Modern cosmology and high energy astrophysics have become interdisciplinary research subjects, and new graduate students are now expected to assimilate a broad range of physical concepts. This innovative book provides a clear and pedagogical introduction to research in these areas through a series of problems and answers. The problems are designed to develop each core topic in a simple and coherent way and full solutions are provided to make this book completely self-contained.

The first half of the book covers the core subjects of astrophysical processes, gravitational dynamics, radiative processes, fluid mechanics and general relativity. The second half uses these concepts to develop modern cosmology: topics include the Friedmann model and thermal history, dynamics of dark matter and baryons in an expanding universe, physics of high redshift objects and the very early universe.

This unique self-study textbook will be of key interest to graduate students and researchers in cosmology, astrophysics, relativity and theoretical physics. It is particularly well suited to graduate-level courses.

Cosmology and astrophysics *through problems*

Cosmology and astrophysics *through problems*

T. Padmanabhan

**Inter-University Centre for Astronomy and Astrophysics,
Pune, India**

CAMBRIDGE UNIVERSITY PRESS
Cambridge, New York, Melbourne, Madrid, Cape Town,
Singapore, São Paulo, Delhi, Tokyo, Mexico City

Cambridge University Press
The Edinburgh Building, Cambridge CB2 8RU, UK

Published in the United States of America by
Cambridge University Press, New York

www.cambridge.org
Information on this title: www.cambridge.org/9780521467834

First published 1996

A catalogue record for this publication is available from the British Library

Library of Congress Cataloguing in Publication data
Padmanabhan, T. (Thanu), 1957–
Cosmology and astrophysics through problems / T. Padmanabhan.
    p. cm.
Includes bibliographical references.
ISBN 0 521 46230 4 – ISBN 0 521 46783 7 (pbk)
1. Cosmology – problems, exercises, etc.
2. Astrophysics – Problems, exercises, etc.
I. Title.
QB981.P244   1996
523.1–dc20   95-49977   CIP

ISBN 978-0-521-46230-3 Hardback
ISBN 978-0-521-46783-4 Paperback

To all my Teachers

Contents

Note: The number within the square bracket denotes the level of difficulty of the problem: 1 is easy, 2 is moderately difficult and 3 is hard. Those problems which require numerical work are marked with the letter N.

Preface

... vadah pravadatamaham

*(... among those who debate I manifest
as the dialectic (the process of logical enquiry
through questions and answers))*

– Gita, Chapter X, Verse 32

This book is an experiment. It is based on the idea that students learning certain subjects can benefit substantially from a book which is structured differently from either a pure problem book or a pure text book but captures the good features of both. Inevitably, such a book will also capture some bad features of both and I hope the good aspects far outweigh the bad ones.

The book is intended for graduate students, researchers and teachers who are working in the broad area of astrophysical cosmology and structure formation. This subject has progressed significantly over the last decade or so and – as a natural consequence – has become fairly interdisciplinary. To do anything worthwhile in this subject a student needs to know not only a fair amount of general relativity and cosmology but also have a basic grasp of several other topics such as radiative processes, fluid mechanics, galactic dynamics and even quantum field theory. Conventionally, good students try to acquire this background from the graduate school and by reading relevant parts of text books in different areas. This process, unfortunately, is slow. What is more, conventional text books *teach* a student rather than *help the student learn*; there is a significant difference between these two approaches, especially in areas of applied physics in which a particular way of thinking about problems is important. In tackling a complex astrophysical problem, the student should be able to discern the important processes from the incidental ones, introduce approximations without killing the essence and estimate the accuracy of the final result. Such a real-life situation is quite different from the sanitized surroundings of text books, and people who have mastered the 'exact results' of text books often have difficulty in making the transition to the new mode of thinking. It will be a considerable help if we can supplement the conventional modes of learning in some way to help the student acquire the background more quickly and efficiently.

The present book is intended to be such a supplement.

Let us ask what should be the ingredients in such a supplement when targeted at students interested in cosmology or structure formation. It seems to me that one can achieve the objective quickly if the student is helped to think on his/her own right from the start. There is no real reason why students cannot learn conventional topics (say e.g. the propagation of sound waves, self-absorption in a synchrotron source, derivation of the Schwarzschild metric...) by working out the steps on their own rather than reading them from a book. If these results are presented as problems in a suitably graded form, it will only take marginally more time to work them through as compared to learning from a text book. This extra time is well spent for two reasons: first, it provides the reader with an opportunity for *active* learning and thinking; second, it gives a clearer logical picture to a student regarding the various assumptions that go into any particular result.

With this philosophy in mind, I have structured this book in the form of a series of problems (with answers) in different topics. Within each topic, the problems proceed in a logical order, deriving the key results which are needed and discussing applications whenever relevant. The student is expected to work through the problems, or at least make sure he/she understands all the key elements. The answers, of course, are sufficiently detailed to help the student in case of any difficulty.

The book is intended for a student with a physics (rather than an astronomy) background and I have tried to explain the astronomical jargon and units whenever this is needed. There is a section at the beginning of the book called 'How to use this book' which provides detailed tips to the reader and also explains the interdependency of various problems. Those students who are seriously interested in the subject are expected to read through this section carefully! This section will also help lecturers to adapt parts of this book for their courses. To provide the student with some idea regarding the level of difficulty of the problems, I have classified them on a scale of three levels: 1 is easy, 2 is moderately difficult and 3 is difficult.

It is impossible to arrive at a consensus regarding the topics that such a book should contain. I have attemped to select the topics along the following lines: Part 1 of the book has five chapters dealing with a set of 'core topics': Dynamics and gravitational physics, Fluid mechanics, Radiative processes and General relativity. (In addition, I have thrown in chapter 1, called 'Astrophysical processes', which gives the reader an overall feel for the entire collection of topics. Processes discussed in this introductory chapter are elaborated upon in the rest of the book). I believe any serious student of cosmology requires a good grasp of these topics, and the time spent learning them will be a good investment for the future. In Part 2, there are five chapters, all dealing with modern cosmology: Friedmann model and thermal history, Dynamics in the expanding universe, Structure formation, High redshift objects and Very early universe. These chapters make use of results from Part 1 and build further on them. I give a more detailed description of the chapters below.

Chapter 1, in some sense, presents the whole book in miniature form. It discusses several processes from radiation theory, fluid mechanics, Newtonian gravity, observational astronomy and cosmology in order to give the reader an overview. In this chapter, the emphasis is on back-of-the envelope calculations and order of magnitude estimates in different contexts. This chapter also reviews several background topics with which the reader should be familiar, and sets the stage for the entire book.

Chapter 2 deals with issues of dynamics and Newtonian gravity. The emphasis is on the 'modern' approach to classical mechanics and statistical physics geared towards understanding the gravitational many body problem. I have included fairly detailed accounts of topics such as surfaces of section, statistical mechanics of gravitating systems, isothermal sphere, collisional and collisionless evolution, etc. These topics are usually picked up by students through an 'intellectual osmosis' and it is important to provide a more methodical introduction for them.

Chapter 3 provides a cross-section from the vast lore of fluid mechanics. After presenting a logical derivation of the Navier–Stokes limit from microscopic theory, I concentrate on those aspects which could be of relevance in extragalactic astronomy. So problems developing supersonic flow, shock waves, particle acceleration mechanisms, two- and three-dimensional accretion, magnetohydrodynamics (MHD) and flux freezing, etc., find a place here rather than more conventional topics such as boundary layer theory.

Chapter 4 discusses the basic radiative processes. I have noticed that, in spite of the existence of several excellent texts, students find this subject somewhat difficult. To ease the difficulty, I have tried to include a brief discussion of most of the radiative processes in chapter 1 and provide detailed derivations in chapter 4. This repetition is *intentional* and serves to transmit the key physical ideas without burdening the reader with too many details (in chapter 1) to be followed by rigorous derivations (in chapter 4). I have also tried to maintain a logical progression of ideas in presenting different processes. By and large, the general principles are worked out first followed by specific applications to various phenomena. This chapter is fairly exhaustive; the reader will find almost all the key processes needed in extragalactic astronomy and cosmology here.

Chapter 5 introduces the reader to general relativity, the most beautiful of all physical theories known to mankind. In spite of its aesthetic value (and its new-found utilitarian value in high-energy astrophysics and cosmology), most graduate students are not as exposed to it as to, say, nuclear physics. I have tried to include problems which will illustrate all the key concepts but this is one area in which the student may really need to gain more practice from a specialist text. After developing the concepts I have applied them to a select set of topics (e.g. gravitational lensing) which are relevant to cosmology.

These five chapters are intended to provide the reader with a background which will be useful in many areas of extragalactic astronomy. The second part of the book uses many of the concepts to develop ideas in structure formation.

Chapter 6 discusses the Friedmann model and the thermal history of the universe. Many of the processes discussed in chapter 4 find application here. Once in a while, I have included problems which ask the reader to provide a broad, qualitative description, but I have also included a more detailed discussion in the answer.

Chapters 7 and 8 develop the subject of structure formation. I have concentrated on the dynamical issues and dark matter in chapter 7 and have discussed baryon-related processes in chapter 8. These two chapters, especially chapter 7, include problems which are at the forefront of current research.

Chapter 9 is titled High redshift objects, which – in conventional jargon – refers to the contents of the universe in the redshift range of $z < 20$. It was quite difficult to decide which aspects to include and which to omit. With the background which has been developed in Part 1, a reader can attack a host of interesting issues in extragalactic astronomy. I have decided – with a heavy heart – to omit the physics of active galactic nuclei and radio galaxies and to concentrate on the intergalactic medium (IGM), ionization history, the abundance of high redshift objects and quasar absorption systems.

Chapter 10 deals with the very early universe. In spite of the glamour (and rate of preprint production) surrounding this subject area, it is probably not necessary to burden a student with too many speculative ideas. I have selected only two themes: the production of initial perturbations in inflationary scenarios and the nature of cosmological defects.

Finally, chapter 11 is a collection of advanced problems which further develop some of the ideas from the ten chapters. By their very nature, some of these are open ended and so I have not given answers to these problems. A reader who has read the preceding ten chapters should not have much difficulty in providing the answers on his/her own.

Because of the nature of the book, I have not given detailed references to original literature except where the solution is not available in standard text books. The annotated list of references appears at the end of the book and I have given – for each chapter – a few 'core text books' which I have found useful. These books, of course, contain extensive bibliographies and references to original literature. The selection clearly reflects the personal bias of the author and I apologize to anyone who feels that their contribution has been overlooked.

Several people have contributed to the making of this book. Many of my colleagues who liked the collection of problems in my earlier book *Structure formation in the universe*, encouraged me to try out a problem book of this kind.

I thank them for convincing me in this matter and I hope they find the final product satisfactory.

This is the second time I have worked with Rufus Neal of CUP, and it has again been a pleasure. Rufus Neal and Adam Black have been very efficient in handling all aspects of the logistics related to this project and I thank them for all the effort they have put in.

I thank J. S. Bagla, Ravi Subrahmanian, K. Subramanian and M. Vivekanand for their comments on some of the chapters in an earlier version.

J. S. Bagla did an excellent job of producing the original diagrams for the book. The photographic work relating to the diagrams was produced by B. Prem Kumar. The initial parts of the book were typed by Manjiri Mahabal. The final round of typing, formatting and proof-correcting was done by Vasanthi Padmanabhan. I thank all of them for their help. I thank IUCAA for the use of the computer facilities in these tasks.

T. Padmanabhan

How to use this book

You may find the following comments useful when deciding on the best possible way of using this book for your purpose:

1. Read through the preface (and this section!), which explains the overall structure of this book. Then browse through the table of contents to get a feel for the topics which are covered.

2. The problems in chapters 2–10 are ordered in such a way as to develop the subject in a logical sequence. Hence the problems show a fair amount of variation in the level of difficulty. Depending on your level of familiarity with the subject you can adopt one of the following strategies:

(a) If you are thoroughly familiar with the basic theoretical development of a particular subject, you could go through the earlier problems of that chapter fairly rapidly. It is enough if you read the question to see what needs to be demonstrated and glance through the answer to make sure you understand the derivation. (I have provided a number of 'non-standard' proofs and derivations even for some standard text book results; so it will be useful to read through the answers even where you know the subject.) The latter parts of the chapter will usually contain applications of the basic formalism. You can attempt these questions on your own and compare your solutions with the ones provided.

(b) If you have only a moderate level of familiarity with the subject, you should try to work through the earlier parts of the chapter yourself. These problems develop the subject in a methodical fashion and are designed to help you arrive at the basic results on your own. If you face any serious difficulty with a particular problem, you could read the answer and then go back to tackling the rest of the set on your own. In general, it is a good idea to read the answer to each problem and compare it with your solution before proceeding to the next one.

(c) If you know nothing about a subject, I am not certain whether this book – by itself – can help you to master it. In such a situation, it is best to use this book as a supplement to a more introductory conventional text book.

3. Very often, you will be interested in certain sets of applications or in some special selection of material. This book is designed in a modular fashion so as to be useful in these contexts. I give a reorganized 'table of contents' of the book below to help you in such circumstances.

In the table given below, the contents of the book are presented in the form of 13 'study-units' labelled A–M. Within each study-unit, subjects are presented in a 'tree-format'. The first levels of each of these sub-units are numbered consecutively from 1 to 29. Such a layered structure tells you at a glance the interdependency of various problems. The problems in each sub-unit require knowledge of the problems in the preceding layer but – by and large – each sub-unit can be studied independently. For example, consider the constraints which can be imposed on astrophysical sources from the study of synchrotron radiation. This problem (4.7) appears in study-unit D, sub-unit 7(ii). The tree structure shows that one should ideally work out (or be able to understand) the problems in D7 and in sub-section (ii) before proceeding to problem 4.7.

There are, of course, exceptions. In some cases, especially in Part 2, one needs to know the material of one study-unit in order to follow another. In such cases, this is clearly indicated by the relevent unit number within brackets. For example, unit G develops the Friedmann model and unit H deals with the derivation of the basic equations for studying structure formation. To study linear perturbation theory in the Friedmann universe, which is unit I, one needs to know most of units G and H. This is indicated by placing the letters [G,H] next to the title of unit I.

Study–units

4. There are no 'plug-in' kinds of elementary problem in this book. In fact, problems in many chapters require you to think up approximation procedures and mathematical techniques on your own. This is intended for you to develop confidence as regards when an approximation is valid (or useful) and when it is not. Such approximations are, of course, clearly spelt out in the *answers but are not always stated explicitly* in the questions. This is intentional since, in a

real research problem, it is up to you to decide how to introduce a suitable mathematical model. You should not approach the questions as mathematical theorems requiring you to proceed from statement A to statement B; they should only be treated as broad indicators of the kind of results one is aiming for.

5. I have indicated the level of difficulty of each problem on a scale of 1–3 in the table of contents. Problems numbered 1 are the easiest and those numbered 3 are the most difficult. The letter N next to an entry indicates that the problem requires numerical work. If the student is not familiar with numerical methods, he/she can skip these problems at the first go and read the answers. This is, of course, not a permanant (or desired) solution and I would strongly urge the student to remedy this handicap as quickly as possible!

6. Chapters 1 and 11 are special. Chapter 1 presents, in condensed form, most of the material in the book and I would expect all students to attempt the problems in it. Many of these problems are taken up for more detailed study in the remaining chapters. When the study-unit gives the number of a problem in chapter 1 as well as from a later chapter, it is better to attempt the chapter 1 problem first.

Chapter 11 contains a set of advanced problems for which no solutions are given. (For some of the problems I have given the reference to the published literature wherein the solution can be found.) This chapter is intended for the 'brave ones' but I strongly recommend that all students should attempt at least a few of the problems contained therein.

Oh Traveller, there are no paths.
Paths are made by walking.
–Antonio Machado

Part one

1

Astrophysical processes

This chapter presents a collection of processes which play a crucial role in astrophysics, and reviews some of the basic concepts in statistical mechanics, plasma physics, etc. The key idea behind each process is brought out in the problems and the reader is only expected to provide a simple, qualitative estimate in the answers. Most of these processes will be discussed in greater detail in the following chapters.

The reader is assumed to be familiar with: (i) quantum mechanics; and (ii) special relativity using the four-vector notation. The number within the square bracket denotes the level of difficulty of the problem: 1 is easy, 2 is moderately difficult and 3 is difficult. Those problems which require numerical work are marked with the letter N.

1.1 Why does an accelerated charge radiate? [2]

(a) The electric field of a stationary point charge falls as r^{-2}. The field of a charge, moving with a uniform velocity, can be obtained from the Coulomb field by a Lorentz transformation; show that this field also decreases as r^{-2}.

(b) But when the charge is accelerating, the electric field picks up a term which falls as r^{-1}, usually called the 'radiation field'. A field with $E \propto r^{-1}$ has an energy flux $S \propto E^2 \propto r^{-2}$. Since the surface area of a sphere increases as r^2, the same amount of energy will flow through spheres of different radii. This fact allows the accelerating charge to transfer energy to large distances and thus provides the radiation field with an independent dynamical existence.

Consider a charged particle which was at rest till $t = 0$ and was accelerated to a velocity $\mathbf{v}$ in a small interval of time Δt. Examine the electric field everywhere at some time $t \gg \Delta t$ and show that the electric field in a shell-like region of thickness $c\Delta t$, located around $r \cong ct$, has an r^{-1} dependence.

Use the above result to show that the radiation field due to a *non-relativistic* charged particle with acceleration $\mathbf{a}(t)$ can be expressed in the form

$$\mathbf{E}(t, \mathbf{r}) = \left[\frac{q}{c^2 r} (\hat{\mathbf{n}} \times (\hat{\mathbf{n}} \times \mathbf{a})) \right]_{\text{ret}}, \qquad \mathbf{B} = \hat{\mathbf{n}} \times \mathbf{E}, \qquad (1.1)$$

where **r** is the vector from the charge to the field point and $\hat{\mathbf{n}} = (\mathbf{r}/r)$. The subscript 'ret' implies that the expression should be evaluated at the time $t' = t - (r/c)$.

(c) Consider a single charged particle moving with acceleration a. Show that the power radiated is $(d\mathcal{E}/dt) \cong (q^2 a^2 / c^3)$.

1.2 Paranoid effect [1]

Consider two Lorentz frames S and S', with S' moving along the positive x-axis of S with velocity **v**. We will call S' the rest frame and S the lab frame and denote quantities in these frames by the subscripts R and L. Assume that $v \lesssim c$ so that

$$\left(1 - \frac{v}{c}\right) \cong \frac{(1 - v^2/c^2)}{(1 + v/c)} \cong \frac{1}{2\gamma^2}, \tag{1.2}$$

with $\gamma = (1 - v^2/c^2)^{-1/2} \gg 1$. Consider a photon with frequency ω_L travelling along the direction (θ_L, ϕ_L) in the lab frame.

(a) What is the direction (θ_R, ϕ_R) in which it will be seen as propagating in the rest frame? What is the frequency ω_R in the rest frame?

(b) Plot $\cos\theta_R$ against $\cos\theta_L$. Mark in the plot the direction of propagation in the two frames corresponding to (i) $\cos\theta_L = 1$; (ii) $\cos\theta_L = (v/c)$; (iii) $\cos\theta_L = 0$; (iv) $\cos\theta_L = -1$. For each of these cases relate ω_R to ω_L.

(c) What is the range of angles for which a photon propagating forward in one frame will appear to propagate forward in the other frame as well? What is the range of angles at which the photon is blueshifted? For what angle does the photon suffer no change in frequency?

(d) An object is emitting photons isotropically in its rest frame. Sketch qualitatively the angular dependence of the emission pattern as seen from the lab frame.

(e) A high energy particle is moving through an isotropic bath of radiation. Sketch the angular pattern of collisions between the charge and photons in the rest frame of the particle.

1.3 Bremsstrahlung and synchrotron radiation [1]

We shall now estimate the radiation emitted in two important astrophysical contexts:

(a) In a fully ionized plasma, collisions between electrons and ions will accelerate the electrons and make them radiate. Estimate the energy radiated per second per unit volume from a hydrogen plasma at temperature T. (This process is called 'thermal Bremsstrahlung'.)

(b) Several astrophysical plasmas host magnetic fields. Charged particles, accelerated by the magnetic field, will emit radiation (called 'cyclotron' or 'synchrotron' radiation depending on whether the charge is non-relativistic or relativistic). Estimate the power radiated in this case.

1.4 Thomson scattering and Eddington limit [1]

(a) An electromagnetic wave with amplitude E hits a free charged particle and makes it oscillate with non-relativistic velocities. The oscillating charge, in turn, emits radiation in all directions. This process may be thought of as the scattering of electromagnetic radiation by a free charged particle. Estimate the scattering cross-section for this process.

(b) Matter falling towards a massive gravitating object gains kinetic energy. If this matter is brought to a halt suddenly – say, at the surface of the central body – then the kinetic energy can be converted into radiation with some efficiency ϵ. This process is invoked in several astrophysical contexts to account for large luminosities of extragalactic objects. Show that the luminosity which can be produced by such accretion is limited by the value (called the Eddington limit)

$$L_{\mathrm{E}} \cong \frac{4\pi G M m_{\mathrm{p}} c}{\sigma_{\mathrm{T}}} \cong 10^{46} \left(\frac{M}{10^8 \ \mathrm{M_\odot}} \right) \mathrm{erg \ s^{-1}}. \tag{1.3}$$

Estimate the corresponding limiting values for (i) accretion rate (dm/dt); (ii) typical time-scale for the accretion; and (iii) peak frequency of the emission of radiation. Take the size of the central object to be $R \cong (GM/c^2)$.

1.5 Inverse Compton scattering and CMBR [1]

(a) Consider a beam of relativistic charged particles moving through a radiation field. Because of the collisions with the photons in the radiation field, the charged particles will lose part of their energy. Estimate the net gain in energy by the radiation field. (This process is called the inverse Compton effect.)

(b) Very high energy electrons with $\mathscr{E} \cong 100$ GeV have been observed at the top of earth's atmosphere. While travelling through space, these electrons must have undergone inverse Compton scattering with the microwave photons which are at a temperature $T \cong 2.7$ K. Estimate the characteristic time-scale in which these electrons lose their energy.

1.6 Interaction of matter and radiation [2]

(a) Consider a gas of electrons at temperature T, interacting with a distribution of photons with mean energy $\langle E \rangle$. Assume that $\langle E \rangle \ll mc^2$ and $kT \ll mc^2$. During the scattering, energy is exchanged between electrons and photons. Show that the average energy transfer $\langle \Delta E \rangle$ from the photons to the electrons is

$$\frac{\langle \Delta E \rangle}{\langle E \rangle} = \frac{\langle E \rangle - 4kT}{mc^2}. \tag{1.4}$$

(b) When electrons and photons coexist in a region of size l, the repeated scattering of the photons by the electrons will distort the original spectrum of the photons. Show that this process – called 'Comptonization' – is important if

$4y \gtrsim 1$, where

$$y = \frac{kT_e}{m_e c^2} \max\left(\tau_e, \tau_e^2\right) \qquad (1.5)$$

and $\tau_e = n_e \sigma_T l$ is called the 'optical depth' of the region under consideration. What is the time-scale over which this process is effective? Also show that the final energy $\mathscr{E}'$ of a photon is related to its initial energy $\mathscr{E}$ by

$$\frac{\mathscr{E}'}{\mathscr{E}} = \exp\left(4y\right). \qquad (1.6)$$

(c) Show that the Comptonization can drive the photons to a distribution function of the form

$$N_\gamma \propto \left\{\exp\left[\left(\epsilon - \mu\right)/kT\right] - 1\right\}^{-1} \qquad (1.7)$$

if the optical depth is larger than a critical value τ_{crit}, with

$$\tau_{\text{crit}} = \left[\left(\frac{m_e c^2}{4kT_e}\right) \ln\left(\frac{4kT_e}{\hbar \omega_i}\right)\right]^{1/2} \qquad (1.8)$$

1.7 Diffusion in energy space [1]

Let $n(E, t, \mathbf{x})$ denote the number density of some species of particles with energy between E and $E + dE$, at time t, in some region of interest. This quantity can change due to several physical processes. In many astrophysical contexts, one is interested in determining the time evolution of $n(E, t, \mathbf{x})$.

(a) The first process which can change the spectrum will be the injection of fresh particles into the region. Let the number of particles injected per second into a small volume dV be $Q(E, t)dV$. Further, the particles inside the volume may lose or gain energy due to radiative and acceleration mechanisms at a rate $b(E)$, that is we assume that the rate of change of energy of a single particle is $(dE/dt) = -b(E)$. The particles could also diffuse out of the region of interest with a diffusion constant D.

Show that the evolution of $n(E, t, \mathbf{x})$ can then be described by a differential equation of the form

$$\frac{\partial n}{\partial t} = D\nabla^2 n + \frac{\partial}{\partial E}\left[b(E)n\right] + Q(E, t). \qquad (1.9)$$

(b) The energy loss for high energy electrons can often be expressed in the form

$$b(E) = A_1 + A_2 E + A_3 E^2, \qquad (1.10)$$

with A_1, A_2, A_3 being constants. The first term A_1 represents a constant rate of loss of energy due to ionization, etc. The term linear in E can represent losses due to uniform adiabatic expansion of the region. The quadratic term describes inverse Compton and synchrotron losses.

The injection of particles could be, for example, due to a uniform distribution of sources, with each one producing the particles with a power law spectrum

$Q(E) = kE^{-p}$, with $p \neq 1$. In such a case, diffusion is often unimportant. Determine the steady state spectrum $n(E)$ for this case.

1.8 Line widths [1]

(a) An atom in an excited state will emit a photon spontaneously and make a transition to the ground state. Estimate the rate of spontaneous transitions by using the correspondence principle and the formula for radiation from an accelerated classical charge.

The finite lifetime of the excited state will lead to a finite width for the spectral line. Estimate the width of the line.

(b) The random motions of the atoms at any finite temperature will also lead to broadening of the spectral line due to Doppler shift. How does the Doppler width compare with the natural line width?

1.9 Photon: particle or wave? [2]

The distribution function for photons, in thermal equilibrium at temperature T, is given by the Planck distribution. The mean number $\bar{n}$ of photons and the mean energy $\bar{E}$ at frequency ω are given by

$$\bar{n}(\omega) = \frac{1}{e^{\hbar\omega/kT} - 1}, \qquad \bar{E}(\omega) = \frac{\hbar\omega}{e^{\hbar\omega/kT} - 1}. \qquad (1.11)$$

Calculate the mean square fluctuation in the number of photons at a temperature T. Use this expression to provide a criterion as to when photons may be thought of as particles and when they should be treated as a classical electromagnetic wave.

1.10 Sensitivities of telescopes [1]

An optical telescope is used to study a source from which $N(v)$ photons cm^{-2} s^{-1} Hz^{-1} are expected. This source is observed for a time interval t in a bandwidth Δv. Let the effective diameter of the telescope be D. One is often interested in knowing how long one should observe the source to obtain a particular signal-to-noise ratio.

(a) Show that the time t needed to obtain a particular signal-to-noise ratio varies as D^{-2} if the main source of noise is the intrinsic fluctuations in the number count of photons.

(b) Show that $t \propto \Omega D^{-2}$ if the main source of noise is the background radiation in the sky. Here Ω is the effective solid angle subtended by the telescope to the sky.

(c) In telescopes using detectors such as CCD there will be a 'dark current' which could contribute to the noise by producing (unwanted) photon noise at a constant rate. (This is the noise produced due to the generation of unwanted electrons in the CCD system contaminating the signal.) Show that, if dark current dominates, $t \propto D^{-4}$.

1.11 Newtonian gravity [2]

When the gravitation field is weak, it can be described by a single function, $\phi(t, \mathbf{x})$, called the gravitational potential. The gravitational force acting on a particle of mass m is given by $(-m\nabla\phi)$. The potential $\phi(t, \mathbf{x})$ at any location can be determined from the distribution of matter $\rho(t, \mathbf{x})$ via the relation

$$\phi(t, \mathbf{x}) = -G \int \frac{\rho(t, \mathbf{x}') \, d^3\mathbf{x}'}{|\mathbf{x} - \mathbf{x}'|}. \tag{1.12}$$

(a) Show that the above equation implies the differential relation $\nabla^2 \phi = 4\pi G \rho$ (called Poisson's equation). Is the converse true? That is, does Poisson's equation imply (1.12)?

(b) It is well known that: (i) an infinite, plane sheet of matter with surface density σ produces a gravitational field which is a constant; and (ii) a spherical shell of matter produces zero gravitational field inside.

These results, however, are not unique in the following sense. Show that it is possible to have a physically reasonable density distribution $\rho(\mathbf{x})$ (which exists in a finite region of space and $\rho(\mathbf{x}) \geq 0$) such that the gravitational force due to $\rho(\mathbf{x})$ in some compact region of space is constant. Also show that there exist physically reasonable density distributions which are *not* spherically symmetric but will produce zero gravitational force in some compact region of space.

(c) Let $\phi[\mathbf{x}; \rho(\mathbf{x})]$ be the gravitational potential at a point $\mathbf{x}$ due to a density distribution $\rho(\mathbf{x})$. Consider a new density distribution $\rho'(\mathbf{x}) = (a/x)^5 \rho(a^2\mathbf{x}/x^2)$ with $x = |\mathbf{x}|$. Show that the gravitational potential due to $\rho'(\mathbf{x})$ is $\phi'(\mathbf{x}) = (a/x)\phi(a^2\mathbf{x}/x^2)$.

(d) Consider a compact region $\mathscr{R}$ in the three-dimensional space outside which the matter density is zero. It is given that the gravitational force due to the matter, distributed inside $\mathscr{R}$, has the form $(-A\mathbf{x}/x^3)$ everywhere outside $\mathscr{R}$, that is, the gravitational force outside $\mathscr{R}$ obeys the inverse square law. This can easily be achieved if the matter distribution in $\mathscr{R}$ is spherically symmetric. Show that, while this is a sufficient condition, it is not necessary. In other words, it is possible to have a mass distribution which, though not spherically symmetric, will produce an inverse square law force outside the distribution.

1.12 Neutral gaseous system [2]

One of the simplest systems we encounter in astrophysics consists of a collection of atoms or molecules interacting via short range ('collisional') forces. It is usual to study such a system using statistical mechanics.

Table 1.1. Fermion gas

State	Region in $\rho-T$ plane (ρ in g cm^{-3}; T in kelvin)	Pressure (dyne cm^{-2})
Radiation dominated	$\rho \lesssim 3 \times 10^{-23} T^3$	$P \cong 2.5 \times 10^{-15} T^4$
Non-degenerate ideal gas	$3 \times 10^{-23} T^3 \lesssim \rho \lesssim 2.4 \times 10^{-8} T^{3/2}$	$P \cong 8.3 \times 10^7 \rho T$
Fully degenerate non-relativistic	$2.4 \times 10^{-8} T^{3/2} \lesssim \rho \lesssim 7.3 \times 10^6$	$P \cong 10^{13} \rho^{5/3}$
Fully degenerate ultrarelativistic	$7.3 \times 10^6 \lesssim \rho$	$P \cong 1.2 \times 10^{15} \rho^{4/3}$

(a) Table 1.1 describes four different regions in the density–temperature plane of a fermion gas. Describe and justify the entries in the table.

(b) Mark these four regimes on a $T-\rho$ plot. On the same diagram, indicate the location of the following systems: (i) stellar core; (ii) white dwarf; (iii) normal metallic solid at room temperature and pressure.

1.13 Transport coefficients for a gas [2]

Estimate (i) the coefficient of viscosity; (ii) the coefficient of thermal conductivity; and (iii) the coefficient of diffusion for a gas. When are the viscous forces important? What is the time-scale over which diffusion becomes effective?

1.14 Time-scale for gravitational collisions [3]

(a) Consider a collection of stars moving under the action of their mutual gravitational forces (say, in a galaxy or in a star cluster). Physical collisions between stars are extremely rare in such systems. The gravitational potential at any given point can be approximated, to reasonable accuracy, as that due to a smooth density distribution of stars. This will lead to systematic orbits for the stars. The actual potential, at any point, will differ from the smooth potential due to the granularity of the system. Some amount of randomness in the motion of the stars is introduced due to the difference between the actual gravitational force acting on any one star and the mean gravitational force calculated from a smoothed-out distribution of mass in the system. Estimate the time-scale t_{gc} over which these deviations can significantly affect the distribution of stars in the system.

(b) Consider a 'test' particle of mass M_t moving through a region containing a cloud of 'field' particles of mass m_f and number density n_f. If the test particle is moving with a speed v which is larger than the velocity dispersion of the field particles, it will experience a drag force. This force (called 'dynamical friction') arises because of the transfer of energy from the test particle to the field particles. Estimate the time-scale for this process.

(c) Consider a system made up of a large number of particles, each of size r_0 and mass m_0. Let the size of the system be R and the number of particles in the system be N. Typical systems include ideal gas in a container, stars in a galaxy or globular cluster, galaxies in a cluster, etc.

We are often interested in the following physical parameters characterizing the system: (i) typical velocity of a particle, v; (ii) the critical impact parameter b_{crit} below which the interactions between two particles can lead to a significant deflection ('collisional range'); (iii) the crossing time, t_{cross}, which is the typical time-scale over which the particle can traverse the system; (iv) collisional relaxation time, t_R; (v) the 'collisionality' of the system which may be defined to be the ratio, (t_{cross}/t_R); (vi) the mean-free-path $l \cong v t_R$; and (vii) the 'structureness' of the system which may be defined to be the ratio between the size of the particle and mean interparticle separation.

Estimate the above quantities for globular clusters, galaxies and clusters of galaxies. Also estimate the same parameters for hydrogen gas at normal temperature and pressure.

1.15 Collisionless evolution [3]

There is considerable evidence to suggest that the visible parts of galaxies are embedded within large massive halos of non-luminous matter. These halos could be composed of fermionic particles which interact among themselves only through gravity. Since the total number N of such particles in any halo will be very large, the collisional relaxation time t_R will be much larger than the age of the universe; so the halo can be considered to be a collisionless system. Assume that, at some initial time $t = t_0$, these particles were described by a distribution function f_0 which was spatially uniform (except for small fluctuations) and was a monotonically decreasing function of $|\mathbf{p}|$. As time goes on, small fluctuations in the density distribution will become amplified by the self-gravity of the system and the particles could form bound systems governed by their self-gravity. It is believed that the halos around the galaxies have formed by such a process. Suppose the typical spatial scale of a galactic halo is R and the velocity dispersion of halo particles is σ. Let the particles be fermions of mass m. Show that the bound

$$m \gtrsim 30 \text{ eV} \left(\frac{\sigma}{220 \text{ km s}^{-1}} \right)^{-1/4} \left(\frac{R}{10 \text{ kpc}} \right)^{-1/2} \tag{1.13}$$

should necessarily hold if the final configuration arises from the initial one through collisionless evolution.

1.16 When is gravity important? [1]

As a body becomes larger, gravity will start to play an increasingly important role. Show that self-gravity is important if the number of atoms in the body is larger than a critical number, $N_{\text{crit}} \cong \alpha^{3/2} \alpha_G^{-3/2} \cong 10^{54}$, where $\alpha \equiv (e^2/\hbar c)$, $\alpha_G \equiv (Gm_p^2/\hbar c)$ and m_p is the proton mass. (We can think of α_G as the 'fine-structure constant' for gravity.) Estimate the corresponding critical mass at which gravity becomes important.

1.17 Making a star [2]

Consider a large mass of gaseous cloud of hydrogen contracting under self-gravity. Assume that the total pressure of the cloud is the sum of the gaseous pressure and degeneracy pressure of the electrons.

(a) Show that the maximum temperature reached by the cloud is about $T_{\text{max}} \cong N^{4/3} \alpha_G^2 m_e$.

(b) If the temperature T_{max} is high enough to trigger nuclear fusion in the centre of the cloud, then the gaseous cloud can become a star. Show that the critical temperature at which a small fraction, say 10^{-5}, of the atomic nuclei can fuse together is given by $T_{\text{max}} \cong \eta \alpha^2 m_p$, with $\eta \cong 0.1$. Combining with the results of (a), show that the minimum mass M_* of gaseous cloud which can turn into a star is given by

$$M_* \cong \eta^{3/4} \left(\frac{m_p}{m_e} \right)^{3/4} \left(\frac{\alpha}{\alpha_G} \right)^{3/2} m_p. \tag{1.14}$$

(c) The nuclear burning will release energy from the star, primarily in the form of photons. These photons will be repeatedly scattered by the ionized gas in the star and will slowly diffuse to the surface of the star. The information about scattering of the photons can be expressed conveniently in terms of a mean-free-path $l = l(\rho, T)$. At very high temperatures, almost all the electrons would have been stripped off from the atoms and scattering of photons is primarily due to Thomson scattering by free electrons. In this case, the mean-free-path is $l \cong (n\sigma_T)^{-1}$, where $\sigma_T \cong \alpha^2 m_e^{-2}$ (in units with $\hbar = c = 1$) is the Thomson cross section. (At lower temperatures, some of the electrons will still be bound to the atoms and most of the scattering of high energy photons is due to photoionization of these electrons. In such a case, it turns out that $l \propto T^{7/2} n^{-2}$. This result will be derived later in problem 4.5.) Assuming that Thomson scattering dominates, estimate: (i) the time taken by a typical photon to escape from the centre of the sun; (ii) the typical luminosity of a star; (iii) the typical lifetime of a star. How does the lifetime of the star compare with the gravitational and thermal time-scales?

(d) The nuclear reactions which take place in the cores of stars depend sensitively on the temperature. When temperature increases, more energy is produced and is radiated away, thereby decreasing the temperature, and vice versa. Hence, we may assume that most of the so-called 'main-sequence' stars have nearly constant central temperature. Show that, for such a class of stars, the surface temperature T_s scales as $T_s \propto L^k$, where k depends on the dominant scattering mechanism and varies between $(1/12)$ and $(3/20)$.

(e) A star will become unstable (and will be blown apart by the radiation pressure) if the radiation energy content of the star is considerably larger than the thermal energy, say, a hundred times larger. Use this criterion to estimate the maximum mass of a stable star.

1.18 Death of stars [2]

Once a star has exhausted its nuclear fuel, it will start contracting under its own weight. For a stellar remnant, supported by the degeneracy pressure of *non-relativistic* electrons, show that $M^{1/3}R =$ constant. On the other hand, if the degeneracy pressure is provided by a *relativistic* gas of particles with mass m_p, then show that $M \lesssim M_{Ch} \cong \alpha_G^{-3/2} m_p$. (This quantity M_{Ch} is called the 'Chandrasekhar mass'.)

1.19 Forming the galaxies [2]

In problem 1.17, it was shown that a gaseous system with $M \gtrsim M_*$ will condense under its own gravity, become heated up and could become self-luminous by triggering nuclear reactions. The evolution of gaseous systems with $M \gg M_*$ is a lot more complicated. In this case, we must take into account the fact that the system may fragment into smaller structures.

(a) Show that if a body can cool rapidly enough, then it could become gravitationally unstable and fragment into smaller objects.

(b) The dominant mechanism for cooling of large ionized gaseous clouds is thermal Bremsstrahlung. Using the results derived in problem 1.3(a), show that efficient cooling can occur for systems with $R < R_g$ and $M > M_g$, where

$$R_g \cong \alpha^3 \alpha_G^{-1} m_e^{-1} \left(\frac{m_p}{m_e} \right)^{1/2} \cong 2.23 \times 10^{23} \text{ cm} = 74 \text{ kpc}, \qquad (1.15)$$

$$M_g \cong \alpha_g^{-2} \alpha^5 \left(\frac{m_p}{m_e} \right)^{1/2} m_p \cong 3 \times 10^{11} M_*. \qquad (1.16)$$

These parameters correspond to the masses and sizes of galaxies.

1.20 Galaxy luminosity function [1]

In the study of galaxies, one is often interested in three key properties:

(1) The redshift z of the distant galaxy which is receding from us with a speed proportional to its distance. The redshift of the galaxy is related to its distance r by $z = (v/c) \cong (H_0 r/c)$, where H_0 is a constant. Observationally, $H_0 = 100h \ \mathrm{km \ s^{-1} \ Mpc^{-1}}$, with $0.5 < h < 1$.

(2) The observed flux of radiation, f, from the galaxy is related to its intrinsic luminosity L by $f = (L/4\pi r^2)$. (The expressions for z and f are not exact; but they are good approximations for $z \ll 1$.)

(3) The luminosity function of the galaxies which describes the distribution of galaxies with different intrinsic luminosities. More precisely, the galaxy luminosity function is defined to be the number dN of galaxies per unit volume with luminosity in the range L to $(L + dL)$. Observations suggest that

$$dN = \phi \left(\frac{L}{L_*} \right) \frac{dL}{L_*}, \tag{1.17}$$

where

$$\phi(y) = \phi_* y^\alpha e^{-y} \tag{1.18}$$

and the parameters α, L_* and ϕ_* have the values

$$\alpha = -1.07 \pm 0.05, \quad L_* = 1.0 \times 10^{10} e^{\pm 0.23} h^{-2} \ \mathrm{L_\odot}, \quad \phi_* = 0.01 e^{\pm 0.4} h^3 \ \mathrm{Mpc^{-3}}. \tag{1.19}$$

Given the above facts, estimate: (i) the number of galaxies per unit redshift interval per unit solid angle with flux densities between f and $f + df$; (ii) the mean redshift of galaxies with a given flux f; (iii) the mean number of galaxies per logarithmic interval in the flux; (iv) the mean luminosity per unit volume; (v) the 'mass-to-light ratio' of galaxies is about $12 e^{\pm 0.2} h (\mathrm{M_\odot / L_\odot})$. Using this, estimate the mass density contributed by galaxies.

1.21 Basic plasma physics [1]

When the gas is not made up of neutral atoms but is decomposed into electrons and positive ions, it is usual to call it a plasma. The simplest of such systems is a completely ionized gas of hydrogen made up of protons and electrons.

(a) At sufficiently high temperatures, the collisions between the atoms can cause ionization. Calculate the fraction of atoms which will be ionized if the gas is in thermal equilibrium at a given temperature T. Plot the temperature T_{crit} at which 50% of the atoms are ionized, as a function of the number density n. Should the temperature T be comparable to the binding energy of hydrogen for significant ionization to occur?

(b) In the study of astrophysical plasmas, it is usual to assume that the following conditions are satisfied: (i) the plasma is significantly ionized; (ii) it has low enough densities to be treated classically; and (iii) the velocities of the particles are low enough to make non-relativistic approximation valid. Each of these conditions will exclude certain regions in the $(n-T)$ plane for the plasma.

Mark out these regions on a graph. Where do the following systems figure in such a graph: intergalactic medium, interstellar medium, stellar core?

(c) A positive charge Q is introduced into a plasma which is at temperature T. This charge will (preferentially) attract negative charges around it. Consequently, the electric field of this charge Q will be 'shielded' at large distances. Show that the potential due to this charge, at a large distance r from the charge, will be of the form

$$\phi \cong \left(\frac{Q}{r}\right) \exp\left(-\frac{\sqrt{2}r}{\lambda_{\mathrm{D}}}\right), \tag{1.20}$$

where $\lambda_{\mathrm{D}} = \left(kT/4\pi q^2 n\right)^{1/2}$. This quantity λ_{D} is called the Debye length.

(d) It is also possible to give an intrinsic meaning to the Debye length. Because of thermal fluctuations, charge neutrality can be violated in a local region of size λ in the plasma. Show that $\lambda^2 \cong \lambda_{\mathrm{D}}^2 \cong (kT/4\pi q^2 n)$. Calculate the corresponding time-scale during which charge neutrality can be violated. What is the condition for the plasma to behave collectively as a gas?

(e) Estimate: (i) the time-scale over which electrons will relax to an equilibrium distribution; (ii) the time-scale for ions to reach equilibrium; (iii) the time-scale for electrons and ions to reach the same temperature; (iv) the electrical conductivity of a fully ionized plasma.

(f) Consider a hot plasma in which there exists a temperature gradient. When photons (which are assumed to be in local thermodynamic equilibrium with matter) move from region to region, they will transport energy. This process is similar to thermal conductivity, discussed earlier. Assume that photons are scattered by electrons with a scattering cross-section σ_{T}. Show that there is a diffusive heat flow $\mathbf{q} = -\kappa \nabla T$ with $\kappa \cong \left(caT^3/\sigma_{\mathrm{T}} n_{\mathrm{e}}\right)$ where n_{e} is the number density of the electrons.

2

Dynamics and gravity

This chapter deals with dynamics of particles. The emphasis is on the gravitational many-body problem as applied to astrophysical systems. The reader is expected to be familiar with the basic concepts of classical mechanics and statistical mechanics, though they are reviewed in some of the problems in order to introduce the necessary perspective. A few of the problems require numerical integration of differential equations.

2.1 Dynamical systems [1]

Dynamical systems which are of interest to us live and move in a $2N$-dimensional space Γ called 'phase space'. Because of historical reasons we use a coordinate system in Γ with components $(q_1, q_2, ..., q_N; p_1, p_2, ..., p_N) \equiv (q_i, p_j)$, where $i, j = 1, 2, ..., N$. The first N quantities q_i are called 'coordinates' and the next N quantities p_j are called momenta.

The above description implicitly assumes that the state of the system at any given instant is completely described by specifying its location in the phase space. The dynamical evolution makes the system trace a parameterized curve $\mathscr{C}(t) = [q_i(t), p_j(t)]$ in Γ, where t is a real parameter called 'time'. The central question in the study of dynamical systems revolves around determining this curve $\mathscr{C}(t)$ and studying the properties of the space spanned by it.

It is usually postulated that this curve $\mathscr{C}(t)$ is determined by the following 'least action principle'. Consider the value of the quantity (called 'action')

$$A = \int_{\mathscr{P}_1}^{\mathscr{P}_2} dt \left(p_i \dot{q}^i - H(p,q) \right) = \int_{\mathscr{P}_1}^{\mathscr{P}_2} \left[p_i \, dq^i - H(p,q) \, dt \right] \tag{2.1}$$

for all possible functions $[p_i(t), q^i(t)]$ which satisfy the condition $q^i(t_1) = q_1^i, q^i(t_2) = q_2^i$. Here $H(p,q)$ is a given function (called the 'Hamiltonian') characterizing the system. The evolution of the system is postulated to be such that it makes the value of A a local extremum.

15

(a) Vary the paths $q_i(t)$ keeping the end points fixed and show that A is extremized when $q_i(t)$ and $p_i(t)$ satisfy the equations

$$\dot{q}_i = \frac{\partial H}{\partial p_i}, \qquad \dot{p}_i = -\frac{\partial H}{\partial q_i}, \quad i = 1, ..., N.$$ (2.2)

These equations can be integrated if the initial values q_i and p_i are given.

(b) For a wide class of Hamiltonians, it is possible to invert the relation $\dot{q}_i = (\partial H / \partial p_i)$ and express p_i in terms of q_j. In that case the quantity $L = p\dot{q} - H$ can be expressed as a function of q and $\dot{q}$, that is, $L = L(q, \dot{q})$. (We shall omit the subscript i on q_i, etc., when it is not relevent.) Then the action principle involves extremizing the quantity

$$A = \int_{\mathscr{P}_1}^{\mathscr{P}_2} L(q, \dot{q}) \, dt,$$ (2.3)

with respect to all functions $q(t)$ satisfying the conditions $q(t_1) = q_1, q(t_2) = q_2$. Thus, for systems for which p_i can be expressed in terms of $\dot{q}_j$, we can study the dynamics either using the Hamiltonian $H(p, q)$ or using the Lagrangian $L(q, \dot{q})$. Show that the variation of $q(t)$ in (2.3) leads to the equations

$$\frac{d}{dt} \left(\frac{\partial L}{\partial \dot{q}} \right) = \frac{\partial L}{\partial q}.$$ (2.4)

(c) Given a Lagrangian $L = L(q, \dot{q})$, one can construct a modified Lagrangian L' defined by the relation

$$L' = L - \frac{d}{dt} \left(q \frac{\partial L}{\partial \dot{q}} \right).$$ (2.5)

Show that varying the path $q(t)$ in L', keeping the momentum $(\partial L / \partial \dot{q})$ fixed at the end points, will lead to the same equations of motion as varying the path $q(t)$ in L, keeping the coordinates fixed at end points. What is the crucial difference between L and L'?

2.2 Hamiltonian dynamics [1]

As far as the derivation of the equations of motion is concerned, most of the physical systems allow us to use either the Hamiltonian approach or the Lagrangian approach. These approaches, however, are very different in their abilities to attack more formal questions of principle. Since the Lagrangian has only N independent variables q_i, the most general transformation of a Lagrangian consists of: (i) changing the set q_i to another set of generalized coordinates $Q_i = F_i(q_j)$; and (ii) adding a total time derivative of an arbitrary function $G(q_i, t)$. Since the Hamiltonian depends on $2N$ independent variables q_i and p_j it is possible to consider a more general class of transformation mixing up qs and ps. It is of central importance to know what is the most general class of transformation from (q, p, H) to (Q, P, H') such that the form of the equations of motion in (2.2) is preserved.

(a) If the same equations have to be obtained from (Q_i, P_j) show that we must have the relation

$$p_i \, dq^i - H dt = P_i \, dQ^i - H' dt + dS , \qquad (2.6)$$

where dS is a total differential of some funtion S (called the 'generating function'). Taking $S = S(q, Q, t)$ show that

$$p_i = \frac{\partial S}{\partial q_i}, \quad P_i = -\frac{\partial S}{\partial Q_i}, \quad H' = H + \frac{\partial S}{\partial t} . \qquad (2.7)$$

(b) The equations of motion in a new coordinate system will be simplest if the new Hamiltonian vanishes, that is, if $H' = 0$. In this case P and Q will be constant and the evolution will be frozen in the new coordinate system. The vanishing of the new Hamiltonian can be achieved by using a generating function S which satisfies the equation

$$\frac{\partial S}{\partial t} + H\left(\frac{\partial S}{\partial q}, q\right) = 0 . \qquad (2.8)$$

Describe how, by integrating this equation, one can determine the trajectory $\mathscr{C}(t)$ in the phase space in terms of the old coordinates, q and p.

2.3 Integrals of motion [2]

If the equations of motion can be integrated, then we can obtain the functions $q_i = q_i\left(t; q_i^0, p_i^0\right)$ and $p_i = p_i\left(t; q_i^0, p_i^0\right)$ in terms of the initial conditions $\left(q_i^0, p_i^0\right)$. In principle we can invert these relations and express the $2N$ quantities q_i^0 and p_i^0 in terms of q_j, p_j and t. By their very construction, these $2N$ functions of q_j, p_j and t are constants, that is, their value remains the same as the system evolves. Out of these $2N$ constants, one constant is trivial since the origin of time is immaterial for a closed system. We can always change t to $(t + t_0)$ in a parametrized curve $\mathscr{C}(t)$ while still maintaining the physics invariant.

Thus any dynamical system has $(2N - 1)$ non-trivial constants of motion which restricts the system to a $[2N - (2N - 1)] = 1$-dimensional surface in phase space. This one-dimensional 'surface' is the trajectory $\mathscr{C}(t)$.

Most of the $(2N - 1)$ constants of motion described above are, however, fairly useless since these constants can be determined only after the equations of motion are integrated. They are, therefore, of no help in studying the dynamics. For certain physical systems there might exist a subset of k constants of motion $f_A(q, p), A = 1, 2, ..., k$, which are *independent* of t. Constants of motion which are independent of time are called 'integrals of motion'.

(a) Consider a system for which $N = 3$. Assume that the Lagrangian for the system can be expressed in the form $L = (1/2) \dot{q}_i \dot{q}^i - V(q_j)$, with $q^i = (x, y, z)$ representing the Cartesian coordinates. Determine the *obvious* integrals of motion for such a system when: (i) the potential $V(q_j)$ depends on all the coordinates non-trivially; (ii) $V(q_j)$ depends only on q_1 and q_2; (iii) $V(q_j)$ is axially symmetric about the axis q_1; (iv) $V(q_j)$ depends only on the radial distance $q_i q^i$.

(b) Even among integrals of motion, not all of them are of equal importance. Certain integrals of motion isolate the time evolution of the system to a restricted region in phase space, thereby simplifying the physics. Such integrals are called 'isolating integrals'. The integrals of motion which do not isolate the motion to a well-defined region in phase space are called 'non-isolating'.

Consider a two-dimensional harmonic oscillator described by the Lagrangian of the form

$$L = \frac{1}{2}\left(\dot{x}^2 + \dot{y}^2\right) - \frac{1}{2}\left(\omega_x^2 x^2 + \omega_y^2 y^2\right). \tag{2.9}$$

Determine for this system all the (i) constants of motion; (ii) isolating integrals of motion; (iii) non-isolating integrals of motion. What is the topology of the surface in the phase space Γ to which $\mathscr{C}(t)$ is confined?

(c) Consider a three-dimensional system described by the Lagrangian in spherical polar coordinates (r, θ, ϕ):

$$L = \frac{1}{2}\left(\dot{r}^2 + r^2\dot{\theta}^2 + r^2\sin^2\theta\,\dot{\phi}^2\right) + \frac{\alpha}{r} + \frac{\beta}{r^2}. \tag{2.10}$$

Determine for this system all the (i) isolating integrals of motion; (ii) non-isolating integrals of motion. While doing so pay special attention to the cases with $\alpha \neq 0, \beta = 0; \alpha = 0, \beta \neq 0$; and $\alpha \neq 0, \beta \neq 0$.

2.4 Surfaces of section [3,N]

Consider the motion of a particle in the x–y plane under the action of a gravitational potential $\phi(x, y)$. The phase space for this system is four dimensional. The conservation of energy will restrict the motion to a three-dimensional surface in phase space. The path of the system $\mathscr{C}(t)$ on this three-dimensional surface will cut the $y = 0$ plane in a two-dimensional surface. This surface can be characterized by the two coordinates (x, p_x).

One can obtain valuable insight into the nature of the motion by plotting the intersection of the curve $\mathscr{C}(t)$ with the $y = 0$ surface. In general, the motion will be spread over a bounded region in the (x, p_x) plane since we expect the motion to span a two-dimensional surface. If the system has any other hidden integrals of motion, then the motion can become one-dimensional and the points of intersection of $\mathscr{C}(t)$ with the $y = 0$ surface will lie on a curve. Study the nature of the surface of section by numerically integrating the equations of motion for the potential

$$\phi(x, y) = \frac{1}{2}v_0^2 \ln\left(R_c^2 + x^2 + \frac{y^2}{q^2}\right). \tag{2.11}$$

Take $q = 0.9, R_c = 0.14$ and the conserved energy to be $E = -0.337$.

2.5 Fully integrable systems [2]

A system with $2N$-dimensional phase space is called 'integrable' if there exist N integrals of motion including the Hamiltonian.

Consider an integrable system with the integrals of motion given by $F_m(p_i, q_i)$, where $m = 1, 2, ..., N$. We now make a transformation to a new set of coordinates and Hamiltonian $(Q, P, \mathcal{H})$ such that $P_m = F_m(q, p)$. Prove that $\mathcal{H}(Q, P) = \mathcal{H}(P)$ and $Q_i = (\partial \mathcal{H}/\partial P_i) t + \delta_i$, where δ_i are constants. Find the generating function for this transformation and show how the trajectory in phase space can be determined.

2.6 Systems with many particles [2]

The behaviour of systems with a large number of particles is complicated by the fact that it is impossible to integrate the dynamical equations governing the trajectory $\mathcal{C}(t)$ in phase space in closed form. It is therefore difficult – and, in some sense, unnecessary – to study the dynamics of the system by following the trajectories when $N \cong 10^{23}$ or so. For such systems we are often interested in the average behaviour of physical quantities. It is possible to obtain insight into the average behaviour along the following lines.

The exact location of the system in the phase space at any given time t will depend on the initial conditions from which the system has begun to evolve. We may assume that, by varying the initial condition over all possible values, one can make the system probe all regions of the phase space which are allowed by the isolating integrals of motion. For any closed system, we will have seven isolating integrals, namely, the total momentum, the angular momentum and the energy. Most often we consider the system to be confined in space by an external box of volume V. Such an external box is equivalent to a potential $U(\mathbf{x})$ such that $U = \infty$ outside the box and zero inside. In the presence of such a potential neither total momentum nor angular momentum is conserved. Thus only energy is a stable conserved quantity.

(a) In statistical mechanics it is usually assumed that the memory of initial conditions is destroyed by random collisions within a short time-scale. It may appear that this is not necessarily true for certain 'special' initial conditions. As an example, consider the following situation. We put a large number (N) of 'molecules' inside a rectangular box. At $t = 0$, all the particles are arranged to move only along the x-axis. We will assume the molecules to be hard spheres of radius r, mean separation l and mean velocity v. The collisions are elastic and occur head-on when two particles, which are moving along the same line in the x-direction, meet; i.e., the particles are distributed in such a way that there are no glancing collisions. The collisions with the walls are also elastic. It may seem that, in this situation, the initial conditions can be preserved for arbitrarily long periods of time and no thermodynamic equilibrium will ensue.

Prove that the above conclusion is erroneous. Consider the gravitational force on our system due to an electron located at the edge of the universe at a distance $D \cong 6000$ Mpc. Show that this perturbation is enough to destroy the orderly state within about 60 collisions if our system is made up of oxygen molecules at room temperature and pressure.

(b) Let $f\left(q_i, p_j\right)$ be some observable characterizing the system. The value of this observable at any given time t will depend not only on t but also on the initial conditions which determine the way the system has been prepared. If the observable f is a 'good', 'macroscopic' observable, then we expect it to approach a steady state value at late times when the system has relaxed into an equilibrium configuration. Show that the average value of such an observable $f(p, q)$ can be expressed in the form

$$\langle f\left(p, q\right)\rangle = \left(\frac{1}{g\left(E\right)}\right) \int f\left(p, q\right) \delta\left(E - H\left(p, q\right)\right) dp\, dq , \qquad (2.12)$$

where the quantity $g\left(E\right)$ (called the 'density of states') is defined to be

$$g\left(E\right) = \int \delta\left(E - H\left(q, p\right)\right) dq\, dp . \qquad (2.13)$$

(c) Consider, as a simple example, an ideal gas of N particles with the Hamiltonian

$$H\left(p, q\right) = \sum_{i=1}^{3N} \frac{p_i^2}{2m} + H_{\text{int}} , \qquad (2.14)$$

where the H_{int} represents the Hamiltonian describing molecular interactions between the particles. We shall assume that the effect of H_{int} is to merely redistribute the energy among the particles without actually contributing to the total energy. Calculate $g\left(E\right)$ for this system.

(d) Consider now a subsystem of the ideal gas consisting of particles $1, 2, 3, ..., k$. Let the probability that these k particles are in a small phase volume $d\Gamma_1 = dp_1...dp_{3k}\, dq_1...dq_{3k}$ be $\mathscr{P}_k\left(p_1...p_{3k}\right) d\Gamma_1$. Show that

$$\mathscr{P}_k \propto g_k\left(\epsilon\right) \exp\left[-\left(\frac{3N}{2E}\right)\epsilon\right] \propto g_k\left(\epsilon\right) \exp\left(-\beta\epsilon\right) , \qquad (2.15)$$

where the quantity β is *defined* by

$$\beta \equiv \frac{3N}{2E} = \frac{\partial}{\partial E} \ln g_N\left(E\right) \qquad (2.16)$$

and ϵ is the energy of the subsystem. Interpret this result. Show that the properly normalized probability distribution can be written as

$$\mathscr{P}\left(\epsilon\right) = \frac{1}{Z\left(\beta\right)} g_k\left(\epsilon\right) \exp\left(-\beta\epsilon\right) , \qquad (2.17)$$

where

$$Z(\beta) = \int_0^\infty \exp(-\beta\epsilon) g_k(\epsilon) \, d\epsilon = \int dp \, dq \, \exp(-\beta H(p,q)) \qquad (2.18)$$

and the integral is over the coordinates and momenta of the subsystem. The quantity Z is called the partition function of the subsystem.

(e) Generalize the above result to an arbitrary system interacting via short range forces. Let a system with energy E be divided into a small subsystem of energy E_1 and another large system of energy $E - E_1$. The probability for such a configuration to arise will be proportional to the phase volume occupied by the configuration, namely, $g_1(E_1) g_2(E - E_1)$. We define a quantity called entropy by the relation $S \equiv \ln g(E)$. Show that the most probable configuration is the one for which

$$\beta_1 \equiv \frac{\partial S_1}{\partial E_1} \equiv \frac{\partial \ln g_1(E_1)}{\partial E_1} = \beta_2 = \frac{\partial S_2}{\partial E_2} = \frac{\partial \ln g_2(E_2)}{\partial E_2}. \qquad (2.19)$$

(f) Show that the probability $\mathscr{P}(\epsilon)$ estimated in part (d) above can be very well approximated to be

$$\mathscr{P}(\epsilon) = \mathscr{P}(\bar{\epsilon}) \exp\left[-\frac{\beta^2}{2C_v}(\epsilon - \bar{\epsilon})^2\right], \qquad (2.20)$$

where the *most probable* energy $\bar{\epsilon}$ is determined by the condition

$$\left.\frac{\partial}{\partial\epsilon}\ln g(\epsilon)\right|_{\epsilon=\bar{\epsilon}} = \beta \qquad (2.21)$$

and $C_v = -\beta^2 (\partial\bar{\epsilon}/\partial\beta) = \beta^2 (\langle E^2 \rangle - \langle E \rangle^2)$. (This result shows that $C_v > 0$ for systems with short range interactions.) Use this result to evaluate Z approximately and show that

$$\ln Z \cong \ln g(\bar{\epsilon}) - \beta\bar{\epsilon} + \mathcal{O}\left(\frac{\ln N}{N}\right). \qquad (2.22)$$

Show also that mean energy and most probable energy match to the same order of accuracy.

(g) One is often interested in physical observables which can be expressed in the form

$$F = \frac{\partial H(p,q;\lambda)}{\partial\lambda}, \qquad (2.23)$$

where λ is some parameter which appears in the Hamiltonian H. Show that the mean values of such observables can be expressed as

$$\langle F \rangle = \frac{1}{\beta}\frac{\partial \ln Z}{\partial(-\lambda)}. \qquad (2.24)$$

Prove that the mean pressure exerted by a system of volume V is $\langle P \rangle = \beta^{-1}(\partial \ln Z / \partial V)$.

2.7 Thermodynamics [1]

Consider a macroscopic system exerting a pressure P. Suppose its volume changes by dV when its energy changes by dE. The central principle of thermodynamics states that the quantity $dE + P\,dV$ can be expressed in the form $\mathscr{T}\,d\mathscr{S}$, where $d\mathscr{S}$ is an exact differential. The quantity $\mathscr{T}$ is called the thermodynamic temperature and $\mathscr{S}$ is called the thermodynamic entropy.

Use the results of the preceding problems to show that we can take $k\mathscr{T} = \beta^{-1} \equiv T$ and $\mathscr{S} = k \ln g(E)$, where k is an (unimportant) constant. This constant may be set to unity by suitable choice of units. Show that, for an ideal gas,

$$\mathscr{S} = N \ln V + \frac{3}{2} N \ln \frac{2}{3} \frac{E}{N} + q(N) . \tag{2.25}$$

The function $q(N)$ can be determined from the condition that the entropy should be an extensive quantity. That is,

$$\mathscr{S}(\lambda N, \lambda V, \lambda E) = \lambda \mathscr{S}(N, V, E) . \tag{2.26}$$

Use this to show that $q(N) = -\ln(N!)$.

2.8 Systems without extensivity of energy [3,N]

The description of statistical mechanics and thermodynamics outlined in the last two problems assumes that the energy of the system is extensive. In other words, if the system is divided into two parts, 1 and 2, the total energy can be expressed as $E = E_1 + E_2$. This is possible only if the interaction energy between parts 1 and 2 is small compared to E_1 or E_2. If this is not the case then the description of statistical mechanics and thermodynamics needs to be modified drastically. We shall study the effects which arise due to non-extensivity of energy by using a simple toy model.

Consider a system with two particles described by a Hamiltonian of the form

$$H(\mathbf{P}, \mathbf{Q}; \mathbf{p}, \mathbf{r}) = \frac{\mathbf{P}^2}{2M} + \frac{\mathbf{p}^2}{2\mu} - \frac{Gm^2}{r} , \tag{2.27}$$

where $(\mathbf{Q}, \mathbf{P})$ are coordinates and momenta of the centre of mass, $(\mathbf{r}, \mathbf{p})$ are the relative coordinates and momenta, $M = 2m$ is the total mass, $\mu = m/2$ is the reduced mass and m is the mass of the individual particles. This system may be thought of as consisting of two particles (each of mass m) interacting via gravity. We shall assume that the quantity r varies in the interval (a, R). This is equivalent to assuming that the particles are hard spheres of radius $a/2$ and that the system is confined to a spherical box of radius R. We will study the 'statistical mechanics' of this simple toy model.

(a) Calculate the density of states $g(E)$ for the system and show that

$$g(E) \propto \begin{cases} \dfrac{1}{3}R^3(-E)^{-1}\left(1 + \dfrac{aE}{Gm^2}\right)^3 & \text{(for } E_1 < E < E_2) \\[4mm] \dfrac{1}{3}R^3(-E)^{-1}\left[\left(1 + \dfrac{RE}{Gm^2}\right)^3 - \left(1 + \dfrac{aE}{Gm^2}\right)^3\right] & \text{(for } E_2 < E < \infty) \end{cases},$$

$$(2.28)$$

with $E_1 = -Gm^2/a$ and $E_2 = -Gm^2/R$. Define the entropy and the temperature of the system by the relations $S(E) = \ln g(E)$ and $T(E) = (\partial S/\partial E)^{-1}$. Plot the function $T(E)$ and show that it is not monotonic in E and that the specific heat is negative in the range $E_1 < E < E_2$. Interpret the behaviour of $T(E)$. What is the physical origin of the negative specific heat region? Discuss how this system behaves when a or R is varied.

(b) It must be noted that most real life systems are indeed found in the range $E_1 < E < E_2$. For astrophysical systems this range is quite wide since $|E_2| \ll |E_1|$. This is precisely the region with negative specific heat. Since systems described by canonical distribution cannot exhibit negative specific heat (see problem 2.6(f)), it follows that canonical distribution will lead to a very different physical picture for this range of (mean) energies. In other words, canonical and microcanonical descriptions are very different for gravitating systems *in the most important range of energies*. It is, therefore, of interest to look at our system from the point of view of canonical distribution.

Compute the partition function for the system by performing the integral

$$Z(\beta) = \int d^3P \, d^3p \, d^3Q \, d^3r \exp(-\beta H).$$

$$(2.29)$$

Define a dimensionless temperature $t = (aT/Gm^2)$ and show that $Z(t)$ can be well approximated by the expression

$$Z \cong \frac{1}{3}t^3\left(\frac{R}{a}\right)^6\left(1 + \frac{3a}{2Rt}\right) \quad \text{(for } t > t_c)$$

$$\cong \left(\frac{R}{a}\right)^3 t^4(1 - 2t)^{-1}\exp\left(\frac{1}{t}\right) \quad \text{(for } t < t_c),$$

$$(2.30)$$

where $t_c^{-1} = 3\ln(R/a)$. Given $Z(\beta)$ one can compute $E(\beta)$ by the relation $E(\beta) = -(\partial \ln Z/\partial \beta)$. Compare the $T(E)$ obtained by this process with the $T(E)$ obtained in part (a) above. Interpret the difference between them. Show that if our system is put in contact with a heat bath it will undergo a rapid phase transition at a critical temperature.

2.9 Mean field equilibrium of gravitating systems [3]

The statistical definition of entropy used in the preceding section involves a $6N$-dimensional integration over the whole phase space. Under certain circumstances, it is possible to approximate this expression by one which involves only a six-dimensional integration. Such an approximation, which we shall call the mean field approximation, reduces the problem to a tractable level.

Consider a system of N particles interacting via a two-body potential $U(\mathbf{x}_i, \mathbf{x}_j)$. The entropy of the system can be defined in terms of the phase volume which is occupied:

$$e^S = g(E) = \frac{1}{N!} \int d^{3N}x \, d^{3N}p \, \delta(E-H) = \frac{A}{N!} \int d^{3N}x \left(E - \frac{1}{2} \sum_{i \neq j} U(\mathbf{x}_i, \mathbf{x}_j) \right)^{3N/2}. \tag{2.31}$$

To evaluate this quantity, we divide the spatial volume V into M cells of equal size, each large enough to contain many particles but small enough so that the variation of the potential U inside the cell can be neglected. Any particular configuration of particles can be characterized by giving the set of integers $(n_1, n_2, n_3, ...)$, where n_a is the number of particles located at $\mathbf{x}_a$.

(a) The mean field approximation consists of estimating $(\exp S)$ as $(\exp S_{\max})$, where $S_{\max}$ is the entropy corresponding to the configuration which maximizes S. Take the continuum limit in the mean field approximation and show that the density distribution is determined by the integral equation

$$\rho(\mathbf{x}) = A \exp[-\beta \phi(\mathbf{x})], \quad \phi(\mathbf{x}) = \int d^3y \, U(\mathbf{x}, \mathbf{y}) \rho(\mathbf{y}), \tag{2.32}$$

where the constants A and β can be determined if the total mass and energy of the system are known. (This is also called 'saddle-point approximation'.)

In the case of gravitational interaction, these equations take the form

$$\rho(\mathbf{x}) = A \exp[-\beta \phi(\mathbf{x})], \quad \phi(\mathbf{x}) = -G \int \frac{\rho(\mathbf{y}) d^3y}{|\mathbf{x} - \mathbf{y}|}. \tag{2.33}$$

(b) We have derived the mean field limit directly from the phase volume by using the saddle-point approximation. It is also possible to obtain the same result in a more intuitive approach which is very useful for studying the dynamical evolution of the system.

The central quantity in this approach is the one-particle distribution function $f(\mathbf{x}, \mathbf{p}, t)$, which may be defined through the (operational) relation

$$dm = f(\mathbf{x}, \mathbf{p}, t) d^3x \, d^3p, \tag{2.34}$$

where dm is the total mass of particles contained in the cell $(\mathbf{x}, \mathbf{p}; \mathbf{x} + d^3\mathbf{x}, \mathbf{p} + d^3\mathbf{p})$. This function is obtained from the full N-body distribution function of the system by integrating out $(N-1)$ variables:

$$f(\mathbf{x}_1, \mathbf{p}_1, t) \equiv \int f_N(\mathbf{x}_1, \mathbf{p}_1; \mathbf{x}_2, \mathbf{p}_2; \mathbf{x}_3, \mathbf{p}_3; \ldots \mathbf{x}_N, \mathbf{p}_N, t) d^3\mathbf{x}_2 \, d^3\mathbf{p}_2 \ldots d^3\mathbf{x}_N \, d^3\mathbf{p}_N. \tag{2.35}$$

We shall assume that the functions which appear in (2.34) and (2.35) are smooth. This excludes, for example, the function f being given as a sum of delta functions. If we use such a smoothed-out description, then it is clear that f contains far less information than f_N. In particular, the correlation between the particles is washed out in $f(\mathbf{x}, \mathbf{p}, t)$. Thus f could provide a useful description of the system only as long as the correlations are not significant.

Given the fundamental Hamiltonian of the system, it should be possible to write down an (approximate) equation satisfied by f. This task, however, is not easy and we shall merely note that the equation obeyed by f has the following generic form:

$$\frac{\partial f}{\partial t} + \mathbf{v} \cdot \frac{\partial f}{\partial \mathbf{x}} - \nabla \phi \cdot \frac{\partial f}{\partial \mathbf{v}} = C(f), \tag{2.36}$$

where $\mathbf{v} = \mathbf{p}/m$ and $\phi(\mathbf{x}, t)$ is the mean gravitational field produced by f:

$$\phi(\mathbf{x}, t) = -G \int \frac{f(\mathbf{y}, \mathbf{v}, t)\, d^3 y\, d^3 v}{|\mathbf{x} - \mathbf{y}|}. \tag{2.37}$$

The right-hand side of (2.36) describes the effect of 'collisions' on the evolution of f. The collisional evolution of the system, driven by $C(f)$, exhibits two reasonable properties: (i) the mean field energy E:

$$E = K + U \equiv \frac{1}{2} \int \mathbf{v}^2 f d^3 x\, d^3 v - \frac{G}{2} \int \frac{f(\mathbf{x}, \mathbf{v}) f(\mathbf{x}', \mathbf{v}')}{|\mathbf{x} - \mathbf{x}'|} dx\, dv\, dx'\, dv'$$

$$= \frac{1}{2} \int \mathbf{v}^2 f d^3 x\, d^3 v + \frac{1}{2} \int \rho(\mathbf{x}) \phi(\mathbf{x}) d^3 x \tag{2.38}$$

and mass M:

$$M = \int f\, d^3 x\, d^3 v \tag{2.39}$$

are conserved in the evolution; (ii) the mean field entropy, defined by

$$S = - \int f \ln f\, d^3 x\, d^3 p \tag{2.40}$$

does not decrease (and, generically, increases) during the evolution. It is, therefore, reasonable to ask: among the class of all f with the same E and M, which f maximizes the mean field entropy S?

Since we are using gravitational potential without a short distance cut-off there will be no *global* maximum for entropy, that is, we can construct configurations in which S is arbitrarily high. It is easy to devise such configurations by distributing part of the matter in a tightly bound core and the rest of the material as a halo at a large distance. The question, therefore, is really the following: are there configurations which are *local* maxima of the entropy? To answer this question we have to consider the change δS in entropy when f is changed by δf:

$$\delta S = \left(\frac{\delta S}{\delta f} \right) \delta f + \frac{1}{2} \left(\frac{\delta^2 S}{\delta f^2} \right) (\delta f)^2 + \ldots. \tag{2.41}$$

The vanishing of the first term defines the extremum condition and the sign of the second term decides whether the extremum is stable or merely a saddle point.

Show that among all $f(\mathbf{x}, \mathbf{v})$ with a given E, M and $\rho(\mathbf{x})$, the entropy is extremized by the Maxwellian distribution function with

$$f(\mathbf{x}, \mathbf{v}) = (2\pi T)^{-3/2} \rho(\mathbf{x}) \exp\left(-v^2/2T\right), \tag{2.42}$$

where $T = (2K/3m)$ and K is the kinetic energy of the system. Consider now a small variation in the spatial density $\delta\rho$ leading to a corresponding variation $\delta\phi$ in the gravitational potential. Show that S is maximized when ρ and ϕ satisfy equations (2.33). Also show that the second variation of S has the form

$$\delta^2 S = -\int d^3\mathbf{x} \left(\frac{\delta\rho\,\delta\phi}{2T} + \frac{(\delta\rho)^2}{2\rho}\right) - \frac{1}{3MT^2}\left(\int d^3\mathbf{x}\,\phi\,\delta\rho\right)^2. \tag{2.43}$$

Whether the extremum solution to (2.33) is stable or unstable is determined by this second variation.

2.10 Isothermal sphere [2,N]

A self-gravitating system is described by equations (2.33) in the mean field limit. The two equations may be combined to obtain

$$\nabla^2\phi = 4\pi G\rho_c \exp\left(-\beta\left[\phi(\mathbf{x}) - \phi(0)\right]\right), \tag{2.44}$$

which describes an 'isothermal sphere'. Introduce length, mass and energy scales by the definitions

$$L_0 \equiv (4\pi G\rho_c\beta)^{1/2}, \quad M_0 = 4\pi\rho_c L_0^3, \quad \phi_0 \equiv \beta^{-1} = \frac{GM_0}{L_0} \tag{2.45}$$

and express the radial distance r, density ρ, mass $M(r)$ contained within a radius r and potential ϕ in terms of these variables in dimensionless form:

$$x \equiv \frac{r}{L_0}, \quad n \equiv \frac{\rho}{\rho_c}, \quad m \equiv \frac{M(r)}{M_0}, \quad y \equiv \beta[\phi - \phi(0)]. \tag{2.46}$$

(a) Show that $n = (2/x^2), m = 2x, y = 2\ln x$ is a solution to (2.44) and that all other solutions to this equation tend to $y = 2\ln x$ for large x.

(b) Introduce the variables $v = (m/x)$ and $u = (nx^3/m)$ and show that the isothermal equations can be expressed as

$$\frac{u}{v}\frac{dv}{du} = -\frac{u-1}{u+v-3}, \tag{2.47}$$

with the boundary conditions $v = 0$ at $u = 3$ and $(dv/du) = -5/3$ at $(u, v) = (3, 0)$. Explain why a second-order differential equation could be reduced to a first-order equation in terms of u and v. Integrate equation (2.47) numerically and plot v as a function of u.

(c) It is clear from the analysis in part (a) that an isothermal sphere extends to infinity and has an infinite amount of mass. In order to obtain a more realistic solution, we may assume that the system is bounded by a spherical box of radius R. We can now characterize the isothermal sphere by a dimensionless parameter $\lambda \equiv (RE/GM^2)$. Show that an isothermal sphere cannot exist if $\lambda < \lambda_{\text{crit}}$, where $\lambda_{\text{crit}} \cong -0.335$. In other words, the entropy of a self-gravitating system has an extremum only if $\lambda > \lambda_{\text{crit}}$.

2.11 Gravity and degeneracy pressure [2]

The description in the last few problems relates to systems of point particles. Another class of systems which are of interest in astrophysics are self-gravitating 'fluid' systems described by an equation of state of the form $P = P(\rho)$. A degenerate Fermi gas of electrons or neutrons, for example, can be described by such an equation of state. If the degeneracy pressure can balance the inward pull of gravity, then it is possible to obtain a stable static configuration. Such configurations are of relevance in the study of the final stages of stellar evolution. The purpose of this exercise is to derive the behaviour of systems supported by the degeneracy pressure.

(a) As the material of the star is compressed, the individual atoms lose their identity and – if the temperature is not too high – the electrons become a degenerate Fermi gas. The equation of state of an ideal, degenerate Fermi gas was obtained in problem 1.12. It can be written as

$$P_{\text{nr}} = \frac{(3\pi^2)^{2/3}}{5} \frac{\hbar^2}{m} \left(\frac{Z}{A} \right)^{5/3} \left(\frac{\rho}{m_{\text{p}}} \right)^{5/3} \equiv \lambda_{\text{nr}} \rho^{5/3} \tag{2.48}$$

in the non-relativistic limit, and

$$P_{\text{r}} = \frac{1}{4} (3\pi^2)^{1/3} \hbar c \left(\frac{Z}{A} \right)^{4/3} \left(\frac{\rho}{m_{\text{p}}} \right)^{4/3} \equiv \lambda_{\text{r}} \rho^{4/3} \tag{2.49}$$

in the relativistic limit. Here m is the mass of the electron and m_{p} is the mass of the proton. Show that such a gas becomes *more and more* ideal as the density *increases* and satisfies the condition:

$$\rho \gg \left(\frac{me^2}{\hbar^2} \right)^3 \left(\frac{m_{\text{p}} A}{Z} \right) Z^2 . \tag{2.50}$$

(b) Further, show that, for a spherically symmetric system in which pressure balances gravity, the density ρ satisfies the equation

$$\frac{1}{r^2} \frac{d}{dr} \left(\frac{r^2}{\rho^{1/3}} \frac{d\rho}{dr} \right) = -\frac{12\pi G}{5\lambda_{\text{nr}}} \rho \tag{2.51}$$

in the non-relativistic case, and

$$\frac{1}{r^2} \frac{d}{dr} \left(\frac{r^2}{\rho^{2/3}} \frac{d\rho}{dr} \right) = -\frac{3\pi G}{\lambda_{\text{r}}} \rho \tag{2.52}$$

in the relativistic case.

(c) Consider the solution to the non-relativistic case first. Argue from dimensional considerations that the solution must have the form $\rho(r) \propto R^{-6}f(r/R)$, where R is the radius of the star. Hence, show that for such systems $R \propto M^{-1/3}$ and $\bar{p} \propto M^2$. With a suitable transformation of variables, reduce the equation and the boundary conditions to the form

$$\frac{1}{x^2}\frac{d}{dx}\left(x^2\frac{dy}{dx}\right) = -y^{3/2}, \quad y'(0) = 0, \quad y(1) = 0. \tag{2.53}$$

Numerical integration shows that $y(0) \cong 178$ and $y'(1) \cong -132$. From these values show that

$$MR^3 \cong \frac{92\hbar^6}{G^3 m^3 m_{\mathrm{p}}^5}\left(\frac{Z}{A}\right)^5. \tag{2.54}$$

(d) In the relativistic case, argue that the solution must have the form $\rho(r) \propto R^{-3}$, $f(r/R)$. Hence, show that $M = M_0$, some fixed constant. Convert the equation into the form

$$\frac{1}{x^2}\frac{d}{dx}\left(x^2\frac{dy}{dx}\right) = -y^3, \quad y'(0) = 0, \quad y(1) = 0. \tag{2.55}$$

Numerical integration gives $y(0) \cong 6.9$ and $y'(1) \cong -2.0$. Using these values show that

$$M_0 \cong \frac{3.1}{m_{\mathrm{p}}^2}\left(\frac{Z}{A}\right)^2\left(\frac{\hbar c}{G}\right)^{3/2} \cong 5.8\left(\frac{Z}{A}\right)^2 M_\odot \tag{2.56}$$

if the gas is made up of neutrons. (For $A \cong 2Z$, $M_0 \cong 1.45\,M_\odot$.)

2.12 Collisional evolution of gravitating systems [3]

The gravitational force acting on any particle in a self-gravitating system can be divided into two parts, $\mathbf{f}_{\mathrm{sm}} + \mathbf{f}_{\mathrm{fluc}}$. The $\mathbf{f}_{\mathrm{sm}}$ is due to the gravitational potential arising from the smooth distribution of matter. Since the matter is made up of individual particles, there will be a deviation from the smooth force $\mathbf{f}_{\mathrm{sm}}$, and this deviation is denoted by the fluctuating part of the force $\mathbf{f}_{\mathrm{fluc}}$. The latter part produces a slow diffusion of particles in the momentum space. This process is called 'soft collisions' and the time-scale for this process was estimated in problem 1.14.

In this exercise, we shall derive the equation satisfied by the distribution function $f(\mathbf{x}, \mathbf{p}, t)$ describing the system, taking into account the slow diffusion in the momentum space due to soft collisions. In general, such a diffusion process can be studied by an equation of the following kind:

$$\frac{df}{dt} = \frac{\partial f}{\partial t} + \mathbf{v} \cdot \frac{\partial f}{\partial \mathbf{x}} - \nabla\phi \cdot \frac{\partial f}{\partial \mathbf{v}} = -\frac{\partial J^\alpha}{\partial p^\alpha}. \tag{2.57}$$

The right-hand side is the divergence of a particle current J^α in the momentum space which is characteristic of diffusive process. We shall now determine the form of J^α.

(a) Consider a collision in which two particles with momenta $\mathbf{p}_1$ and $\mathbf{p}_2$ scatter to momenta $\mathbf{p}_1'$ and $\mathbf{p}_2'$. Let the momentum transfer in the collision be $\mathbf{q}$ so that the scattering is from $(\mathbf{p}_1, \mathbf{p}_2)$ to $(\mathbf{p}_1 + \mathbf{q}, \mathbf{p}_2 - \mathbf{q})$. It is, however, more convenient to characterize the scattering in terms of the mean momenta of the particles $\bar{\mathbf{p}}_1 = \mathbf{p}_1 + (1/2)\mathbf{q}$ and $\bar{\mathbf{p}}_2 = \mathbf{p}_2 - (1/2)\mathbf{q}$. Let the quantity $w\left(\mathbf{p} + \mathbf{q}/2, \mathbf{p}' - \mathbf{q}/2; \mathbf{q}\right) f(\mathbf{p}) f(\mathbf{p}')\, d\mathbf{p}'\, d\mathbf{q}$ represent the number of collisions which take place in unit time between some particle of momentum $\mathbf{p}$ and all particles with momentum in the range $(\mathbf{p}', \mathbf{p}' + d\mathbf{p}')$ such that the momentum transfer is in the range $(\mathbf{q}, \mathbf{q} + d\mathbf{q})$. (To simplify the notation, we will omit the x- and t-dependence of $f(\mathbf{x}, \mathbf{p}, t)$ and write it as $f(\mathbf{p})$ if no confusion is likely to arise.) From time reversibility of the scattering, we have the condition $W(\bar{\mathbf{p}}_1, \bar{\mathbf{p}}_2; \mathbf{q}) = W(\bar{\mathbf{p}}_1, \bar{\mathbf{p}}_2; -\mathbf{q})$.

Show that the x-component of the current in momentum space can be written in the form

$$J_x = \int\limits_{\text{all}} d\mathbf{l} \int\limits_{q_x > 0} d\mathbf{q} \int\limits_{p_{1x} - q_x}^{p_{1x}} dl_x\, W \left(\mathbf{l} + \frac{1}{2}\mathbf{q}, \mathbf{l}' - \frac{1}{2}\mathbf{q}, \mathbf{q}\right) \left[f(\mathbf{l}) f(\mathbf{l}') - f(\mathbf{l} + \mathbf{q}) f(\mathbf{l} - \mathbf{q})\right] .$$

$$(2.58)$$

Since the momentum transfer in each collision is expected to be small, we can expand all the quantities in a Taylor series in $\mathbf{q}$ and retain up to the first non-vanishing order. By this process reduce the current to the form

$$J_\alpha = \int d\mathbf{l}'\, B_{\alpha\beta}(\mathbf{l}, \mathbf{l}') \left(f \frac{\partial f'}{\partial l'_\beta} - f' \frac{\partial f}{\partial l_\beta}\right) , \qquad (2.59)$$

where

$$B_{\alpha\beta}(\mathbf{l}, \mathbf{l}') = \frac{1}{2} \int q_\alpha q_\beta\, W(\mathbf{l}, \mathbf{l}'; \mathbf{q})\, d\mathbf{q} . \qquad (2.60)$$

(b) Show that

$$B_{\alpha\beta} = \frac{1}{2} \frac{B_0}{|\mathbf{k}|} \left(\delta_{\alpha\beta} - \frac{k_\alpha k_\beta}{k^2}\right) , \qquad \mathbf{k} = \mathbf{l} - \mathbf{l}' , \qquad (2.61)$$

with $B_0 = 4\pi G^2 m^5 \ln(b_{\max}/b_{\min})$. The logarithmic term arises due to the long range nature of the interaction and has been estimated in problem 1.14 to be

$$\ln\left(\frac{b_{\max}}{b_{\min}}\right) \cong \ln N . \qquad (2.62)$$

(c) Show that the Maxwellian distribution with $f(\mathbf{l}) \propto \exp\!-\left(\sigma l^2\right)$ is a solution to the collisional Boltzmann equation.

(d) Manipulate the expression for J_α to reduce it to the form

$$J_\alpha(\mathbf{l}) = a_\alpha(\mathbf{l}) f(\mathbf{l}) - \frac{1}{2} \frac{\partial}{\partial l_\beta}\left(\sigma_{\alpha\beta}^2 f\right) , \qquad (2.63)$$

where $a_\alpha (\mathbf{l}) = B_0 \left(\partial \eta / \partial l_\alpha \right)$, $\sigma_{\alpha\beta}^2 = B_0 \left(\partial^2 \psi / \partial l_\alpha \, \partial l_\beta \right)$ and the potentials η and ψ satisfy the equation

$$\nabla_l^2 \eta (\mathbf{l}) = -8\pi f (\mathbf{l}) , \qquad \nabla^2 \psi = \eta . \tag{2.64}$$

(e) To understand the physics behind this equation, consider a simpler equation of similar form

$$\frac{\partial f}{\partial t} = \frac{\partial}{\partial v} \left[(\alpha v) f + \frac{1}{2} \sigma^2 \frac{\partial f}{\partial v} \right] \equiv -\frac{\partial J}{\partial v} , \tag{2.65}$$

where α and σ are constants. Argue that the αv-term provides a 'dynamical friction' which reduces the mean bulk velocity as time goes on, while the σ^2-term makes the velocity dispersion Δv^2 increase to a value (σ^2 / α) at late times. Find the solution to this equation with the initial condition $f (v, 0) = \delta (v - v_0)$.

By analogy the a_α-term in (2.63) may be interpreted as providing a dynamical friction and the $\sigma_{\alpha\beta}$-term as characterizing the diffusion in the velocity space. Explain the physical origin of the two terms and the necessity for the existence of both. Estimate $\alpha = a/v$ for a Maxwellian distribution of velocities using (2.63) and (2.64).

(f) Consider a system characterized by a distribution function of the form $f (\mathbf{l}) = f (|\mathbf{l}|)$. Show that, in such a case, the dynamical friction on a particle with momentum $\mathbf{l}$ is only due to particles with lower momentum.

2.13 Collisionless relaxation [3]

Consider the evolution of a self-gravitating system at time-scales short compared to the collisional relaxation time-scale. Assume that at $t = 0$ the particles were distributed in some finite region in the phase space occupying a phase volume Γ. The evolution of the system will distort the shape of the occupied region in phase space. We are interested in knowing the properties of the distribution function at late times given the fact that the evolution is collisionless.

(a) Argue that as the evolution proceeds the volume occupied in the phase space will remain constant but the phase elements will become mixed at finer and finer scales. Hence, show that the distribution function, when averaged ('coarse grained') over phase cells of some finite volume will reach a steady state at late times, even though the evolution continues to occur at finer and finer scales. Determine the form of the coarse grained distribution function at late times if: (i) the initial phase space density were constant at some regions of phase space and zero elsewhere; (ii) the evolution conserved the total energy of the system; and (iii) the relaxation process reached completion. What is the time-scale for this process?

(b) Consider a collisionless system with a coarse grained distribution function $f_c (\mathbf{x}, \mathbf{v})$. The volume of the phase space with phase density larger than q is defined by the relation

$$V(q) = \int d\mathbf{x}\, d\mathbf{v}\, \theta\, (f_c(\mathbf{x}, \mathbf{v}) - q)\,, \tag{2.66}$$

where $\theta(z) = 1$ for $z > 0$ and zero otherwise. Similarly, the mass contained in a region with a phase density larger than q is defined as

$$M(q) = \int d\mathbf{x}\, d\mathbf{v}\, f_c \theta\, (f_c(\mathbf{x}, \mathbf{v}) - q)\,. \tag{2.67}$$

Show that

$$M(V) = \int_0^V q(V')\, dV'\,, \tag{2.68}$$

where $q(V)$ is the inverse of the function $V(q)$.

Prove that collisionless evolution makes the distribution function f_1 evolve to f_2 only if $M_2(V) \leq M_1(V)$ for all V. In particular, show that the maximum value of the phase density can only decrease during collisionless evolution.

2.14 Equilibria of collisionless gravitating systems [2]

Systems like galaxies appear to be in the steady state as far as their macroscopic properties are concerned. They are also collisionless since the collisional time-scale for a galactic system is much larger than the age of the universe. Under such circumstances the steady state distribution function $f(\mathbf{x}, \mathbf{v}, t) = f(\mathbf{x}, \mathbf{v})$ can be constructed in a simple manner.

(a) Let $c_i(\mathbf{x}, \mathbf{v})$, with $i = 1, 2, ...$, be a set of isolating integrals of motion for orbits in some gravitational potential ϕ. Show that the distribution function which satisfies collisionless, steady state equations may be taken to be a function of the c_is. That is, $f(\mathbf{x}, \mathbf{v}) = f[c_i(\mathbf{x}, \mathbf{v})]$. The gravitational potential has to be obtained in a self-consistent manner using this distribution function.

(b) The simplest distribution function for a steady state collisionless Boltzmann equation is obtained when f depends only on the energy E of the particle. Let $\psi = -\phi + \phi_0$, where ϕ is the gravitational potential, and let $\epsilon = -E + \phi_0$, where E is the energy of the individual particle. We assume that $\phi < 0$ and $\phi \to 0$ as $r \to \infty$; similarly $E < 0$. The constant ϕ_0 is chosen so that ϵ and ψ are positive. Show that given any function $f(\epsilon)$ one can construct a valid solution to a collisionless Boltzmann equation by choosing $\psi(r)$ to be the solution to the equation

$$\frac{1}{r^2}\frac{d}{dr}\left(r^2\frac{d\psi}{dr}\right) = -16\pi^2 G \int_0^\psi d\epsilon\, f(\epsilon)\, \sqrt{2(\psi - \epsilon)}. \tag{2.69}$$

(c) One possible choice for the distribution function is a power law: $f(\epsilon) = A\epsilon^{n-3/2}$ for $\epsilon > 0$ and zero otherwise. What is the range of values of n for which this leads to sensible gravitational potential? In particular, solve for the gravitational potential and the density of matter when $n = 5$.

(d) In the above example, a particular form of $f(\epsilon)$ was chosen and the density distribution $\rho(r)$ was determined by integrating the equations. Show that, given $\rho(r)$, it is possible to determine an $f(\epsilon)$ along the following lines: (i) given $\rho(r)$ we can determine $\psi(r)$; eliminating r between $\rho(r)$ and $\psi(r)$ we can obtain the function $\rho(\psi)$; (ii) the distribution function is then given by

$$f(\epsilon) = \frac{1}{\sqrt{8\pi^2}} \frac{d}{d\epsilon} \int_0^\epsilon \left(\frac{d\rho}{d\psi}\right) \frac{d\psi}{\sqrt{\epsilon - \psi}}. \tag{2.70}$$

Does this always give a physically acceptable $f(\epsilon)$?

(e) A wider class of models can be obtained if one assumes that f can depend on the energy ϵ as well as the total angular momentum J^2. Construct a distribution function $f(\epsilon, J^2)$, such that the density distribution has a given form $\rho(r)$, by making f depend on ϵ and J^2 only through the quantity $Q = (\epsilon - J^2/2R^2)$ with some constant R. Show that if $f = f(\epsilon)$ then $\langle v_r^2 \rangle = \langle v_\theta^2 \rangle = \langle v_\phi^2 \rangle$, while if $f = f(\epsilon, J^2)$ then $\langle v_\theta^2 \rangle = \langle v_\phi^2 \rangle \neq \langle v_r^2 \rangle$.

2.15 Axisymmetric systems and halos [2]

(a) Consider a system with the density distribution $\rho(r, \theta) = \rho_0 S(\theta) (r_0/r)^2$ which is axisymmetric and falls as r^{-2}. Let $S(\theta)$ be normalized such that $\int_0^\pi S(\theta) \sin\theta \, d\theta = 2$. In general, we would have expected the radial and angular components of the force

$$F_r = -\frac{\partial\Phi}{\partial r}, \quad F_\theta = -\frac{1}{r}\frac{\partial\Phi}{\partial\theta} \tag{2.71}$$

to depend on θ as well as r. Show that $(\partial F_r/\partial\theta) = 0$, that is, the radial force is independent of the latitude. Hence, show that $F_r(r, \theta) = F_r(r) = -4\pi G\rho_0 r_0^2/r \equiv -v_0^2/r$.

(b) Prove that the potential must have the form

$$\Phi(r, \theta) = v_0^2 [\ln(r/r_0) + P(\theta)]. \tag{2.72}$$

Determine the equation connecting $P(\theta)$ and $S(\theta)$.

(c) What is the potential, if $S(\theta) = \delta_{\text{Dirac}}(\theta - \pi/2) + b$, which corresponds to the disc embedded in a spherical halo?

3
Fluid mechanics

This chapter deals with some aspects of fluid mechanics which are of relevance to high energy astrophysics and extragalactic astronomy. The problems are somewhat harder than those in the last two chapters and some of them involve a fair amount of algebra. Familiarity with the basic concepts of thermodynamics is needed.

3.1 Basic equations of fluid mechanics [2]

At the microscopic level, a fluid can be thought of as a collection of 'molecules'. Ignoring the internal structure of the molecules, we can specify the state of any molecule of mass m by giving its position $\mathbf{x}$ and momentum $\mathbf{p} = m\mathbf{v}$. Let $dN = f(\mathbf{x}, \mathbf{p}, t) d^3\mathbf{x}\, d^3\mathbf{p}$ denote the number of molecules in a phase volume $d^3\mathbf{x}\, d^3\mathbf{p}$ at time t. We are interested in the form and evolution of this distribution function.

(a) The distribution function changes due to two kinds of physical process. Macroscopic external force fields, for example the gravitational field, can exert forces on the molecules and influence their motion. Such a force, $\mathbf{F}_{sm}(\mathbf{x}, t) \equiv -\nabla U_{sm}(\mathbf{x}, t)$, will vary smoothly over the microscopic scales and can be derived from a suitable potential U_{sm}. A particular molecule will also experience the force, $\mathbf{F}_{coll}$, due to 'collision' with another molecule whenever it is close to another molecule. The collisions are assumed to be uncorrelated in space and time and conserve the total momentum and energy of the colliding particles. The dynamics of such a collision can be completely characterized by a differential cross-section $\sigma(\Omega)$, where Ω denotes the two angular coordinates θ and ϕ.

Show that, under these conditions, the evolution of the distribution function is governed by the equation

$$\frac{\partial f}{\partial t} + \mathbf{v} \cdot \nabla f - \nabla U_{sm} \cdot \frac{\partial f}{\partial \mathbf{p}} = C[f], \tag{3.1}$$

where

$$C[f] = \int |\mathbf{v} - \mathbf{v}_1| \left(f' f_1' - f f_1 \right) \sigma(\Omega)\, d\Omega\, d^3\mathbf{p}_1. \tag{3.2}$$

33

Here $f = f(\mathbf{x}, \mathbf{p}, t), f_1 = f(\mathbf{x}, \mathbf{p}_1, t), f' = f(\mathbf{x}, \mathbf{p}', t)$ and $f'_1 = f(\mathbf{x}, \mathbf{p}'_1, t)$. The collision described by $\sigma(\Omega)$ changes the initial momenta $(\mathbf{p}, \mathbf{p}_1)$ to the final momenta $(\mathbf{p}', \mathbf{p}'_1)$, conserving the total momentum and energy. Contrast the structure of this equation with the one derived in problem 2.12.

(b) Multiplying equation (3.1) by m, p^i and ϵ (where ϵ is the energy of the molecule) and integrating over $d^3\mathbf{p}$, show that

$$\frac{\partial \rho}{\partial t} + \frac{\partial}{\partial x^i}\left(\rho V^i\right) = 0, \tag{3.3}$$

$$\frac{\partial \left(\rho V^i\right)}{\partial t} + \frac{\partial T^{ik}}{\partial x^k} + \rho \frac{\partial U_{\text{sm}}}{\partial x^i} = 0, \tag{3.4}$$

$$\frac{\partial \left(N\bar{\epsilon}\right)}{\partial t} + \nabla \cdot \mathbf{q} + \rho \frac{\partial U_{\text{sm}}}{\partial x^k} V^k = 0, \tag{3.5}$$

where

$$N = \frac{\rho}{m} = \int d^3\mathbf{p}\, f(\mathbf{x}, \mathbf{p}, t), \quad \mathbf{V} = \frac{1}{N}\int \mathbf{v} f\, d^3\mathbf{p}, \quad \bar{\epsilon} = \frac{1}{N}\int \epsilon f\, d^3\mathbf{p} \tag{3.6}$$

and

$$T^{ik} = \int m v^i v^k f\, d^3\mathbf{p}, \quad \mathbf{q} = \int \epsilon \mathbf{v} f\, d^3\mathbf{p}. \tag{3.7}$$

(The indices $i, j, \ldots$, etc., take values 1,2,3 and summation over repeated indices is assumed. The positioning of indices as superscripts or subscripts has no special significance in this chapter and will be used interchangeably.) Interpret each of the terms in these equations.

(c) Consider a closed system without any external forces, that is, with $U_{\text{sm}} = 0$. To first approximation, we may assume that the fluid in different regions of space is in local thermodynamic equilibrium, even though the full system has not reached equilibrium. Then the distribution function in each volume element can be set equal to the equilibrium function f_0 corresponding to the density, temperature and macroscopic velocity $\mathbf{V}$ which exist at that region. At this level of approximation we shall ignore the gradients in the temperature and velocity in the fluid, which is equivalent to ignoring dissipative processes such as viscosity, thermal conduction, etc. Evaluate T^{ik} and $\mathbf{q}$ to this order of accuracy and show that

$$T^{ik} = \rho V^i V^k + P\delta^{ik}, \quad q^i = \rho \left(\frac{1}{2}V^2 + w\right) V^i, \tag{3.8}$$

where P is the pressure and $w = \epsilon + P/\rho$ is the enthalpy density of the fluid.

3.2 Viscosity and heat conduction [3]

(a) To study the dissipative processes, it is necessary to consider the next order of approximation. Assume that $f = f_0 + \delta f$, where δf is a small correction. Here $f_0 \propto \exp(-\epsilon/T)$ corresponds to the distribution function in local thermodynamic equilibrium, discussed in the last question. We shall write δf in the form $\delta f \equiv -\left(\partial f_0/\partial \epsilon\right) h = \left(f_0 h/T\right)$, separating out the factor $\left(f_0/T\right)$ from δf (which is done purely for later convenience). Show that

$$\langle h \rangle = 0, \qquad \langle h\epsilon \rangle = 0, \qquad \langle h\mathbf{p} \rangle = 0, \tag{3.9}$$

where the symbol $\langle\ \rangle$ stands for the averaging procedure

$$\langle a \rangle \equiv N^{-1} \int d^3\mathbf{p}\, f a. \tag{3.10}$$

To study the evolution of h we will linearize equation (3.1) with respect to h. Show that, to the lowest order, the linearized equation is

$$\frac{\epsilon\left(\mathbf{p}\right) - c_p T}{T}\mathbf{v} \cdot \nabla T + \left[m v^i v^k - \delta^{ik}\frac{\epsilon\left(\mathbf{p}\right)}{c_v} \right] V^{ik} = I\left(h\right), \tag{3.11}$$

where c_p and c_v are the specific heats at constant pressure and constant volume, respectively, and

$$V^{ik} = \frac{1}{2}\left(\frac{\partial V^i}{\partial x^k} + \frac{\partial V^k}{\partial x^i} \right), \tag{3.12}$$

$$I\left(h\right) = \int |\mathbf{v} - \mathbf{v}_1| f_{01}\left(h' + h'_1 - h - h_1\right)\sigma\left(\Omega\right) d\Omega\, d^3\mathbf{p}_1. \tag{3.13}$$

The quantity f_{01} represents the equilibrium distribution function f_0 with $\mathbf{p} = \mathbf{p}_1$.

(b) The first-order correction to f_0 in the form of δf will change the expressions for T_{ik} and q_i from the results obtained in problem 3.1(c), in which the gradients of T and $\mathbf{V}$ were ignored. In the next order, T_{ik} and q_i will contain terms which are proportional to the gradients $\left(\partial T/\partial x^i\right)$ and $\left(\partial V_k/\partial x^i\right)$. We will write these expressions, correct to linear order, as

$$T_{ik} = \rho V_i V_k + P\delta_{ik} - \sigma_{ik}, \qquad q_i = \left(\frac{1}{2}\rho V^2 + w\right) V_i + Q_i. \tag{3.14}$$

The first-order approximation to the Boltzmann transport equation (BTE, for short), derived in part (a) above, can be used to obtain the corrections to T_{ik} and q_i due to the spatial gradients. To solve the linearized BTE we will use the ansatz

$$h = g_1^i \frac{\partial T}{\partial x^i} + g^{ik} V_{ik} + g_3 \frac{\partial V_i}{\partial x_i}, \tag{3.15}$$

where g_1^i, g^{ik} and g_3 depend only on v^i and are to be determined by solving the linearized BTE. Note that, by definition, g_{ik} is symmetric; it may also be taken to be traceless since any part of g_{ik} proportional to δ_{ik} can be absorbed into the third term by redefining $g_3\left(v\right)$.

(i) Use symmetry considerations to show that g_1^i and g_{ik} must have the forms

$$g_1^i = \frac{v^i}{v} g_1(v), \qquad g^{ik} = \left(v^i v^k - \frac{1}{3}\delta^{ik}v^2\right) g_2(v).$$ (3.16)

(ii) Using these in the definition of T_{ik} and q_i show that

$$\sigma_{ik} = 2\eta\left(V_{ik} - \frac{1}{3}\delta_{ik}\nabla\cdot\mathbf{V}\right) + \zeta\delta_{ik}\nabla\cdot\mathbf{V}, \qquad Q_i = -\kappa\frac{\partial T}{\partial x^i} - V_k\sigma^{ik},$$ (3.17)

with η, ζ, and κ defined by the relations

$$\kappa = -\frac{1}{3T}\langle\epsilon v g_1(v)\rangle, \qquad \eta = -\left(\frac{m}{15T}\right)\langle v^4 g_2(v)\rangle, \qquad \zeta = -\frac{m}{3T}\langle v^2 g_3\rangle,$$ (3.18)

irrespective of the detailed nature of the collisions. The collisional cross-section $\sigma(\Omega)$ determines the actual form of $g_1(v), g_2(v)$ and $g_3(v)$.

(iii) Interpret the form of σ_{ik} and explain why no other term is expected to arise.

3.3 Macroscopic flow of fluid [1]

The last problem shows that the flow of fluids can be described in terms of some well-defined macroscopic variables and certain coefficients which describe the dissipative processes. It is conventional to choose these variables to be: (i) the macroscopic velocity $\mathbf{v}(\mathbf{x},t)$; (ii) the density $\rho(\mathbf{x},t)$; and (iii) the pressure $p(\mathbf{x},t)$. All the other thermodynamic quantities are determined once ρ and p are given. (The velocity field and pressure were denoted in problems 3.1 and 3.2 by capital letters $\mathbf{V}(\mathbf{x},t)$ and $P(\mathbf{x},t)$ to avoid confusion with microscopic velocity $\mathbf{v}$ and momentum $\mathbf{p}$. Since the latter quantities will not be needed any longer we shall use the lowercase letters $\mathbf{v}(\mathbf{x},t)$ and $p(\mathbf{x},t)$ to denote fluid velocity and pressure from now on.) We also found that the dissipative processes can be characterized by the coefficient of thermal conduction κ, and the coefficients of viscosity η and ζ.

(a) Show that the equations for fluid flow, correct to first order in the gradients of $\mathbf{v}$ and T, can be expressed as

$$\frac{d\rho}{dt} + \rho\nabla\cdot\mathbf{v} = 0,$$ (3.19)

$$\rho\frac{d\mathbf{v}}{dt} = -\nabla p + \eta\nabla^2\mathbf{v} + \left(\zeta + \frac{1}{3}\eta\right)\nabla(\nabla\cdot\mathbf{v}),$$ (3.20)

$$\rho T\left(\frac{ds}{dt}\right) = \sigma^{ik}\frac{\partial v_i}{\partial x^k} + \nabla\cdot(\kappa\nabla T),$$ (3.21)

where

$$\frac{d}{dt} = \frac{\partial}{\partial t} + \mathbf{v}\cdot\nabla$$ (3.22)

and s is the entropy density. Argue that the variations of s can be assumed to take place at constant pressure, so that $T \, (ds/dt) = c_p \, (dT/dt)$. In this case, we can write an equation for the temperature variation of the fluid as

$$c_p \rho \frac{dT}{dt} = \sigma^{ik} \frac{\partial v_i}{\partial x^k} + \nabla \cdot (\kappa \nabla T). \tag{3.23}$$

The temperature $T(\mathbf{x}, t)$ can be expressed in terms of $p(\mathbf{x}, t)$ and $\rho(\mathbf{x}, t)$ by the usual thermodynamic relations. These five equations will allow one to determine the five variables ρ, p and $\mathbf{v}$.

(b) Interpret these equations physically. Are they consistent with energy conservation? Increase of entropy? How should one define thermodynamic variables when the system has not reached full equilibrium and there exist gradients in macroscopic variables?

3.4 Visualization of flow [1]

Consider a small fluid element which is moving along with the fluid. Let $\mathbf{x}_0 (t)$ be the trajectory of some fiducial fluid particle located inside this element. Consider the separation vector $\mathbf{l}$ between this fiducial fluid particle and some other particle located a short distance away. The change in the shape of the fluid element is essentially determined by the rate of change of the vector $\mathbf{l}$. This quantity $(d\mathbf{l}/dt)$ will be the difference in the velocities of the two particles. For an infinitesimal fluid element this difference can be written as $-l_k \, (\partial \mathbf{v}/\partial x_k)$, which gives the change in velocity in the direction of $\mathbf{l}$. So the deformation of a fluid element is governed by the gradient of the velocity field $(\partial v_i/\partial x^k)$.

(a) Show that the velocity gradient can be expressed in the form

$$\frac{\partial v_i}{\partial x^k} = \frac{1}{3} \theta \delta_{ki} + S_{ki} + \frac{1}{2} \epsilon_{kil} \Omega^l, \tag{3.24}$$

where

$$\theta = \nabla \cdot \mathbf{v}, \quad \mathbf{\Omega} = \nabla \times \mathbf{v}, \quad S_{ki} = \frac{1}{2} \left[\frac{\partial v_i}{\partial x^k} + \frac{\partial v_k}{\partial x^i} \right] - \frac{1}{3} \theta \delta_{ki}. \tag{3.25}$$

Interpret each of the terms in (3.24) by arguing that: (i) θ (called 'expansion') produces a volume change without rotation or deformation of the fluid; (ii) S_{ik} (called 'shear') deforms the fluid element without volume change or rotation; and (iii) $\mathbf{\Omega}$ (called 'vorticity') rotates the fluid element without volume change or deformation.

(b) Let $\mathbf{A}(\mathbf{x}, t)$ be a vector field; it could be, for example, $\mathbf{v}(\mathbf{x}, t)$ or $\mathbf{\Omega}(\mathbf{x}, t)$. Interpret the equations

$$\frac{d\mathbf{A}}{dt} = \left[\frac{\partial}{\partial t} + (\mathbf{v} \cdot \nabla) \right] \mathbf{A} = 0, \tag{3.26}$$

$$\frac{d\mathbf{A}}{dt} - (\mathbf{A} \cdot \nabla) \mathbf{v} = 0, \tag{3.27}$$

for the vector field geometrically and contrast them with the simpler equation $(\partial \mathbf{A}/\partial t) = 0$.

(c) Consider any vector field $\mathbf{A}(\mathbf{x}, t)$ which has zero divergence, $\nabla \cdot \mathbf{A} = 0$. Suppose the fluid moves in such a way that the fluid lines of $\mathbf{A}$ are 'frozen' into the fluid. In other words, each fluid particle remains attached to the same field line as time goes on. (The fluid particles, of course, can move up and down the field line like beads on a string; but they cannot move from one field line to another.) Show that $\mathbf{A}$ must satisfy the equation

$$0 = \frac{\partial \mathbf{A}}{\partial t} - \nabla \times (\mathbf{v} \times \mathbf{A}) = \frac{d\mathbf{A}}{dt} - (\mathbf{A} \cdot \nabla)\mathbf{v} + (\nabla \cdot \mathbf{v})\mathbf{A}. \qquad (3.28)$$

Also show that the flux of $\mathbf{A}$ through any surface $\mathscr{S}$ bounded by a closed curve $\mathscr{C}$ remains constant as the surface moves through the fluid.

3.5 Ideal fluids [1]

The simplest (and most drastic) approximation which one can make in the study of fluids is to set $\kappa = 0, \eta = 0, \zeta = 0$, thereby ignoring all dissipative processes. Such a fluid is called 'ideal'. We shall now explore some of the properties of such a fluid.

(a) Show that $(ds/dt) = 0$ in an ideal fluid. This equation, along with the continuity and Euler equations, determines the motion of the fluid. In this case, show that the Euler equation can be transformed to the form

$$\frac{\partial \mathbf{\Omega}}{\partial t} = \nabla \times (\mathbf{v} \times \mathbf{\Omega}) + \frac{1}{\rho^2} \nabla \rho \times \nabla p, \qquad (3.29)$$

where $\mathbf{\Omega} = \nabla \times \mathbf{v}$ is called the vorticity. Using this, prove that any fluid particle carries with it a constant value of the quantity

$$\mathscr{S} = \frac{1}{\rho}(\nabla s \cdot \mathbf{\Omega}). \qquad (3.30)$$

(b) If, further, the pressure can be expressed as a function of density alone, $p = p(\rho)$, then the flow is called 'barotropic'. Give two examples of barotropic flow.

Consider an (imaginary) closed curve $\mathscr{C}$ drawn somewhere in the fluid. As the fluid moves, the curve will move in some fashion due to the motion of the fluid particles making up the curve. We define a quantity called the 'circulation', by the line integral

$$\Gamma = \oint_{\mathscr{C}} \mathbf{v} \cdot d\mathbf{l} \qquad (3.31)$$

taken around the closed contour $\mathscr{C}$ in the fluid. Show that the circulation Γ is conserved during the flow in an ideal, barotropic fluid. By conservation we mean that the value of the circulation defined through some contour $\mathscr{C}$ has the same value as the contour moves through the fluid. Equivalently, the vorticity, $\mathbf{\Omega}$, is frozen on to the fluid.

(c) The equations of motion for the ideal fluid take still simpler forms if more assumptions are made. A flow is called 'steady' if $(\partial/\partial t)$ (anything) $= 0$. In this case, prove Bernoulli's theorem: the quantity $B = \left[(1/2)\,v^2 + w\right]$ remains constant along any streamline, where $w = \epsilon + (p/\rho)$ is the enthalpy of the fluid.

(d) Consider next the situation in which the flow is not necessarily steady but is irrotational (i.e. $\nabla \times \mathbf{v} = 0$) and isentropic. In this case we can express the velocity field as a gradient: $\mathbf{v} = \nabla \psi$. Show that the first integral to the equations of motion is

$$\frac{\partial \psi}{\partial t} + \frac{1}{2}v^2 + w = 0. \qquad (3.32)$$

3.6 Viscous fluids [2]

A more realistic description of fluids will be possible if one assumes that the viscosity is not zero.

(a) Show that the transport of vorticity is now governed by the equation

$$\frac{\partial \mathbf{\Omega}}{\partial t} - \nabla \times (\mathbf{v} \times \mathbf{\Omega}) = \frac{\eta}{\rho}\nabla^2\mathbf{\Omega} + \frac{1}{\rho^2}(\nabla\rho) \times \left[\nabla p - \eta\nabla^2\mathbf{v} - \left(\zeta + \frac{1}{3}\eta\right)\nabla\theta\right], \qquad (3.33)$$

with $\theta = \nabla \cdot \mathbf{v}$. This is a generalization of equation (3.29).

(b) The general equations for a viscous fluid are extremely complicated to solve and progress can be made only by resorting to simplifying assumptions. What condition should be satisfied for the flow of a viscous fluid to be considered incompressible? To be treated as adiabatic?

(c) Consider the flow of an incompressible viscous fluid. Show that its flow is governed by the equation

$$\frac{\partial \mathbf{\Omega}}{\partial t} - \nabla \times (\mathbf{v} \times \mathbf{\Omega}) = \frac{d\mathbf{\Omega}}{dt} - (\mathbf{\Omega} \cdot \nabla)\mathbf{v} = \frac{\eta}{\rho}\nabla^2\mathbf{\Omega} \qquad (3.34)$$

and

$$\nabla^2 p = -\rho\frac{\partial^2}{\partial x_i \partial x_k}(v_i v_k). \qquad (3.35)$$

(d) It is possible to obtain some general results regarding steady, incompressible viscous flow of fluids past solid bodies. Such a problem is of relevance in several practical situations. In analysing such a flow, one introduces a quantity called the Reynolds number, R, defined to be

$$R = \frac{\rho u l}{\eta} \equiv \frac{ul}{v}, \qquad (3.36)$$

where l is a characteristic scale in the problem (which could, for example, be related to the dimension of the solid body moving through the fluid), u is the characteristic velocity and $v = (\eta/\rho)$. Show that the velocity and the pressure can be expressed in the form

$$\mathbf{v} = u\mathbf{f}_1\left(\mathbf{x}/l, R\right), \quad p = \rho u^2 f_2\left(\mathbf{x}/l, R\right), \qquad (3.37)$$

where $\mathbf{f}_1$ and f_2 are functions which can be determined only by solving the equation of motion. Further, show that the force acting on the solid body moving through the fluid can be expressed as $F = \rho u^2 l^2 f_3(R)$.

3.7 Instabilities and turbulence [3]

The analysis of viscous fluid flow, based on the equations derived above, becomes very complicated in realistic situations. To begin with, exact solutions can be found only with very special assumptions. What is more, these solutions are usually unstable to small perturbations. The growth of small perturbations in a fluid makes the flow complicated and turbulent in realistic situations. In this problem we shall study one of the simplest kinds of instability called Kelvin–Helmholtz instability.

(a) Consider two layers of an incompressible fluid with velocities $\mathbf{v}_1$ and $\mathbf{v}_2$. We assume that one layer is attempting to 'slide' on top of another, along the common surface which they share. We shall consider a small region of the fluid on both sides of the surface of separation with fluid velocities tangential to the surface; the surface itself is assumed to be a plane. This kind of a flow, which occurs very often in nature (e.g. wind blowing on the surface of a lake), happens to be unstable to small perturbations. Consider a perturbation in which the surface of discontinuity becomes slightly distorted from the planar shape. Show that, in an ideal incompressible fluid, such an instability will grow. Determine the time-scale for the growth.

(b) Instabilities like the one mentioned above make the fluid flow turbulent in any realistic situation. Explain quantitatively how a complicated turbulent velocity field can arise due to the repeated action of instability like the one described in (a) above.

(c) A highly turbulent flow cannot be analysed quantitatively from the basic equations of fluid mechanics. It is, however, possible to make some progress by introducing reasonable physical assumptions regarding the transport of energy in the turbulent flow. It is usual to assume that fully developed turbulence can be understood in terms of 'eddies' of different sizes coexisting at the same time. Let λ be the size of a generic eddy and let v_λ and ϵ_λ denote the typical change in the velocity (across the eddy) and energy per unit mass contained in the eddy. At very large scales, viscosity does not play any crucial role and the energy which is put into the system is merely transferred to the next level of smaller eddies. We will assume that the rate $\dot{\epsilon}$ at which energy per unit mass is fed from the larger scale to the smaller one is a constant independent of the scale. Let the velocity at the largest scale, L, be U. Show that

$$v_\lambda \cong U \left(\frac{\lambda}{L} \right)^{1/3}. \tag{3.38}$$

This process of energy transfer can go on till the viscous effects become important. Let this happen at some small scale λ_s. Show that $\lambda_s \cong (L/R^{3/4})$, where R is the Reynold's number. Also show that the power spectrum of the velocity field in the range $L^{-1} \ll k \ll \lambda_s^{-1}$ is

$$S(k) \propto \dot{\epsilon}^{1/3} k^{-5/3}. \tag{3.39}$$

This is known as the Kolmogorov spectrum.

3.8 Sound waves [2]

(a) Show that a *compressible* ideal fluid supports the propagation of small disturbances in density and pressure, in the form of a longitudinal wave, with velocity

$$c_s = \left(\frac{\partial p}{\partial \rho}\right)_s^{1/2} = \left(\frac{\gamma p}{\rho}\right). \tag{3.40}$$

What happens to the perturbations in vorticity or entropy?

Also show that: (i) the velocity perturbation is related to the pressure and density perturbation by $v' = (p'/\rho c_s) = (c_s \rho'/\rho)$, where the perturbations are indicated by primed quantities; (ii) the temperature perturbation is related to the velocity perturbation by $T' = c_s \beta T v'/c_p$, where $\beta = (1/V)(\partial V/\partial T)_p$ is the coefficient of thermal expansion.

(b) The dissipative processes will dampen the amplitude of a sound wave propagating through a fluid. Show that, in the lowest order of approximation, the amplitude is attenuated by a factor $\exp(-\gamma x)$, where

$$\gamma = \frac{\omega^2}{\rho c_s^3} \left[\left(\frac{4}{3}\eta + \zeta\right) + \kappa\left(\frac{c_p - c_v}{c_p c_v}\right)\right]. \tag{3.41}$$

(c) If a small disturbance is generated inside the fluid, it will propagate in all directions with the speed of sound, relative to the fluid. When viewed from a fixed coordinate system, the disturbance will move with a velocity which is the sum of sound velocity and the velocity of gas flow **v**. Different phenomena can arise depending on whether $|\mathbf{v}| > c_s$ or $|\mathbf{v}| < c_s$. Show that (i) if $|\mathbf{v}| < c_s$, then the disturbance can eventually reach all points in the fluid; while (ii) if $|\mathbf{v}| > c_s$, then the disturbance is propagated downstream only inside a conical region with opening angle $\alpha = \sin^{-1}(c_s/|\mathbf{v}|)$.

(d) The analysis in part (a) above assumed that the amplitude of the disturbance was small. But since c_s was proportional to $\rho^{1/3}$, regions of the fluid with higher density will be travelling faster. This will necessarily distort the shape of the wave and cause it to steepen. Such a distortion of shape cannot be understood in linearized theory and we need to study the exact equations.

Consider a one-dimensional fluid flow which is isentropic. All quantities depend only on x and t and we take $v_x = v(x,t), v_y = v_z = 0$. Show that the equations for isoentropic flow can be solved to give

$$v = \pm \int c_s \frac{d\rho}{\rho} = \pm \int \frac{dp}{\rho c_s}. \tag{3.42}$$

$$x = t\,[v \pm c_s(v)] + f(v). \tag{3.43}$$

Here $f(v)$ is an arbitrary function of velocity and these two relations implicitly determine $v(x, t)$. (Given $c_s(\rho)$, equation (3.42) determines $v = v(\rho)$; eliminating ρ between $c_s(\rho)$ and $v(\rho)$ one can obtain $c_s(v)$.) For the case of a polytropic gas simplify these equations to the form

$$v = F(q), \quad q = x - t\left\{\frac{1}{2}(\gamma + 1)v \pm c_0\right\} \tag{3.44}$$

and

$$c_s = c_0 \pm \frac{1}{2}(\gamma - 1)v \tag{3.45}$$

where F is another arbitrary function. Describe how the profile of the wave changes with time in this case.

(e) A gas in a semi-infinite cylindrical pipe ($x > 0$) is terminated by a piston at one end. The piston begins to move with a uniformly accelerated velocity $v_{\text{pist}} = \pm at$ at $t = 0$. Determine the gas flow, assuming the gas to be polytropic.

3.9 Supersonic flow of gas [3]

(a) Consider the flow of gas through a tube with variable area of cross-section $A(x)$. Assume that the flow is steady and adiabatic. Show that

$$\frac{dA}{A} = \frac{c_s^2}{v^2}\left(1 - \frac{v^2}{c_s^2}\right)\frac{d\rho}{\rho} = \frac{p}{\rho v^2}\left(1 - \frac{v^2}{c_s^2}\right)\frac{dp}{p} = -\left(1 - \frac{v^2}{c_s^2}\right)\frac{dv}{v}, \tag{3.46}$$

where A is the area of cross-section of the tube. Interpret the nature of gas flow in the regions with $v^2 < c_s^2$ and $v^2 > c_s^2$.

(b) Assume that the gas is polytropic with an index γ and constant specific heats. Integrate the equations of motion to obtain the T, p and ρ at some arbitrary location in the streamline in terms of the velocity v.

(c) Several astrophysical radio sources exhibit jet-like structures which, presumably, are made up of matter flowing out from a central source. These jets are fairly well collimated even when they propagate a long distance from the central source. Assume that the gas in the jet is ideal and polytropic (with an index 5/3) and that the flow is adiabatic and stationary. We may also assume that the jet is in pressure equilibrium with the outside gas, the pressure of which diminishes as r^{-2} as we go away from the central source. Estimate the cross-sectional area of the jet as a function of the outside pressure, as well as the distance from the source. Describe qualitatively the nature of the flow. Show that the angle subtended by the jet at the central source, $\theta = (A^{1/2}/r)$, diminishes as $r^{-2/5}$ so that a supersonic jet can remain well collimated.

3.10 Shock waves [2]

In the last two problems we considered situations in which gas flow can result in nearly discontinuous changes in the parameters. This occurs generically when the flow velocity changes from supersonic to subsonic values in some regions. Such a nearly discontinuous transition in density, pressure, etc., is called a shock.

Shocks arise in several astrophysical contexts in which there is a supersonic flow of gas relative to a solid obstacle. (In the laboratory frame, this could arise either because of an object moving through a fluid with supersonic velocities or because gas flowing with supersonic velocities suddenly encounters an obstacle.) Because of the boundary conditions near the solid body, the flow has to change from supersonic to subsonic somewhere near the object. Further, this information cannot reach upstream where the flow is supersonic. Hence, the pattern of flow changes very rapidly over a short region, resulting in a shock wave. In the frame in which the gas is at rest, we may consider the solid body to be moving with supersonic speed and pushing the gas. This necessarily leads to steepening of density profile at some place ahead of the body. We shall consider some of the properties of shock waves in this problem.

(a) Idealize the surface of discontinuity as a plane with gas flow occurring perpendicular to it. Let p_1, ρ_1, v_1 denote the state of the gas to the left of the shock front and let p_2, ρ_2, v_2 give the corresponding values on the right side. Assume that the orientation of gas flow is such that the flow is supersonic on the left side. Using the equations of flow for an ideal polytropic fluid with index γ, relate the ratios $(p_2/p_1), (\rho_2/\rho_1), (v_2/v_1)$ and (T_2/T_1) to the index γ and the Mach number $M_1 = (v_1/c_s)$ on one side.

(b) Show that, under the conditions of the above problem, $p_2 > p_1, \rho_2 > \rho_1, T_2 > T_1, v_2 < v_1$ and that the maximum possible increase in the density is by a factor $(\gamma + 1)/(\gamma - 1)$.

(c) In the above analysis, it was assumed that the discontinuity occurs at a very short region, so that it can be treated as infinitesimal for mathematical purposes. Estimate the order of magnitude of various terms in the equations of fluid motion with dissipation, and show that the discontinuity occurs over a region of the order of mean-free-path.

(d) A gas in a semi-infinite cylindrical pipe is terminated by a piston at one end. At $t = 0$, the piston begins to move towards the positive x-axis with a constant speed U. Determine the resulting gas flow.

3.11 Particle acceleration mechanisms [2]

Observations suggest that the energy spectrum of particles in many astrophysical contexts has a power law form. To generate and maintain such a power law, one often requires mechanisms which will accelerate the particles. We shall explore some general features of such acceleration mechanisms in this problem.

(a) Consider a process in which the energy of a particle is increased by a factor β each time the particle undergoes a basic acceleration process. (For the sake of definiteness we shall call the process a 'collision'.) Let P be the probability that the particle remains within the accelerating region after one collision. Show that such a process will lead to a power law distribution in energy with

$$n(E)\, dE \propto E^{-p}\, dE; \qquad p = 1 - \frac{\ln P}{\ln \beta}. \tag{3.47}$$

(b) As a first example of such a process, consider the scattering of relativistic particles by, say, large molecular clouds. Show that, in this case, $\beta = 1 + (8/3)(V/c)^2$ where V is the typical velocity of the scatterer and we ignore terms which are higher order in (V/c). It is assumed that the clouds are sufficiently massive not to undergo changes in velocities during the collisions. If the escape probability of the particle from the accelerating region is τ, estimate the index of the power law spectrum.

(c) As a second example, consider the acceleration of particles by a moving shock front. A shock front in, say, a supernova explosion will move through the surrounding interstellar medium (which has density ρ_1, pressure P_1 and temperature T_1) with a supersonic velocity U. A flux of high energy particles is assumed to be present both in front and behind the shock. As these particles pass through the shock in either direction, they are scattered by the turbulence behind the shock front and irregularities ahead of it, so that their velocity distribution rapidly becomes isotropic in the fluid rest frame on either side. Argue that the high energy particles gain energy when they cross the shock front from either direction and that, for this process,

$$\beta = 1 + \frac{U}{c}, \qquad P = 1 - \frac{U}{c}. \tag{3.48}$$

Hence show that, in this case, the energy spectrum has the form $n(E)\, dE \propto E^{-2}\, dE$ if $U \ll c$. (It is assumed that the particles are ultrarelativistic while the shock front is non-relativistic.)

3.12 Spherical accretion [3]

Consider a spherical object of mass M and radius R embedded inside a gas cloud. We expect the gas to flow towards the central mass due to gravitational attraction. In a highly idealized situation, we may consider the flow to be radially inwards at some steady rate $\dot{M}$. Further, assume that the gas is ideal and polytropic with an index γ where $1 < \gamma < 5/3$.

(a) Solve the equations of fluid flow and show that there are six different types of solution which are possible. Also show that (among the six) there is only one solution which satisfies both the following constraints: (i) the flow is subsonic at large distances and supersonic at small radius; (ii) the velocity $v(r)$ is a single-valued function of r.

(b) Relate the accretion rate to the density and sound speed of the gas at infinity and show

$$\dot{M} = \pi G^2 M^2 \frac{\rho(\infty)}{c_s^3(\infty)} \left[\frac{2}{5-3\gamma}\right]^{(5-3\gamma)/2(\gamma-1)}. \qquad (3.49)$$

Also show how one can determine the velocity and density profiles everywhere.

3.13 Accretion disc [3]

Spherically symmetric accretion can occur only when the angular momentum of the infalling matter is zero. In a more realistic situation one expects the gas to form a disc around the central object and flow in a spiralling pattern. For this process to occur it is necessary for the gas to lose the angular momentum efficiently due to some viscous process.

(a) Consider a viscous accretion disc with an inner radius r_{min} and outer radius r_{max}. It is assumed that: (i) there exists some suitable form of viscous stress which transports angular momentum radially outward; and (ii) the gas flows in nearly Keplerian orbits and spirals inwards slowly. The same viscous stress will also transport energy, some of which will be dissipated locally. Show that, in a steady state, the energy dissipated in an annular ring between r and $(r + dr)$ is

$$|\,dE\,| = \frac{3}{2}\dot{M}v_\phi^2(r)\frac{dr}{r}, \qquad (3.50)$$

where $\dot{M}$ is the mass flowing through the disc per unit time.

(b) Assume that this energy is radiated from each annular ring in the form of a blackbody with a temperature $T(r)$. The net spectra of the accretion disc will be a superposition of blackbody radiation with different temperatures. Show that the intensity scales approximately as $I_v \propto v^{1/3}$ in the Rayleigh–Jeans limit.

3.14 The Sedov solution for a strong explosion [3,N]

An explosion can release a large amount of energy E into an ambient gas on a time-scale which is extremely small. This will result in the propagation of a spherical shock wave centred at the location of the explosion. The resulting flow pattern can be determined everywhere if we make the following assumptions:

(i) The explosion is idealized as one which releases energy E *instantaneously* at the origin. The shock wave is taken to be so strong that the pressure p_2 behind the shock is far greater than the pressure p_1 of the undisturbed gas.

(ii) We neglect the original energy of the ambient gas in comparison with the energy E which it acquires as a result of the explosion.

(iii) We take the gas flow to be governed by the equations appropriate for an adiabatic polytropic gas with index γ.

Argue, from dimensional considerations, that the position of the shock front at time t is given by $R(t) \propto (Et^2/\rho_1)^{1/5}$. Solve the equations of gas flow by assuming that $R(t)$ is the only relevant length scale in the problem and that the solution must be self-similar.

3.15 Basics of magnetohydrodynamics [2]

The discussion so far has assumed that the fluid is electrically neutral. If the fluid is significantly ionized, with free electrons and ions being present, then the dynamics become much more complicated. Under certain conditions, it is possible to study such a fluid, taking only the magnetic field generated by the motion into account. This is called the 'magnetohydrodynamic approximation' or MHD for short. We shall now explore the basic features of this approximation. We shall assume that the fluid satisfies the following conditions:

(a) The electrons and ions in the fluid are 'locked' together by electrostatic forces so that the fluid can be described by a single velocity field $\mathbf{v}(\mathbf{x}, t)$. In reality, of course, there is an extremely tiny velocity difference $\mathbf{v}_{\text{diff}} = \mathbf{v}_e - \mathbf{v}_i$ between the ionic velocity $\mathbf{v}_i$ and the electronic velocity $\mathbf{v}_e$. This difference produces an electric current $\mathbf{j}$ which, in turn, produces a magentic field. In many astrophysical contexts, this difference $\mathbf{v}_{\text{diff}}$, is very small. Show, for example, that, in the outer zones of the Sun with $L \cong 2 \times 10^{10}$ cm, $n_e \cong 10^{23}$ cm^{-3}, $B \cong 10^3$ Gauss, the drift velocity is extremely tiny: $\mathbf{v}_{\text{diff}} \cong 10^{-12}$ cm s^{-1}. Thus we ignore the $\mathbf{v}_{\text{diff}}$ in discussing the macroscopic flow but retain it while calculating the electromagnetic field.

(b) In the local rest frame, S' of the plasma, the current $\mathbf{j}'$ is described by Ohm's law $\mathbf{j}' = \sigma \mathbf{E}'$; σ is asumed to be large so that even a tiny electric field can cause significant currents. We shall also assume that the velocity of the fluid in the lab frame is small compared to c so that we need to retain only terms of order (v/c). Show that these asumptions imply

$$\frac{1}{c}\left|\frac{\partial \mathbf{E}}{\partial t}\right| \ll \frac{|\mathbf{j}|}{c}, \quad \rho \ll \frac{|\mathbf{j}|}{c}, \quad |\mathbf{E}| \ll |\mathbf{B}|. \tag{3.51}$$

(c) Simplify Maxwell's equations and the equations of fluid dynamics using the above assumptions and show that MHD can be described by the following equations:

$$\frac{\partial \rho}{\partial t} + \nabla \cdot \rho \mathbf{v} = 0. \tag{3.52}$$

$$\rho \frac{d\mathbf{v}}{dt} = -\nabla p + \frac{1}{4\pi}(\nabla \times \mathbf{B}) \times \mathbf{B}. \tag{3.53}$$

$$\frac{\partial \mathbf{B}}{\partial t} - \nabla \times (\mathbf{v} \times \mathbf{B}) = \frac{c^2}{4\pi\sigma}\nabla^2 \mathbf{B}. \tag{3.54}$$

$$\rho T \frac{ds}{dt} = \sigma_{ik}\frac{\partial v_i}{\partial x^k} + \nabla \cdot (\kappa \nabla T) + \frac{j^2}{\sigma}. \tag{3.55}$$

(d) The magnetic Reynolds number R_M is defined by the relation

$$R_M = \frac{4\pi\sigma v L}{c^2}. \tag{3.56}$$

Show that if $R_M \gg 1$, then the magnetic field is frozen in the fluid and is carried away by the flow. On the other hand, if $R_M \ll 1$ the field diffuses in space on a time-scale $t_{\text{diff}} \cong 4\pi\sigma L^2/c^2$. Estimate R_M and t_{diff} for (i) intergalactic medium; and (ii) laboratory plasma.

3.16 Landau damping [1]

It is possible for charged particles to be accelerated through a resonant interaction with the waves in a plasma. This process has the effect of transferring energy from the plasma waves to the particles (and is called Landau 'damping', since it damps the energy of the waves). Consider a particle with charge q and velocity v_x which is suddenly acted upon by an electric field of the form $E_z = E_0 \cos(kx - \omega t)$. This is a transverse wave varying along the x-axis with electric field pointing in the z-direction. Show that the kinetic energy, contributed by the z-component of the velocity of the particle at time t, is

$$\frac{1}{2}mv_z^2 = \frac{q^2 E_0^2}{2m(kv - \omega)^2} \sin^2\left[(kv_x - \omega)t\right]. \tag{3.57}$$

Let $f(v_x)\,dv_x$ denote the number of particles in the system with velocities between v_x and $v_x + dv_x$ and assume that $f(v_x)$ varies slowly near $v_x = (\omega/k)$. Show that the rate of transfer of energy from the wave to the particle is

$$P \cong \frac{\pi q^2 E_0^2}{2mk} f(v_x)\Bigg|_{v_x = \omega/k}. \tag{3.58}$$

4

Radiation processes

The key radiative processes which are needed in extragalactic astronomy and high energy astrophysics are reviewed in this chapter. The reader is assumed to be familiar with four-vector notation and Maxwell's equations in tensor form. Familiarity with quantum mechanical time-dependent perturbation theory is also needed for some problems. Previous exposure to quantum theory of radiation will help but it is not essential. Some of the problems in this chapter elaborate on the concepts developed in chapter 1.

4.1 Feynman formula for classical radiation [2]

Consider a charge q moving along a trajectory $\mathbf{z}(t)$. The charge and current densities due to this particle are given by

$$\rho(\mathbf{x}, t) = q\delta\left(\mathbf{x} - \mathbf{z}(t)\right), \quad \mathbf{j} = q\mathbf{v}\delta\left(\mathbf{x} - \mathbf{z}(t)\right), \tag{4.1}$$

where $\mathbf{v}$ is the three-velocity of the charge. Maxwell's equations connecting the electromagnetic potentials to the charge and current densities can be written (in a suitable gauge) as

$$\Box\,\phi(\mathbf{x}, t) = 4\pi\rho(\mathbf{x}, t), \qquad \Box\mathbf{A}(\mathbf{x}, t) = \left(\frac{4\pi}{c}\right)\mathbf{j}(\mathbf{x}, t). \tag{4.2}$$

Here, $\Box$ stands for the d'Alembertian operator:

$$\Box \equiv \frac{1}{c^2}\frac{\partial^2}{\partial t^2} - \nabla^2. \tag{4.3}$$

(a) Solve these equations and show that the potentials can be expressed in the form

$$\phi(\mathbf{x}, t) = \int d^3\mathbf{y}\, dt'\, \frac{q}{R}\delta\left(\mathbf{y} - \mathbf{z}(t')\right)\delta\left(t' - t + \frac{R}{c}\right), \tag{4.4}$$

$$\mathbf{A}(\mathbf{x}, t) = \frac{1}{c}\int d^3\mathbf{y}\, dt'\, \frac{q\mathbf{v}(t')}{R}\delta\left(\mathbf{y} - \mathbf{z}(t')\right)\delta\left(t' - t + \frac{R}{c}\right), \tag{4.5}$$

where $R = |\mathbf{x} - \mathbf{y}|$.

(b) Given the electromagnetic potentials, **E** and **B** can be computed using the relations

$$\mathbf{E} = -\nabla\phi - \frac{1}{c}\frac{\partial\mathbf{A}}{\partial t}, \quad \mathbf{B} = \nabla \times \mathbf{A}. \tag{4.6}$$

Show that electric field is

$$\mathbf{E} = \frac{\hat{\mathbf{n}}}{R^2}\frac{q}{(1-v_R)} + \frac{q}{(1-v_R)}\frac{d}{cdt'}\left[\frac{\mathbf{R} - (\mathbf{v}/c)R}{R^2(1-v_R)}\right], \tag{4.7}$$

with $R = |\mathbf{R}|$, $\hat{\mathbf{n}} = (\mathbf{R}/R)$, $v_R = (\mathbf{v}\cdot\mathbf{n}/c)$ and all the quantities on the right are evaluated at the retarded time $t' = t - R/c$. Manipulate this expression to arrive at

$$\mathbf{E} = \frac{q}{R^2}\hat{\mathbf{n}} + \frac{qR}{c}\frac{d}{dt}\left(\frac{\hat{\mathbf{n}}}{R^2}\right) + \frac{q}{c^2}\frac{d^2\hat{\mathbf{n}}}{dt^2}. \tag{4.8}$$

All quantities on the right are evaluated at the retarded time. Interpret each of the three terms. (This formula was originally derived by Feynman.)

The conventional form for **E** and **B**, found in text books, can be obtained by direct differentiation of (4.7); this gives

$$\mathbf{E} = \frac{q}{R^2}\frac{(1-v^2/c^2)}{(1-v_R)^3}\left(\hat{\mathbf{n}} - \frac{\mathbf{v}}{c}\right) + \frac{q}{R^2(1-v_R)^3}\mathbf{R} \times \left[\left(\hat{\mathbf{n}} - \frac{\mathbf{v}}{c}\right) \times \frac{1}{c^2}\frac{d\mathbf{v}}{dt'}\right]. \tag{4.9}$$

The magnetic field is given by $\mathbf{B} = \hat{\mathbf{n}} \times \mathbf{E}$.

4.2 Radiation field in simple cases [2]

The formulas obtained in the last problem take simple forms in several special cases. We will discuss some of these cases now.

(a) Let a system of charges, confined to a finite region of size a, execute non-relativistic motion with $v \ll c$. We consider the radiation field of this system at a distance R with $R \gg a$. Also assume that $R \gg \lambda$, where λ is the typical wavelength of radiation. Show that the energy radiated per second into a solid angle $d\Omega$ is

$$\frac{dE}{dt\,d\Omega} = \frac{|\ddot{\mathbf{d}}|^2}{4\pi c^3}\sin^2\theta, \tag{4.10}$$

where $\mathbf{d} = \sum q_i\mathbf{x}_i$ is the dipole moment of the system and θ is the angle between $\mathbf{d}$ and $\hat{\mathbf{n}}$. What is the total energy radiated by the system of charges?

(b) Show that the energy radiated in the frequency interval $d\omega$ is

$$\frac{dE}{d\omega} = \frac{2}{3\pi}\left(\frac{\omega^4}{c^3}\right)|\mathbf{d}(\omega)|^2, \tag{4.11}$$

where $\mathbf{d}(\omega)$ is the Fourier transform of $\mathbf{d}(t)$.

(c) The velocity of a charged particle changes from $\mathbf{v}_1$ to $\mathbf{v}_2$ on a short time-scale τ. Consider the radiation emitted at frequencies $\omega \ll \tau^{-1}$. Show that the energy emitted into a solid angle $d\Omega$ in the frequency interval $d\omega$ is

$$\frac{dE}{d\omega\,d\Omega} = \frac{q^2}{4\pi^2 c^3}\left|\left\{\frac{\mathbf{v}_2 \times \hat{\mathbf{n}}}{1 - \hat{\mathbf{n}}\cdot\mathbf{v}_2/c} - \frac{\mathbf{v}_1 \times \hat{\mathbf{n}}}{1 - \hat{\mathbf{n}}\cdot\mathbf{v}_1/c}\right\}\right|^2$$

$$\cong \frac{q^2}{4\pi^2 c^3}\left[(\mathbf{v}_2 - \mathbf{v}_1) \times \hat{\mathbf{n}}\right]^2 = \frac{q^2}{4\pi^2 c^3}(\Delta v)^2 \sin^2\theta, \qquad (4.12)$$

where the expression in the second line is valid for non-relativistic motion. Also show that the total energy emitted in all directions is

$$\frac{dE}{d\omega} \cong \frac{2}{3\pi}\left(\frac{q^2}{c^3}\right)(\Delta v)^2. \qquad (4.13)$$

(d) The radiation in a frame in which the system of charges is at rest on the whole is given by $(dE'/dt'd\Omega') = f(\theta', \phi')$. Another frame is moving along the polar axis with velocity v. Find the angular distribution of radiation $(dE/dt\,d\Omega) = f(\theta, \phi)$ in this frame.

4.3 Radiation reaction [3]

(a) Show that the four-momentum dP^k radiated by a charge in a propertime interval ds can be written in a Lorentz invariant way as

$$dP^k = -\frac{2}{3}\frac{q^2}{c}\left(a^i a_i\right)u^k ds, \qquad (4.14)$$

where $a^i(s)$ is the four-acceleration of the charge and $u^i(s)$ is the four-velocity.

(b) A charged particle is accelerated by an electromagnetic field $(\mathbf{E}, \mathbf{B})$. Show that the instantaneous energy radiated by the charge in an interval Δt is

$$\Delta\mathscr{E} = \frac{2}{3}\frac{q^4}{m^2}\gamma^2\left[(\mathbf{E} + \mathbf{v}\times\mathbf{B})^2 - (\mathbf{E}\cdot\mathbf{v})^2\right]\Delta t \qquad (4.15)$$

in units with $c = 1$. The corresponding four-dimensional formula is

$$\Delta P^k = -\frac{2}{3}\left(\frac{q^4}{m^2}\right)\left(F^{ia}F_{ib}\right)u_a u^b u^k \Delta s. \qquad (4.16)$$

(c) Show that, for non-relativistic velocities, the radiation of energy will lead to a damping force on the charge given by

$$\mathbf{f}_{\text{damp}} = \frac{2}{3}q^2\ddot{\mathbf{v}} \qquad (4.17)$$

for any bounded motion. If the charge is accelerated by an electromagnetic field $(\mathbf{E}, \mathbf{B})$, then show that

$$\mathbf{f}_{\text{damp}} = \frac{2}{3}\frac{q^3}{m}\dot{\mathbf{E}} + \frac{2}{3}\frac{q^4}{m^2}(\mathbf{E}\times\mathbf{B}) \qquad (4.18)$$

in the instantaneous rest frame of the charge.

(d) For a charge moving with relativistic velocities, the radiation reaction force **f** can be expressed in terms of a four-force g^i with components $(\gamma \mathbf{f} \cdot \mathbf{v}, \gamma \mathbf{f})$, where $\gamma = (1 - v^2)^{-1/2}$. Show that

$$
\begin{aligned}
g^i &= \frac{2}{3} q^2 \left(\frac{d^2 u^i}{ds^2} - u^i u^k \frac{d^2 u_k}{ds^2} \right) \\
&= \frac{2}{3} q^2 \left[\frac{d^2 u^i}{ds^2} + u^i \left(a^k a_k \right) \right].
\end{aligned} \tag{4.19}
$$

(e) Consider a charged particle which is accelerated by an electromagnetic field F_{ab} which is constant in space and time. Let the energy–momentum tensor of this electromagnetic field be $T_b^a = -(4\pi)^{-1} \left[F^{ak} F_{bk} - (1/4)\delta_b^a \left(F_{ik} F^{ik} \right) \right]$. Show that the radiation reaction force on the charged particle can be written as

$$
g^i = \left(\frac{\sigma_T}{c} \right) \left[T^{ik} u_k - \left(T^{ab} u_a u_b \right) u^i \right], \quad \sigma_T = \frac{8\pi}{3} \left(\frac{q^2}{mc^2} \right)^2. \tag{4.20}
$$

4.4 Radiation in ultrarelativistic case [2]

The radiation field in the relativistic case is given by the second term in (4.9). Using this, show that the angular distribution of radiation is given by

$$
\frac{dE}{dt\, d\Omega} = \frac{q^2}{4\pi c^3} \left[2\mu^5 \frac{(\mathbf{n} \cdot \mathbf{a})(\mathbf{a} \cdot \mathbf{v})}{c} + \mu^4 a^2 - \mu^6 \gamma^{-2} (\mathbf{n} \cdot \mathbf{a})^2 \right], \tag{4.21}
$$

where $\mu = (1 - \mathbf{v} \cdot \mathbf{n}/c)^{-1}; \gamma = (1 - v^2/c^2)^{-1/2}$. From this expression conclude that an ultrarelativistic particle radiates mainly along the direction of motion, within a small range of angles given by

$$
\Delta\theta \cong \gamma^{-1} = \left(1 - \frac{v^2}{c^2} \right)^{1/2}. \tag{4.22}
$$

Also show that when **a** and **v** are parallel or perpendicular, one obtains simpler expressions:

$$
\begin{aligned}
\frac{dE}{dt\, d\Omega} &= \frac{q^2 a^2}{4\pi c^3} \left[\mu^4 - \mu^6 \gamma^{-2} \sin^2 \theta \cos^2 \Phi \right] \quad (\mathbf{v} \perp \mathbf{a}), \\
\frac{dE}{dt\, d\Omega} &= \frac{q^2 a^2}{4\pi c^3} \mu^6 \sin^2 \theta \quad (\mathbf{v} \parallel \mathbf{a}),
\end{aligned} \tag{4.23}
$$

where $\mu = (1 - v \cos\theta/c)^{-1}$ and ϕ is the azimuthal angle of $\hat{\mathbf{n}}$ relative to the plane containing **v** and **a**.

4.5 Bremsstrahlung [2]

(a) In a plasma, electrons are constantly accelerated during their collisions with ions. This leads to emission of radiation by the plasma, called thermal Bremsstrahlung. Show that the amount of energy emitted by the plasma, per unit volume per second per unit frequency interval, due to ion–electron collisions with impact parameter in the range $(b, b + db)$ is

$$\frac{dE}{dV \, d\omega \, dt} \cong \frac{16z^2 e^6 n^2}{3c^3 m^2 \langle v \rangle} \frac{1}{b} db, \tag{4.24}$$

where n is the number density of electrons and $\langle v \rangle$ is the mean velocity $(kT/m)^{1/2}$. Integrate this expression over b from b_1 to b_2. What are the physically reasonable choices for b_1 and b_2?

From the final expression one can obtain several quantities of astrophysical interest. We are often interested in the limit $\hbar\omega \ll kT$; in this limit, calculate:

(i) The amount of energy emitted by the plasma per unit volume per second in a frequency interval $d\omega$. This quantity,

$$j_\omega = \frac{dE}{dt \, dV \, d\omega}, \tag{4.25}$$

is called the 'specific emissivity' of the plasma. Show that $j_\omega \propto n^2 T^{-1/2}$.

(ii) The amount of energy emitted per second per unit volume of plasma over all frequencies. This quantity

$$J = \int_0^\infty d\omega \, j_\omega = \frac{dE}{dt \, dV} \tag{4.26}$$

is called the 'volume emissivity' of the plasma. Show that $J \propto n^2 T^{1/2}$.

(iii) The cooling time-scale for the plasma due to thermal Bremsstrahlung. This is defined to be

$$t_{\text{cool}} = \left[\frac{V}{E} \frac{dE}{dt \, dV} \right]^{-1}. \tag{4.27}$$

Show that $t_{\text{cool}} \propto n^{-1} T^{1/2}$.

(b) The electrons can also absorb photons while being accelerated by the ions. This process, called free–free absorption, can be characterized by a frequency-dependent cross-section $\sigma_{\text{ffa}}(\omega)$. Show that, in a plasma, $\sigma_{\text{ffa}}(\omega) \propto n\omega^{-2} T^{-3/2}$ for $\hbar\omega \lesssim kT$.

(c) In the calculations performed above, we have ignored the scattering of electrons by electrons and ions by ions. Is this justifiable?

4.6 Magnetic fields and synchrotron radiation [2]

(a) Most plasmas host magnetic fields and the charged particles move in helical paths around the magnetic field lines. Find the angular frequency of motion (ω_B, called the 'gyro frequency') and the orbital radius (r_B, called the gyro radius) of an electron in a magnetic field B. Estimate ω_B and r_B for (i) ionosphere; (ii) solar chromosphere; (iii) solar corona; (iv) interstellar medium; (v) neutron star magnetosphere; and (vi) intergalactic medium. How do ω_B and r_B compare with the plasma frequency ω_P and mean-free-path l in these systems? What conclusions can be drawn from this comparison?

(b) The charged particles, say electrons, radiate energy when accelerated by the magnetic field. Show that the net power radiated by a particle is

$$P_{syn} = \frac{4}{3}\sigma_T c\gamma^2\beta^2 U_B, \qquad (4.28)$$

where $U_B = (B^2/8\pi)$ is the energy density of the magnetic field. This loss of energy leads to the electrons 'cooling' in a time-scale

$$t_{syn\,cool} \cong \frac{\gamma mc^2}{P_{syn}} \cong 5 \times 10^8 \mathrm{s}\left[\gamma^{-1}B_{Gauss}^{-2}\right]. \qquad (4.29)$$

(c) Estimate the angular region into which most of the radiation is emitted if the motion is ultrarelativistic. Show that the power emitted per unit frequency interval per orbit has the frequency dependence

$$\frac{dE}{dt\,d\omega} = (\text{constant})F\left(\frac{\omega}{\omega_c}\right), \qquad (4.30)$$

where $\omega_c = (3/2)\left(\gamma^2\sin\alpha\right)(qB/mc)$ and α is the angle between the magnetic field and the velocity vector. What can one say about the polarization of synchrotron radiation?

(d) Show that if synchrotron radiation is emitted by a bunch of electrons with a power law spectrum $n(\epsilon)\,d\epsilon = k\epsilon^{-p}d\epsilon$ (for $\epsilon_1 < \epsilon < \epsilon_2$), then the total power emitted by the system varies with ν and B as $P_{tot} \propto \nu^{-\alpha}B^{1+\alpha}$, where $\alpha = (p-1)/2$.

4.7 Energy content of synchrotron sources [3]

Consider a source of volume V from which synchrotron radiation is detected. Assume that the radiation is emitted by relativistic electrons with a power law distribution in energy: $n(E)\,dE \propto E^{-p}\,dE$. The total energy content of the system will be the sum of the magnetic energy and the energy of the relativistic particles. Assume that the energy content in protons is β times that of relativistic electrons. Show that, for a source with a given luminosity, the total particle energy scales as $B^{-3/2}$ while the magnetic energy scales as B^2. Hence argue that the system must have a minimum magnetic field in order to produce a given luminosity.

4.8 Thomson scattering [2]

(a) Show that the cross-section for scattering of unpolarized electromagnetic waves by a charged particle – in the non-relativistic limit – is given by

$$\left(\frac{d\sigma}{d\Omega}\right) = \frac{1}{2}\left(\frac{e^2}{mc^2}\right)^2\left[1+(\hat{\mathbf{n}}\cdot\hat{\mathbf{n}}')^2\right], \qquad (4.31)$$

where $\hat{\mathbf{n}}$ is a unit vector in the direction of propagation of the wave and $\hat{\mathbf{n}}'$ is the scattered direction. Also show that the total cross-section is $\sigma_T = (8\pi/3)\left(e^2/mc^2\right)^2$ if $\hbar\omega \ll mc^2$. What is the mean-free-path for photons due to this scattering in a fully ionized plasma? If the charged particle was originally at rest, will the scattering transfer a significant amount of energy to the particle? What about momentum?

(b) What is the force exerted by the electromagnetic wave on the charged particle?

(c) Consider a charge which is moving with a velocity $\mathbf{v}$ through an *isotropic* bath of photons with energy density U_{rad}. Show that the particle will feel a drag force

$$\mathbf{f}_{\mathrm{drag}} = -\frac{4}{3}\sigma_T U_{\mathrm{rad}}\gamma^2\left(\frac{\mathbf{v}}{c}\right) \cong \frac{4}{3}\sigma_T\left(aT^4\right)\left(\frac{\mathbf{v}}{c}\right). \tag{4.32}$$

The second expression is valid when $v \ll c$ and U_{rad} is due to a thermal bath of photons at temperature T.

4.9 Compton and inverse Compton effects [3]

Consider a plasma embedded in a radiation field of temperature T. The scattering of photons by the electrons in the plasma will continuously exchange energy between the two. The high energy photons with $m_e v^2 \ll \hbar\omega$ will transfer energy to the low energy electrons, but will gain energy from high energy electrons (with $\hbar\omega \ll m_e v^2$). In thermal equilibrium, the net transfer of energy will be zero. But if the electron temperature T_e is very different from the photon temperature, there can be a net transfer of energy. When $T_e \gg T_{\mathrm{rad}}$, the electrons 'cool' by transferring energy to the photons. The spectrum of the photons is distorted by this process.

(a) Consider a scattering of a photon by an electron which was originally at rest. Show that the frequencies of the final and initial photons are related by

$$\frac{\omega_f}{\omega_i} = \left[1 + \left(\frac{\hbar\omega_i}{m_e c^2}\right)(1 - \cos\theta)\right]^{-1}, \tag{4.33}$$

where θ is the angle between the initial and final photon directions. Using this, show that the average energy lost by the photon (and gained by the electron) is

$$\langle\Delta\epsilon\rangle = -\left(\frac{\hbar\omega_i}{m_e c^2}\right)\hbar\omega_i \tag{4.34}$$

if $\hbar\omega_i \ll m_e c^2$. Thus the photon loses energy while scattering off low velocity ($v \cong 0$) electrons.

(b) The photons can gain energy when they scatter off high energy electrons. The scattering of a photon by a high energy electron is best studied in the rest frame of the electron. Show that the net power lost by an electron (and gained by a photon) is

$$\frac{dE}{dt} = \frac{4}{3}\sigma_T c U_{rad} \gamma^2 \left(\frac{v}{c}\right)^2 \equiv P_{inComp}, \qquad (4.35)$$

where $U_{rad} = aT^4$ is the energy density of the radiation field. In the non-relativistic limit, for electrons with temperature T_e, this gives

$$\left(\frac{dE}{dt}\right)_{nr} \cong \frac{4}{3}\sigma_T c U_{rad} \left(\frac{3kT_e}{m_e c^2}\right). \qquad (4.36)$$

Show that the average fractional energy gained by a photon per collision is

$$\left\langle \frac{\Delta \epsilon}{\epsilon} \right\rangle = \frac{\langle \Delta \epsilon \rangle}{\hbar \omega_i} = \frac{4}{3}\gamma^2 \left(\frac{v}{c}\right)^2 = \begin{cases} (4/3)\gamma^2 & (\text{if } v \cong c) \\ (4kT_e/m_e c^2) & (\text{if } v \ll c) \end{cases}. \qquad (4.37)$$

Combining with (4.34) we find that the net energy change of photons in Compton scattering with a thermal distribution of electrons with temperature T_e is

$$\left\langle \frac{\Delta \epsilon}{\epsilon} \right\rangle = -\frac{\hbar \omega_i}{m_e c^2} + \frac{4kT_e}{m_e c^2}, \qquad (4.38)$$

for $v \ll c$.

The process described above acts as a major source of cooling for relativistic plasma as well as a mechanism for producing high energy photons. The time-scale for 'Compton cooling' of individual relativistic particle is

$$t_{Compcool} \cong \frac{\gamma m c^2}{P_{inComp}} \cong 4 \times 10^{-3} \gamma^{-1} \beta^{-2} \left(\frac{T_{rad}}{10^6 K}\right)^{-4} \text{s}, \qquad (4.39)$$

where T_{rad} is the radiation temperature. In the case where electrons are non-relativistic with temperature T_e, this time-scale is

$$t_{Compcool} \cong \frac{kT_e}{P_{inComp}} \cong \frac{1}{n_\gamma \sigma_T} \left(\frac{m}{T_{rad}}\right) = 1.3 \times 10^{-3} \left(\frac{T_{rad}}{10^6 \text{ K}}\right)^{-4} \text{s}. \qquad (4.40)$$

4.10 Comptonization [3]

The energy transfer between electrons and photons takes place through Compton scattering. This process dominates when (i) scattering is more important than true absorption; and (ii) the temperature of the low density electron gas is far higher than the Planck distribution with the same energy. In that case, there is net energy transfer from the electrons to the photons which distorts the original photon spectrum. We will now work out the details of this process.

(a) In a scattering of a photon of frequency v with an electron of energy E, resulting in a photon of frequency v' and an electron of energy E', the energy transfer is

$$h(v' - v) \equiv h\Delta \cong -\left(\frac{hv}{mc}\right) \mathbf{p} \cdot (\hat{\mathbf{n}} - \hat{\mathbf{n}}'), \qquad (4.41)$$

where $\mathbf{p}$ is the initial momentum of the electron and $\hat{\mathbf{n}}$ and $\hat{\mathbf{n}}'$ are the directions of the initial and final photons. This is the lowest order result for $(\Delta v/v)$ and is $\mathcal{O}(v/c)$. Prove this result.

(b) The evolution equation for photon number density in a homogeneous medium is

$$\frac{\partial n(v)}{\partial t} = \int d^3p \int d\Omega \left(\frac{d\sigma}{d\Omega}\right) c \left[n(v)\left(1 + n(v')\right)N(E) - n(v')\left(1 + n(v)\right)N(E')\right],$$
(4.42)

where $(d\sigma/d\Omega)$ is the electron–photon scattering cross-section, $n(v)$ is the photon distribution function and $N(E)$ is the electron distribution function. Explain the origin of each of the terms in this equation.

(c) The integro-differential equation in (4.42) can be approximated as a differential equation when $\Delta \ll 1$. (This is similar to the approach used in problem 2.12.) Expand $n(v') = n(v + \Delta)$ and $N(E') = N(E - h\Delta)$ in a Taylor series in Δ, retaining up to quadratic order. Assuming that $N(E)$ is a Maxwellian distribution with temperature T, show that equation (4.42) can now be written as

$$\frac{\partial n}{\partial t} = \left(\frac{h}{kT}\right)\left(\frac{\partial n}{\partial x} + n(n+1)\right)I_1$$
$$+ \frac{1}{2}\left(\frac{h}{kT}\right)^2\left(\frac{\partial^2 n}{\partial x^2} + 2(1+n)\frac{\partial n}{\partial x} + n(n+1)\right)I_2,$$
(4.43)

where $x = (hv/kT)$ and I_1 and I_2 are the integrals

$$I_1 = \int d^3p\, d\Omega \left(\frac{d\sigma}{d\Omega}\right) cN(E)\,\Delta,$$
(4.44)

$$I_2 = \int d^3p\, d\Omega \left(\frac{d\sigma}{d\Omega}\right) cN(E)\,\Delta^2.$$
(4.45)

(d) Use the expression for average energy transfer $\Delta E = (hv/mc^2)(hv - 4kT)$ (see equation (4.38)) to show that

$$I_1 = (\sigma_T n_e)\left(\frac{hv}{mc^2}\right)(4 - x) = (\sigma_T n_e)\left(\frac{kT}{mc^2}\right)x(4 - x).$$
(4.46)

(e) Using (4.41) evaluate I_2 and show that

$$I_2 = 2\left(\frac{v}{mc}\right)^2 (kT)(mc)\, n_e \sigma_T.$$
(4.47)

Hence, arrive at the final form of the equation (called Kompaneet's equation):

$$\frac{\partial n}{\partial y} = \frac{1}{x^2}\frac{\partial}{\partial x}\left[x^4\left(\frac{\partial n}{\partial x} + n^2 + n\right)\right],$$
(4.48)

where

$$y = t\left(\frac{kT}{mc^2}\right)n_e\sigma_T c.$$
(4.49)

(f) In most of the astrophysically interesting cases one can ignore the n^2 term with respect to n. The evolution of the total energy of the photons,

$$E_{\text{pho}}(y) = \frac{2(kT)^4}{h^3 c^3} \int_0^\infty n(x) x 4\pi x^2 dx, \qquad (4.50)$$

can be then determined using (4.48). Show that, if $n(x)$ vanishes suficiently fast for large x, then

$$\frac{dE_{\text{pho}}}{dy} = \frac{8\pi (kT)^4}{(hc)^3} \left[4 \int_0^\infty nx^3 dx - \int_o^\infty nx^4 dx \right]. \qquad (4.51)$$

If one can further assume that most of the photon energy is concentrated at $x \ll 1$, then we can ignore the second term with respect to the first on the right-hand side. Show that, in that case, the photon energy, E_{pho}, increases as

$$E_{\text{pho}}(y) \cong E_{\text{pho}}(0) \exp(4y). \qquad (4.52)$$

The characteristic e-folding time is

$$t_{\text{Comp}} = \left(\frac{mc^2}{4kT} \right) \left(\frac{1}{n_e \sigma_T c} \right). \qquad (4.53)$$

(g) Solve (4.48) when both n and n^2 are ignorable compared to $(\partial n/\partial x)$. Show that the solution is

$$n(x,y) = \frac{1}{(4\pi y)^{1/2}} \int_0^\infty \frac{d\mu}{\mu} n(\mu, 0) \exp \left[-\frac{1}{4y} \left(3y + \ln \frac{x}{\mu} \right)^2 \right]. \qquad (4.54)$$

Consider an initial distribution of photons with $n(x,0) \cong x^{-1}$, which corresponds to the Rayleigh–Jeans limit of the Planck spectrum. Show that this distribution will evolve to

$$n(x,y) \cong x^{-1} e^{-2y} \qquad (4.55)$$

due to Comptonization. Hence, the temperature of the radiation will appear to have been diminished by a factor e^{-2y} at the Rayleigh–Jeans end. Such an effect has been observed when microwave background radiation passes through hot intergalactic gas in clusters.

(h) Equation (4.51) can also be used to study the change in the electron energy. Since $(E_{\text{pho}} + E_e)$ should be a constant $(dE_e/dy) = -(dE_{\text{pho}}/dy)$. Calculate (dE_e/dy) when $n(x) = n_0 \exp(-x/\alpha)$. What is the cooling rate of electrons for $\alpha \ll 1$?

4.11 Quantum theory of radiation [3]

In the quantum mechanical treatment, an electromagentic field is treated as a bunch of photons. The emission and absorption of radiation are to be represented as processes which cause transitions between states with different numbers of photons. In general, such a transition will make the atomic system change from some state i to some other state f.

(a) Consider a free electromagnetic field, represented by a vector potential $A^i = (0, \mathbf{A})$, with $\nabla \cdot \mathbf{A} = 0$. (This is called the 'transverse gauge'.) Let us assume that the field is confined to a cubic box of volume L^3 and vanishes on the boundaries of the box. Expand $\mathbf{A}(t, \mathbf{x})$ as

$$\mathbf{A}(t, \mathbf{x}) = \sum_{\mathbf{k}} \mathbf{q_k}(t) \exp(i\mathbf{k} \cdot \mathbf{x}) = \sum_{\mathbf{k}} \left(\mathbf{a_k} \exp(i\mathbf{k} \cdot \mathbf{x}) + \mathbf{a_k^*} \exp(-i\mathbf{k} \cdot \mathbf{x}) \right), \quad (4.56)$$

with $\mathbf{k} = (2\pi/L)\,\mathbf{n}$, where $\mathbf{n}$ is a vector with integer components. Show that $\mathbf{k} \cdot \mathbf{a_k} = \mathbf{k} \cdot \mathbf{a_k^*} = 0$. This allows the vector $\mathbf{a_k}$ to have two components $\mathbf{a}_{\mathbf{k}d}\,(d = 1, 2)$ in the plane perpendicular to $\mathbf{k}$, with $\{a_{\mathbf{k}1}, a_{\mathbf{k}2}, \mathbf{k}\}$ forming an orthogonal system. Show that the Hamiltonian for the electromagnetic field

$$H = \frac{1}{8\pi} \int d^3x \left(E^2 + B^2 \right) \quad (4.57)$$

can be expressed in the form

$$H = \sum_{\alpha=1}^{2} \sum_{\mathbf{k}} \frac{1}{2} \left(P_{\mathbf{k}\alpha}^2 + \omega_{\mathbf{k}}^2 Q_{\mathbf{k}\alpha}^2 \right), \quad \omega_{\mathbf{k}}^2 = |\mathbf{k}|^2, \quad (4.58)$$

with

$$Q_{\mathbf{k}\alpha} = \left(\frac{L^3}{4\pi} \right)^{1/2} \left(a_{\mathbf{k}\alpha} + a_{\mathbf{k}\alpha}^* \right), \quad (4.59)$$

$$P_{\mathbf{k}\alpha} = -i\omega_{\mathbf{k}} \left(\frac{L^3}{4\pi} \right)^{1/2} \left(a_{\mathbf{k}\alpha} - a_{\mathbf{k}\alpha}^* \right). \quad (4.60)$$

These expressions shows that the quantum state of the electromagnetic field can be described in terms of the quantum states of harmonic oscillators, with each oscillator being labelled by $\mathbf{k}$ and α. It is convenient to express the vector potential in quantum theory in the form

$$\mathbf{A}(t, \mathbf{x}) = \sum_{\mathbf{k}\alpha} \left(c_{\mathbf{k}\alpha} A_{\mathbf{k}\alpha} + c_{\mathbf{k}\alpha}^\dagger A_{\mathbf{k}\alpha}^* \right), \quad (4.61)$$

with

$$A_{\mathbf{k}\alpha} = \left(\frac{2\pi}{L^3 \omega} \right)^{1/2} \hat{\mathbf{e}}_\alpha \exp i(\mathbf{k} \cdot \mathbf{x} - \omega t). \quad (4.62)$$

Show that (i) the time dependence of $A_{\mathbf{k}\alpha}$ is correct; (ii) the set of vectors $\left\{ \hat{\mathbf{e}}_1, \hat{\mathbf{e}}_2, \hat{\mathbf{k}} \right\}$ forms an orthonormal basis; and (iii) the quantities $c_{\mathbf{k}\alpha}$ satisfy the commutation rule

$$\left[c_{\mathbf{k}\alpha}, c_{\mathbf{p}\beta}^\dagger \right] = \delta_{\mathbf{k}\mathbf{p}} \delta_{\alpha\beta}. \quad (4.63)$$

(b) The coupling between the electromagnetic field and the atomic system is described by the Hamiltonian

$$H_{\text{int}} = \int d^3x \, \mathbf{J} \cdot \mathbf{A}, \tag{4.64}$$

where $\mathbf{J} = q\mathbf{v} = (q/m)\,\mathbf{p} = (q/m)\,(-i\nabla)$. To the lowest order in the perturbation theory the matrix elements

$$\langle E_f, n_{\mathbf{k}\alpha} + 1 | H_{\text{int}} | E_i, n_{\mathbf{k}\alpha} \rangle, \quad \langle E_f, n_{\mathbf{k}\alpha} | H_{\text{int}} | E_i, n_{\mathbf{k}\alpha} + 1 \rangle \tag{4.65}$$

govern the atomic transition from state i to state f accompanied by emission/absorption of a photon. Show that the probability of emission of a photon with $(\mathbf{k}, \mathbf{e}_\alpha)$ is given by

$$\begin{aligned}
\frac{dP_\alpha}{d\Omega \, dt \, d\omega} &= \frac{2\pi}{\hbar} \cdot \left(\frac{2\pi\hbar}{V\omega} \right) (n_{\mathbf{k}\alpha} + 1) \, \mathbf{e}_{\mathbf{k}\alpha} \cdot |\mathbf{M}_{fi}|^2 \cdot \frac{V}{(2\pi)^3} \frac{\omega^2}{\hbar c^3} \delta\left(\omega - \omega_{fi}\right) \\
&= \left(\frac{\omega}{2\pi\hbar c^3} \right) (n_{\mathbf{k}\alpha} + 1) \, |\mathbf{e}_{\mathbf{k}\alpha} \cdot \mathbf{M}_{fi}|^2 \, \delta\left(\omega - \omega_{fi}\right),
\end{aligned} \tag{4.66}$$

where $\omega_{fi} = (E_f - E_i)/\hbar$ and

$$|\mathbf{e}_{\mathbf{k}\alpha} \cdot \mathbf{M}_{fi}|^2 = \frac{q^2}{m^2} |\langle E_f | \exp\left(-i\mathbf{k} \cdot \mathbf{x}\right) \mathbf{e}_{\mathbf{k}\alpha} \cdot \mathbf{p} | E_i \rangle|^2. \tag{4.67}$$

Similarly, show that the corresponding probability for absorption is

$$\frac{dP_\alpha}{d\Omega \, dt \, d\omega} = \left(\frac{\omega}{2\pi\hbar c^3} \right) n_{\mathbf{k}\alpha} |\mathbf{e}_{\mathbf{k}\alpha} \cdot \mathbf{M}_{fi}|^2 \, \delta\left(\omega - \omega_{fi}\right). \tag{4.68}$$

We shall apply the results to different processes in the next few problems.

4.12 Classical limit and dipole approximation [1]

In problem 4.2, we developed the *classical* theory of radiation from non-relativistic sources. Since atomic systems are non-relativistic, one should be able to recover the results of problem 4.2 by taking the classical limit of the quantum theory developed in the last problem. This will provide the correspondence between the classical and quantum theories of radiation for non-relativistic sources.

(a) Consider the emission of radiation by an atom when (i) $n_{\mathbf{k}\alpha} = 0$; and (ii) the wavelength of the radiation λ is much larger than the atomic size. By approximating $\exp(i\mathbf{k} \cdot \mathbf{x}) \cong 1$ in (4.67), show that

$$\frac{dP_\alpha}{d\Omega \, dt \, d\omega} = \frac{\omega^3}{2\pi\hbar c^3} |\mathbf{e}_{\mathbf{k}\alpha} \cdot \mathbf{d}_{fi}|^2 \, \delta\left(\omega - \omega_{fi}\right), \tag{4.69}$$

where $\mathbf{d}_{fi} = \langle \phi_f | q\mathbf{x} | \phi_i \rangle$ is the transition element of the dipole moment. Perform an average over the polarization directions and multiply by $\hbar\omega$ to obtain the energy radiated at frequency ω_{fi}, in the direction $\hat{\mathbf{k}}$:

$$\frac{dE}{dt \, d\Omega} = \frac{\omega_{fi}^4}{4\pi c^3} |\hat{\mathbf{k}} \times \mathbf{d}_{fi}|^2 = \frac{\omega_{fi}^4}{4\pi c^3} |\mathbf{d}_{fi}|^2 \sin^2 \theta. \tag{4.70}$$

This agrees with the classical result in (4.10), provided we identify $|\omega^2 \mathbf{d}_{fi}|^2$ with $|\ddot{\mathbf{d}}|^2$ of the classical theory.

(b) Consider now the corresponding absorption probability between states i and f when $n_{\mathbf{k}\alpha}$ photons are originally present. Assume that the initial radiation field is isotropic and unpolarized so that $n_{\mathbf{k}\alpha} = (n(\omega)/2)$. Sum the probability of absorption over all directions of $\mathbf{k}$ and polarization states to obtain

$$\frac{dP_{fi}}{dt} = \frac{2}{3} \frac{q^2}{\hbar c^3} n(\omega_{fi}) \omega_{fi}^3 |\mathbf{x}_{fi}|^2. \tag{4.71}$$

(c) The 'bound–bound' absorption cross-section $\sigma_{bb}(\omega)$ for the atom is defined by the relation

$$\begin{bmatrix} \text{fraction of photons} \\ \text{absorbed per second} \end{bmatrix} = \int_0^\infty \sigma_{bb}(\omega) \, [\text{flux of photons}] \, [\text{density of states}] \, d\omega \tag{4.72}$$

or, equivalently,

$$\frac{dP_{if}}{dt} = \int_0^\infty \sigma_{bb}(\omega) \, [cn(\omega)] \, \frac{4\pi (\omega/c)^2 \, d(\omega/c)}{(2\pi)^3}. \tag{4.73}$$

It is convenient to write σ_{bb} as

$$\sigma_{bb}(\omega) = \frac{\pi q^2}{mc} f \, [2\pi \, \delta(\omega - \omega_{fi})] \equiv \frac{\pi q^2}{mc} f \phi_\omega, \tag{4.74}$$

where f is called the 'oscillator strength' and ϕ_ω is called the 'line profile function'. Show that, in the dipole approximation,

$$f = \frac{2}{3} \frac{m\omega}{\hbar} |\mathbf{x}_{fi}|^2 = \frac{2}{3} \left[\frac{m\omega^2 |\mathbf{x}_{fi}|^2}{\hbar \omega} \right]. \tag{4.75}$$

Interpret this result. Write the total emission and absorption rates for isotropic, unpolarized radiation as

$$\left(\frac{dP_{\text{abs}}}{dt} \right) = \left(\frac{2q^2}{mc^3} \right) n_\alpha(\omega) \omega^2 f, \tag{4.76}$$

$$\left(\frac{dP_{\text{emis}}}{dt} \right) = \left(\frac{2q^2}{mc^3} \right) [1 + n_\alpha(\omega)] \omega^2 f, \tag{4.77}$$

where $n_\alpha(\omega) = [n(\omega)/2]$ is the number of photons in each polarization state.

(d) Compute f_{ij} for the hydrogen atom when the initial state is the $n = 1$ ground state and the final state is any one of the three states $\{n = 2, l = 1, m = -1, 0 \text{ or } 1\}$. Sum over all the three final states and show that

$$f = \frac{2^{14}}{3^9} \cong 0.83. \tag{4.78}$$

4.13 Photoionization [3]

A sufficiently energetic photon can ionize the hydrogen atom, that is, the electron can absorb the photon and make a transition from the $n = 1$ ground state to a free particle state. Let the rate of such transitions be (dP/dt) when the number of incident photons is $n(\omega)$.

(a) The cross-section σ_{PI} for photoionization is defined by the relation

$$\left(\frac{dP}{dt}\right) dN_e = \sigma_{PI}(\omega) \cdot cn(\omega) \cdot \frac{4\pi \left(\omega^2/c^2\right) d\left(\omega/c\right)}{(2\pi)^3} , \qquad (4.79)$$

where $dN_e = \left[V d^3 p / (2\pi\hbar)^3\right]$ is the density of states available for the final electron. Explain the factors in this definition.

(b) To evaluate $\sigma_{PI}(\omega)$, one needs to compute (dP/dt) which – essentially – involves computing the matrix element of $(\exp(i\mathbf{k} \cdot \mathbf{x})\mathbf{p})$ between the initial and final states of the electron. The initial state is the, $n = 1$, ground state of a hydrogen atom with energy $-E_0$. Assume, for simplicity, that the final state is a plane wave state. Show that, if $\hbar\omega \gg E_0$, then

$$\sigma_{PI}(\omega) = \frac{2^8 \pi}{3} \left(\frac{q^2}{\hbar c}\right) a_0^2 \left(\frac{E_0}{\hbar\omega}\right)^{7/2} , \qquad (4.80)$$

with $a_0 = \left(\hbar^2/mq^2\right)$.

(c) Strictly speaking, the final state of the electron is a scattering state in the field of a nucleus rather than a plane wave state. Use a more exact representation of the final wave function of the electron to obtain the result

$$\sigma_{PI} = \frac{2^9 \pi^2}{3} \left(\frac{q^2}{\hbar c}\right) a_0^2 \left(\frac{E_o}{\hbar\omega}\right)^4 \left[\frac{\exp\left(-4\mu \cot^{-1}\mu\right)}{1 - \exp\left(-2\pi\mu\right)}\right] , \qquad (4.81)$$

with $\mu = \left(q^2/\hbar c\right)\left(c/v\right)$, where v is the velocity of the electron in the final state. Show that this result reduces to that of (b), if $\mu \ll 1$.

(d) Show that the cross-section σ_{rec} for recombination, in which a free electron and a proton combine to form a bound state emitting a photon, is related to σ_{PI} by $\sigma_{rec} = 2\left(k/p\right)^2 \sigma_{PI}$, where $\mathbf{p}$ is the momentum of incident electron and $\mathbf{k}$ is the momentum of the emitted photon.

(e) The coefficient of radiative recombination is defined to be $\alpha = <v_e\sigma_{rec}>$, where v_e is the speed of the electrons and the averaging is performed over a Maxwellian distribution with temperature T. With this definition, the rate of change of number density of electrons due to recombination will be $-\alpha n_e n_i$, where n_e and n_i are the electron and ion number densities. Estimate α.

4.14 Einstein's *A, B* coefficients [1]

(a) In classical radiation theory one often uses a quantity called the intensity of radiation. Intensity I_v is defined to be the amount of radiation energy flowing normal to a unit area, per second, per solid angle per frequency interval. That is,

$$I_v = \frac{dE}{dv \, dt \, (\hat{\mathbf{n}}dA) \cdot \left(\hat{\mathbf{k}} \, d\Omega\right)}, \tag{4.82}$$

where $\hat{\mathbf{n}}$ is the unit normal to the area dA and $\hat{\mathbf{k}}$ is a unit vector in the direction of the solid angle $d\Omega$. Show that

$$I_v = \sum_{\alpha=1}^{2} n_\alpha (\mathbf{x}, \mathbf{p}, t) \frac{hv^3}{c^2} = n (\mathbf{x}, \mathbf{p}, t) \frac{2hv^3}{c^2}, \tag{4.83}$$

where $n_\alpha (\mathbf{x}, \mathbf{p}, t)$ is the number of photons at $\mathbf{x}$ with momentum $\mathbf{p}$ and polarization α and the second relation is valid for unpolarized radiation.

(b) The rate of absorption and emission in (4.76) and (4.77) can now be written in terms of I_v. Show that

$$\mathcal{R}_{abs} \equiv \left(\frac{dP_{abs}}{dt}\right) = BI_v, \quad \mathcal{R}_{emis} \equiv \left(\frac{dP_{emis}}{dt}\right) = A + BI_v, \tag{4.84}$$

with

$$B = 4\pi^2 \left(\frac{q^2}{mc}\right) \left(\frac{f}{hv}\right), \quad A = 4\pi^2 \left(\frac{q^2}{mc}\right) \left(\frac{2v^2}{c^2}\right) f. \tag{4.85}$$

These are conventionally called 'Einstein's A, B, coefficients'. It follows from the definitions that $(A/B) = \left(2hv^3/c^2\right)$ and $\sigma_{bb} (v) = (B/4\pi) (\phi_v hv)$.

(c) Consider two particular quantum states of an atomic system with energies E_1 and E_2, with $E_2 > E_1$ and $E_2 - E_1 = hv$. Let the number of photons with frequency v be $n(v)$ and N_1 and N_2 be the number of atoms in the two states. Show that, if the system is in steady state, then we must have

$$n_{eq} (v) = \frac{1}{e^{\beta hv} - 1}, \tag{4.86}$$

where β is *defined* by the relation $(N_2/N_1) = \exp{-\beta (E_2 - E_1)}$. Interpret this result. Let the radiation intensity and energy density corresponding to $n_{eq} (v)$ be $I_{eq} (v)$ and $U_{eq} (v)$. Show that

$$I_{eq} (v) = \frac{2hv^3}{c^2 \left(e^{\beta hv} - 1\right)}, \quad U_{eq} (v) = \frac{8\pi h}{c^3} \frac{v^3}{e^{\beta hv} - 1} \tag{4.87}$$

and

$$U = \int_0^\infty dv \, U_{eq} (v) = \frac{\pi^2}{15 \, (\hbar c)^3} \frac{1}{\beta^4}. \tag{4.88}$$

(d) The emissivity j_v of a material is defined to be the amount of radiation energy emitted by the material spontaneously per second per unit volume per unit frequency range per solid angle. Show that

$$j_v = n_2^{\text{atom}} hv_{21} \phi_v (4\pi)^{-1} A_{21} = \frac{dE}{dt \, d^3x \, dv \, d\Omega}, \tag{4.89}$$

where $\phi_v = \delta (v - v_{21})$, n_2^{atom} is the number density of atoms in the upper state, A_{21} is the Einstein coefficient for transition from state 2 to state 1 which are separated by energy hv_{21}. The emissivity per atom is

$$\frac{dE}{dt\, dv\, \Omega} = J_v \equiv \frac{j_v}{n_2} = hv_{21}\phi_v\,(4\pi)^{-1}\,A_{21}. \tag{4.90}$$

(e) The absorption coefficient α_v for matter is defined in a similar manner via the equation

$$\alpha_v I_v = \left(\frac{dE}{d^3x\, dt\, dv\, d\Omega}\right)_{\text{abs}}. \tag{4.91}$$

Show that $\alpha_v = n_1 B_{12} hv \phi_v\,(4\pi)^{-1} = n_1 \sigma_v$. The coefficient of induced emission α_v^{ind} is defined in an identical manner with

$$\alpha_v^{\text{ind}} I_v = \left(\frac{dE}{d^3x\, dt\, dv\, d\Omega}\right)_{\text{ind emis}}. \tag{4.92}$$

Show that $\alpha_v^{\text{ind}} = n_2 B_2 hv \phi_v\,(4\pi)^{-1}$. The net absorption coefficient is

$$\alpha_v^{\text{net}} = \frac{hv}{4\pi}\,(n_1 - n_2)\,B_{12}\phi_v. \tag{4.93}$$

Find the ratio $\left(j_v/\alpha_v^{\text{net}}\right)$ and show that

$$\alpha_v = \frac{c^2}{2hv^3}\left(e^{hv/kT} - 1\right) j_v, \tag{4.94}$$

when matter is in thermal equilibrium. It is also usual to define a quantity κ_v by the relation $\kappa_v = (\alpha_v/\rho)$.

(f) Show that I_v/v^3, $\rho J_v/v^2$ and $\rho\kappa_v v$ and are Lorentz invariant.

4.15 Absorption and emission in continuum case [2]

In several astrophysical situations, photons are emitted by charged particles (say, electrons) in 'free' motion; that is, their energies form a continuous range rather than discrete levels. Let $(d\epsilon/dt\, dv) = P\,(v, E)$ be the energy radiated per second per frequency interval at frequency v by an electron with energy E. Show that the emissivity is

$$j_v = \frac{d\epsilon}{dt\, d^3x\, d\Omega\, dv} = \int d^3p\, P\,(v, E)\frac{f\,(p)}{4\pi} = \int_0^{\infty} dE\, n(E) P(v, E), \tag{4.95}$$

where $n\,(E)\,dE = f\,(p)\,4\pi p^2 dp$ is the number of electrons per unit volume with energies between E and $E + dE$. Show that the net absorption coefficient is

$$\rho\kappa_v = \frac{c^2}{8\pi hv^3}\int d^3p_2 P\,(v, E_2)\left[f\,(p_2') - f\,(p_2)\right], \tag{4.96}$$

where p_2' is the momentum corresponding to energy $(E_2 - hv)$.

Verify that if the electrons are in thermal equilibrium at temperature T, then $\rho\kappa_v I_v^{\text{eq}}\,(T) = j_v$.

4.16 Self-absorption of continuum radiation [1]

Consider a system of charged particles which is emitting radiation by some process (e.g. Bremsstrahlung, synchrotron, etc.). It is possible for some of the photons emitted by the system to be reabsorbed by the charged particles in the system itself. Since such a process will be frequency-dependent, the final spectrum (observed from the source) could be quite different from the spectrum emitted by individual charged particles. This effect can be investigated as follows:

(a) Let the emissivity, absorption coefficient and the size of the system be $\rho j_v, \alpha_v = \rho \kappa_v$ and R respectively. Show that: (i) if $\tau_v \cong \rho \kappa_v R \ll 1$ then the observed intensity is $I_v \cong R \rho j_v$; and (ii) if $\tau_v \gg 1$, then the observed intensity is $I_v = (\rho j_v / \alpha_v)$.

(b) Apply the above result to a plasma which is in thermal equilibrium and is emitting Bremsstrahlung radiation. Using the results derived in problem 4.5, show that $I_v \propto v^2$ for small v, $I_v \cong$ constant for intermediate values and $I_v \propto \exp(-hv/kT)$ for large v.

(c) Apply the results of (a) to a source of synchrotron radiation consisting of a power law spectrum of relativistic electrons $N(E)\,dE = CE^{-p}dE$. Assume that the electrons are ultrarelativistic, with $E \cong pc$, and that $hv \ll \bar{E}$, where $\bar{E}$ is the typical energy of the electron. Calculate the absorption coefficient $\alpha_v = \rho \kappa_v$ for synchrotron radiation under these conditions and show that $\alpha_v \propto v^{-(p+4)/2}$. Hence, show that $I_v \propto v^{5/2}$ for small v and $I_v \propto v^{-(p-1)/2}$ for large v.

(d) How does the synchrotron loss of a relativistic electron compare with the inverse Compton loss?

5
General relativity

General relativity is developed through problems in this chapter. All the key ideas are introduced *ab initio* and no previous exposure is assumed. It requires, however, a fair amount of practice to become fully conversant with the manipulation of tensor indices. Exposure to a more conventional course in general relativity will help in this matter.

We use a line element with the signature $(+ - - -)$. The Latin indices $(i, k, ...,$ etc.) go over 0,1,2,3, while the Greek indices $(\alpha, \beta, ...,$ etc.) go over 1,2,3. Summation over repeated indices is assumed.

5.1 Accelerated frames, special relativity and gravity [2]

Combining special relativity with Newtonian gravity turns out to be far more difficult than one would have imagined *a priori*. Any attempt to do so inevitably suggests a geometrical description for gravity. In this problem we shall examine some simple attempts to combine Newtonian gravity and special relativity and see how a geometrical description arises in the process.

(a) Let (T, X, Y, Z) be an inertial coordinate system. Consider an observer moving along the X-axis of this frame with a constant acceleration g. We will construct a coordinate system (t, x, y, z) for the accelerated observer by the following procedure. (i) The observer will use the proper time τ shown by a clock carried by him (her) for his (her) measurements. (ii) Let $\mathscr{P}$ be some event in the spacetime to which the observer has to attribute coordinates (t, x, y, z). Since the motion is along the X-axis the observer can set $y = Y$, $z = Z$. To attribute the (t, x) coordinates he (she) proceeds as follows. He (she) sends a light signal at time $\tau = t_A$ to this event. The light signal is reflected at the event $\mathscr{P}$ and the observer receives it back at time $\tau = t_B$. The observer then attributes to the event $\mathscr{P}$ the time coordinate $t = (t_A + t_B)/2$ and space coordinate $x = (t_B - t_A)/2$.

Use this criterion to determine the coordinate transformation between the accelerated and inertial observers to be

$$X = g^{-1}e^{gx}\cosh gt, \quad T = g^{-1}e^{gx}\sinh gt. \tag{5.1}$$

Show that one can introduce another space coordinate $\bar{x}$ such that the transformation becomes

$$(1 + gX) = (1 + g\bar{x}) \cosh gt, \qquad gT = (1 + g\bar{x}) \sinh gt, \quad Y = y; \quad Z = z.$$
$$(5.2)$$

These transformations are clearly non-linear and hence do not preserve the form of the line element ds^2. Show that the line elements in the inertial and accelerated frames are related by

$$ds^2 = dT^2 - dX^2 - dY^2 - dZ^2 = e^{2gx} \left(dt^2 - dx^2 \right) - dy^2 - dz^2$$
$$= (1 + g\bar{x})^2 \, dt^2 - d\bar{x}^2 - dy^2 - dz^2. \tag{5.3}$$

This result shows that non-inertial frames are described by intervals of the form $ds^2 = g_{ik}(x) \, dx^i dx^k$, where $g_{ik}(x)$ – called the metric tensor – will, in general, depend on t and $\mathbf{x}$.

(b) The action for a particle of mass m, in an external gravitational field characterized by the Newtonian potential $\phi(t, \mathbf{x})$, can be taken to be

$$A = \int dt \, \frac{1}{2} mv^2 - \int dt \, m\phi \tag{5.4}$$

in Newtonian gravity. The simplest generalization of this action, in the context of special relativity, will be

$$A = -mc^2 \int dt \sqrt{1 - \frac{v^2}{c^2}} - \int dt \, m\phi. \tag{5.5}$$

Argue that an action of this form leads to the following conclusions:

(i) The gravitational field is locally indistinguishable from a suitably chosen non-inertial frame of reference.

(ii) The gravitational field affects the rate of flow of clocks in such a way that the clocks slow down in strong gravitational fields. To the lowest order in (ϕ/c^2),

$$\Delta t' = \Delta t \left(1 - \frac{\phi}{c^2} \right), \tag{5.6}$$

where Δt is the time interval measured by a clock in the absence of the gravitational field and $\Delta t'$ is the corresponding interval measured by a clock, located in the gravitational potential.

(c) We know that annihilation of electrons and positrons can lead to γ-rays and that under suitable conditions one can produce $e^+ e^-$-pairs from the radiation. Devise a suitable thought experiment based on the above facts to conclude that gravitational field must necessarily affect the rate of flow of clocks. Hence, argue that global inertial frames cannot exist in the presence of gravity.

(d) Show that the dynamics of a particle in an external, weak $(\phi \ll c^2)$ gravitational field can be derived from a purely 'geometrical' action of the form

$$A = -mc \int ds = -mc \int \sqrt{g_{ik} \, dx^i \, dx^k},$$ (5.7)

where

$$g_{ik} \, dx^i \, dx^k \cong \left(1 + \frac{2\phi}{c^2}\right) c^2 dt^2 - d\mathbf{x}^2.$$ (5.8)

This result suggests that the Newtonian gravitational field can be thought of as modifying the line interval ds^2. Compare this result with the one obtained in (a) above.

The above results imply that a weak gravitational field cannot be distinguished from a modified spacetime interval as far as mechanical phenomena are concerned.

5.2 Gravity and the metric tensor [2]

The last problem shows that weak gravitational fields can be provided with a geometrical description as far as mechanical phenomena are concerned. Einstein made a bold generalization of this result by postulating that: *all aspects of gravitational physics allow a geometrical description.* In other words, we generalize the tentative conclusion of the preceding problem to include *arbitrarily strong gravitational fields* and *all* possible physical phenomena. We will now explore some elementary consequences of this postulate.

(a) To begin with, we shall generalize the result obtained in problem 5.1(d) and postulate that the gravitational field will modify the spacetime interval to the form

$$ds^2 = g_{ik}(x) \, dx^i \, dx^k,$$ (5.9)

where the metric tensor $g_{ik}(x)$ characterizes a specific gravitational field. Show that: (i) it is not possible, in general, to change the coordinates from x^i to some other x'^i such that ds^2 in the above equation reduces to the Lorentz form; (ii) it is, however, possible to choose coordinate systems such that, around any event $\mathscr{P}$, the following conditions are satisfied: (1) $g'_{ik}(\mathscr{P}) = \eta_{ik}$; (2) $(\partial g'_{ik}/\partial x^{k'})_{\mathscr{P}} = 0$. Can one make the second derivatives $(\partial^2 g_{ik}/\partial x^a \partial x^b)$ at $\mathscr{P}$ vanish as well?

(b) Since g_{ik} cannot be reduced to any preassigned form, no coordinate system has a preferred status in the presence of a gravitational field. This implies that all laws of physics must be expressed in terms of physical quantities which are coordinate-independent. The simplest such entity is a tangent vector v^i to any curve $x^i(\lambda)$ defined by $v^i = (dx^i/d\lambda)$. All other four-vectors will be *defined* to be quantities which transform similar to the tangent vector v^i. Determine how a vector v^i transforms when the coordinate system is changed from x^i to x'^i. What is the natural transformation law for such higher order tensors as T^{ik}, S^{ijk}, etc.?

(c) Treating the metric tensor g_{ik} as a matrix we can define its inverse g^{ik} such that $g^{ik} g_{kl} = \delta^i_l$. Derive the transformation law for g^{ik}. Let g denote the determinant of g_{ik}. Show that

$$\frac{\partial g}{\partial x^i} = g g^{ab} \left(\frac{\partial g_{ab}}{\partial x^i} \right) = -g g_{ab} \left(\frac{\partial g^{ab}}{\partial x^i} \right). \tag{5.10}$$

(d) We define the operation of 'raising and lowering an index' of a tensor by using the appropriate form of metric tensor. That is, given T^{ik} we define two new tensors $T^i{}_k$ and T_{ik} by the relations

$$T^i{}_k \equiv g_{ka} T^{ia}, \quad T_{ik} \equiv g_{ia} g_{kb} T^{ab}. \tag{5.11}$$

Show that these are valid tensor operations in the sense that if these equations are true in one frame they will be true in all frames.

5.3 Particle trajectories in a gravitational field [3]

(a) In special relativity, the equation of motion for a free particle can be obtained by varying the action

$$A = -m \int ds = -m \int \left(\eta_{ab} \frac{dX^a}{d\lambda} \frac{dX^b}{d\lambda} \right)^{1/2} d\lambda. \tag{5.12}$$

The variation of A gives $\left(d^2 X^a / d\lambda^2 \right) = \left(du^a / d\lambda \right) = 0$, where $u^a = \left(dX^a / d\lambda \right)$ is the four-velocity which is a tangent vector to the curve $X^a(\lambda)$. This equation can be written equivalently as

$$\frac{du^a}{d\lambda} = \left(\frac{dX^b}{d\lambda} \right) \left(\frac{\partial u^a}{\partial X^b} \right) = u^b \left(\frac{\partial u^a}{\partial X^b} \right) \equiv u^b \left(u^a{}_{,b} \right) = 0. \tag{5.13}$$

Show that this is *not* a valid tensor equation in the sense that the form of this equation will not remain the same in an arbitrary coordinate system.

(b) To obtain the correct equation of motion, which is generally covariant, we could proceed as follows. In any small region around an event, one can always choose an inertial coordinate system. Hence, in such a small region, the form of the action in equation (5.12) remains valid. In that small region $\eta_{ab} = g_{ab}$ and so we can also write the action in the form

$$A = -m \int ds = -m \int \sqrt{g_{ab} \, dx^a \, dx^b} = -m \int \left(g_{ab} \frac{dx^a}{d\lambda} \frac{dx^b}{d\lambda} \right)^{1/2} d\lambda. \tag{5.14}$$

Since *this* action is made up of proper tensorial quantities it will remain valid in any coordinate system. Hence, the equations derived from this action will provide us with the correct generalization of equation (5.13).

Vary this action and show that the equation of motion for a particle in an arbitrary spacetime is given by

$$\frac{du^i}{d\lambda} + \Gamma^i_{kl} u^k u^l = 0, \tag{5.15}$$

where

$$\Gamma^i_{kl} = \frac{1}{2}g^{ia}\left(-\frac{\partial g_{kl}}{\partial x^a} + \frac{\partial g_{ka}}{\partial x^l} + \frac{\partial g_{al}}{\partial x^k}\right). \tag{5.16}$$

Show, further, that this equation can be written in the form $u^i u^k{}_{;i} = 0$, where the 'covariant derivative' of the vector u^i is defined by the relation

$$u^a{}_{;b} \equiv u^a{}_{,b} + \Gamma^a_{bc} u^c. \tag{5.17}$$

By its very construction, the covariant derivative will be a tensor, and defines a natural generalization of the ordinary derivative to curved spacetime. How does Γ^i_{kl} transform under a coordinate transformation?

(c) We shall use the definition of covariant derivative in equation (5.17) for any vector field $v^i(x)$. We can generalize the notion of covariant derivative to other tensorial quantities by assuming that (1) the 'chain rule' of differentiation should remain valid; and (2) the covariant derivative of a scalar should be the same as the ordinary derivative. Show, using these criteria, that (i) the covariant derivative of v_i differs from that of v^i only in a sign:

$$v_{i;k} = v_{i,k} - \Gamma^l_{ik}v_l ; \tag{5.18}$$

(ii) for a mixed tensor the covariant derivative will appear with positive sign for upper indices and negative sign for lower indices:

$$T^a_{b;c} = T^a_{b,c} + \Gamma^a_{dc} T^d_b - \Gamma^d_{bc} T^a_d . \tag{5.19}$$

Similar definitions can be used for any higher rank tensor.

(d) Show that covariant derivatives do not commute. More specifically, show that

$$A^a{}_{;b;c} - A^a{}_{;c;b} = -R^a_{ibc} A^i , \tag{5.20}$$

where

$$R^a_{ibc} = \frac{\partial \Gamma^a_{ic}}{\partial x^b} - \frac{\partial \Gamma^a_{ib}}{\partial x^c} + \Gamma^a_{kb}\Gamma^k_{ic} - \Gamma^a_{kc}\Gamma^k_{ib} . \tag{5.21}$$

Is this quantity a tensor?

5.4 Parallel transport [1]

Consider a curve $x^i(\lambda)$ passing through an event $\mathscr{P}$ in spacetime. Let the coordinates of $\mathscr{P}$ be $x^i(0)$. If $v^j(x)$ is a vector field, then we can interpret the quantity $v^j{}_{;i}\,(dx^i/d\lambda)$ to be the change of the vector field along a direction specified by the tangent vector $(dx^i/d\lambda)$. Show how this concept can be used to generate a vector *field* from a *given* vector at $\mathscr{P}$. (This process is called 'parallel transport'.)

5.5 Formulas for covariant derivative [1]

The calculation of covariant derivatives can be simplified using certain identities which we shall derive below.

(a) Show that

$$\Gamma^a_{ba} = \frac{\partial}{\partial x^b} \left(\ln \sqrt{-g} \right) , \quad g^{cd} \Gamma^a_{cd} = -\frac{1}{\sqrt{-g}} \frac{\partial \left(\sqrt{-g} g^{ab} \right)}{\partial x^b} . \tag{5.22}$$

(b) Using the above results show that

$$A^i_{;i} = \frac{1}{\sqrt{-g}} \frac{\partial \left(\sqrt{-g} A^i \right)}{\partial x^i} , \tag{5.23}$$

$$\Box \phi = \frac{1}{\sqrt{-g}} \frac{\partial}{\partial x^i} \left(\sqrt{-g} g^{ik} \frac{\partial \phi}{\partial x^k} \right) . \tag{5.24}$$

(c) If A^{ik} is an antisymmetric tensor, show that

$$A^{ik}_{;k} = \frac{1}{\sqrt{-g}} \frac{\partial \left(\sqrt{-g} A^{ik} \right)}{\partial x^k} . \tag{5.25}$$

(d) If T^{ik} is a symmetric tensor, show that

$$T^k_{i;k} = \frac{1}{\sqrt{-g}} \frac{\partial \left(\sqrt{-g} \, T^k_i \right)}{\partial x^k} - \frac{1}{2} \frac{\partial g_{kl}}{\partial x^i} T^{kl} . \tag{5.26}$$

5.6 Physics in curved spacetime [2]

The description of a gravitational field in terms of g_{ik} assumes that all the laws of physics can be suitably generalized from flat spacetime to curved spacetime. The usual procedure for doing this is as follows. Around any event $\mathscr{P}$ we choose a locally inertial frame in which the laws of special relativity are valid. By writing these laws in a covariant manner using suitable tensors, we will have a description which is valid in any arbitrary coordinate system around $\mathscr{P}$. Since the form of the equations is expected to be same (locally) in an accelerated frame and in the presence of gravity, the above procedure allows us to describe physics in curved spacetime. (This was the procedure we followed in problem 5.3 to describe the motion of a particle in curved spacetime.)

(a) Carry out this procedure for the case of a free electromagnetic field and obtain the form of Maxwell's equation which is valid in an arbitrary spacetime.

(b) Introduce the electric and magnetic fields by the definitions $E_\alpha = F_{0\alpha}$ and $B^\alpha = -(1/2\sqrt{\gamma})\epsilon^{\alpha\beta\gamma} F_{\beta\gamma}$, where γ is the determinant of the metric of three-dimensional space. Write down Maxwell's equations in terms of **E** and **B**. Show that if $g_{0\alpha} = 0$ and other components of g_{ik} are independent of time then the gravitational field can be interpreted as equivalent to a dielectric medium with $\epsilon = \mu = (1/\sqrt{g_{00}})$.

5.7 Curvature [2]

(a) Consider two curves $x_A^i(\lambda)$ and $x_B^i(\lambda)$ which intersect at the events $\mathscr{P}$ and $\mathscr{Q}$. Given any vector at event $\mathscr{P}$ we can parallel transport it to $\mathscr{Q}$ along either of these two curves. Show that: (i) the resultant vector at $\mathscr{Q}$ will not, in general, be the same; (ii) we obtain the same vector at $\mathscr{Q}$, irrespective of the curve chosen to transport it from $\mathscr{P}$ to $\mathscr{Q}$, only if $R_{klm}^i = 0$, where

$$R_{klm}^i = \frac{\partial \Gamma_{km}^i}{\partial x^l} - \frac{\partial \Gamma_{kl}^i}{\partial x^m} + \Gamma_{bl}^i \Gamma_{km}^b - \Gamma_{bk}^i \Gamma_{lm}^b \tag{5.27}$$

is defined in problem 5.3(d). Also show that the vanishing of this tensor is the necessary and sufficient condition for the spacetime to be flat.

(b) From the definition of R_{klm}^i one can construct the tensors $R_{iklm} \equiv g_{ia} R_{klm}^a$, $R_{ik} \equiv R_{ijk}^j$ and the scalar $R \equiv g^{ik} R_{ik}$. Show that

$$R_{klm;n}^i + R_{kmn;l}^i + R_{knl;m}^i = 0 \tag{5.28}$$

and

$$\left(R_k^i - \frac{1}{2} \delta_k^i R \right)_{;i} = 0. \tag{5.29}$$

(c) How many independent components does the Riemann tensor R_{iklm} have in a four-dimensional spacetime?

5.8 Practice with metrics [3]

Given a metric g_{ik} one can compute all other geometrical quantities such as Γ_{kl}^i, R_{iklm}, etc. The following metrics will provide us with some practise in manipulating various tensorial quantities. For each of the following metrics compute all non-trivial components of $G_k^i = R_k^i - \frac{1}{2} \delta_k^i R$.

(a) The spherically symmetric metric:

$$ds^2 = e^\nu dt^2 - e^\lambda dr^2 - r^2 \left(d\theta^2 + \sin^2 \theta \, d\phi^2 \right), \tag{5.30}$$

with $\lambda = \lambda(t, r)$ and $\nu = \nu(r, t)$.

(b) The metric describing a homogeneous and isotropic spacetime:

$$ds^2 = dt^2 - a^2(t) \left[\frac{dr^2}{1 - kr^2} + r^2 \left(d\theta^2 + \sin^2 \theta \, d\phi^2 \right) \right], \tag{5.31}$$

where k is a constant.

(c) The metric describing a weak gravitational field:

$$ds^2 = \left(1 + \frac{2\phi}{c^2} \right) c^2 dt^2 - dx^2 - dy^2 - dz^2, \tag{5.32}$$

with $\phi = \phi(t, \mathbf{x}) \ll c^2$. In this case, evaluate various components only to the lowest non-trivial order in (ϕ/c^2).

5.9 Dynamics of gravitational field [3]

In Newtonian gravity, the gravitational field is described by one scalar function $\phi(t, \mathbf{x})$ which can be determined if the density distribution $\rho(t, \mathbf{x})$ is given. The action for the gravitational field may be taken to be

$$A = \frac{1}{8\pi G} \int (\nabla \phi)^2 \, d^3\mathbf{x} \, dt + \int \rho \phi \, d^3\mathbf{x} \, dt. \tag{5.33}$$

We shall now determine the corresponding generalization in the case of general relativity.

We can take the total action A_{tot} to be a sum of two terms, A_g representing the gravitational part and A_m which depends on the metric as well as matter variables. We have already seen in problem 5.6 that the matter part of the action in curved spacetime can be obtained by generalizing the corresponding special relativistic action. We therefore only have to determine A_g.

The action principle must be such that the equations of motion obtained from it involve only up to second derivatives of g_{ik}. Further, the action should reduce to the Newtonian action for the metric given in equation (5.32).

(a) This would suggest that A_g should contain up to quadratic terms in the first derivatives of g_{ik}. Show that no non-trivial scalar quantity can be constructed from g_{ik} and their first derivatives alone. Hence, conclude that one must look for a scalar quantity which contains the metric, first derivatives *and* second derivatives. Argue that the equations of motion will still involve only up to second derivatives of g_{ik}, provided the action is linear in the second derivatives of the metric.

(b) Show that one can take the action to be proportional to

$$A_g = \int R\sqrt{-g}\, d^4x \tag{5.34}$$

by demonstrating that $R\sqrt{-g}$ can be written in the form

$$R\sqrt{-g} = \sqrt{-g}\, g^{ik} \left(\Gamma^m_{il}\Gamma^l_{km} - \Gamma^l_{ik}\Gamma^m_{lm} \right) + \frac{\partial \sqrt{-g}\, w^i}{\partial x^i} = \sqrt{-g}\,\mathscr{G} + \frac{\partial \sqrt{-g}\, w^i}{\partial x^i}, \tag{5.35}$$

where the first term is quadratic in the derivatives of g_{ik} and the second term is a four-divergence of some four-vector w^i. Identify w^i.

(c) Explain the nature of variation in g_{ik} which will be appropriate for this Lagrangian. In problem 2.1 it was shown that, given any Lagrangian of the form $L(\dot{q}, q)$, it is possible to construct another Lagrangian $L'(\ddot{q}, \dot{q}, q)$, which is linear in $\ddot{q}$ such that both L and L' lead to the same equations of motion; while performing the variation in L the coordinates are kept fixed at the end points while with L' the canonical momenta are kept fixed. Show that if we take $L = \sqrt{-g}\,\mathscr{G}$ then L' turns out to be $R\sqrt{-g}$.

(d) Take the full action for the system containing gravity and matter to be

$$A = \frac{1}{c_1} \int R\sqrt{-g}\, d^4x + \int \mathscr{L}_m \left(q^A, g_{ik} \right) \sqrt{-g}\, d^4x, \tag{5.36}$$

where q^A denotes the matter variables. Evaluate the first term in the action for the weak gravitational field using (5.32) and show that we obtain the correct Newtonian limit only if $c_1 = 16\pi G$. Show that under suitable variation of g^{ik}, we will obtain the field equations for the gravitational field:

$$R_{ik} - \frac{1}{2}g_{ik}R = 8\pi G T_{ik},\qquad\qquad (5.37)$$

where the quantity T_{ik} is *defined* via the relation

$$\delta A_{\mathrm{m}} = \frac{1}{2}\int T_{ik}\,\delta g^{ik}\,\sqrt{-g}\,d^4x.\qquad\qquad (5.38)$$

(e) Interpret the quantity T_{ik} as the stress-tensor of the matter field. Show that $T^i_{k;i} = 0$ for any, arbitrary, matter Lagrangian. Work out T_{ik} for a (i) massless scalar field; (ii) electromagnetic field.

(f) Suppose that the matter action is invariant under the transformation $g_{ik} \rightarrow \Omega^2(x)g_{ik}$ called 'conformal transformation'. Show that $T^i_i = 0$ for such a matter Lagrangian.

(g) How many independent equations are there in (5.37)? What are the appropriate initial conditions which need to be specified to solve these equations?

5.10 Geodesic deviation [2]

(a) Consider a bunch of particles moving along geodesics in a spacetime. Let $x^i = x^i(s,v)$ denote the family of geodesics where s represents the proper time along the geodesics and v labels a particular geodesic. The vector $n^i = (\partial x^i/\partial v)\delta v \equiv v^i\delta v$ denotes the deviation between two neighbouring geodesics parametrized by the values v and $v + dv$. Show that

$$\frac{D^2v^i}{Ds^2} \equiv \left(v^i{}_{,k}u^k\right)_{;l}u^l = R^i_{klm}u^ku^lv^m,\qquad\qquad (5.39)$$

where $u^i = (\partial x^i/\partial s)$ is the tangent vector to the geodesic. This is called the 'geodesic deviation equation'.

(b) Consider two observers moving along neighbouring geodesics in a given spacetime. Let **g** be the relative, three-dimensional acceleration of one observer with respect to the other. Show that, in the locally inertial frame of the first observer,

$$\nabla\cdot\mathbf{g} = -4\pi G(\rho + T^\alpha_\alpha) = -4\pi G(\rho + 3p),\qquad\qquad (5.40)$$

where T^α_α is summed over $\alpha = 1,2,3$ and the second expression is applicable if the source is an ideal fluid with pressure p.

5.11 Schwarzschild metric [2]

(a) Consider the gravitational field produced by a spherically symmetric mass distribution confined to the interior of a spherical region of radius R_1. Show that the metric outside this region (i.e. for $r > R$) has the form

$$ds^2 = \left(1 - \frac{2GM}{r}\right) dt^2 - \left(1 - \frac{2GM}{r}\right)^{-1} dr^2 - r^2 \left(d\theta^2 + \sin^2\theta \, d\phi^2\right). \quad (5.41)$$

Also determine the metric due to a spherical shell of matter both inside and outside the shell.

(b) Consider the motion of a material particle of mass m in this metric. Show that the orbit of the particle can be determined from the equations

$$\left(1 - \frac{2GM}{r}\right)^{-1} \frac{dr}{dt} = \frac{1}{\mathscr{E}} \left[\mathscr{E}^2 - V^2\left(r\right)\right]^{1/2}, \quad r^2\dot\phi = \left(\frac{L}{\mathscr{E}}\right) \left(1 - \frac{2GM}{r}\right), \quad (5.42)$$

where $\mathscr{E}$ and L are constants and

$$V_{\text{eff}}^2\left(r\right) = m^2 \left[\left(1 - \frac{2GM}{r}\right) \left(1 + \frac{L^2}{m^2r^2}\right)\right]. \quad (5.43)$$

Plot the effective potential $V_{\text{eff}}\left(r\right)$ for different values of L and describe qualitatively the nature of orbits for various values of $\mathscr{E}$ and L.

(c) Consider an ultrarelativistic particle which is moving past a body of mass M with velocity $v \lesssim c$. Let the impact parameter be R and the angle of deflection be $\delta\phi$. Calculate $\delta\phi$. Show that $\delta\phi$ has twice the value it would have in the case of Newtonian gravity in the ultrarelativistic limit of $v = c$. Also estimate the corresponding deflection if the force of interaction between the two particles is electromagnetic rather than gravitational. What happens to the deflection as $v \to c$ in the case of electromagnetic interaction?

5.12 Gravitational lensing [2]

Light rays reaching us from distant cosmological sources are deflected by the gravitational field of intervening masses. This could lead, under favourable conditions, to the formation of multiple images of the source. Assume that all the deflection takes place when the light crosses the 'deflector plane', which is defined to be the plane containing the deflecting object and is perpendicular to the line connecting the source and the observer. It is convenient to project all relevent quantities on to this two-dimensional plane. Let **s** and **i** denote the two-dimensional vectors giving the source and image positions on this two-dimensional plane; and let **d(i)** be the (vectorial) deflection produced by the lens.

(a) Show that the image and source positions are related by the equation

$$\mathbf{s} = \mathbf{i} - \frac{D_{\text{LS}}}{D_{\text{OS}}} \mathbf{d(i)}, \quad (5.44)$$

where L, S and O stand for lens, source and observer and D_{LS} is the distance between the lens and the source, etc.

(b) The deflection produced by the lens can be related to the surface density $\Sigma(\mathbf{x})$ of the intervening mass distribution. Show that

$$\mathbf{d}(\mathbf{i}) = \frac{4GD_{OL}}{c^2} \int d^2x \, \Sigma(\mathbf{x}) \frac{(\mathbf{i} - \mathbf{x})}{|\mathbf{i} - \mathbf{x}|^2} \,. \tag{5.45}$$

Describe qualitatively the behaviour of the function $\Sigma(x)$ when the mass distribution is spherically symmetric. Argue that, for spherically symmetric, non-singular mass distributions there will be either one or three images.

(c) Show that the images can be amplified (or de-amplified) with respect to the source. Find an expression for the amplification in terms of the position s of the source.

(d) Relate the source and image position when the lensing is due to (i) a uniform sheet with surface density Σ_0; (ii) a point mass M; (iii) an isothermal sphere with a small core radius. Calculate the amplification in each of these cases.

(e) Let us choose a coordinate system in the lensing plane such that the source is at the origin. Show that the location of the images can be obtained by calculating the extremum of the function

$$P(\mathbf{i}) = \frac{1}{2}i^2 - \psi(\mathbf{i}), \tag{5.46}$$

where

$$\psi(\mathbf{i}) = \frac{4GD_{OL}D_{LS}}{c^2 D_{OS}} \int d^2x \, \Sigma(\mathbf{x}) \ln \left(|\mathbf{i} - \mathbf{x}| \right). \tag{5.47}$$

Part two

6
Friedmann model and thermal history

This chapter introduces the 'standard model' of cosmology which is based on the Friedmann universe. Several geometrical and physical characteristics of this model are explored and the stage is set to discuss structure formation in the following two chapters.

6.1 Maximally symmetric universe [2]

The simplest model for the universe, called the Friedmann model, is based on the assumption that the matter distribution in the universe is homogeneous and isotropic. Since the geometrical properties of the spacetime are determined by the matter distribution via Einstein's equations, it follows that three-dimensional space must be homogeneous and isotropic. That is, three-dimensional space has geometrical properties which are independent of location (homogeneous) and are the same in all directions (isotropic). The observed universe, of course, is inhomogeneous at the scale of galaxies, clusters, etc., and does not exhibit the symmetry invoked above. In constructing the model for the universe, we shall ignore this complication and assume that when averaged over sufficiently large scales the universe will be isotropic and homogeneous.

(a) Show that, in any universe which is homogeneous and isotropic, it is possible to choose a coordinate system in which the metric has the form

$$ds^2 = dt^2 - a^2(t) \left[\frac{dr^2}{1 - kr^2} + r^2 \left(d\theta^2 + \sin^2 \theta \, d\phi^2 \right) \right], \qquad (6.1)$$

where $k = 0, 1$ or -1.

(b) Further, show that this metric can be re-expressed in the forms

$$ds^2 = a^2(\eta) \left[d\eta^2 - d\chi^2 - f^2(\chi) \left(d\theta^2 + \sin^2 \theta \, d\phi^2 \right) \right]$$
$$= e^\nu \, dT^2 - e^\lambda \, dR^2 - R^2 \left(d\theta^2 + \sin^2 \theta \, d\phi^2 \right). \qquad (6.2)$$

Determine the coordinate transformations from (t, r, θ, ϕ) to $(\eta, \chi, \theta, \phi)$ and (T, R, θ, ϕ) and the functions $f(x), \nu(R, T)$ and $\lambda(R, T)$.

(c) Interpret the *spatial* part of the $(\eta, \chi, \theta, \phi)$ coordinate system geometrically.

(d) We saw in chapter 5 that the Newtonian limit of a metric has the form

$$ds^2 \cong \left(1 + \frac{2\phi_N}{c^2}\right) c^2 \, dT^2 - dR^2 - R^2(d\theta^2 + \sin^2\theta \, d\phi^2). \tag{6.3}$$

Find a suitable transformation from the Friedmann coordinates (t, r, θ, ϕ) to a new set of coordinates (T, R, θ, ϕ) such that *for small R*, the metric has the form

$$ds^2 \cong \left(1 - \frac{\ddot{a}}{a}\frac{R^2}{c^2}\right) c^2 \, dT^2 - \left(1 + \frac{k}{a^2}R^2 + \frac{\dot{a}^2}{a^2}R^2\right) dR^2 - R^2\left(d\theta^2 + \sin^2\theta \, d\phi^2\right). \tag{6.4}$$

What is the effective Newtonian potential to $\mathcal{O}\left(R^2\right)$?

6.2 Kinematics in Friedmann universe [2]

(a) Observers located at constant values of (r, θ, ϕ) in a Friedmann universe are called comoving observers. Consider two comoving observers located with an infinitesimal separation δx. A narrow pencil of electromagnetic radiation crosses both the observers at different times. By studying the frequencies observed by the two observers, conclude that the frequency $\omega(t)$ of electromagnetic radiation decreases with the expansion of the universe according to the law $\omega(t) \propto a(t)^{-1}$.

Consider next the wave equation for the electromagnetic wave (described by a four-vector potential A_a) in a $k = 0$ Friedmann universe. Show that the solution to the wave equation has the time dependence

$$A_a \propto \exp\left[-ik \int \frac{dt}{a(t)}\right]. \tag{6.5}$$

Argue that the instantaneous frequency $\omega(t)$ decreases as $a(t)^{-1}$.

(b) Suppose the metric of the spacetime is slightly different from that of a Friedmann model and is given by

$$ds^2 = dt^2 - a^2(t) \left[\delta_{\alpha\beta} - h_{\alpha\beta}(t, x)\right] dx^\alpha dx^\beta. \tag{6.6}$$

A photon emitted by a distant source propagates through this spacetime and is received by an observer at the present epoch. Show that, to the lowest order in $h_{\alpha\beta}$, the quantity ωa will vary by an amount

$$\ln\left(\frac{\omega_2 a_2}{\omega_1 a_1}\right) = \frac{1}{2} \int_{\text{along path}} \dot{h}_{\alpha\beta} n^\alpha n^\beta \, dt, \tag{6.7}$$

where the trajectory of the photon is taken to be

$$x^\alpha = n^\alpha \int_{t_1}^{t_2} \frac{dt'}{a(t')}. \tag{6.8}$$

The integral in (6.7) is taken along the trajectory of the photon.

(c) Consider a material particle which crosses two comoving observers separated by a distance δx. By studying the momentum of the particle, as measured by the two observers, conclude that the magnitude of the momentum decreases with expansion as a^{-1}.

(d) A Friedmann universe is filled with radiation with intensity distribution of the form $I(\omega) = \omega^3 G(\omega/T)$, where T is some parameter characterizing the spectrum. Show that as the universe expands the intensity distribution retains this form with $T \propto a^{-1}$. What does this result imply for a Planck spectrum?

(e) Most of our information in cosmology arises from the study of the radiation received from distant sources. Suppose a source emits radiation at time t and the radiation is received by us at $t = t_0$. From the result of part (a) above, we know that the frequencies of radiation at emission and at reception are related by

$$\frac{\omega_{em}}{\omega_{rec}} = \frac{a(t_0)}{a(t)} \equiv 1 + z(t). \tag{6.9}$$

The last relation defines the redshift z in terms of the expansion factor $a(t)$.

(i) Consider a distant cosmological source located at (r_1, θ, ϕ) which is emitting radiation towards an observer. If the source has an intrinsic luminosity L and the observer receives a flux l, then the 'luminosity distance' $d_L(z)$ to the source is defined by the relation, $l = L/(4\pi d_L^2)$. Show that

$$d_L(z) = a_0 r_1(t_1)(1 + z), \tag{6.10}$$

where $a_0 = a(t_0)$. In these expressions t_1 refers to the time at which the radiation is emitted, t_0 is the instant at which the radiation is detected and z is the redshift. The coordinates r_1 and t_1 are related by

$$\int_0^{r_1} \frac{dr}{\sqrt{1 - kr^2}} = -\int_{t_0}^{t_1} \frac{dt}{a(t)}. \tag{6.11}$$

(ii) Consider a cosmic source of radiation which has a physical size D and subtends a small angle δ to the observer. The 'angular diameter distance' $d_A(z)$ is defined by the relation $\delta = (D/d_A)$. Show that $d_A = (1 + z)^{-2} d_L$.

6.3 Dynamics of the Friedmann model [3]

(a) The geometrical properties of the Friedmann metric were calculated in problem 6.1. Using these results show that: (i) the energy–momentum tensor of the source must have the form $T_k^i = \text{dia}\,[\rho(t), -p(t), -p(t), -p(t)]$; and (ii) the variables $a(t)$, $\rho(t)$ and $p(t)$ are related by the equation

$$\frac{\dot{a}^2 + k}{a^2} = \frac{8\pi G}{3}\rho, \tag{6.12}$$

$$\frac{2\ddot{a}}{a} + \frac{\dot{a}^2 + k}{a^2} = -8\pi G p. \tag{6.13}$$

If the source is an ideal fluid then ρ will be the energy density and p will be the pressure. In general, it is not necessary for ρ and p to have such a simple interpretation.

(b) The above set gives two equations connecting three variables $a(t)$, $\rho(t)$ and $p(t)$. To integrate these equations, we will need an equation of state $p = p(\rho)$ and suitable boundary conditions. The boundary condition may be taken to be the value of $a(t)$ and $\dot{a}(t)$ at the present instant $t = t_0$.

It is, however, conventional to specify the boundary conditions in terms of two other parameters directly connected with observations. The first one is the Hubble constant $H_0 \equiv (\dot{a}/a)_0$, which is observed to have the value $100\,h$ km s^{-1} Mpc^{-1} with $0.5 < h < 1$. The parameter h quantifies the observational uncertainty in the measurement of the Hubble constant. One can specify H_0 instead of $\dot{a}(t_0)$ as one initial condition.

To define the second parameter, we will proceed as follows. Using H_0 one can define a quantity – called the critical density – by the relation

$$\rho_{\text{crit}} \equiv \frac{3}{8\pi G} \left(\frac{\dot{a}}{a}\right)_0^2 \equiv \frac{3H_0^2}{8\pi G} \cong 1.88 \times 10^{-29} h^2 \text{g cm}^{-3}. \tag{6.14}$$

We shall define the density of different constituents of the universe in terms of the critical density ρ_{crit}. The ratio $(\rho/\rho_{\text{crit}})$ will be denoted by the symbol Ω. For example, radiation density in the universe will be written as $\Omega_{\text{rad}}\rho_{\text{crit}}$, baryonic density in the universe will be written as $\Omega_B \rho_{\text{crit}}$, etc. The total density due to all constituents will be taken to be $\Omega \rho_{\text{crit}}$. Express $\dot{a}(t_0)$ and $a(t_0)$ in terms of H_0 and Ω, if $k \neq 0$. What happens when $k = 0$?

(c) From the equations for the Friedmann model argue that, if $(\rho + 3p) > 0$, then the universe must have had a singular beginning at some time $t = t_{\text{sing}}$ such that

$$t_0 - t_{\text{sing}} < t_{\text{univ}} \equiv H_0^{-1} = 9.8 \times 10^9 \, h^{-1} \text{ yr}. \tag{6.15}$$

Does this conclusion sound reasonable ?

(d) Rewrite the Friedmann equations in the form

$$\frac{d}{dt}\left(\rho a^3\right) = -p\frac{da^3}{dt}. \tag{6.16}$$

How does ρ vary with a if the equation of state has the form $p = w\rho$? Integrate Einstein's equation for a $\Omega = 1$ universe when $p = w\rho$. In particular, consider the cases $w = -1, 0, 1/3$ and 1. Which physical systems can give rise to stress-tensors with such equations of state?

(e) Observations suggest that the present universe is made of (i) radiation, with an equation of state $p = (1/3)\rho$; (ii) baryons, with an equation of state $p \cong 0$; and (iii) non-baryonic dust-like matter, possibly made of weakly interacting elementary particles with an equation of state $p \cong 0$. It is also possible that the universe contains a 'cosmological constant' which is equivalent to a fluid with the equation of state $p = -\rho$.

Estimate the energy density of radiation from the observed microwave background temperature of $T_\gamma \cong 2.73$ K and obtain Ω_{rad}. It is also known from observations that: (i) $\Omega_B \cong 0.02 h^{-1}$; (ii) $\Omega_{\text{DM}} \lesssim 1$; and (iii) the fraction of critical density Ω_V contributed by the cosmological constant is less than 0.8. Show that the evolution of such a universe will be described by the equation

$$\frac{\dot{a}^2}{a^2} + \frac{k}{a^2} = H_0^2 \left[\Omega_{\text{rad}} \left(\frac{a_0}{a}\right)^4 + (\Omega_B + \Omega_{DM}) \left(\frac{a_0}{a}\right)^3 + \Omega_V \right]. \qquad (6.17)$$

The form of the right-hand side shows that different kinds of matter will be influencing the dynamics of the universe at different epochs. Explain qualitatively how the evolution will proceed for the universe described by the above equation. Obtain explicit forms of $a(t)$ when the universe: (i) is radiation dominated; (ii) has matter and radiation but $k = 0$ and $\Omega_V = 0$; (iii) is matter dominated with $k \neq 0$ but $\Omega_V = 0$; and (iv) is matter dominated with $\Omega_V \neq 0$ but $k = 0$.

(f) The characteristic length scale at which the expansion of the universe is important is given by $d_H(t) \equiv (\dot{a}/a)^{-1}$. How does d_H change with the expansion factor $a(t)$ for the universe described in part (e) above? The physical length scale, characterizing a region which has a size λ_0 today, will change with expansion as $\lambda(a) = \lambda_0 (a/a_0)$. How does $\lambda(a)$ compare with $d_H(a)$ for small a and large a? In particular, compute the redshift at which $\lambda(a) = d_H(a)$ for different values of λ_0, if $\Omega_V = 0$ and $k = 0$.

6.4 Optical depth due to galaxies [2]

(a) Suppose the number of galaxies per unit proper volume within the redshift range $(z, z + dz)$ is $n(z)$. (If the galaxies are neither created nor destroyed, $n(z) = n(0)(1 + z)^3$.) Let the effective cross-section of a galaxy be πr_G^2. Estimate the probability that a galaxy with a redshift in the range $(z, z + dz)$ intervenes in our line of sight to a distant source of redshift z_s. Also calculate the optical depth due to galaxies for a source located at a redshift of z_s.

Observations suggest that the effective radius of a galaxy is $r_G \cong 10h^{-1}$ kpc and the comoving number density at $z = 0$ is $n_G \cong 0.02h^3$ Mpc^{-3}. Estimate the fraction of the sky which is covered by galaxies at $z = 1$ in a matter dominated universe. At what redshift does the optical depth due to galaxies become unity? (Take $\Omega = 1$ and assume that galaxies are conserved.)

(b) Calculate the expected number density of galaxies per unit redshift interval, per unit solid angle along any line of sight.

(c) Distant objects in the universe can be affected by the gravitational lensing phenomena discussed in problem 5.12. Assume that the density distribution of each lens is that of an isothermal sphere with velocity dispersion σ. Estimate the probability that a distant source produces multiple images due to a lens located within the redshift range $(z, z + dz)$. Also compute the optical depth for gravitational lensing if the population of lenses is conserved.

6.5 Radiative processes in an expanding universe [2]

(a) Most of our knowledge about the local universe arises from studying the radiation emitted by sources in the redshift range of $z = 0.5$ to 5.0. In relating the observed parameters to the intrinsic properties of the source, it is necessary to take various cosmological factors into account. Consider some emission process which takes place in a redshift interval $z_1 < z < z_2$ with a specific emissivity $J_\omega(z)$ (with units erg $cm^{-3}\,s^{-1}\,Hz^{-1}$), which could explicitly depend on z. Show that the flux of radiation observed today from these sources will be given by

$$F(\omega_0; z = 0) = \frac{c}{4\pi H_0} \int_{z_1}^{z_2} \frac{J[\omega_0(1+z); z]\, dz}{(1+z)^5 (1+\Omega z)^{1/2}}. \tag{6.18}$$

(b) The $n = 1$ state of the hydrogen atom, in which the electron and the proton have parallel spins, has a slightly higher energy than the state with antiparallel spins. The radiation emitted when the atom makes the transition between these two states has a wavelength of 21 cm (corresponding to a frequency of $v_H = c/\lambda = 1420$ MHz). Ignoring the natural and Doppler widths, the emissivity can be written as

$$J(v) = \frac{3}{4} A n_H h v\, \delta(v - v_H), \tag{6.19}$$

where $A = 2.85 \times 10^{-15}\,s^{-1}$ is the rate of spontaneous transition (which is rather small because this is a magnetic dipole transition) and n_H is the number density of hydrogen atoms. Estimate the flux of 21 cm radiation from cosmological neutral hydrogen.

(c) If the ionized gas has a temperature higher than 10^6 K, the most dominant radiation loss is through Bremsstrahlung. The emissivity for this process (in units erg $cm^{-3}\,s^{-1}\,Hz^{-1}$) is

$$J[v] \cong A T_e^{-1/2} n_e^2 \left[\exp\left(-\frac{hv}{kT_e} \right) \right], \tag{6.20}$$

where T_e is the electron temperature, n_e is the number density of electrons and A is a constant. Estimate the Bremsstrahlung radiation from a cosmological source of ionized gas.

(d) Cosmological considerations also modify the absorption of radiation in an expanding universe. Assume that radiation emitted by a source with a redshift z is absorbed by matter in the redshift range 0 to z. Let the number density of absorbers be $n(z)$ and the absorption cross-section be $\sigma(v)$, where v is the frequency at the time of emission. Show that the total optical depth due to intervening matter will be

$$\tau[v_0; z] = \frac{c}{H_0} \int_0^z \frac{\sigma[v_0(1+z)]\, n(z)}{(1+z)^2 (1+\Omega z)^{1/2}}\, dz. \tag{6.21}$$

As an example of the use of this formula, consider the absorption of radiation by neutral hydrogen at a wavelength of 21 cm. The absorption cross-section for this process is

$$\sigma(\nu) = \frac{A}{4\pi} \left(\frac{3}{4}\right) \left(\frac{h\nu}{2kT_{sp}}\right) \left(\frac{c}{\nu}\right)^2 \delta(\nu - \nu_H),$$
(6.22)

where T_{sp} is the so-called 'spin-temperature.' It is defined by the relation

$$\frac{n_{up}}{n_{down}} = \exp\left(-\frac{h\nu_H}{kT_{sp}}\right),$$
(6.23)

where n_{up} and n_{down} denote the number of atoms in the upper and lower energy levels. Estimate the optical depth due to 21 cm absorption.

6.6 Statistical mechanics in an expanding universe [3]

The discussion in problem 6.3 shows that the universe was radiation dominated at redshifts higher than $z_{eq} \cong 3.9 \times 10^4 \Omega h^2$. In the radiation dominated phase, the temperature will be greater than $T_{eq} \cong 9.2 \Omega h^2$ eV and will be increasing as $(1 + z)$. As we go to earlier phases of the universe the radiation will produce particle – antiparticle pairs of different kinds. These elementary particles will be interacting with each other via different processes which will try to maintain statistical equilibrium between the particles. In order to study the early phases of the universe, one would like to understand how the material content of the universe changes as it expands.

(a) Consider some early epoch at which the universe was populated by different species of particles such as electrons, positrons, muons, neutrinos, etc. Let the distribution function for the particle species A be $f_A(\mathbf{x}, \mathbf{p}, t)$. Determine the conditions under which we may assume these particles to be in statistical equilibrium with (i) other particles and (ii) radiation.

If the species A is in statistical equilibrium, then we can take the distribution function to be

$$f_A(\mathbf{p}, t)\, d^3\mathbf{p} = \frac{g_A}{(2\pi)^3} \left\{ \exp\left[(E_p - \mu_A)/T_A(t) \right] \pm 1 \right\}^{-1} d^3\mathbf{p},$$
(6.24)

where g_A is the spin degeneracy factor of the species, $\mu_A(T)$ is the chemical potential, $E(\mathbf{p}) = (\mathbf{p}^2 + m^2)^{1/2}$ and $T_A(t)$ is the temperature characterizing this species at time t. The upper sign ($+1$) corresponds to fermions and the lower sign (-1) is for bosons. Should the temperature be the same for all species and photons, that is, should $T_A = T_B = \ldots = T$?

(b) Express the number density of particles, (n), energy density (ρ) and pressure (p) as integrals over the distribution function $f(\mathbf{p})$. Using these expressions show that

$$d\left(sa^3\right) \equiv d\left\{ \frac{a^3}{T} (\rho + P - n\mu) \right\} = \left(\frac{\mu}{T}\right) d\left(na^3\right).$$
(6.25)

Argue that the quantity s, defined by the first equation can be interpreted as the entropy density of the universe. Under what conditions will the expansion of the universe be treated as adiabatic?

(c) Show that the energy density, number density and entropy density contributed by highly relativistic ($T \gg m$) and non-degenerate ($T \gg \mu$) particles can be written as

$$\rho_{\rm rad} = g_{\rm rad}\left(\frac{\pi^2}{30}\right)T^4, \quad n = \lambda\left(\frac{\zeta(3)}{\pi^2}\right)T^3, \quad s = q\left(\frac{2\pi^2}{45}\right)T^3, \tag{6.26}$$

where $\zeta(m)$ is a Riemann zeta function of order m and

$$g_{\rm rad} = \sum_{\rm B} g_{\rm B}\left(\frac{T_{\rm B}}{T}\right)^4 + \sum_{\rm F} \frac{7}{8}g_{\rm F}\left(\frac{T_{\rm F}}{T}\right)^4. \tag{6.27}$$

$$\lambda = \sum_{\rm B} g_{\rm B}\left(\frac{T_{\rm B}}{T}\right)^3 + \frac{3}{4}\sum_{\rm F} g_{\rm F}\left(\frac{T_{\rm F}}{T}\right)^3. \tag{6.28}$$

$$q = \sum_{\rm B} g_{\rm B}\left(\frac{T_{\rm B}}{T}\right)^3 + \frac{7}{8}\sum_{\rm F} g_{\rm F}\left(\frac{T_{\rm F}}{T}\right)^3. \tag{6.29}$$

Here $T_{\rm B}$, $T_{\rm F}$ refer to the temperatures characterizing the distribution functions of specific species of boson or fermion. What is the number density when $T \ll m$?

Also show that, during the radiation dominated era, the temperature of the universe at time t is given by the equation

$$t \cong 0.3g^{-1/2}\left(\frac{m_{\rm pl}}{T^2}\right) \cong 1{\rm s}\left(\frac{T}{1\,{\rm MeV}}\right)^{-2}g^{-1/2}. \tag{6.30}$$

(d) Consider a species of particles which was in equilibrium with the rest of the matter in the universe up to a time $t = t_{\rm D}$ and 'decouples' at $t = t_{\rm D}$. For $t > t_{\rm D}$ we may assume that each of the particles of this species moves along a geodesic. Determine the distribution function for such a species at $t \gg t_{\rm D}$. In particular, discuss the form of the distribution function if (i) the particles decouple when they are extremely relativistic ($T_{\rm D} \gg m$); or (ii) the particles decouple when they are non-relativistic ($T_{\rm D} \ll m$). What is the energy density contributed by the decoupled species in either of the above cases?

6.7 Relics of relativistic particles [2]

Consider the composition of the universe at a temperature slightly lower than 10^{12} K. We will assume that the universe is populated by electrons (e), positrons ($\bar{\rm e}$), three species of massless neutrino $(\nu_{\rm e}, \nu_\mu, \nu_\tau)$, neutrons (n), protons (p) and photons (γ).

(a) Estimate the number density of p, n, e, $\bar{\rm e}$, $\nu_{\rm e}$, ν_μ and $\nu_{\rm e}$ relative to the photons γ at this temperature.

(b) Since neutrinos have no electric charge they have no direct, strong coupling with photons. Their interaction with baryons can be ignored because of the low density of baryons. So they are kept in equilibrium essentially through reactions like $v\bar{v} \leftrightarrow e\bar{e}$, $ve \leftrightarrow ve$, etc. The cross-section, $\sigma(E)$, for these weak interaction processes is of the order of $(\alpha^2 E^2/m_x^4)$, where $\alpha \cong 2.8 \times 10^{-2}$ and $m_x \cong 50$ GeV is the mass of the gauge vector boson mediating the weak interaction. Show that neutrinos decouple from the rest of the matter at temperatures below $T_D \cong 1.4$ MeV.

(c) At the time of v-decoupling, the photons, neutrinos and the rest of the matter had the same temperature. As long as the photon temperature decreases as a^{-1}, neutrinos and photons will continue to have the same temperature even though the neutrinos have decoupled. However, the photon temperature will decrease at a slightly lower rate if the g-factor is changing. In that case, T_γ will become higher than T_v as the universe cools. Such a change in the value of g occurs when the temperature of the universe falls below $T \cong m_e$. The electron rest mass $m_e \cong 0.5$ MeV corresponds to a temperature of 5×10^9 K. When the temperature of the universe becomes lower than this value the mean energy of the photons will fall below the energy required to create $e\bar{e}$-pairs. Thus the backward reaction in $e\bar{e} \leftrightarrow \gamma\gamma$ will be severely suppressed. The forward reaction will continue to occur, resulting in the disappearance of the $e\bar{e}$ pairs.

Show that after $e - \bar{e}$ annihilation is complete, the temperature of the photons T_γ will be related to the temperature of the neutrinos T_v by $T_\gamma = (11/4)^{1/3} T_v$. Use this to estimate the fraction of critical density Ω_R contributed by relativistic particles today and show that $\Omega_R h^2 \cong 4.3 \times 10^{-5}$. Calculate the epoch at which non-relativistic and relativistic particles were contributing equal amounts of energy density and show that $(1 + z_{eq}) = 2.3 \times 10^4 \Omega h^2$. What was the temperature T_{eq} and time t_{eq} at this epoch ?

6.8 Relics of massive particles [1]

The scenario described in the preceding problem will change if the neutrinos have a mass. Consider the situation in which the neutrino has a mass of about 30 eV and decouples at $T_D \cong 1$ MeV. Show that the energy density contributed by such neutrinos today corresponds to

$$\Omega_v h^2 = \left(\frac{m_v}{91.5\,\text{eV}}\right). \tag{6.31}$$

Use this result to put a bound on the mass of the neutrino.

6.9 Decoupling of matter and radiation [3]

(a) As the universe cools, there will be a tendency for electrons and protons to combine and form neutral hydrogen atoms. Show that if we assume the process of recombination to take place in thermodynamic equilibrium with each recombination producing a H-atom in its ground state, then the recombination occurs at a redshift of $z_{rec} \cong 1300$. This epoch is defined to be the redshift at which 90% of electrons and protons have combined to form neutral atoms.

(b) For the above process to be described by equilibrium thermodynamics, the reaction rate $p + e \rightarrow H + \gamma$ should be higher than the expansion rate. Estimate the temperature T_D below which this condition fails if $\Omega \cong 1$ and $\Omega_B \cong 0.1$. What fraction of electrons (and protons) will remain as free charged particles at $T \ll T_D$?

(c) When most of the electrons and protons combine to form neutral atoms, the mean-free-path of photons (due to Thompson scattering) will become larger than $d_H(t)$. Estimate the temperature T_{dec} at which this happens.

(d) After decoupling, the free electrons will continue to exchange energy with radiation. Collisions between e, p and H will distribute this energy among the constituents of the (nearly neutral) plasma. Estimate the redshift z_{crit} up to which this process can keep the temperature of matter and radiation equal. (Assume that $\Omega \cong 1$ and $\Omega_B h^2 \cong 0.015$.)

(e) The above analysis was based on the assumption that recombination produces a hydrogen atom in the ground state. Is this reasonable? How will the results be modified in a more realistic scenario?

6.10 Perturbed Friedmann universe [3]

The Friedmann model describes a homogeneous and isotropic universe. It is generally believed that structures in the universe grew out of small perturbations which existed in the very early universe. To study the formation of such structures, one needs to understand how small perturbations around the Friedmann model evolve. We shall consider several aspects of this issue in this problem.

(a) Consider a spacetime described by the line interval

$$ds^2 = dt^2 - a^2(t) \left(\delta_{\alpha\beta} - h_{\alpha\beta} \right) dx^\alpha dx^\beta , \qquad (6.32)$$

where $\alpha, \beta = 1, 2, 3$ and $h_{\alpha\beta}$ is treated as a small perturbation. We shall confine our attention to the lowest order in all perturbed quantities and assume that matter is made up of an ideal fluid with the stress-tensor

$$T^{ik} = (\rho + p)u^i u^k - g^{ik} p \qquad (6.33)$$

and equation of state $p = w\rho$ with a constant w. The coordinate system in which the line element has the form (6.32) is taken to be the one in which the fluid is at rest with four-velocity $u^i = (1, 0, 0, 0)$. We also introduce the density contrast $\delta(t, \mathbf{x})$ by the relation $\rho(t, \mathbf{x}) = \rho_b(t)(1 + \delta)$.

Calculate the components of the stress-tensor in this spacetime. Show that, to the lowest order in $h_{\alpha\beta}$ and δ, we have the equation

$$\dot{\delta} = \frac{1}{2}(1 + w)\dot{h}, \qquad h = h^\alpha_\alpha. \qquad (6.34)$$

(When greek indices are repeated, summation is implied over the range 1,2,3.)

(b) To proceed further, we need to write down the perturbed Einstein equation for the metric (6.32). It turns out that the evolution of the density contrast can be determined using the time–time part of Einstein's equation *and* the equation derived in part (a) above.

Evaluate R_{00} for the perturbed Friedmann universe to linear order in $h_{\alpha\beta}$. Write down the time–time component of Einstein's equation and separate out the zeroth-order and first-order contributions. Show that the zeroth order terms give the standard Friedmann equations (as to be expected). The first-order term gives

$$\ddot{h} + 2\frac{\dot{a}}{a}\dot{h} = 8\pi G\rho_{\mathrm{b}}(1+3w)\delta. \tag{6.35}$$

Combining the equations derived in parts (a) and (b), show that the evolution of the density contrast is determined by the equation

$$\ddot{\delta} + 2\frac{\dot{a}}{a}\dot{\delta} = 4\pi G\rho_{\mathrm{b}}(1+w)(1+3w)\delta. \tag{6.36}$$

(c) Solve this equation and show that the two independent solutions are

$$\delta_{\mathrm{g}} \propto t^{n}, \quad \delta_{\mathrm{d}} \propto t^{-1}, \quad n = \frac{2}{3}\frac{(1+3w)}{(1+w)}, \tag{6.37}$$

if $w \neq -1$. What happens when $w = -1$?

7

Dynamics in the expanding universe

This chapter and the next deal with the formation of structures in the universe. This chapter concentrates on the gravitational dynamics in an expanding background and studies the evolution of perturbations in the dark matter. The physics of the baryonic component is taken up in the next chapter. Some of the problems are adapted from current research in this area and are fairly difficult. Quite a few of the problems require numerical integration of differential equations.

7.1 Trajectories in perturbed Friedmann universe [3]

The observed universe contains structures such as galaxies, clusters, etc., which are distributed very inhomogeneously at small scales. It is believed that these structures have formed due to the growth of small perturbations which originated in the very early universe. Observations also suggest that visible astrophysical systems such as galaxies, clusters, etc., are embedded in the halos of dark matter which are ten to a hundred times more massive than the visible matter. The dark matter is usually assumed to be made up of elementary particles interacting with each other only through gravity.

To the first approximation, the problem of structure formation reduces to understanding the formation of dark matter halos in an expanding universe. We need to understand how small perturbations in dark matter density will evolve as the universe expands and whether they can grow sufficiently fast to form the dark matter potential wells.

The evolution of *dark matter* perturbations in the universe can be analysed at three different levels. (i) To begin with, one can study the evolution of the individual trajectories $\mathbf{x}_i(t)$ of each dark matter particle in an expanding universe. This provides the most detailed level of description. (ii) At the next level, one can try to describe the dark matter by a distribution function $f(\mathbf{x}, \mathbf{p}, t)$ such that a small element in phase volume $d^3\mathbf{x}\, d^3\mathbf{p}$ contains $dN = f d^3\mathbf{x}\, d^3\mathbf{p}$ particles. At this level we ignore the individual trajectories but content ourselves with an average description. (iii) At the third level, we treat the system as made up of pressureless dust and describe the dark matter using a smoothed density $\rho(t, \mathbf{x})$ and mean velocity $\mathbf{v}(t, \mathbf{x})$.

The key difference between (ii) and (iii) is the following. The distribution function f allows for particles with different velocities to exist at some given location $\mathbf{x}$; in other words, there exists a velocity dispersion at each location $\mathbf{x}$. In the pressureless dust limit, we ignore the velocity dispersion and assume that at any given point there is a single velocity of flow $\mathbf{v}(t, \mathbf{x})$. We shall now study each of these descriptions.

Let us consider the trajectories of particles in a slightly perturbed Friedmann universe with the metric

$$ds^2 = a^2(\eta) \left[(1 + 2\phi)\, d\eta^2 - (1 - 2\phi) \left(dx^2 + dy^2 + dz^2 \right) \right] , \tag{7.1}$$

where ϕ is the perturbed gravitational potential arising from the perturbed energy density.

(a) Explain why this metric provides an approximate description of the perturbed universe at scales $\lambda \ll d_{\mathrm{H}}$. What is the effective Newtonian potential for this metric?

(b) The action for a particle of unit mass in this metric is

$$A = -\int ds = \int a \left[(1 + 2\phi) - (1 - 2\phi)\, |\dot{\mathbf{x}}|^2 \right]^{1/2} d\eta , \tag{7.2}$$

with $\dot{\mathbf{x}} = (d\mathbf{x}/d\eta)$. Linearize this action in ϕ and assume that $\phi \left(1 - v^2 \right)^{-1/2} \ll 1$ (even though v^2 need not be small compared to unity). Determine the perturbed Hamiltonian describing the motion of the particle to the linear order. Hence, obtain the equations of motion for a particle. Show that these equations take simple forms in the extreme relativistic (ER) and non-relativistic (NR) limits. In the ER limit we obtain

$$\dot{\mathbf{x}} \cong (1 + 2\phi) \frac{\mathbf{p}}{|\mathbf{p}|} , \qquad \dot{\mathbf{p}} \cong -2|\mathbf{p}|\nabla\phi , \tag{7.3}$$

while in the NR limit we have

$$\dot{\mathbf{x}} = \frac{\mathbf{p}}{a} , \qquad \dot{\mathbf{p}} = -a\nabla\phi . \tag{7.4}$$

How do these equations compare with the equations of motion in Newtonian gravity?

(c) The equations obtained above assumed that $\phi \ll c^2$ but made no assumptions regarding the magnitude of density inhomogeneities. In other words, these equations can be used to study *highly non-linear perturbations* provided the scales which are of interest to us satisfy $\lambda \ll d_{\mathrm{H}}$. In this case, we can assume that matter is non-relativistic and use the NR limit of the equations.

Introduce the time coordinate t with $a(\eta)\,d\eta = dt$. Recast the NR equation into the form

$$\frac{d^2 \mathbf{x}_i}{dt^2} + \frac{2\dot{a}}{a}\frac{d\mathbf{x}_i}{dt} = -\frac{1}{a^2}\nabla_{\mathbf{x}}\phi\,, \tag{7.5}$$

$$\nabla_{\mathbf{x}}^2 \phi = 4\pi G a^2 \rho_{\mathrm{bm}}\delta\,, \tag{7.6}$$

where $\rho_{\mathrm{bm}}(t)$ is the smooth background density of matter and δ is the density contrast of the matter defined by

$$\delta = \frac{\rho(\mathbf{x},t) - \rho_{\mathrm{bm}}(t)}{\rho_{\mathrm{bm}}(t)}\,. \tag{7.7}$$

Provide a simpler derivation of these equations using the effective Newtonian potential of the perturbed Friedmann metric.

(d) The equation for the particle trajectories can be recast into simpler form by using other coordinate systems. For example, one can introduce a new time coordinate $b = b(t)$ instead of t and the corresponding 'velocity' $\mathbf{w} = (d\mathbf{x}/db) = (\dot{\mathbf{x}}/\dot{b}) = (\mathbf{v}/a\dot{b})$, where $\mathbf{v} = a\dot{\mathbf{x}}$ is the original peculiar velocity. It turns out to be useful to choose $b(t)$ to be the growing solution to the equation

$$\ddot{b} + \frac{2\dot{a}}{a}\dot{b} = 4\pi G\rho_{\mathrm{bm}}(t)b\,. \tag{7.8}$$

Show that, in terms of $\mathbf{w}$ and b, the equations of motion for particles can be written as

$$\frac{d\mathbf{w}}{db} = -\frac{3A}{2b}(\mathbf{w} + \nabla\psi)\,, \quad \nabla^2\psi = \left(\frac{\delta}{b}\right)\,, \quad A = \left(\frac{\rho_{\mathrm{bm}}}{\rho_{\mathrm{c}}}\right)\left(\frac{b\dot{a}}{a\dot{b}}\right)^2\,. \tag{7.9}$$

7.2 Density contrast from the trajectories [3]

One is often interested in the evolution of the density contrast $\delta(t,\mathbf{x})$ rather than the trajectories. Since the density contrast can be expressed in terms of the trajectories of the particles, it should be possible to write down a differential equation for $\delta(t,\mathbf{x})$ based on the equations for the trajectories $\mathbf{x}_i(t)$ derived above. It is, however, somewhat easier to write down an equation for $\delta_{\mathbf{k}}(t)$ which is the Fourier transform of $\delta(t,\mathbf{x})$. Show that $\delta_{\mathbf{k}}(t)$ can be expressed in terms of the positions of the individual particles via the relation

$$\delta_{\mathbf{k}}(t) = \frac{1}{N}\sum_i \exp\left[i\mathbf{k}\cdot\mathbf{x}_i(t)\right]\,, \tag{7.10}$$

where N is the total number of particles in a large volume V under consideration. Using the equations of motion for $\mathbf{x}_i(t)$ show that $\delta_{\mathbf{k}}$ satisfies the equation

$$\ddot{\delta}_{\mathbf{k}} + 2\frac{\dot{a}}{a}\dot{\delta}_{\mathbf{k}} = 4\pi G\rho_{\mathrm{b}}\delta_{\mathbf{k}} + A - B\,, \tag{7.11}$$

with

$$A = 2\pi G \rho_b \sum_{\mathbf{k}' \neq 0, \mathbf{k}} \delta_{\mathbf{k}} \delta_{\mathbf{k}-\mathbf{k}'} \left\{ \frac{\mathbf{k} \cdot \mathbf{k}'}{k'^2} + \frac{\mathbf{k} \cdot (\mathbf{k} - \mathbf{k}')}{|\mathbf{k} - \mathbf{k}'|^2} \right\},$$

$$B = \frac{1}{N} \sum_j (\mathbf{k} \cdot \dot{\mathbf{x}}_j)^2 \exp\left[i\mathbf{k} \cdot \mathbf{x}_j(t)\right].$$

(7.12)

Here the overdot represents the derivative with respect to t.

7.3 Vlasov equation and the fluid limit [2]

(a) The description in the last two problems is exact but intractable. Some progress can be made if one is content to describe the system using a distribution function $f(\mathbf{x}, \mathbf{p}, t)$. This quantity is defined in such a way that the number of particles dN in a phase volume $d^3x \, d^3\mathbf{p}$ is

$$dN = f(\mathbf{x}, \mathbf{p}, t) d^3x \, d^3\mathbf{p}.$$

(7.13)

The mass density corresponding to f will be

$$\rho(\mathbf{x}, t) = \frac{m}{a^3} \int d^3\mathbf{p} \, f(\mathbf{x}, \mathbf{p}, t) = \rho_b(t) \left[1 + \delta(\mathbf{x}, t)\right],$$

(7.14)

where the last equation relates $\delta(\mathbf{x}, t)$ to $\rho(\mathbf{x}, t)$. Write down the Vlasov equation (also called the 'collisionless Boltzmann equation') in the non-relativistic limit using the time coordinates t and τ, where $d\tau = dt/a^2$.

(b) The crudest form of aproximation to study the evolution of density inhomogeneities in an expanding universe is obtained by taking the moments of the Vlasov equation and ignoring the velocity dispersion. Show that the equations describing such a 'pressureless dust' are

$$\frac{\partial \delta}{\partial b} + \partial_i \left[u^i(1 + \delta)\right] = 0, \quad \nabla^2 \psi = \left(\frac{\delta}{b}\right),$$

(7.15)

$$\frac{\partial u^i}{\partial b} + u^k \partial_k u^i = -\frac{3A}{2b} \left[\partial^i \psi + u^i\right], \quad A = \left(\frac{\rho_{bm}}{\rho_c}\right) \left(\frac{\dot{a}b}{a\dot{b}}\right)^2.$$

(7.16)

Here $u^i = u^i(b, \mathbf{x})$ is the velocity of the fluid at 'time' b and location $\mathbf{x}$; since the velocity dispersion has been ignored, u^i satisfies the same equation as the velocity w^i of the individual particles.

(c) Consider the evolution of density contrast in the fluid limit. Decompose the derivative of the velocity $\partial_a u_b$ into shear σ_{ab}, rotation Ω^c and expansion θ by writing

$$\partial_a u_b = \sigma_{ab} + \epsilon_{abc} \Omega^c + \frac{1}{3} \delta_{ab} \theta.$$

(7.17)

Show that the fluid equations can be combined to give

$$\frac{d^2\delta}{db^2} + \frac{3A}{2b} \frac{d\delta}{db} - \frac{3A}{2b^2} \delta(1 + \delta) = \frac{4}{3} \frac{1}{(1 + \delta)} \left(\frac{d\delta}{db}\right)^2 + (1 + \delta)(\sigma^2 - 2\Omega^2),$$

(7.18)

$\sigma^2 = \sigma_{ab}\sigma^{ab}$ and $\Omega^2 = \Omega_a\Omega^a$. Also show that the equation for Ω^a can be written as

$$\frac{\partial \mathbf{\Omega}}{\partial b} = -\frac{3A}{2b}\mathbf{\Omega} + \nabla \times (\mathbf{u} \times \mathbf{\Omega}). \qquad (7.19)$$

(d) If the flow is irrotational then one can write u^i as a gradient of some function Φ, that is, $u^i = \partial^i \Phi$. Show that the equations of motion for the irrotational flow can be written in terms of two coupled scalar fields (ψ, Φ):

$$\frac{\partial \Phi}{\partial b} + \frac{1}{2}(\partial_i \Phi)^2 = -\frac{3A}{2b}(\psi + \Phi),$$

$$\partial_i \left[\frac{\partial}{\partial b}(b\partial^i \psi) + (1 + b\nabla^2 \psi)\partial^i \Phi\right] = 0. \qquad (7.20)$$

7.4 Measures of inhomogeneity [3]

Consider some physical quantity $f(\mathbf{x}, t)$ which characterizes the inhomogeneities in the universe. This could be, for example, the density contrast or gravitational potential or a component of peculiar velocity. The simplest way of quantifying the variation of $f(\mathbf{x}, t)$ is to use the amplitude of the Fourier transform $|f_{\mathbf{k}}(t)|^2$. In the literature one often uses several other related measures to represent the density inhomogeneity.

(a) Define the two-point correlation function $\xi_f(\mathbf{x})$ of $f(\mathbf{x})$ by

$$\xi_f(\mathbf{x}) = \langle f(\mathbf{x} + \mathbf{y}) f(\mathbf{y})\rangle, \qquad (7.21)$$

where $\langle\,\rangle$ denotes averaging over a large volume V in space. (Here, and in what follows, we shall suppress the t-dependence to simplify the notation.) Show that

$$\xi_f(\mathbf{x}) = \int \frac{d^3\mathbf{k}}{(2\pi)^3} P_f(\mathbf{k}) e^{i\mathbf{k}\cdot\mathbf{x}}, \quad P_f(\mathbf{k}) = |f_k|^2 V^{-1}, \qquad (7.22)$$

where $P_f(\mathbf{k})$ is called the 'power spectrum' of $f(\mathbf{x})$.

(b) Given a variable $f(\mathbf{x})$ we can smooth it over some scale by using window functions $W(\mathbf{x})$ of suitable radius and shape. Let the smoothed function be

$$f_{\mathrm{W}}(\mathbf{x}) \equiv \int f(\mathbf{x} + \mathbf{y}) W(\mathbf{y})\, d^3\mathbf{y}. \qquad (7.23)$$

Show that

$$\langle f_{\mathrm{W}}^2\rangle = \int \frac{d^3k}{(2\pi)^3} P(\mathbf{k})|W_{\mathbf{k}}|^2, \qquad (7.24)$$

where $W_{\mathbf{k}}$ is the Fourier transform of the window function $W(\mathbf{y})$. Also show that

$$\langle f_{\mathrm{W}}^2\rangle = \int d^3\mathbf{x}\, \xi_f(\mathbf{x})\xi_{\mathrm{W}}(-\mathbf{x})V. \qquad (7.25)$$

(c) Consider, as an example, the mass contained within a sphere of radius R centred at some point $\mathbf{x}$ in the universe. As we change $\mathbf{x}$, keeping R constant, the mass enclosed by the sphere will vary randomly around a mean value

$M_0 = (4\pi/3)\rho_b R^3$, where ρ_b is the matter density of the background universe. The mean square fluctuation in this mass $\langle(\delta M/M)^2_R\rangle$ is a good measure of the inhomogeneities present in the universe at the scale R. In this case, $f(\mathbf{x}) = \delta(\mathbf{x})$ and the window function is $W(\mathbf{y}) = 1$ for $|\mathbf{y}| \leq R$ and zero otherwise. Show that

$$\sigma^2(R) \equiv \left\langle \left(\frac{\delta M}{M}\right)^2_R \right\rangle = \int_0^\infty \frac{dk}{k} \frac{k^3 P(k)}{2\pi^2} \left[\frac{3\sin kR}{(kR)^3} - \frac{3\cos kR}{(kR)^2}\right]^2. \tag{7.26}$$

What is the corresponding expression $\sigma^2_G(R)$ if the window function is a Gaussian with scale length R?

(d) The average value of the correlation function within a sphere of radius R is defined to be

$$\bar{\xi} = \frac{3}{R^3} \int_0^R \xi(x) x^2 \, dx. \tag{7.27}$$

Relate $\bar{\xi}(x)$ to $P(k)$. Show that $\sigma(R)$, $\xi(R)$ and $\bar{\xi}(R)$ are connected by the equations

$$\begin{aligned}
\sigma^2(R) &= \frac{3}{R^3} \int_0^{2R} x^2 \, dx \, \xi(x) \left(1 - \frac{x}{2R}\right)^2 \left(1 + \frac{x}{4R}\right) \\
&= \frac{3}{2} \int_0^{2R} \frac{dx}{2R} \bar{\xi}(x) \left(\frac{x}{R}\right)^3 \left[1 - \left(\frac{x}{2R}\right)^2\right].
\end{aligned} \tag{7.28}$$

7.5 Gaussian random fields [2]

A physical quantity $f(t, \mathbf{x})$ – which could be, for example, $\delta(t, \mathbf{x})$ or $\mathbf{v}(t, \mathbf{x})$ – evolves in time in accordance with some differential equation. To determine any such quantity at time t, we need to know its exact space dependence at some initial instant $t = t_i$ (e.g. to determine $\delta(t, \mathbf{x})$, we need to know $\delta(t_i, \mathbf{x})$). Often, we are not interested in the exact form of $\delta(t, \mathbf{x})$ but only in its 'statistical properties' in the following sense. We may assume that, for sufficiently small t_i, each Fourier mode $\delta_\mathbf{k}(t_i)$ was a Gaussian random variable with

$$\langle \delta_\mathbf{k}(t_i)\delta^*_\mathbf{p}(t_i)\rangle = (2\pi)^3 P(\mathbf{k}, t_i)\delta_\mathrm{Dirac}(\mathbf{k} - \mathbf{p}), \tag{7.29}$$

where $P(\mathbf{k}, t_i)$ is the power spectrum of $\delta(t_i, \mathbf{x})$ and $\langle\cdots\rangle$ now denotes an ensemble average.

(a) Define (at $t = t_i$) the correlation function $\xi_\delta(\mathbf{x})$ as the *ensemble* average $\langle\delta(\mathbf{x} + \mathbf{y})\delta(\mathbf{y})\rangle$. Does it agree with the result in problem 7.4?

(b) Consider the density distribution $\delta_W(\mathbf{x})$ which is obtained by smoothing $\delta(t_i, \mathbf{x})$ by some window function $W(\mathbf{y})$. What is the probability that $\delta_W(\mathbf{x})$ has a value q at some location $\mathbf{x}$?

(c) Equation (7.29) treats each mode $\delta_\mathbf{k}$ as an independent Gaussian variable at $t = t_i$. Will evolution preserve this property?

7.6 Linear evolution in the smooth fluid limit [3]

The equations for the evolution of perturbations described in the above problems are fairly complicated and exact solutions to these equations are nearly impossible. One can make progress by using one of the three possible techniques:

(i) When the density contrast δ is small, it is possible to linearize the equations in δ and obtain a solution which describes the growth of linear perturbations. Since the density contrasts are expected to be small in the early universe this approximation will be valid for a considerable period of time.

(ii) When the density contrast is comparable to unity, these equations can be solved only if specific assumptions (e.g. spherical symmetry) are made regarding the solutions. The usefulness of such solutions, of course, depends on the validity of the underlying assumptions.

(iii) In the non-linear stage ($\delta \gg 1$) the equations for particle trajectories have to be integrated numerically. Such an integration is time consuming and it is usually not easy to obtain physical insight about the evolution, even when the numerical integration is performed successfully. In order to tackle this difficulty one often introduces some simplifying ansatz while integrating the equations of motion. By comparing the results of exact numerical integration with the simplified picture, it is often possible to gain more physical insights into the dynamics.

We shall now explore the simplest of the above techniques, namely, the linear limit of perturbation equations. More complicated approaches will be discussed in the later problems.

(a) When the density contrasts are small, each Fourier mode $\delta_{\mathbf{k}}(t)$ evolves independently. The physical wavelength $\lambda(t) = (2\pi/|\mathbf{k}|)a(t)$ of this perturbation grows as $t^{1/2}$ in the radiation dominated phase and as $t^{2/3}$ in the matter dominated phase. The Hubble radius of the universe, $d_{\mathrm{H}}(t)$, grows as $d_{\mathrm{H}}(t) \propto t$ both in the matter and radiation dominated epochs. It follows that the wavelength of any perturbation will be bigger than the Hubble radius at sufficiently early times. It is conventional to say that 'the mode was outside the Hubble radius' when $\lambda(t) > d_{\mathrm{H}}(t)$ and that 'the mode was inside the Hubble radius' when $\lambda(t) < d_{\mathrm{H}}(t)$; the mode 'enters' the Hubble radius at a time t_{enter} when $\lambda = d_{\mathrm{H}}$. (See problem 6.3.)

Consider the growth of the density contrast $\delta_{\mathbf{k}}(t)$ when the mode is outside the Hubble radius. Since $\lambda \gg d_{\mathrm{H}}$ we cannot use Newtonian perturbation theory. Nevertheless, it is easy to determine the evolution of the density perturbation along the following lines. Consider an $\Omega = 1$ Friedmann universe with a smooth background density $\rho_{\mathrm{b}}(t)$. Let a spherical region of size $R \cong \lambda$ in this universe be perturbed to have a density contrast $\rho_{\mathrm{b}}(t) + \delta\rho(t)$. Because of spherical symmetry, the outside universe will not affect the evolution of the perturbed spherical region. This region will evolve as a Friedmann universe with a different density contrast, while the background universe continues to expand as an unperturbed universe.

Show that such an evolution is indeed possible with

$$\frac{\delta\rho}{\rho} \propto \frac{1}{\rho_b a^2} = \begin{cases} a^2 & \text{(radiation dominated phase)} \\ a & \text{(matter dominated phase).} \end{cases} \tag{7.30}$$

(The same result was obtained more rigorously by perturbing the Friedmann model in problem 6.10.)

(b) Perturbations which are of interest in cosmology have comoving wavelengths less than $\lambda_{eq} \cong 13(\Omega h^2)^{-1}$ Mpc. They enter the Hubble radius in the radiation dominated phase. We have seen above, in part (a), that these perturbations were growing as a^2 before entering the Hubble radius. We shall next consider their behaviour once they are inside the Hubble radius.

When $\lambda \ll d_H$, we can use Newtonian equations to study their growth. In this case, the equations are (see (7.11); in the linear order $A = B = 0$)

$$\ddot{\delta}_k + \frac{2\dot{a}}{a}\dot{\delta}_k \cong 4\pi G\rho_{DM}\delta_k, \tag{7.31}$$

$$\frac{\dot{a}^2}{a^2} = \frac{8\pi G}{3}(\rho_{rad} + \rho_{DM}), \tag{7.32}$$

where ρ_{DM} denotes the background dark matter density and ρ_{rad} is the background radiation density. Solve this equation and show that the two independent solutions are

$$\delta_g = 1 + \frac{3}{2}x,$$

$$\delta_d = \left(1 + \frac{3x}{2}\right)\ln\left[\frac{(1+x)^{1/2}+1}{(1+x)^{1/2}-1}\right] - 3(1+x)^{1/2}. \tag{7.33}$$

Here $x = a/a_{eq}$ and we have suppressed the subscript k in δ_k for simplicity of notation. How does δ behave for small x? Large x? Why?

(c) Combining the results of parts (a) and (b) above, we can determine the evolution of density perturbations in dark matter during all relevent epochs. Assume that the dark matter perturbation was given by $\delta(x) = x^2$ for $x \ll 1$. Use appropriate matching conditions when the mode enters the Hubble radius and determine the total growth for the perturbation from the time it enters the Hubble radius *until* the universe becomes matter dominated. Assume that $a_{enter} \ll a_{eq}$ and show that

$$\frac{\delta_{final}}{\delta_{enter}} \equiv \frac{\delta(a_{eq})}{\delta(a_{enter})} \cong 5\ln\left(\frac{\lambda_{eq}}{\lambda}\right). \tag{7.34}$$

(d) In the discussion above, we have assumed that $\Omega = 1$, which is a valid assumption in the early phases of the universe. During the later stages of evolution in a matter dominated phase, we have to take into account the actual value of Ω. Solve equation (7.31) for a matter dominated universe with arbitrary Ω and demonstrate that the decaying solution $\delta_d(t)$ and the growing solution $\delta_g(t)$ are given by

$$\delta_d(t) = H(t), \qquad \delta_g(t) = H(t) \int \frac{dt}{a^2 H^2(t)} \equiv b(t), \qquad (7.35)$$

where $H(t) = (\dot{a}/a)$. (The $b(t)$ is the growing solution used earlier.) Hence, show that, in the x-space, $\delta(t,\mathbf{x})$, $\psi(t,\mathbf{x})$ and $\mathbf{u}(t,\mathbf{x})$ (defined in problem 7.3) evolve according to

$$\delta(t,\mathbf{x}) = b(t)q(\mathbf{x}), \quad \nabla^2\psi(\mathbf{x}) = q(\mathbf{x}), \quad \mathbf{u}(\mathbf{x}) = -\nabla\psi(\mathbf{x}). \qquad (7.36)$$

Here $q(\mathbf{x})$ is a given function determined by the initial conditions. Thus $\mathbf{u}$ and ψ are constant in time in the linear theory.

(e) If $\mathbf{v} = a\dot{\mathbf{x}}$ is the peculiar velocity of the fluid and $\mathbf{g}$ is the acceleration produced by the perturbed matter, show that

$$\mathbf{v} = \frac{2f}{3H\Omega}\mathbf{g}, \qquad (7.37)$$

where $f = (a/\delta)\,(d\delta/da)$, $H(t) = (\dot{a}/a)$ and $\Omega(t) = (\rho/\rho_c)$.

(f) Calculate δ_g explicitly for a matter dominated universe with arbitrary Ω and show that it can be expressed as

$$\delta_g(z) = \frac{1 + 2\Omega + 3\Omega z}{(1-\Omega)^2} - \frac{3}{2}\frac{\Omega(1+z)(1+\Omega z)^{1/2}}{(1-\Omega)^{5/2}}\ln Q(z), \qquad (7.38)$$

where

$$Q(z) = \frac{(1+\Omega z)^{1/2} + (1-\Omega)^{1/2}}{(1+\Omega z)^{1/2} - (1-\Omega)^{1/2}}. \qquad (7.39)$$

The redshift z is related to the expansion factor $a(z) = a_0(1+z)^{-1}$. How does this solution behave for large and small z?

7.7 Linear evolution of distribution function [3]

The linear approximation can also be used to study the solutions to collisionless Boltzmann equations. For this purpose, it is convenient to use the time coordinate $d\tau = dt/a^2$. The linear approximation is obtained by writing $f(\mathbf{x}, \mathbf{p}, \tau) = f_0(\mathbf{p}) + f_1(\mathbf{x}, \mathbf{p}, \tau)$ and linearizing the equation in f_1. The resulting equation can be solved by standard Laplace transform techniques.

(a) Show that the solution f_1 can be expressed as

$$f_1(\mathbf{x}, \mathbf{p}, \tau) = \int \frac{d^3\mathbf{k}}{(2\pi)^3} f_\mathbf{k}(\mathbf{p}, \tau) e^{i\mathbf{k}\cdot\mathbf{x}}, \qquad (7.40)$$

where

$$f_\mathbf{k}(\mathbf{p}, \tau) = f_\mathbf{k}(\mathbf{p}, \tau_i)\exp(-i(\mathbf{k}\cdot\mathbf{p}/m)(\tau - \tau_i))$$
$$+ m\left(i\mathbf{k}\cdot\frac{\partial f_0}{\partial\mathbf{p}}\right)\int_{\tau_i}^{\tau} ds\, a^2(s)\phi_\mathbf{k}(s)\exp(-i(\mathbf{k}\cdot\mathbf{p}/m)(\tau - s)), \qquad (7.41)$$

$f_{\mathbf{k}}(\mathbf{p}, \tau_i)$ is the initial perturbation and $\phi_{\mathbf{k}}(\tau)$ is the Fourier transform of the perturbed gravitational potential.

(b) Given the distribution function, one can compute the density contrast by integrating over the velocities (see equation (7.14)). Show that the Fourier transform of the density contrast satisfies the equation

$$\delta_{\mathbf{k}}(\tau) = -k^2 \int_{\tau_i}^{\tau} ds\, a^2(s)\phi_{\mathbf{k}}(s)(\tau - s)\mathscr{G}\left[k(\tau - s)/m\right], \tag{7.42}$$

where

$$\mathscr{G}(\mathbf{q}) = \frac{1}{n_0} \int d^3\mathbf{p}\, f_0(\mathbf{p})e^{-i\mathbf{p}\cdot\mathbf{q}}. \tag{7.43}$$

This is an integral equation for $\delta_{\mathbf{k}}(\tau)$ since $\phi_{\mathbf{k}}(\tau)$ is related to $\delta_{\mathbf{k}}(\tau)$ by the Poisson equation.

(c) Assume that the $f_0(\mathbf{p}) = f_0(|\mathbf{p}|)$. Further, let the gravitational field be due to the particles described by f_1, as well as some external sources. Then

$$\nabla^2\phi = 4\pi G\rho_b a^2\left(\delta + \delta^{\mathrm{ext}}\right), \tag{7.44}$$

where $\rho_b\delta^{\mathrm{ext}}$ is the (given) mass density due to external sources. (We assume that $\delta^{\mathrm{ext}} \ll 1$ to allow for perturbation theory.) Show that $\delta_{\mathbf{k}}(\tau)$ satisfies the equation

$$\delta_{\mathbf{k}}(\tau) = \frac{16\pi^2 Gm^2}{k} \int_{\tau_i}^{\tau} d\tau'\, a(\tau')\left[\delta_{\mathbf{k}}(\tau') + \delta_{\mathbf{k}}^{\mathrm{ext}}(\tau')\right] I_{\mathbf{k}}(\tau - \tau'), \tag{7.45}$$

where $\delta_{\mathbf{k}}^{\mathrm{ext}}(\tau)$ is the Fourier transform of $\delta^{\mathrm{ext}}(\tau, \mathbf{x})$ and

$$I_{\mathbf{k}}(z) = \int_0^{\infty} q\, dq\, f_0(q) \sin\left[\frac{kq}{m}z\right]. \tag{7.46}$$

(d) Solve the equation for $\delta_{\mathbf{k}}$ in the long wavelength limit with $\delta^{\mathrm{ext}} = 0$.

7.8 Free streaming [3]

(a) Consider a system of particles described by the perturbed distribution function $f_1(\tau, \mathbf{x}, \mathbf{p})$ studied in the last problem. Let us assume that at the initial epoch the perturbation had a reasonable amount of power at all scales, that is, for all values of k. The velocity dispersion will cause particles to move from the 'crests' of a perturbation to the 'troughs'. Show that the amplitude of the perturbations is rapidly destroyed for modes with small wavelengths. (This process is called 'free streaming'.)

(b) In the case of pressureless dust discussed in problem 7.6 one could obtain a differential equation for $\delta_{\mathbf{k}}(t)$. In the present case, when the velocity dispersion is non-zero we are led to the integral equation (7.42) and it is not possible (in general) to obtain a differential equation for $\delta_{\mathbf{k}}$. There is, however, a special case in which one can manipulate the various factors to obtain a closed differential equation for $\delta_{\mathbf{k}}$. This occurs when the background distribution function has the form $f_0(\mathbf{p}) \propto \left(\alpha^2 + p^2\right)^{-2}$.

Show that in this particular case, the integral equation for $\delta_{\mathbf{k}}$ can be expressed as a differential equation:

$$\frac{d^2\delta_{\mathbf{k}}}{dt^2} + 2\left(\frac{\dot{a}}{a} + c_s\frac{k}{a}\right)\frac{d\delta_{\mathbf{k}}}{dt} + \left(\frac{c_s^2 k^2}{a^2} - 4\pi G\rho_b\right)\delta_{\mathbf{k}} = 0, \qquad (7.47)$$

with $c_s = \alpha/a$. (We have assumed that $\delta_{\mathbf{k}}^{\text{ext}} = 0$.) When compared to equation (7.31) we see that there are two extra terms, one in the damping and one in the restoring force. Interpret these terms.

(c) In the limit of $(c_s k/a) \gg H(t)$ we can ignore the $(\dot{a}/a)$ and $4\pi G\rho_b$ terms in the above equation. In this limit solve the equation and show that the solutions decay at late times.

(d) The free-streaming scale can also be obtained more rigorously by analysing the motion of the particles. Estimate the proper distance which a dark matter particle can travel during the time interval $(0, t_0)$. Assume that the particle is relativistic with $v \cong c$ for $t < t_{\text{nr}}$ and that it is non-relativistic with $v \propto a^{-1}$ for $t > t_{\text{nr}}$. Show that the maximum comoving distance the particle could have travelled is given by

$$\lambda_{\text{FS}} = (2ct_{\text{nr}})\left(\frac{a_0}{a_{\text{nr}}}\right)\left(\frac{5}{2} + \ln\frac{a_{\text{eq}}}{a_{\text{nr}}}\right) \cong (2ct_{\text{nr}})\left(\frac{a_0}{a_{\text{nr}}}\right). \qquad (7.48)$$

Consider a universe populated by neutrinos of mass $m_\nu \cong 30\,\text{eV}$ which decouple at $T_{\text{D}} \cong 1\,\text{MeV}$. After decoupling, neutrinos can be described adequately by a collisionless Boltzmann equation. The neutrinos will be relativistic at temperatures $T \gtrsim m_\nu$ and will become non-relativistic at $T \ll m_\nu$. Let the epoch at which neutrinos become non-relativistic be t_{nr}. Estimate the free-streaming scale for neutrinos.

7.9 Power spectra for dark matter [2,N]

(a) Most models for structure formation assume that in the very early universe, density perturbations were described by power spectra of the form $P(k) = Ak^n$. Using the results of the preceding problems evolve such a power spectrum forward in time and determine its approximate shape at the time when $a = a_{\text{dec}}$.

(b) Calculate numerically and plot the measures of inhomogeneity $\Delta(k) = (k^3 P(k)/2\pi^2)^{1/2}$, $\sigma(R)$ and $[\bar{\xi}(R)]^{1/2}$ for the following power spectra: (i) $P(k) = Ak\exp[-\lambda(k/k_0)^{3/2}]$ with $A = (24h^{-1}\,\text{Mpc})^4$, $\lambda = 4.61, k_0 = 0.16\,\text{Mpc}^{-1}$. (ii) $P(k) = Ak(1 + Bk + Ck^{3/2} + Dk^2)^{-2}$ with $B = 1.7h^{-2}\,\text{Mpc}$, $C = 9h^{-3}$ $\text{Mpc}^{3/2}$ and $D = 1\,h^{-4}\,\text{Mpc}^2$. Take $h = 0.5$.

The spectrum in (i) corresponds to 'hot' dark matter. For $k > k_0$, the power is exponentially suppressed. The spectrum in (ii) is that of 'cold' dark matter with a free-streaming scale which is too small for cosmological interest. For both these cases, the initial spectrum was taken to be a power law with $n = 1$.

7.10 Spherically symmetric evolution [2]

In the non-linear regime when $\delta \gtrsim 1$ it is not possible to solve equations for pressureless dust exactly. Some progress, however, can be made if we assume that the solutions are spherically symmetric. We consider, at some initial epoch t_i, a spherical region of the universe which has a slight constant overdensity compared to the background. As the universe expands, the overdense region will expand more slowly compared to the background, will reach a maximum radius, contract and virialize to form a bound non-linear system. Such a model is called 'spherical top-hat'.

(a) Show that in the case of spherical symmetry the density contrast δ satisfies the equations

$$\delta'' + \frac{3A}{2b}\delta' = \frac{4}{3}\frac{(\delta')^2}{(1+\delta)} + \frac{3A}{2b^2}\delta(1+\delta), \quad A = \left(\frac{\rho_b}{\rho_c}\right)\left(\frac{\dot{a}b}{a\dot{b}}\right)^2, \tag{7.49}$$

or

$$\ddot{\delta} - \frac{4}{3}\frac{\dot{\delta}^2}{(1+\delta)} + \frac{2\dot{a}}{a}\dot{\delta} = 4\pi G\rho_b\delta(1+\delta), \tag{7.50}$$

where the dot denotes (d/dt) and the prime denotes (d/db); here $b(t)$ is the growing solution to the linear perturbation equation. Further, show that the solution to (7.50) can be expressed in the form

$$\delta(t) = \frac{2GM}{\Omega_m H_0^2 a_0^3}\left[\frac{a(t)}{R(t)}\right]^3 - 1 \equiv \lambda\left(\frac{a}{R}\right)^3 - 1, \tag{7.51}$$

where M is a constant and $R(t)$ satisfies the equation

$$\ddot{R} = -\frac{GM}{R^2} - \frac{4\pi G}{3}(\rho + 3p)_{\text{rest}}R, \tag{7.52}$$

in which $(\rho + 3p)_{\text{rest}}$ is due to components other than dust-like matter. Interpret this equation.

(b) Assume that the background is described by a $\Omega = 1$, matter dominated universe for which $b = a = (t/t_0)^{2/3}$. In this case, show that the evolution of density contrast can be expressed in the form $1 + \delta(a) = (2GM/H_0^2)\left[a^3/R^3(a)\right]$, where $R(t)$ satisfies the equation

$$\frac{d^2R}{dt^2} = -\frac{GM}{R^2} \tag{7.53}$$

and M is a constant. Let $R(t_i) = r_i$ where t_i is the time at which initial conditions are specified. Let $M = (4\pi/3)r_i^3\rho_b(t_i)(1+\delta_i)$, with $\delta_i > 0$. Further assume that $\left(\dot{R}^2/2\right)_{t=t_i} = (1/2)H_i^2 r_i^2$.

(i) Show that the evolution of the density contrast $\delta(z)$ is given implicitly by the functions $[\delta(\theta), z(\theta)]$ with

$$(1+z) = \left(\frac{4}{3}\right)^{2/3} \frac{\delta_i (1+z_i)}{(\theta - \sin \theta)^{2/3}} = \left(\frac{5}{3}\right)\left(\frac{4}{3}\right)^{2/3} \frac{\delta_0}{(\theta - \sin \theta)^{2/3}}, \qquad (7.54)$$

$$\delta = \frac{9}{2}\frac{(\theta - \sin \theta)^2}{(1 - \cos \theta)^3} - 1 \qquad (7.55)$$

where δ_i is the density contrast at the initial redshift z_i and $\delta_0 = (3/5)\delta_i(1+z_i)$ is the corresponding density contrast at the present epoch if the density perturbation grows according to linear theory. Also show that the density contrast at any epoch, predicted by the linear theory, is

$$\delta_L = \frac{3}{5}\left(\frac{3}{4}\right)^{2/3}(\theta - \sin \theta)^{2/3}. \qquad (7.56)$$

(ii) Compare the results for δ and δ_L at different epochs to estimate the accuracy of linear theory.

(iii) Prove that the spherical region reaches a maximum radius of $r_m = (3x/5\delta_0)$ at a redshift z_m such that $(1 + z_m) \cong (\delta_0/1.062)$ and the density contrast at maximum expansion is about 4.6.

(c) After reaching the maximum expansion, the matter will collapse inwards. It is likely that the system will become virialized during the collapse and form a gravitationally bound object. (i) Estimate the redshift z_{coll} at which the virialization occurs and (ii) the density of the collapsed object.

7.11 Scaling laws for spherical evolution [3]

(a) Explain qualitatively how the evolution described in parts (c) and (d) of the last problem becomes modified if the initial density is not constant in the spherical region but is given by some profile $\rho_i(r)$ at $t = t_i$.

(b) Assume that the initial density profile is such that the energy $E(M)$ of a shell containing mass M is given by a power law $E(M) = E_0(M/M_0)^{2/3-\epsilon}$. Describe the final density profile assuming that the evolution is self-similar.

7.12 Self-similar evolution of Vlasov equation [2]

At the next level of approximation, one would like to obtain non-linear solutions to the collisionless Boltzmann equation. Needless to say, this is far more difficult than in the case of pressureless dust. Some amount of progress can be made by assuming that the evolution is self-similar in a finite region.

(a) Show that the Vlasov equation (see problem 7.3) admits self-similar solutions of the form

$$f = t^{-(3\alpha+1)}\hat{f}\left(\frac{\mathbf{x}}{t^\alpha}, \frac{\mathbf{p}}{t^{\alpha+1/3}}\right). \qquad (7.57)$$

(b) This solution can be used to relate the behaviour of the density contrast at different scales. In many models of structure formation, there will be more power (per decade) at small scales compared to the large scales. In that case, small scales will go non-linear first and will form gravitationally bound objects while the large scales are still evolving in accordance with the linear perturbation equation. (It is assumed that the non-linearity of the small scales does not affect the large scale evolution.) At very large scales (i.e. for small k), the power spectrum is still described by the linear theory and the k-dependence of $\delta_k(t)$ will be given by the initial conditions. Assume that $P(k) \propto k^n$ in the linear theory. We have no idea, *a priori*, about the k-dependence of $\delta_k(t)$ for large k. However, *if* we *assume* that the evolution of density perturbation in a region of interest is self-similar, then we can use the results of part (a) to relate $\delta_k(t)$ at large k with $\delta_k(t)$ at small k. Show that at large k the density contrast can be expressed as $\delta_k(t) \propto t^{(2/3)(3-m)}k^m$, with $m = (9 + 3n)/(5 + n)$, if the evolution is self-similar. Provide a geometrical description for this result.

7.13 Non-linear evolution from a scaling ansatz [3,N]

Given an initial density contrast, one can trivially obtain the density contrast at any later epoch in the *linear* theory. If there is a procedure for relating the non-linear density contrast and the linear density contrast (even approximately), then one can make considerable progress in understanding non-linear clustering. It is actually possible to make one such ansatz along the following lines.

Let $v_{\rm rel}(a, x)$ denote the relative pair velocities of particles separated by a distance x, at an epoch a, averaged over the entire universe. This relative velocity is a measure of gravitational clustering at the scale x at the epoch a. Let $h(a, x) \equiv -[v_{\rm rel}(a, x)/\dot{a}x]$ denote the ratio between the relative pair velocity and the Hubble velocity at the same scale.

(a) Argue that $h \cong 1$ in the extreme non-linear limit (when $\bar{\xi} \gg 1$), and $h \cong (2/3)\,\bar{\xi}$ when $\bar{\xi} \ll 1$. (Assume $\Omega = 1$.)

(b) This suggests the ansatz that h depends on a and x only through the density contrast at the epoch a at the scale x. As a measure of the density contrast we shall use $\bar{\xi}(a, x)$. In other words, we assume that $h(a, x) = H[\bar{\xi}(a, x)]$.

Show that the assumption $h(a, x) = H[\bar{\xi}(a, x)]$ implies that the non-linear density contrast $\bar{\xi}(x, a)$ is a universal function of the linear density contrast $\bar{\xi}_{\rm L}(l, a)$ with $l^3 = x^3(1 + \bar{\xi})$. The relationship between $\bar{\xi}_{\rm L}(l, a)$ and $\bar{\xi}(x, a)$ is given by

$$\bar{\xi}_{\rm L}(l, a) = \exp\left(\frac{2}{3} \int^{\bar{\xi}} \frac{d\mu}{H(\mu)(1 + \mu)}\right). \tag{7.58}$$

The functional relationship $\bar{\xi} = F(\bar{\xi}_L)$ clearly depends on the form of the function $h(\bar{\xi})$. Numerical simulations suggest that

$$\bar{\xi}(a,x) = \begin{cases} \bar{\xi}_L(a,l) & (\text{for } \bar{\xi}_L < 1.2, \ \bar{\xi} < 1.2) \\ 0.7[\bar{\xi}_L(a,l)]^3 & (\text{for } 1.2 < \bar{\xi}_L < 6.5, \ 1.2 < \bar{\xi} < 195), \\ 11.7[\bar{\xi}_L(a,l)]^{3/2} & (\text{for } 6.5 < \bar{\xi}_L, \ 195 < \bar{\xi}) \end{cases} \qquad (7.59)$$

with $l^3 = x^3(1 + \bar{\xi})$. This result allows one to determine the non-linear density contrast given the linear density contrast.

(c) Using the above relationship we can evaluate $\bar{\xi}(x,a)$ using $\bar{\xi}_L(x,a)$. This requires a non-local manipulation since the relationship in part (b) only relates $\bar{\xi}$ and $\bar{\xi}_L$ at different scales. Evaluate numerically and plot the true $\bar{\xi}$ for the CDM power spectrum of problem 7.8 at the redshifts $z = 0, 3, 5, 20, 50$ and 100 from the knowledge of $\bar{\xi}_L(x,a)$.

(d) In most situations of interest, different measures of inhomogeneity will be of the same order and can be used interchangeably. Hence, we may assume that the relations obtained above for $\bar{\xi}$ remain approximately valid for $\sigma(R)$ defined by some window function. We can obtain approximate information about several physical quantities from the above result.

As an example, consider a linear power spectrum with index n, for which $\sigma_L^2(a,x) \propto a^2 x^{-(n+3)}$. Determine $\sigma_{NL}^2(a,x)$ in the quasilinear and non-linear epochs. Show that the spectrum with $n = -1$ follows linear evolution even in the quasilinear phase, up to $\bar{\xi} \cong 200$.

7.14 Zeldovich approximation and pancakes [2]

The equations of motion for the particles are most difficult to integrate in closed form. It is nevertheless possible to invoke some approximations at the level of individual trajectories, which gives an insight about the motion. One such approximation, called Zeldovich approximation, provides an interesting picture about the formation of structures in the universe.

We begin by noting that, in the linear theory, the velocities w^i are related to the gradients of the perturbed gravitational potential by $\mathbf{w} = -\nabla\psi$. Let us assume that at very early times $t = t_i$, all the particles in the universe had the velocities as given by linear theory. If the location of a particle at this time is $\mathbf{q}$ it may be attributed a specific velocity $\mathbf{u}(\mathbf{q})$. The Zeldovich approximation consists of moving the particles by this velocity even into the non-linear region. That is, we take the position of the particle at time t to be $\mathbf{x}(b) = \mathbf{q} + b\mathbf{u}(\mathbf{q})$, where $b(t)$ is the growing mode of linear perturbation theory. This relation may be interpreted as giving the Eulerian position $\mathbf{x}$ of a particle which originally had a Lagrangian position $\mathbf{q}$.

(a) Show that this ansatz is equivalent to assuming

$$\frac{\partial \Phi}{\partial b} + \frac{1}{2}(\partial_i \Phi)^2 = 0 \qquad (7.60)$$

in equation (7.20). Is this consistent with the trajectories which are assumed?

(b) Show that the density at any time in Zeldovich approximation is given by

$$\rho(\mathbf{r}, t) = \frac{\rho_b(t)}{(1 - b(t)\lambda_1(\mathbf{q}))(1 - b(t)\lambda_2(\mathbf{q}))(1 - b(t)\lambda_3(\mathbf{q}))}, \tag{7.61}$$

where the $\lambda(\mathbf{q})$s are the eigenvalues of the matrix $(\partial^2 \psi / \partial q_i \partial q_j)$. Conclude from this expression that the first non-linear structures which form are planar.

(c) Describe the evolution of a spherical overdense region using Zeldovich approximation. How does it compare with the spherical top-hat model discussed earlier?

(d) Show that Zeldovich approximation is exact in one dimension until the trajectories of the particles cross.

7.15 Accuracy of Zeldovich approximation [1]

The description based on Zeldovich approximation uses trajectories which are built out of the linear theory. Such a description, of course, is not exact in more than one dimension. To understand the nature of the approximation, we may proceed as follows. Given the acceleration field $\ddot{\mathbf{x}}$, we can compute the density distribution ρ as

$$\nabla_x \cdot \ddot{\mathbf{x}} = -4\pi G\rho, \tag{7.62}$$

provided the potential producing the acceleration ($\ddot{\mathbf{x}} = -\nabla \phi_{\text{acc}}$) is the same as the potential generated by the density ($\nabla^2 \phi_N = 4\pi G\rho$). For the *exact* trajectories, equation (7.62) will be an identity; for the approximate ansatz we are using for $\mathbf{x}(t)$, we can compute the quantity $Q = (\nabla \cdot \ddot{\mathbf{x}} + 4\pi G\rho)$ explicitly. The smallness of this quantity will provide a measure for the validity of the approximation.

Since $\mathbf{x}(t)$ is known, we can evaluate this quantity Q explicitly. The result is most easily expressed in terms of the 'invariants' of the tensor $D_{ij} = (\partial^2 \psi / \partial q_i \partial q_j)$:

$$I_1 = \lambda_1 + \lambda_2 + \lambda_3, \quad I_2 = \lambda_1\lambda_2 + \lambda_2\lambda_3 + \lambda_3\lambda_1, \quad I_3 = \lambda_1\lambda_2\lambda_3. \tag{7.63}$$

where the λs are the eigenvalues. Show that the fractional inaccuracy is of the order of

$$\frac{(b^2 I_2 - 2b^3 I_3)}{(1 - bI_1 + b^2 I_2 - b^3 I_3)} = \frac{b^2 I_2 - 2b^3 I_3}{(1 - b\lambda_1)(1 - b\lambda_2)(1 - b\lambda_3)}. \tag{7.64}$$

In a planar collapse with $\lambda_2 \cong \lambda_3 \cong 0$, the approximation is excellent. If the collapse first occurs when $b\lambda_1 = 1$, the approximation remains quite good till that event.

8
Structure formation

The formation of baryonic structures and the radiative coupling between photons and baryons are studied in this chapter. The discussion includes the anisotropies of CMBR and an elementary analysis of the COBE results.

8.1 Perturbations in baryons [1]

The baryons in the universe are tightly coupled to the photons until matter and radiation decouple at $z \cong 10^3$. This tight coupling provides the baryonic fluid with a pressure which could resist the force of gravity. Hence, small scale perturbations in a baryonic fluid cannot grow in amplitude at $z > z_{\text{dec}}$. But at sufficiently large scales, one would expect gravity to overpower the pressure gradient, thereby allowing the perturbation to grow.

(a) The equations describing baryons may be taken to be those of an ideal self-gravitating fluid. Set up these equations and linearize them around the uniform Friedmann background. Assume that the perturbed pressure δp and the perturbed density $\delta \rho$ are related by $\delta p = c_s^2 \delta \rho$, where c_s is the speed at which disturbances propagate in the fluid. Show that modes with

$$k^2 > k_{\text{J}}^2, \qquad k_{\text{J}}^2 \equiv \frac{4\pi G \rho a^2}{c_s^2}, \tag{8.1}$$

oscillate as accoustic vibrations, while modes with $k < k_{\text{J}}$ can grow in amplitude. Here ρ is the density of the most dominant, perturbed component which is contributing to gravity. In particular, large scale modes with $k \ll k_{\text{J}}$ grow like the corresponding dark matter modes.

The condition for the growth of modes could also be expressed in terms of the physical wavelength, $\lambda_{\text{phy}} \equiv (2\pi/k)a(t)$. Modes with $\lambda_{\text{phy}} < \lambda_{\text{J}}$ do not grow, while long wavelength modes with $\lambda > \lambda_{\text{J}}$ can grow where

$$\lambda_{\text{J}} = \left(\frac{\pi c_s^2}{G \rho} \right)^{1/2}. \tag{8.2}$$

The λ_{J} is called the Jeans length and k_{J} is the corresponding Jeans wavenumber.

(b) It is convenient to define a quantity called the 'Jeans mass', M_J, as the amount of baryonic mass contained within a sphere of radius $(\lambda_J/2)$. Describe how the Jeans mass scales with a for $a > a_{eq}$. Hence, provide a qualitative description of the growth of baryonic perturbations at different epochs.

8.2 Acoustic oscillations of baryons [1]

(a) Consider modes with $\lambda_{phy} \ll \lambda_J$ for which the pressure term dominates over gravity. We may assume that, for $z \gtrsim 10^2$, the temperature of matter is approximately equal to that of radiation, which scales as a^{-1}, that is, we can take $c_s^2 = v_0^2(a_0/a)$ (see problem 6.9). Solve the perturbation equation for baryons for this case (ignoring $4\pi G\rho\delta$ terms) and interpret the solutions.

(b) The analysis above shows that baryonic perturbations at moderate scales do not grow when baryons and photons are coupled. However, dark matter perturbations, which are uncoupled to photons, have been growing since $a = a_{eq}$. During the period $a_{eq} < a < a_{dec}$, dark matter perturbations would have grown by a factor $(a_{dec}/a_{eq}) \cong 20\Omega h^2$. For $a > a_{dec}$, baryonic perturbations can grow. During this epoch, baryons will feel the potential wells of the dark matter perturbations which will have already grown by a larger factor. It follows that baryonic perturbations at $a > a_{dec}$ will be essentially driven by the dark matter perturbations. We can then ignore the $\rho_B\delta_B$ term compared to the $\rho_{DM}\delta_{DM}$ term. Solve the baryonic perturbation equation in this limit and show that $\delta_B \cong \delta_{DM}$ for small k, and $\delta_B \propto \delta_{DM}/k^2$ for large k.

8.3 Silk damping [2]

In the study of dark matter, we found that small scale perturbations are wiped out due to the free streaming of the dark matter particles. A somewhat similar phenomenon occurs for baryons as well. Photons could also undergo free streaming if only they were not tightly coupled to baryons. In the limit of infinitely tight coupling, no free streaming can occur. However, in reality, photons *can* diffuse from high density to low density regions, dragging baryons with them. This process can wipe out small scale fluctuations in baryons. Estimate the mass scale M_S (called the 'Silk mass') below which this damping is effective and show that

$$M_S \cong 6 \times 10^{12}\, M_\odot \left(\frac{\Omega}{\Omega_B}\right)^{3/2} (\Omega h^2)^{-5/4}. \tag{8.3}$$

8.4 Overall evolution of baryonic perturbations [1]

In problem 7.9, we worked out the power spectrum of dark matter by putting together the dynamical evolution in various phases. Repeat the same analysis for the baryonic perturbations and provide a qualitative picture of the inhomogeneities in the baryonic component at some time well after decoupling.

8.5 Non-linear collapse of baryons [3,N]

The linear perturbation theory, described in the preceding problems breaks down when $\delta_B \cong 1$. The study of the non-linear regime for the baryons is far more complicated than that of dark matter because of the need to take pressure gradients and radiative processes into account. Numerical simulations are also more complex for baryons since one needs to use hydrodynamic codes. In this problem we shall examine some simple models which allow the estimation of non-linear baryonic structures.

(a) Describe the evolution of the baryonic component located within a spherically symmetric dark matter potential well using the spherical top-hat model discussed in problem 7.10. (Ignore the effects of radiative cooling.) Estimate the virial radius, r_{vir}, the velocity dispersion, v, and the virial temperature of the baryonic component in terms of the collapse redshift, z_{coll} and the total mass M.

(b) The baryonic component will be heated to a rather high temperature if the above analysis is correct. At these temperatures, it is important to take into account the cooling processes operational in the gas. If the gas can radiate energy and cool effectively, it will be able to collapse further and form a more tightly bound object. The evolution will depend critically on the relative values of the dynamical time-scale:

$$t_{\mathrm{dyn}} \cong \frac{\pi}{2} \left[\frac{2GM}{R^3} \right]^{-1/2} = 5 \times 10^7 \, \mathrm{yr} \left(\frac{n}{1 \, \mathrm{cm}^3} \right)^{-1/2} \tag{8.4}$$

and the cooling time-scale:

$$t_{\mathrm{cool}} = \frac{E}{\dot{E}} \cong \frac{3\rho k T}{2\mu \Lambda(T)}, \tag{8.5}$$

where μ is the mean molecular weight of the gas and $\Lambda(T)$ is the cooling rate. Several different processes can contribute to cooling depending on the temperature range. For the sake of simplicity, one may assume that the cooling is dominated by plasma Bremsstrahlung and recombination. Using the expressions derived in problems 4.5 and 4.13, we can estimate $\Lambda(T)$ to be

$$\Lambda(T) = (A_B T^{1/2} + A_{\mathrm{rec}} T^{-1/2}) \rho^2, \tag{8.6}$$

where the $A_B \propto (e^6/m_e^{3/2})$ term represents the cooling due to Bremsstrahlung and the $A_{\mathrm{rec}} \cong e^4 m A_B$ term arises from the cooling due to recombination. This expression is valid for temperatures above 10^4 K; for lower temperatures, the cooling rate drops drastically since hydrogen can no longer be significantly ionized by collisions. Introducing the numerical values appropriate for a hydrogen–helium plasma (with a helium abundance $Y = 0.25$ and some admixture of metals) the expression for t_{cool} becomes

$$t_{\mathrm{cool}} = 8 \times 10^6 \, \mathrm{yr} \left(\frac{n}{1 \, \mathrm{cm}^{-3}} \right)^{-1} \left[\left(\frac{T}{10^6 \, \mathrm{K}} \right)^{-\frac{1}{2}} + 1.5 \left(\frac{T}{10^6 \, \mathrm{K}} \right)^{-\frac{3}{2}} \right]^{-1} . \tag{8.7}$$

Here n is the number density of gas particles. One can see from (8.7) that there is a transition temperature $T^* \cong 10^6$ K. For $T > T^*$ Bremsstrahlung dominates, while for $T < T^*$, line cooling dominates. Describe qualitatively the evolution of baryons taking the cooling processes given above into account.

(c) The baryons can also – in principle – cool through inverse Compton scattering with MBR photons. Show that this process is not very efficient at $z \lesssim 10$ irrespective of the mass of the baryonic structure.

8.6 Angular momentum of galaxies [2]

One important feature of galactic systems, especially the disc-like systems, is their angular momentum. The angular momentum of a galaxy can be conveniently expressed in terms of the dimensionless parameter

$$\lambda \equiv \frac{LE^{1/2}}{GM^{5/2}}. \tag{8.8}$$

This parameter is the ratio between the *actual* angular velocity ω of the system and the *hypothetical* angular velocity ω_0 that is needed to support the system purely by rotation. It follows that a self-gravitating system with appreciable rotational support has a λ comparable to unity. Observations suggest that disc galaxies have $\lambda \cong 0.4$ to 0.5.

One possible way of explaining the angular momentum of galaxies is as follows. During the initial collapse of the baryonic structures, tidal forces will be exerted on each protogalaxy by its neighbours. The tidal torque can spin up the protogalaxies, thereby providing them with some angular momentum. Numerical simulations suggest that this process can provide an initial λ of about $\lambda_i \cong 0.05$, which is only about 10% of the observed value. During the collapse of the gas due to cooling, the binding energy increases while mass and angular momentum remain the same. This will allow λ to increase as $\lambda \propto |E|^{1/2}$ and (possibly) reach observed values.

(a) Examine this idea in the absence of dark matter halos and show that it is not viable.

(b) Now consider the same idea in the presence of the dark matter halo. Show that, in this case, it is possible to circumvent the difficulties encountered in (a) above and produce a sufficiently high value of λ if $M_{\text{disc}} \cong 0.1 M_{\text{halo}}$.

8.7 Angular pattern of MBR anisotropies [3]

The photons, which decouple from matter at $z \cong z_{\text{dec}}$, will propagate freely during $t > t_{\text{dec}}$. The microwave background radiation with temperature $T_0 \cong 2.7$ K which is observed around us is usually interpreted as a relic of this decoupling era. In a strictly Friedmann universe, this radiation will be isotropic and the temperature in any direction in the sky will be constant. Any anisotropy in the temperature can, therefore, contain valuable information about the inhomogeneities in the universe. There are several possible sources of anisotropy in the microwave background radiation:

(1) The energy density of radiation in the last scattering surface can have an intrinsic inhomogeneity $\delta_{\text{rad}} = (\delta\rho_{\text{rad}}/\rho_{\text{rad}})$. (In most models of structure formation, $\delta_{\text{rad}} \cong \delta_{\text{B}}$ around $z \cong z_{\text{dec}}$.) This will appear as a temperature fluctuation $(\delta T/T)$.

(2) If the matter which scattered the radiation had a peculiar velocity (with respect to the comoving frame) when the scattering occurred, then the photon would undergo a Doppler shift. Since the peculiar velocity of matter would be different at different locations in the last scattering surface, corresponding to different directions in the sky today, this would lead to an anisotropy in the temperature distribution in the sky.

(3) If the local gravitational potential on the last scattering surface was different at different locations, then the photons would be climbing out of different gravitational potential wells and, hence, would experience different amounts of redshift. This would lead to an angular dependence for the observed temperature.

(4) If the observer is moving with respect to the comoving frame, then the photons reaching the observer from different directions with respect to the direction of motion will be redshifted (or blueshifted) by different amounts. This will lead to a dependence of T on $\cos\theta$, where θ is the angle between the direction of motion and the direction of the photon.

(5) Processes which take place along the path of the photon reaching us can also produce a $(\delta T/T)$. For example, consider a photon which travels through a large mass concentration which is collapsing. Since the depth of the gravitational potential well of a collapsing structure is increasing with time, the blueshift suffered by the photon as it travels towards the centre of the condensation will be lower than the redshift which occurs while it is emerging from the centre. Similarly, photons travelling through hot gas in a cluster will be affected by Compton scattering with the charged particles in the gas. These two processes can cause temperature distortions in specific directions.

Of these, the anisotropies due to (4) and (5) are not intrinsic to the conditions at $z = z_{\text{dec}}$ and hence cannot directly give us any information about the physics at that epoch. The other three anisoptropies contain valuable information about the processes taking place at $z = z_{\text{dec}}$. We shall concentrate on these anisotropies.

Make an order of magnitude estimate of the anisotropy $(\Delta T/T)$, which is expected at different angular scales due to processes $1, 2, 3$. Take into account the fact that decoupling was not an instantaneous process but took an interval of $\Delta z \cong 80$.

8.8 Derivation of MBR anisotropies [3]

The propagation of light in a perturbed Friedmann metric was discussed in problem 6.2, where we related the frequency shift of the photons $(\Delta\omega/\omega)$ to the time derivative of the metric perturbation $\dot{h}_{\alpha\beta}$. Use this result to show that the anisotropy of the MBR (in an $\Omega = 1$ model) can be expressed as

$$\frac{\Delta T}{T_0} = \left(\frac{\Delta T}{T_0}\right)_{\text{int}} + \mathbf{n} \cdot (\mathbf{v}_{\text{ob}} - \mathbf{v}_{\text{em}}) - \frac{1}{3}\left[\phi(0) - \phi(\mathbf{x}_{\text{em}})\right], \qquad (8.9)$$

where $\mathbf{n}$ is the unit vector from the observer to a point on the last scattering surface (LSS); the $\mathbf{v}_{\text{ob}}$ is the velocity of the observer, $\mathbf{v}_{\text{em}}$ is the velocity of the matter on LSS; ϕ is the perturbed gravitational potential and

$$\mathbf{x}_{\text{em}} = \mathbf{n}\eta(t) = \mathbf{n}\int_t^{t_0} \frac{dt'}{a(t')}. \qquad (8.10)$$

Interpret each of the terms.

8.9 Sachs–Wolfe effect and COBE [2]

The temperature distribution of MBR in the sky can be described by a function $(\Delta T/T) = S(\theta, \psi) = S(\mathbf{n})$, where the (θ, ψ) are the angular coordinates in the sky and $\mathbf{n}$ is the unit vector in the direction of (θ, ψ). It is convenient to expand S in terms of the spherical harmonics $Y_{lm}(\theta, \psi)$ with some coefficients a_{lm}:

$$\frac{\Delta T}{T} = S(\theta, \psi) = \sum_{l,m}^{\infty} a_{lm} Y_{lm}(\theta, \psi). \qquad (8.11)$$

Since absolute measurement $S(\mathbf{n})$ along a given direction is technically very difficult, one is often interested in the angular correlation functions defined out of $S(\mathbf{n})$. To define them, we consider the product

$$S(\mathbf{n})S(\mathbf{m}) = \sum\sum a_{lm}a_{l'm'}^* Y_{lm}(\mathbf{n})Y_{l'm'}^*(\mathbf{m}). \qquad (8.12)$$

This quantity can be averaged, either over the product $(a_{lm}a_{l'm'}^*)$ treating them as Gaussian random fields, or over the product $Y_{lm}(\mathbf{n})Y_{l'm'}^*(\mathbf{m})$ by integrating all over the sky (with $(\mathbf{n} \cdot \mathbf{m}) = \cos\alpha$ kept fixed). Observers use the latter average, while, theoretically, it is easier to predict the former.

(a) Assume that these two averages can be treated as statistically equivalent and derive an expression for the angular correlation function $\mathscr{C}(\alpha)$ of the MBR fluctuation (defined as $\mathscr{C}(\alpha) = \langle S(\mathbf{n})S(\mathbf{m})\rangle$), in terms of the power spectrum of density fluctuations. For the purpose of this calculation, assume that *only* the Sachs–Wolfe term contributes to the anisotropy.

(b) The cosmic background explorer (COBE) was a special purpose satellite which was put into orbit to investigate the properties of MBR. The analysis of COBE suggests that the root-mean-square fluctuations in the MBR temperature, when averaged (approximately) over about $10°$, is

$$\left(\frac{\Delta T}{T}\right)_{\text{rms}} \cong 1.1 \times 10^{-5}. \qquad (8.13)$$

The contribution from the dipole ($l = 1$ term) has been subtracted out in obtaining the above result. The dominant contribution to $(\Delta T/T)_{\text{rms}}$ is from the quadrapole which is estimated to be

$$\left(\frac{\Delta T}{T}\right)_Q \cong 0.48 \times 10^{-5}. \qquad (8.14)$$

Are these observations consistent with the claim that the primordial fluctuation has a power spectrum $P(k) = Ak$? Estimate the amplitude A of the fluctuations, if $n = 1$.

9
High redshift objects

The physical processes which take place in the universe at $z < 20$ or so are discussed in this chapter. The main focus is on the intergalactic medium (IGM), the abundance of high redshift objects and the quasar absorption systems. The constraints on cosmological models arising from these observations are also covered in this chapter.

9.1 Gunn–Peterson effect [2]

We have seen in chapter 6 that neutral matter was formed and the radiation decoupled from matter at $z \cong 10^3$. Since it is unlikely that the formation of structures at lower redshifts could have been 100% efficient, we would expect at least a fraction of the neutral hydrogen to remain in the intergalactic medium (IGM) with nearly uniform density. This neutral hydrogen could, in principle, be detected by examining the spectrum of a distant quasar. Neutral hydrogen absorbs Lyman-α photons, which are photons of wavelength 1216Å. (This wavelength corresponds to the energy difference between the ground state and the first excited state of the hydrogen atom.) Because of the cosmological redshift, the photons which are absorbed will have a shorter wavelength at the source and the signature of the absorption will be seen at longer wavelengths at the observer. We expect the spectrum of the quasar to show a dip at wavelengths on the blue side (shortwards) of the Lyman-α emission line if neutral hydrogen is present between the source and the observer. The magnitude of this dip depends on the neutral hydrogen density and can be calculated, using the optical depth, τ, for such absorption.

No such dip has been seen. Observations suggest that $\tau < 0.05$ for quasars with a mean redshift of about 2.6. What does this bound imply for the number density of neutral hydrogen in the intergalactic medium? It is reasonable to assume that $\tau < 0.1$ in the entire range $0 < z \lesssim 5$. Convert this into a bound on Ω contributed by neutral hydrogen at these redshifts.

9.2 Ionization of IGM [1]

The Gunn–Peterson effect described in the last problem shows that IGM has to be predominantly ionized at redshift $z \lesssim 5$. One can think of collisional ionization or photo-ionization as two possible means of achieving the ionization. Let the number densities of neutral hydrogen atoms, protons and electrons in the IGM be n_H, n_p and n_e, with $n_p = n_e$. If the ionization process is in equilibrium, we can relate n_p to n_H. The bound on n_H can be used to put a bound on n_p.

(a) The cross-section for collisional ionization is maximum when the electron energy is about 100 eV, that is, when the temperature is about 10^6 K. Around this temperature, we can approximate the cross-section by $\sigma_{coll} \cong 2a_0^2$ where a_0 is the Bohr radius. Estimate Ω_{IGM} if collisional ionization is the dominant source and show that

$$\Omega_{IGM} \lesssim 0.4(1+z)^{-3/2}h^{-1}. \tag{9.1}$$

(Ω_{IGM} is contributed by all the *baryons*; both neutral and ionized atoms contribute to Ω_{IGM}.)

(b) The photoionization of IGM could be due to the UV photons emitted by the first generation of quasars. The spectrum of such a metagalactic ionizing flux may be taken to be

$$J_\nu = J_0 \frac{\nu_0}{\nu} \times 10^{-21} \text{ erg cm}^{-2} \text{ s}^{-1} \text{ Hz}^{-1} \text{ ster}^{-1}, \tag{9.2}$$

where $\nu_0 = 3 \times 10^{15}$ Hz is the ionization threshold corresponding to $\lambda \cong 912$ Å. It is usually assumed that J_0 is of order unity. Does this spectrum have a sufficient number of photons to effectively ionize the IGM? What is the bound on Ω_{IGM} if photoionization is the main process for ionization of the IGM?

9.3 Re-ionization and CMBR [1]

Since there is no observational evidence for smoothly distributed neutral hydrogen in the IGM, one is led to believe that the IGM is very efficiently ionized by some physical process at a redshift of $z \gtrsim 5$. Let us assume that the ionization occurred at a redshift $z = z_{ion}$ and that the IGM has remained fully ionized since then. In that case, the radiation will be scattered by the ionized particles in the IGM. What effect will this have on the microwave background anisotropies?

9.4 Mass functions [3,N]

(a) One is often interested in knowing what fraction of the mass, $f(M, z) \, dM$, in the universe exists as gravitationally bound objects with masses between M and $M + dM$ at any given redshift z. We saw in problem 7.10 that a virialized object is formed due to non-linear processes when the linear density contrast is larger than a critical value $\delta_{crit} \cong 1.68$. This fact can be used to estimate $f(M, z)$.

Consider the density fluctuation $\delta(x, t_i)$ at some very early epoch when linear theory was valid. We smooth this density field using a spherical window function of radius R. Let $P(\delta, z)$ be the probability that the smooth density field has a value δ at some redshift z. All regions of space which have a density greater than $\delta_{crit}(1 + z)(1 + z_i)^{-1}$ at an initial epoch z_i could have evolved to a density contrast greater than δ_{crit} at a later redshift z. It seems reasonable to assume that such regions would have collapsed and formed gravitationally bound structures at a redshift of z. Estimate $f(M, z)$ using this criterion. Show that the resulting $f(M, z)$ is not correctly normalized. Explain the origin of this difficulty.

(b) Repeat the above analysis by using a window function which is sharply truncated in the Fourier space. That is, assume that the Fourier transform of the window function is $W_k(R) = \theta(R^{-1} - k)$, where $\theta(x) = 1$ for $x > 0$ and zero otherwise. Show that we now obtain a properly normalized mass function.

(c) Rich Abell clusters have a mass of about $M_{clus} \cong 5 \times 10^{14} h^{-1} M_\odot$ and an abundance of about $4 \times 10^{-6} h^3$ Mpc^{-3}. Consider a class of models in which the linear power spectrum has the form

$$P(k) = \frac{A^4 k}{\left[1 + \left(ak + (bk)^{3/2} + (ck)^2\right)^\nu\right]^{2/\nu}}, \tag{9.3}$$

with $A = 24 h^{-1}$ Mpc, $a = (6.4/\Gamma) h^{-1}$ Mpc, $b = (3.0/\Gamma) h^{-1}$ Mpc, $c = (1.7/\Gamma) h^{-1}$ Mpc, $\nu = 1.13$ and Γ is an adjustable parameter. What kind of constraints can one impose on Γ based on the observed cluster abundance?

9.5 Abundance of quasars [2,N]

Quasar observations over the years have allowed astronomers to construct a luminosity function, $\phi(M_B, z)$ for quasars. This function gives the number density of quasars with magnitudes between M_B and $M_B + dM_B$ at a redshift z. This function has the form

$$\phi(M_B, z)\, dM_B = \frac{\phi^* \, dM_B}{10^{0.4[M_B - M_B^*(z)](\alpha + 1)} + 10^{0.4[M_B - M_B^*(z)](\beta + 1)}}, \tag{9.4}$$

where

$$M_B^*(z) = \begin{cases} A - 2.5k \, \log(1 + z) & \text{(for } z < z_{max}) \\ M_B^*(z_{max}) & \text{(for } z > z_{max}) \end{cases}, \tag{9.5}$$

with $\alpha = -3.9$, $\beta = -1.5$, $A = -22.4$, $k = 3.45$, $z_{max} = 1.9$ and $\phi^* = 6.5 \times 10^{-7}$ Mpc^{-3} mag^{-1}. (It is, of course, possible to give the function in terms of luminosities using the relation $L = L_\odot 10^{0.4[M_\odot - M]}$ with $M_\odot = 5.48$, $L_\odot = 2.36 \times 10^{33}$ erg s^{-1}.)

(a) It is believed that quasars are powered by a central engine consisting of a supermassive blackhole surrounded by an accretion disc, located at the core of a host galaxy. Relate the total mass surrounding the host galaxy of a quasar to its luminosity L by making the following series of assumptions: (i) the accretion produces the maximum possible luminosity, namely, the Eddington luminosity; (ii) the efficiency of conversion of rest mass energy to radiation is $\epsilon \cong 0.1$; (iii) when a galaxy forms, it retains a fraction $f_{ret} \cong 0.1$ of its baryons and loses the rest as a galactic wind. Of these, a fraction $f_{hole} \cong 0.1$ forms the central engine.

(b) Use the above results to calculate the fraction of mass in the universe which must have collapsed to form virialized objects which could host quasars. What does this imply for the models with the power spectra given in problem 9.4(c)?

9.6 Lyman-alpha forest [2]

Even though quasar spectra do not show any absorption due to smoothly distributed neutral hydrogen in the IGM, they do exhibit several other absorption features due to hydrogen. To begin with, quasars show many narrow absorption lines at wavelengths shortward of the Lyman-α emission line. These are thought to be Lyman-α absorption lines arising from clumps of gas along the line-of-sight, with each line being contributed by one gas cloud. Since the lines are very numerous, it is usual to call it a 'Lyman-α forest'. Observations suggest that the number of lines per unit redshift range is given by

$$\frac{dN}{dz} = B(1+z)^{\gamma}, \qquad (9.6)$$

where $\gamma \cong 2.75$ and B is in the range of 3 to 4; we will take $B = 3.5$. This fitting function is valid for clouds with a column density of hydrogen $\Sigma > 10^{14}$ cm^{-2} in the range $2 < z < 4$. The typical temperature of the gas clouds (which is obtained from the Doppler width of the lines) is about 5×10^4 K. The sizes of the clouds are very uncertain and we may parameterize this by writing $l = 10\, l_0 h^{-1}$ kpc. It was originally thought that l_0 would be of order unity though some recent observations suggest that l_0 can be much larger.

Observations also give an estimate of the distribution of column densities in the absorption clouds at different redshift ranges. The probability dP that a line-of-sight intercepts a cloud with a column density Σ in the range $d\Sigma$ at the redshift interval $(z, z + dz)$ is given by

$$dP = g\,(\Sigma, z)\, d\Sigma\, dz \cong 91.2 \left(\frac{\Sigma}{\Sigma_0}\right)^{-\beta} \frac{d\Sigma}{\Sigma_0}\, dz, \qquad (9.7)$$

with $\Sigma_0 \cong 10^{14}$ cm^{-2}, $\beta \cong 1.46$. This formula is valid around $z = 3$ for 10^{13} cm$^{-2} \lesssim \Sigma \lesssim 10^{22}$ cm^{-2}.

Given the above facts, estimate the following quantities: (i) the mean proper distance between the clouds along the line-of-sight; (ii) the characteristic density of hydrogen atoms and the mass of neutral hydrogen in a cloud at $z \cong 3$ if the column density is $\Sigma \cong 10^{14}$ cm^{-2}; (iii) the fraction of the space occupied by the clouds and the mean number density of clouds in space; (iv) the fraction of neutral hydrogen in the cloud if the cloud is in ionization equilibrium with respect to the metagalactic flux; (v) the characteristic total mass of the cloud; (vi) the Ω contributed by the neutral hydrogen and total baryonic mass of the cloud.

9.7 Abundance of damped Lyman-alpha clouds [2,N]

Damped Lyman-alpha systems have a column of $\Sigma \cong 10^{21}\text{--}10^{22}$ cm^{-2}. The number density of these systems per unit redshift interval does not show any significant z-dependence and is given by $(dN/dz) \cong 0.3$ around $z \cong 2.5$.

(a) Repeat the analysis of parts (i)–(vi) of the previous problem for damped Lyman-alpha systems taking $(dN/dz) \cong 0.3$, $\Sigma \cong 10^{22}$ cm^{-2} at $z \cong 3$.

(b) Assuming these clouds are progenitors of present-day disc galaxies, estimate the amount of mass which should have collapsed to host the damped Lyman-alpha systems. How does this constraint compare with the one obtained from quasars?

10

Very early universe

This chapter studies two key ideas in the physics of the very early universe. The first one is the possible generation of density inhomogeneities from the inflationary model and the second is the production of cosmological defects from the phase transitions in the early universe. Both these ideas are at the forefront of current research and – consequently – have a fair amount of speculative ingredients. Some familiarity with quantum field theory will be of help in tackling these problems, though it is not essential.

10.1 Flattening the Hubble radius [1]

We saw in problem 6.3 that the physical wavelength of any mode characterizing the density perturbation will be larger than the Hubble radius at sufficiently early epochs. The proper wavelength $\lambda(M)$ of a perturbation containing mass M will be bigger than the Hubble radius at all redshifts $z > z_{\text{enter}}(M)$, where

$$
z_{\text{enter}}(M) \\
\cong \begin{cases} 1.41 \times 10^5 (\Omega h^2)^{\frac{1}{3}} (M/10^{12}\,\text{M}_\odot)^{-1/3}, & M < M_{\text{eq}} \cong 3.2 \times 10^{14}\,\text{M}_\odot (\Omega h^2)^{-2} \\ 1.10 \times 10^6 (\Omega h^2)^{-\frac{1}{3}} (M/10^{12}\,\text{M}_\odot)^{-2/3}, & M > M_{\text{eq}} \cong 3.2 \times 10^{14}\,\text{M}_\odot (\Omega h^2)^{-2}. \end{cases}
$$

$$(10.1)$$

Note that a galactic mass perturbation is bigger than the Hubble radius for redshifts larger than a moderate value of about 10^6.

This result leads to a major difficulty in conventional cosmology. Normal physical processes can only act coherently over sizes smaller than the Hubble radius. Thus any physical process leading to density perturbations at some early epoch, $t = t_i$, could only have operated at scales smaller than $d_H(t_i)$. But most of the relevant astrophysical scales (corresponding to clusters, groups, galaxies, etc.) were much bigger than $d_H(t)$ at early epochs. Thus, if we want the seed perturbations to have originated in the early universe, then it is difficult to understand how any physical process could have contributed to it.

To tackle this difficulty, we must arrange matters such that, at sufficiently small t, $\lambda(t) < d_H(t)$. If this can be done, then the physical processes can lead to an initial density perturbation.

(a) Show that $d_H(t)$ will rise faster than $a(t)$ as $t \to 0$ if the source of expansion of the universe has positive energy density and positive pressure. Thus the above problem *cannot* be tackled in any Friedmann model with $\rho > 0$ and $p > 0$.

(b) Consider, next, models in which ρ or p is negative. In order to satisfy the first Friedmann equation with $k = 0$ we need $\rho > 0$. Thus the only models in which we can have $d_H > \lambda$ for small t are the ones for which $p < 0$. In such models one can make $a(t)$ increase rapidly (e.g. exponentially) with t for a brief period of time. Such rapid growth is called 'inflation'.

Consider a model for the universe in which the universe was radiation dominated up to, say, $t = t_i$, but expanded exponentially in the interval $t_i < t < t_f$:

$$a(t) = a_i \exp H(t - t_i), \qquad t_i \le t \le t_f. \tag{10.2}$$

For $t > t_f$, the evolution is again radiation dominated [$a(t) \propto t^{1/2}$] until $t = t_{eq} \cong 5.8 \times 10^{10} (\Omega h^2)^{-2}$ s. The evolution becomes matter dominated for $t_{eq} < t < t_{now} = t_0$. Typical values for t_i, t_f, and H, may be taken to be

$$t_0 \cong 10^{-35}\,\text{s}, \quad H \cong 10^{10}\,\text{GeV}, \quad t_f \cong 70H^{-1}, \tag{10.3}$$

which give an overall 'inflation' by a factor of about $A \equiv \exp N \cong \exp(70) \cong 2.5 \times 10^{30}$ to the scale factor during the period $t_i < t < t_f$. (In this chapter we shall use the symbol H to denote the Hubble constant *during* the inflationary phase.) At $t = t_i$, the temperature of the universe is about 10^{14} GeV. During this exponential inflation, the temperature drops drastically; however, the matter is expected to be reheated to the initial temperature of about 10^{14} GeV at $t \cong t_f$ by various high energy processes. Thus, inflation effectively changes the value of $S = T(t)a(t)$ by a factor $A = \exp(70) \cong 10^{30}$. Note that this quantity S is conserved during the non-inflationary phases of the expansion.

What kind of source can lead to the exponential expansion? How does a wavelength, of, say, $\lambda_0 \cong 2$ Mpc evolve in such a model, *vis-à-vis*, the Hubble radius?

10.2 Horizon and flatness problems [1]

The exponential expansion described in the last problem can also help to alleviate some other difficulties in the conventional Friedmann models.

(a) It seems reasonable to assume that the present features of the universe were determined by the initial conditions at some epoch $t = t_i$ when the temperature of the universe was very high, say $T \cong 10^{14}$ GeV (which corresponds to the epoch at which most 'grand unified theories' have significant influence on the evolution of the universe). Estimate the proper distance $R_H(t)$ a light signal could have travelled during the time interval $(0, t)$ assuming the universe was radiation dominated. Physical processes could have made the universe homogeneous over a sphere of radius $R_{\mathrm{H}}(t_i)$ at the initial epoch $t = t_i$. If this sphere could expand to encompass the observed universe, then we would have a simple explanation for the observed homogeneity of the universe. Show that this is not possible in the conventional scenarios in which the universe is either radiation dominated or matter dominated.

Examine the situation in models with inflation, in which the universe expands exponentially within a short time. Show that in these models it is possible to produce the entire observed universe from a region which was causally connected at $t = t_i$.

(b) Consider the ratio $\Omega(t) = \rho(t)/\rho_{\mathrm{crit}}(t)$ between the actual energy density of the universe and the critical density at any given time t. Show that in order to keep this ratio at order unity at the present epoch, its value has to be extremely fine-tuned in the early universe, at $t = t_i$, if the universe was radiation dominated. Explain how an inflationary phase of the universe can eliminate the need for this fine-tuning.

10.3 Origin of density perturbations [3]

In conventional models of inflation, the energy density during the inflationary phase is provided by a scalar field with a potential $V(\phi)$. We saw in problem 6.3 that when the potential energy dominates over the kinetic energy, such a scalar field can act like an ideal fluid with the equation of state $p = -\rho$. Since the field is assumed to be inherently quantum mechanical, it will have characteristic quantum fluctuations. It is possible for these quantum fluctuations to eventually manifest as classical density perturbations. In this problem, we shall work out the basic mechanism behind this phenomena.

(a) Assume that, during the inflationary phase, the universe can be described as a Friedmann model with small inhomogeneities. The source of expansion can be assumed to be some classical scalar field $\Phi(t, \mathbf{x})$ which can be split as $\phi_0(t) + f(t, \mathbf{x})$, where ϕ_0 is the average homogeneous part of the field which drives the inflation and $f(t, \mathbf{x})$ is the space-dependent, fluctuating part. Estimate the Fourier transform of the density fluctuations during the inflationary phase and show that it is reasonable to take it to be

$$\delta(\mathbf{k}, t) = \frac{\dot{\phi}_0(t)}{V_0} \dot{\sigma}_{\mathbf{k}}(t), \tag{10.4}$$

where $\sigma_{\mathbf{k}}(t)$ is defined via the relation

$$\sigma_{\mathbf{k}}^2(t) = \int d^3\mathbf{x} <\psi|\hat{\phi}(t, \mathbf{x} + \mathbf{y})\hat{\phi}(t, \mathbf{y})|\psi> e^{i\mathbf{k}\cdot\mathbf{x}}. \tag{10.5}$$

(b) During the inflationary phase the proper wavelengths of perturbations increase exponentially, while the Hubble radius remains constant. It follows that a mode will exit the Hubble radius at some time during the inflation. After the inflationary phase ends, the universe becomes radiation dominated and the Hubble radius grows faster than the wavelength of the mode. Eventually, the mode will re-enter the Hubble radius at some epoch. Show that, at the time of re-entry, the amplitude of the density perturbation will be of the order of

$$\delta(\mathbf{k}, t_{\text{enter}}) \cong \left(\frac{\dot{\sigma}_{\mathbf{k}}}{\dot{\phi}_0} \right)_{t=t_{\text{exit}}}. \tag{10.6}$$

The right-hand side can be evaluated from the study of the quantum fields in the inflationary phase. This equation provides the spectrum of perturbation at the time when the mode enters the Hubble radius.

10.4 Quantum fluctuations in inflationary phase [3]

During the inflationary phase, the potential $V(\phi)$ remains approximately constant. Hence, for the study of quantum fluctuations in this field, we may ignore the potential and treat ϕ as a massless scalar field. By introducing the Fourier transform $q_{\mathbf{k}}(t)$ of the fluctuating part of the field, $\phi(t, \mathbf{x})$, through the relation

$$\phi(t, \mathbf{x}) = \int \frac{d^3 \mathbf{k}}{(2\pi)^3} q_{\mathbf{k}}(t) e^{i\mathbf{k}\cdot\mathbf{x}}. \tag{10.7}$$

We can decompose the action for the scalar field into that of a bunch of harmonic oscillators:

$$\mathscr{A} = \frac{1}{2} \int d^3\mathbf{x}\, dt \sqrt{-g}(\phi^i \phi_i) = \frac{1}{2} \int dt\, d^3\mathbf{x}\, a^3 \left(\dot{\phi}^2 - \frac{|\nabla\phi|^2}{a^2} \right)$$

$$= \frac{1}{2} \int \frac{d^3 \mathbf{k}}{(2\pi)^3} \int dt\, a^3 \left\{ |\dot{q}_{\mathbf{k}}|^2 - \frac{\mathbf{k}^2}{a^2} |q_{\mathbf{k}}|^2 \right\}. \tag{10.8}$$

To quantize this field, we have to quantize each independent oscillator $q_{\mathbf{k}}$. For the kth oscillator, we have the Schrödinger equation:

$$i\frac{\partial \psi_{\mathbf{k}}}{\partial t} = -\frac{1}{2a^3} \frac{\partial^2 \psi_{\mathbf{k}}}{\partial q_{\mathbf{k}}^2} + \frac{1}{2} a^3 \omega_{\mathbf{k}}^2 q_{\mathbf{k}}^2 \psi_{\mathbf{k}}, \quad \omega = \frac{|\mathbf{k}|}{a}. \tag{10.9}$$

The quantum state of the field can be expressed by a wavefunction ψ, which is the product of the $\psi_{\mathbf{k}}$s for all k.

(a) The Schrödinger equation can be solved by the ansatz

$$\psi_{\mathbf{k}} = A_{\mathbf{k}}(t) \exp\left\{ -B_{\mathbf{k}}(t)[q_{\mathbf{k}} - f_{\mathbf{k}}(t)]^2 \right\}. \tag{10.10}$$

Obtain the equation satisfied by $B_{\mathbf{k}}(t)$, $A_{\mathbf{k}}(t)$ and $f_{\mathbf{k}}(t)$. Show that the quantum fluctuations in the field are Gaussian and determined by the dispersion $\sigma_{\mathbf{k}}^2$ with

$$\sigma_{\mathbf{k}}^2 \propto |Q_{\mathbf{k}}|^2, \tag{10.11}$$

where $Q_{\mathbf{k}}$ satisfies the equation

$$\frac{1}{a^3} \frac{d}{dt} \left(a^3 \frac{dQ_{\mathbf{k}}}{dt} \right) + \frac{k^2}{a^2} Q_{\mathbf{k}} = 0. \tag{10.12}$$

(b) Solve the differential equation for $Q_{\mathbf{k}}(t)$ during the inflationary phase of the universe. Determine the boundary conditions for the solution by assuming that the solution should go over to the ground state of the harmonic oscillator when $a \to 1$ and $H \to 0$. Evaluate the quantity $|\dot{\sigma}_{\mathbf{k}}|$ at the time of $t = t_{\text{exit}}$ and show that

$$|\dot{\sigma}_{\mathbf{k}}| \propto \frac{H^2}{k^{3/2}}. \tag{10.13}$$

What does this imply for the spectral shape of the spectrum?

(c) To determine the amplitude of the perturbations, we also need to calculate $\dot{\phi}_0$ when the mode leaves the Hubble radius. This quantity can be determined by solving the equation of motion for the scalar field

$$\ddot{\phi} + 3\frac{\dot{a}}{a}\dot{\phi} + V'(\phi) = 0, \tag{10.14}$$

with a given $V(\phi)$. To make a simple estimate, one can proceed as follows. (i) Assume that $V(\phi)$ can be approximated as

$$V(\phi) \cong V_0 - \frac{\lambda}{4}\phi^4, \tag{10.15}$$

and that one can ignore the $\ddot{\phi}$ term in the equation of motion. (ii) Suppose the mode which is of interest to us leaves the Hubble radius at some time $t = t_{\text{exit}}$. Assume that the universe is inflated by a factor $(\exp N)$ during the time interval $t_{\text{exit}} < t < t_f$. Show that

$$k^{3/2}\delta_k \cong 12 \left(\frac{2}{3}\right)^{3/2} \lambda^{1/2} N^{3/2}. \tag{10.16}$$

Does this give a reasonable amplitude?

10.5 Cosmological defects [3]

Models for particle physics use different kinds of scalar field to describe high energy interactions. These fields are dynamically significant in the very early universe at temperatures of the order of $T \cong 10^{14}$ GeV. As the universe cools, one expects these fields to relax to the minima of the potentials which govern their evolution. However, it often happens that the minima are not unique. In that case, different regions of the universe may have the scalar field relaxing to different values of the minima. The regions where the field makes a transition from one value of the minima to another will exhibit strong gradients and hence finite energy densities. Such objects may be loosely called 'cosmological defects'. The nature of the defects depends on the original interaction Lagrangian describing the scalar field. We shall examine some simple models for these defects in this problem.

Consider a set of N real scalar fields $\phi_A(\mathbf{x}, t)$ with $A = 1, 2, \ldots, N$, described by a Lagrangian:

$$\mathscr{L} = \frac{1}{2} \frac{\partial \phi_A}{\partial x^i} \frac{\partial \phi^A}{\partial x_i} - V(\phi), \tag{10.17}$$

with

$$V(\phi) = \frac{\lambda}{4} \left(\phi^2 - \eta^2 \right)^2, \qquad \phi^2 = \phi_A \phi^A. \tag{10.18}$$

We assume summation over indices i, A, etc.

(a) The simplest such Lagrangian corresponds to $N = 1$. What are the minima of the potential in this case? Calculate the energy–momentum tensor for the field configuration which minimizes the energy locally but is different at different locations. (Such a 'defect' is called a 'domain wall'.) Calculate the gravitational attraction in the neighbourhood of the wall.

(b) Consider next the case of $N = 2$. The scalar fields ϕ_1 and ϕ_2 can be expressed in terms of two new scalar fields η and θ, with $\phi_1 = \eta \cos \theta$ and $\phi_2 = \eta \sin \theta$. Show that line-like 'defects' can form in this situation (these are called 'cosmic strings').

(c) The stress-tensor for the cosmic string described above may be taken to be of the form

$$T_k^i = \rho(r) \text{dia}[1, 0, 0, -1], \tag{10.19}$$

with $\rho(r) = \rho_0$ for $r < r_0$ and zero otherwise. The quantity r_0 defines the core region of the cosmic string. In writing the above expression for T_k^i we have assumed that the string is along the z-axis and infinitely long. Show that the effective gravitational mass of such a string is zero. Further, show that the metric due to such a string at $r \gg r_0$ has the form

$$ds^2 \cong dt^2 - dr^2 - (1 - 8G\mu) r^2 d\theta^2 - dz^2, \tag{10.20}$$

where $\mu = \pi r_0^2$. What kind of geometry is described by this metric?

11

Advanced problems

This section contains somewhat more advanced problems based on the material covered in chapters 1–10. There is significant variation in the level of difficulty, though most of the problems require a fair amount of algebraic manipulation or numerical work.

11.1 Interior potential of a homoeoid [3]

We saw in problem 1.11 that different density distributions may produce the same gravitational potential in a limited region of space. In that problem, we constructed several density distributions which produce zero gravitational force in some compact region of space. This problem explores yet another 'classical' result leading to constant gravitational potential.

We consider a thin shell of matter of constant density ρ, located between two surfaces with equations

$$\frac{x^2 + y^2}{a^2} + \frac{z^2}{b^2} = q^2, \quad \frac{x^2 + y^2}{a^2} + \frac{z^2}{b^2} = q^2(1 + \delta)^2. \tag{11.1}$$

Such surfaces are called homoeoids. Show that the gravitational potential inside a thin homoeoid is a constant.

11.2 Torus in phase space [3]

Consider an N-dimensional dynamical system which is fully integrable (in the sense of problem 2.5). Show that the trajectory in phase space is confined to the surface of an N-dimensional torus.

11.3 Gravitating systems in two dimensions [2]

In the study of self-gravitating systems in problems 2.8 and 2.9 it was necessary to put a short distance cut-off to the potential. This cut-off is absolutely essential for the existence of the integrals defining the microcanonical distribution (for

$N > 2$) and the canonical distribution (for any N). If this cut-off is removed, then the phase volume – and, consequently, the entropy – will increase without bound.

The situation is different in two dimensions. It turns out that the microcanonical distribution exists for a two-dimensional gravitating system, even in the absence of a cut-off. However, the canonical description exists only above a critical temperature. Consider a system of N-point particles interacting via gravity in two dimensions. Show that the correct Hamiltonian for such a system is

$$H = \sum_{i=1}^{N} \frac{\mathbf{p}_i^2}{2m} + \frac{1}{2} \sum_{i \neq j} 2Gm^2 \ln |\mathbf{x}_i - \mathbf{x}_j|. \tag{11.2}$$

Assume that all the particles are confined within a large square box of size $2L$ so that the range of each x_i is $(-L, +L)$. Evaluate the partition function $Z(\beta)$ for this system and show that the system can exist only if

$$\beta < \beta_{\text{crit}} = 2 \left[Gm^2 (N - 1) \right]^{-1}. \tag{11.3}$$

Also show that the equation of state for such a system is given by

$$\frac{PV}{T} = N \left[1 - (N - 1) \frac{Gm^2}{2T} \right]. \tag{11.4}$$

Interpret this result.

11.4 Antonov instability [3]

We have seen in problem 2.10 that isothermal spheres cannot exist if $(RE/GM^2) < -0.335$. Even when $(RE/GM^2) > -0.335$, the isothermal solution need not be stable. The stability of this solution can be investigated by studying the second variation of the entropy. Show that the following results hold:

(i) Systems with $(RE/GM^2) < -0.335$ cannot evolve into isothermal spheres. Entropy has no extremum for such systems.

(ii) Systems with $(RE/GM^2) > -0.335$ and $\rho(0) > 709\,\rho(R)$ can exist in a meta-stable (saddle-point state) isothermal sphere configuration. Here $\rho(0)$ and $\rho(R)$ denote the densities at the centre and edge respectively. The entropy extrema exist but they are not local maxima.

(iii) Systems with $(RE/GM^2) > -0.335$ and $\rho(0) < 709\,\rho(R)$ can form isothermal spheres which are local maxima of entropy.

11.5 Degenerate self-gravitating systems [2,N]

In problem 2.11 we studied self-gravitating systems supported by degeneracy pressure in two extreme limits. In the non-relativistic limit, the equation of state has the form $P \propto \rho^{5/3}$ and, in the relativistic limit, we have $P \propto \rho^{4/3}$. Obtain the equation of state $P = P(\rho)$ for a fully degenerate system without making non-relativistic or ultrarelativistic approximation. Integrate the equations describing the system numerically, and obtain the relation between the mass and the radius of the system. Show that $M \propto R^{-3}$ in the non-relativistic limit but becomes independent of R in the relativistic limit.

11.6 Thermal conductivity of a fluid [3]

In problem 3.2 we obtained an integral equation describing the first-order perturbation in the distribution function of a gas. This equation is correct to linear order in the spatial gradients of temperature, velocity and density. By solving this equation, one should be able to determine coefficients of thermal conductivity and viscosity in terms of the molecular scattering cross-section σ. Develop a suitable perturbation theory to solve this integral equation and show, for example, that the conductivity of a fluid can be expressed as

$$\kappa = \frac{75}{64}\beta^4 \left(\frac{\beta}{2\pi}\right)^{1/2} \int\limits_0^\pi \int\limits_0^\infty \left(\frac{d\sigma}{d\theta}\right) \exp\left(-\frac{1}{2}\beta v_{\mathrm{r}}^2\right) v_{\mathrm{r}}^7 \sin^2\theta \, dv_{\mathrm{r}} \, d\theta, \qquad (11.5)$$

where $\beta = (m/kT)$.

11.7 Ionization fronts [3]

In the discussion of shocks in problems 3.10 and 3.14, we only considered simple hydrodynamical processes. In many astrophysical scenarios, several other physical processes will interfere with the picture developed in these problems. It will then be necessary to analyse the dynamical evolution of the system stage-by-stage, making appropriate simplifications in each stage.

Such a situation arises in the case of ionization fronts propagating through an ambient gas. Consider, for example, a central source of ultraviolet photons ('star') embedded inside a spherically symmetric ambient gas cloud. The photons propagate outward ionizing the medium. The ionization front initially propagates outward by converting HI to HII fairly quickly and thus heating up the inside region. Very soon the pressure imbalance between the hot inner HII region and the cold outer HI region will cause the gas outside to be compressed. This could drive a shock front through the outer medium. We will now analyse the physics of this process using a simplified model.

(a) Consider first the initial propagation of the ionization front through the ambient medium which has an initial density $\rho = m_{\mathrm{H}} n_0$. The production of electrons by ionization will be governed by the equation

$$\frac{\partial n_{\mathrm{e}}}{\partial t} + \nabla \cdot (n_{\mathrm{e}} \mathbf{u}) = -\alpha n_{\mathrm{e}}^2 - \nabla \cdot \left[\frac{N_*}{4\pi r^2}\hat{\mathbf{r}}\right], \qquad (11.6)$$

where n_{e} is the number density of electrons, α is the recombination coefficient and N_* is the number of ionizing photons released by the central source per second. Assume that the ionization front $R(t)$ moves very rapidly into the medium. That is, (dR/dt) is far greater than the thermal velocity of the ionized region. Show that the radius of the newly created HII region is given by

$$R(t) = R_1 \left(1 - e^{-t/t_R}\right)^{1/3}, \tag{11.7}$$

with $R_1^3 = \left(3N_*/4\pi n_0^2 \alpha\right)$ and $t_R = (\alpha n_0)^{-1}$

(b) The above analysis shows that the ionization front is moving into the medium with a speed $U = (dR/dt)$. Initially, this motion will not cause any bulk flow in the medium and will merely convert the neutral hydrogen into a plasma at a temperature $T \cong 10^4\,\mathrm{K}$. The ambient medium is, however, at a temperature $T \cong 10^2\,\mathrm{K}$. Very soon, this temperature difference will make the gas flow relative to the ionization front. We will assume that radiative processes maintain the temperature constant in both the regions. Determine the junction conditions at the ionization front using the usual conservation equations. Show that

$$v_I = \frac{m_H}{\rho_I} J_{II}, \tag{11.8}$$

where the subscripts I and II denote conditions in the HI and HII regions, $J_{II} = \left(N_*/4\pi R^2\right)$ and $J_I = 0$. The velocities are measured in the frame in which the shock front is at rest. Assume that the thermal velocities $\sigma_I = \left(kT_I/m_H\right)^{1/2}$, $\sigma_{II} = \left(2kT_{II}/m_H\right)^{1/2}$ satisfy the inequality $\sigma_{II} \gg \sigma_I$. Under this condition argue that physically meaningful solutions can exist only if $v_I > v_A$ or $v_I < v_B$, where $v_A \cong 2\sigma_{II}$ and $v_B \cong \left(\sigma_I^2/2\sigma_{II}\right)$. Describe the nature of the ionization front in both these cases.

(c) Use the above results to study the late stage evolution of an ionization front. Assume that the physical situation can be described by three layers of matter. The inner layer of ionized HII gas is separated from a middle layer of compressed HI gas by an ionization front; the middle layer is separated from an outer layer by an isothermal shock. Work out the junction conditions and describe qualitatively the propagation of the ionization front and the shock front through the medium.

11.8 Radiation in classical theory [3]

(a) Consider a charged particle moving along a trajectory $z^i(\tau)$. Let $u^k = \dot{z}^k$ be the four-velocity of the charge. Show that the electromagnetic field at any location x^i can be expressed in a fully Lorentz invariant form as

$$F_{ki}(x) = \left[\frac{q}{\rho}\frac{d}{d\tau}\left(\frac{u_i R_k - u_k R_i}{\rho}\right)\right]_{\mathrm{ret}}, \tag{11.9}$$

where $R^k = (x^k - z^k)$ and $\rho = R^k u_k$.

(b) Is there a Lorentz invariant generalization to the Feynman formula?

11.9 Cerenkov radiation [2]

There are a few situations in which a charge can radiate energy even when it is moving with uniform velocity. One such case corresponds to a charged particle moving with a velocity $v > (c/n)$ in a material with refractive index n. Calculate the energy emitted by a charge moving with velocity v for a unit length of its path. Show that radiation arises for $v > (c/n)$ but not for $v < (c/n)$ and is confined along a cone with opening angle $\theta = \cos^{-1}(c/vn)$. Provide a physical explanation for this phenomenon,

11.10 Lensing and caustics [3]

In problem 5.12 we found that the location of the images formed by gravitational lensing can be obtained by calculating the extremum of a given function. The detailed nature of the image configuration depends on the explicit form of this function. It is, however, possible to study the properties of gravitational lensing by considering a generic class of functions $P(x_I, y_I)$. Choose a useful parametrization for this function and analyse the nature of the images for different (generic) cases.

11.11 Inhomogeneous universe [3]

In the study of cosmology, one is interested in the dynamics of a region which has density higher than the average density of the background universe. One would expect such a region to eventually contract and form gravitationally bound systems such as galaxies, etc. It is therefore of interest to consider a metric which describes a spherically symmetric overdense region embedded in a smooth background universe. This exercise provides a derivation of such a general relativistic, inhomogeneous model. Consider a metric of the form

$$ds^2 = dt^2 - e^{2\alpha}dx^2 - e^{2\beta}(d\theta^2 + \sin^2\theta\, d\phi^2),\qquad(11.10)$$

with $\alpha = \alpha(x,t)$, $\beta = \beta(x,t)$. The components of the Einstein tensor $G^i_k = R^i_k -(1/2)\delta^i_k R$ for this metric are given by

$$
\begin{aligned}
G^0_0 &= \dot\beta^2 + 2\dot\alpha\dot\beta + e^{-2\beta} - e^{-2\alpha}(2\beta'' + 3\beta'^2 - 2\alpha'\beta')\\
G^1_1 &= 2\ddot\beta + 3\dot\beta^2 + e^{-2\beta} - (\beta')^2 e^{-2\alpha}\\
G^2_2 &= \ddot\alpha + \dot\alpha^2 + \ddot\beta + \dot\beta^2 + \dot\alpha\dot\beta - e^{-2\alpha}(\beta'' + \beta'^2 - \alpha'\beta')\\
G^1_0 &= e^{-2\alpha}\left[2\dot\beta' + 2\dot\beta\beta' - 2\dot\alpha\beta'\right]
\end{aligned}
\qquad(11.11)
$$

where the dot denotes differentiation with respect to time and the prime denotes differentiation with respect to x. The source is taken to have a stress-tensor $T^i_k = \mathrm{dia}[\rho(x,t), 0, 0, 0]$.

(a) Show that it is possible to express e^α and e^β in the form

$$e^\beta = xa(x,t),\qquad e^\alpha = \frac{(ax)'}{[1 - x^2/R^2(x)]^{1/2}},\qquad(11.12)$$

where a is a function of x and t, while $R(x)$ is a function of x alone.

(b) Show that the functions $a(x,t)$ and $R(x)$ satisfy the equations

$$\dot a^2 a + aR^{-2} = F(x),\qquad \frac{8\pi G}{3}\rho\frac{\partial}{\partial x}(ax)^3 = \frac{\partial}{\partial x}(x^3 F),\qquad(11.13)$$

where $F(x)$ is arbitrary and needs to be specified by initial conditions. Given $F(x)$, these equations determine $R(x)$ and $a(x,t)$.

(c) As a special case, consider a universe made up along the following lines. We take a uniform density Friedmann universe and evacuate a spherical region of radius r. The mass removed from this region is put at the centre of the sphere. Find the metric due to such a configuration of matter.

11.12 Collapse of dust sphere [3]

Consider a sphere of dust with a constant density ρ and initial radius R which is collapsing under its own weight in a spherically symmetric manner. The spacetime outside the collapsing sphere is described by a Schwarszchild metric, while that inside can be described by a Friedmann model in a suitable coordinate system. Show that it is indeed possible to match these two metrics smoothly at the surface of the dust sphere. Describe the nature of the collapse in both the outside and inside coordinate systems as the radius of the sphere approaches $(2GM/c^2)$.

11.13 Geometrical properties of the Friedmann model [2,N]

Consider a Friedmann model parametrized by Ω_{nr}, Ω_r, Ω_V and $\Omega = \Omega_{nr} + \Omega_r + \Omega_V$. Integrate the Friedmann equations numerically and determine $a(t)$ for the following sets of parameters: (i) $\Omega = 1$, $\Omega_V = 0.7$, $\Omega_r \cong 10^{-5}$; (ii) $\Omega = 0.3$, $\Omega_V = 0$, $\Omega_r \cong 10^{-5}$; (iii) $\Omega = 0.3$, $\Omega_V = 0.1$, $\Omega_r \cong 10^{-5}$.

In each of these cases determine the following quantities as a function of redshift: (A) the age of the universe, $t(z)$; (B) the angular diameter distance $d_A(z)$; (C) the Hubble radius $d_H(z)$; (D) the optical depth $\tau(z)$ due to galaxies if the number densities of galaxies scale as $(1+z)^3$; (E) the optical depth for gravitational lensing due to galaxies, modelled as isothermal spheres.

11.14 Nucleosynthesis [3]

As the universe cools to temperatures lower than the binding energy of the nuclei, stable nuclei can form. In the standard big bang model, the nucleosynthesis proceeds in two steps: neutron and proton combine to form deuterium nuclei which combine to form ^{4}He. In the second reaction almost all the deuterium is cooked into helium. Given that (i) the binding energy of helium is about 0.28 MeV; (ii) that of deuterium is 0.07 MeV; (iii) the weak interaction rate converting neutron to proton and (vice versa) falls below the expansion rate at $T_D \cong 0.75$ MeV; (iv) the lifetime of a free neutron is $\tau = 915.4$; and (v) the neutron–proton mass difference $Q = m_N - m_p = 1.3$ MeV, describe qualitatively the process of nucleosynthesis in the universe.

In the conventional big bang model, the primordial abundance of helium is about 0.25 by mass and that of deuterium is about 10^{-5}. Discuss whether the abundance of helium and deuterium would have gone up or down under the following conditions: (i) the baryon density of the universe was increased; (ii) the decay time of the free neutron was decreased; (iii) the number of neutrino species in the universe was increased; (iv) there were many more neutrinos than antineutrinos in our universe; (v) the gravitational constant G was larger in the past than today.

11.15 Velocity space caustics [2]

Galaxy redshifts are often used as 'spatial coordinates' in denoting the position of a galaxy. However, since galaxies have peculiar velocities, the redshift space will not match properly with coordinate space. Consider, for example, the formation of caustics in the redshift space.

(a) The net velocity at any location, in Zeldovich approximation (see problem 7.14), is given by $\mathbf{u} = (\dot{a}/a)\mathbf{x} - a\dot{b}\mathbf{V}\,(\mathbf{q})$, where $\mathbf{V} = \nabla_q\Phi(\mathbf{q})$ in an obvious notation. Show that the redshift of a galaxy at Lagrangian coordinate $\mathbf{q}$ will be

$$z = aq_3 - (1+f)abV_3, \qquad (11.14)$$

where we have taken the line-of-sight to be along the third axis and $f = (\dot{b}/b)/(\dot{a}/a) \cong \Omega^{0.6}$.

(b) Let the redshift space coordinates be $(x = x_1, y = x_2, z = x_3 + v/H)$. On transforming from the Lagrangian system to the redshift coordinates, the entries in the third row of the Jacobian matrix will be multiplied by $(1+f)$ factors. Show that, in the simplest case, this will give the density

$$\rho(x, y, z, t) = \frac{\rho_0}{a^3} \frac{1}{(1 - b\lambda_1)(1 - b\lambda_2)(1 - b(1+f)\lambda_3)}. \qquad (11.15)$$

Hence, show that redshift-space-caustic occurs at $b\lambda_3 = (1+f)^{-1} = (1/2)$ for $\Omega = 1$. In general, show that the 'redshift caustics' will precede the 'density caustics' if a region is collapsing.

11.16 Probability distribution of matrix eigenvalues [3]

In the Zeldovich approximation a central role is played by the matrix $M_{ij}(\mathbf{q}) \equiv (\partial^2\psi/\partial q^i\partial q^j)$, where $\psi(\mathbf{q})$ is the initial gravitational potential. At any location $\mathbf{q}$, we can diagonalize this matrix and obtain the eigenvalues $\lambda_1(\mathbf{q}), \lambda_2(\mathbf{q})$ and $\lambda_3(\mathbf{q})$. Assume that the gravitational potential ψ is a Gaussian random field with a power spectrum $P_\psi(k)$. Determine the probability distribution $P(\lambda_1, \lambda_2, \lambda_3)$ for the distribution of eigenvalues λ_i at any given location in space.

This distribution $P(\lambda_1, \lambda_2, \lambda_3)$ cannot be factorized as a product of the kind $A(\lambda_1)B(\lambda_2)C(\lambda_3)$. Find a new set of variables μ_1, μ_2, μ_3 which are functions of $\lambda_1, \lambda_2, \lambda_3$ such that the probability distribution is factorizable in the new set of variables.

11.17 Numerical simulations of Zeldovich approximation [3,N]

Zeldovich approximation is fairly simple to implement in a simulation and can help one to understand some of the nuances of cosmological N-body simulations.

(a) The first step in implementing the Zeldovich approximation is to generate the initial gravitational potential $\psi(\mathbf{x})$ as a Gaussian random field. We can express $\psi(\mathbf{x})$ as

$$\psi(\mathbf{x}) = \int \frac{d^3\mathbf{k}}{(2\pi)^3} \, (a_\mathbf{k} + ib_\mathbf{k}) \, e^{i\mathbf{k}\cdot\mathbf{x}}, \tag{11.16}$$

where $a_\mathbf{k}$ and $b_\mathbf{k}$ are real Gaussian random variables with zero mean and given variance: $< a_\mathbf{k}^2 > = < b_\mathbf{k}^2 > = P_\psi(\mathbf{k})/2$. Given the power spectrum in the theory, one can generate a set of Gaussian random numbers $a_\mathbf{k}$ and $b_\mathbf{k}$ with the above property for each value of $\mathbf{k}$. (In numerical work, of course, one has to use a grid in $\mathbf{k}$-space.) By using a standard fast-Fourier-transform (FFT) routine, one can then construct $\psi(\mathbf{x})$ in real space.

What conditions should one impose on $P(\mathbf{k})$ if the resulting potential has to be physically meaningful and useful? In doing an FFT, one will approximate the universe as a cubical box of size L and grid spacing l. How does the accuracy of the numerical realization depend on l and L? It is also usual to impose periodic boundary conditions on the surfaces of the cube while performing FFT. Is this meaningful?

(b) Given an initial gravitational potential on a grid, one can compute the gradient $\nabla\psi$ on the grid using a suitable numerical differentiation technique. What kind of errors will be introduced by such a numerical differentiation?

(c) In Zeldovich approximation, one starts with the particles initially located on the grid in real space. These particles are then moved by the amount $\mathbf{D} \equiv a\nabla\psi$ (where $\nabla\psi$ is evaluated at the initial location of the particle) to find the configuration at an epoch with expansion factor a. As a increases, more and more particles will tend to cross each other and will form characteristic pancake-like structures. Study the evolution of structures in Zeldovich approximation as a increases and determine when the approximation breaks down for different kinds of power spectra.

11.18 CMBR anisotropies in a general cosmology [2,N]

In problems 8.8 and 8.9, we calculated the CMBR anisotropy due to the Sachs–Wolfe effect in a cosmological model with $\Omega = 1$. Repeat this analysis for a general Friedmann model parametrized by Ω_r, Ω_{nr} and Ω_V.

When $\Omega \neq 1$, the analysis needs to be changed for several reasons. Among them are the following. (i) The gravitational potential is not independent of time in a matter dominated universe if $\Omega \neq 1$. Because of this the Sachs–Wolfe effect receives contributions all along the path of the photon. (ii) Several kinematical relations, including that between the length scale and the angle subtended in the sky, change when $\Omega \neq 1$. Hence, a physical phenomena characterized by a particular length scale will lead to an effect that is visible at different angular scales in the sky when Ω is changed. (iii) When $\Omega \neq 1$ there is a preferred length scale in the background universe. The concept of the scale invariant spectrum requires a much more careful definition in this case.

Answers

1

Astrophysical processes

1.1 Why does an accelerated charge radiate?

(a) Consider a charge which is moving along the x-axis with velocity v. In the rest frame of the charge, the scalar and vector potentials are given by

$$\phi' = \frac{q}{r'}, \quad \mathbf{A}' = 0, \tag{1.1}$$

where r' is the distance between the location of the charge and the point where the field is measured. Making a Lorentz transformation to the lab frame and using the fact that $(\phi, \mathbf{A})$ are components of a four-vector, we obtain

$$\phi = \gamma \frac{q}{r'}, \quad \mathbf{A} = \phi \left(\frac{\mathbf{v}}{c}\right), \quad \gamma = \frac{1}{\sqrt{1 - v^2/c^2}}. \tag{1.2}$$

The distance r' can be expressed in terms of the lab coordinates by using the standard Lorentz transformation:

$$r' = \left[x'^2 + y'^2 + z'^2\right]^{1/2} = \gamma \left[(x - Vt)^2 + \frac{1}{\gamma^2}\left(y^2 + z^2\right)\right]^{1/2}. \tag{1.3}$$

The electric field in the lab frame can be obtained from ϕ and $\mathbf{A}$ by differentiation. The final answer is

$$\mathbf{E} = \frac{q\mathbf{r}}{r^3} \frac{\left(1 - v^2/c^2\right)}{\left[1 - \left(v^2/c^2\right)\sin^2\theta\right]^{3/2}}, \tag{1.4}$$

where $\mathbf{r} \equiv (x - Vt, y, z)$ is the vector from the point at which the field is measured to the *instantaneous* position of the charge, and θ is the angle between the direction of motion and the radius vector $\mathbf{r}$.

Expression (1.4) shows that: (i) the field falls as r^{-2}; and (ii) for $v \ll c$, the electric field is a *Coulomb field* pointing towards the *instantaneous* position of the particle. (Even when $v \lesssim c$, the field points towards the instantaneous position of the particle but it is not a pure Coulomb field.)

(b) The result that the electric field of an accelerated charge should have an r^{-1} term can be seen along the following lines. From Maxwell's equations it is clear that $\mathbf{A}$ is linear in the velocity $\mathbf{v}$ of the charge. Since $\dot{\mathbf{A}}$ contributes to $\mathbf{E}$, there will be one term in **E** which is linear in the acceleration **a**. (This is, of course, in addition to the usual Coulomb term which is independent of **a** and falls as r^{-2}.) The electric field is also linear in the charge q. Consider this electric field in the instantaneous rest frame of the charge. It has to be constructed from q, a, c^2 and r and hence *must* have the general form

$$E = C(\theta) \frac{qa}{c^n r^m} = C(\theta) \left(\frac{q}{r^2}\right)\left(\frac{a}{c^n r^{m-2}}\right), \tag{1.5}$$

where C is a dimensionless factor, depending only on the angle θ between **r** and **a**, and n and m need to be determined. (Since **v** = 0 in the rest frame, the field cannot depend on the velocity.) From dimensional analysis, it immediately follows that $n = 2, m = 1$. Hence,

$$E = C(\theta) \frac{qa}{c^2 r}. \tag{1.6}$$

Thus, dimensional analysis, plus the fact that **E** must be linear in q and a, implies the r^{-1} dependence for the radiation term. The factor $C(\theta)$ can be determined by studying a special case.

Consider a charge which was at rest, at the origin, from $t = -\infty$ to $t = 0$ and which undergoes constant acceleration **a** along the x-axis for a short time Δt. For $t > \Delta t$, it moves with constant velocity $v = a\Delta t$ along the x-axis. Let us look at the electric field produced by this charge at some time $t \gg \Delta t$. Since Δt is arbitrarily small, we have $a\Delta t \ll c$ and we can use non-relativistic approximation throughout (see Figure 1.1(a,b)).

The 'news', that the charge was accelerated at $t = 0$, could have only travelled a distance $r = ct$ in time t. Thus, at $r > ct$, the electric field should be that due to a charge located at the origin:

$$\mathbf{E} = \frac{q}{r^2}\hat{\mathbf{r}} \quad (\text{for } r > ct). \tag{1.7}$$

At $r \lesssim ct$ the field is that due to a charge moving with velocity v along the x-axis. This will be a Coulomb field radially directed from the *instantaneous* position of the charge (see part (a) above):

$$\mathbf{E} = \frac{q}{r'^2}\hat{\mathbf{r}}' \quad (\text{for } r < ct). \tag{1.8}$$

Around $r = ct$ there exists a small shell of thickness $(c\Delta t)$ in which neither result holds good. From figure 1.1(a) it is clear that the electric field in the transition region should interpolate between the two Coulomb fields. Let $E_{\parallel}$ and $E_{\perp}$ be the magnitudes of the electric fields parallel and perpendicular to the direction $\hat{\mathbf{r}}$. From the geometry, we have

$$\frac{E_{\perp}}{E_{\parallel}} = \frac{v_{\perp}t}{c\Delta t}. \tag{1.9}$$

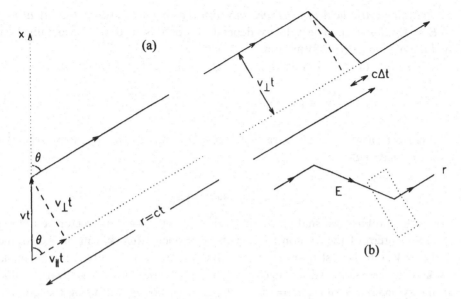

Fig. 1.1. (a) The electric field due to a charged particle which was accelerated for a small time interval Δt. For $t > \Delta t$, the particle is moving with a uniform non-relativistic velocity v along the x-axis. At $r > ct$, the field is that of a charge at rest in the origin. At $r < c(t-\Delta t)$ the field is directed towards the instantaneous position of the particle. The radiation field connects these two Coulomb fields in a small region of thickness $c\Delta t$. (b) Pill box construction to relate the normal component of the electric field around the radiation zone.

But $v_\perp = a_\perp \Delta t$ and $t = (r/c)$, giving

$$\frac{E_\perp}{E_\parallel} = \frac{(a_\perp \Delta t)\,(r/c)}{c\Delta t} = a_\perp \left(\frac{r}{c^2}\right). \tag{1.10}$$

The value of $E_\parallel$ can be determined by applying Gauss' theorem to a small pillbox, as shown in figure 1.1(b). This gives $E_\parallel = E_r = (q/r^2)$; thus we find that

$$E_\perp = a_\perp \left(\frac{r}{c^2}\right) \cdot \frac{q}{r^2} = \frac{q}{c^2}\left(\frac{a_\perp}{r}\right). \tag{1.11}$$

This is the radiation field located in a shell at $r = ct$, which is propagating with a velocity c. The above argument clearly shows that the origin of the r^{-1} dependence lies in the necessity to interpolate between two Coulomb fields. One can express this result in vector notation as

$$\mathbf{E}_{\text{rad}}(t, \mathbf{r}) = \frac{q}{c^2}\left[\frac{1}{r}\hat{\mathbf{n}} \times (\hat{\mathbf{n}} \times \mathbf{a})\right]_{\text{ret}}, \tag{1.12}$$

where $\mathbf{n} = (\mathbf{r}/r)$ and the subscript 'ret' implies that the expression in square brackets should be evaluated at $t' = t - r/c$. Comparison with (1.6) shows that $C(\theta) = \sin\theta$.

The full electric field *in the frame in which the charge is instantaneously at rest* is $\mathbf{E} = \mathbf{E}_{\text{Coul}} + \mathbf{E}_{\text{rad}}$. The velocity dependence of this field can be obtained by making a Lorentz transformation.

(c) The flux of radiation flowing through a sphere of radius r will be

$$\frac{d\mathscr{E}}{dt} \cong \left(4\pi r^2\right) \cdot \frac{c}{4\pi} E^2 \cong cr^2 \cdot \frac{q^2 a^2}{c^4 r^2} \cong \frac{q^2}{c^3} a^2. \tag{1.13}$$

The correct answer has an extra factor of $(2/3)$ due to the averaging of $\sin^2 \theta$ over all solid angles.

1.2 Paranoid effect

(a) In this problem we shall use units with $c = 1$. Let the four-vector describing the momentum of the photon be k^i with components $(\omega_L, \mathbf{k}_L)$ in the lab frame and $(\omega_R, \mathbf{k}_R)$ in the rest frame. For motion along the x-axis, only the angle made by $\mathbf{k}$ with the x-axis will change; the azimuthal angle is clearly invariant due to the symmetry. Writing $k_L^x = \omega_L \cos\theta_L, k_R^x = \omega_R \cos\theta_R$ and using the Lorentz transformations we find that

$$\omega_R = \gamma\omega_L \left(1 - v\cos\theta_L\right), \tag{1.14}$$

$$\omega_R \cos\theta_R = \gamma\omega_L \left(\cos\theta_L - v\right). \tag{1.15}$$

These equations relate the directions of propagation and frequencies in the two frames.

(b) From equations (1.14) and (1.15), we can express $\mu_R \equiv \cos\theta_R$ in terms of $\mu_L \equiv \cos\theta_L$:

$$\mu_R = \frac{\mu_L - v}{1 - v\mu_L}. \tag{1.16}$$

Figure 1.2 shows a plot of μ_R against μ_L when $v \lesssim 1$. Note that when $\mu_L = 1$ we have $\mu_R = 1$ and when $\mu_L = -1$ we have $\mu_R = -1$. In other words, photons travelling along the x-axis appear to do so in both frames. The frequencies, however, are different in the two cases. When $\mu_L = 1$, $\omega_R = \gamma\omega_L (1 - v)$, indicating that $\omega_R < \omega_L$; the radiation is blueshifted, as seen in the lab frame. For $\mu_L = -1$, $\omega_R = \gamma\omega_L (1 + v)$ and the radiation is redshifted.

When $\mu_L = v$, we have $\mu_R = 0$. Since $v \lesssim 1$, we expect $\mu_L = \cos\theta_L \cong \left(1 - \theta_L^2/2\right)$, with $\theta_L \ll 1$. Comparing with the relation $1 - v \cong \left(2\gamma^2\right)^{-1}$ we find that $\theta_L \cong \gamma^{-1} \ll 1$. Hence, this direction of propagation is almost along the positive x-axis in the lab frame. But, since $\mu_R = 0$, it is clear that $\theta_R = \pi/2$ and the photon was propagating *perpendicular* to the x-axis in the rest frame. In other words, a photon emitted in the direction perpendicular to the direction of motion will be 'turned around' to (almost) forward direction by the relativistic motion of the source.

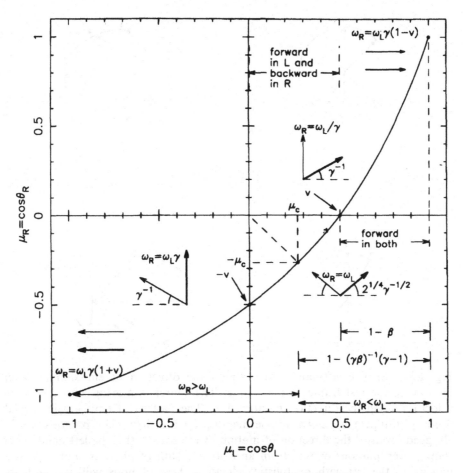

Fig. 1.2.   Relation between $\cos\theta_R$ and $\cos\theta_L$. The directions of two photons in the rest frame and the lab frame are indicated by thin and thick arrows respectively.

When $\mu_L = 0, \mu_R = -v$. In this case, a photon which has been travelling almost along the negative x-axis has been 'turned around' to travel orthogonal to the x-axis.

In the two cases discussed above, the frequency of the photon changes. When $\mu_L = v, \omega_R = \omega_L \gamma^{-1}$ and the photon is still blueshifted. When $\mu_L = 0, \omega_R = \gamma\omega_L$ and the photon is redshifted.

(c) It is clear from the diagram that the photon will appear to propagate forward in both frames only if $\theta_L < \gamma^{-1}$. To find the angle of propagation at which there is no frequency change, we have to set $\omega_L = \omega_R$ and solve for μ. This gives $\theta_{crit} \cong \gamma^{-1/2}$. For $\theta < \theta_{crit}$ the photon is blueshifted. Note that a photon propagating along $\theta = \theta_{crit}$ appears to make the 'same' angle with respect to the x-axis in both frames.

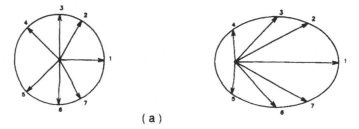

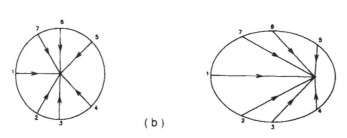

Fig. 1.3. (a) The left-hand frame shows the direction of emission of seven photons in the rest frame of the source. The source is moving with relativistic velocities along the direction numbered 1. The pattern of emission in the lab frame will appear as shown in the diagram on the right. The photons will be 'dragged towards' the direction of motion of the source. (b) The left-hand frame shows seven photons chosen from an isotropic bath of photons. For a particle moving to the left with relativistic velocities, these photons will appear to be distributed as shown in the right-hand frame. More photons will hit the particle face-on than from behind. Note that the directions marked 1,2,...,7 are the same in both (a) and (b).

(d) The discussion above shows that the motion of a source 'drags' the photons forward. This is shown in figure 1.3(a), which depicts the direction of propagation of seven selected photons in the lab and the rest frame. In other words, a charged particle, moving relativistically, will 'beam' most of the radiation in the forward direction.

(e) In the rest frame of the fast moving particles, most of the photons will be seen as coming from the forward direction. This is shown in figure 1.3(b).

1.3 Bremsstrahlung and synchrotron radiation

(a) Consider a collision between an electron of mass m and a proton at an impact parameter b. If the relative velocity is v, then the acceleration of the electron is $a \cong \left(q^2/mb^2 \right)$ and lasts for a time (b/v). This encounter will result in the radiation of energy $\mathscr{E} \cong \left(q^2 a^2/c^3 \right) (b/v) \cong \left(q^6/c^3 m^2 b^3 v \right)$. The amount of energy radiated per unit volume can be found by multiplying this expression by the number density of electrons n_e.

Since each collision lasts for a time $\tau \cong (b/v)$, there will be very little radiation at frequencies greater than $\omega_{crit} \simeq (v/b)$. For frequencies $\omega < \omega_{crit}$, we may take the energy emitted per unit frequency interval to be almost constant. In that case,

$$\left(\frac{d\mathscr{E}}{d\omega \, dt \, dV} \right) \cong n_e \mathscr{E} \cong \left(\frac{q^6}{m^2 c^3} \right) \left(\frac{n_e}{b^3 v} \right). \tag{1.17}$$

(Note that the combination $d\omega \, dt$ is dimensionless.) If the number density of ions is n_i, then $b \cong n_i^{-1/3}$; further, $v = (kT/m)^{1/2}$ if T is the temperature of the plasma. Hence, we can write the result as

$$\left(\frac{d\mathscr{E}}{d\omega \, dt \, dV} \right) \cong \left(\frac{q^6}{m^2 c^3} \right) \left(\frac{m}{kT} \right)^{1/2} n_e n_i. \tag{1.18}$$

To find the total energy radiated we have to integrate this expression over ω in the range $(0, \omega_{max})$, where ω_{max} is the typical frequency up to which we expect most of the radiation. We can estimate ω_{max} by the following argument. In a thermal plasma at temperature T, typical energies of electrons will be kT. An electron with such an energy cannot emit a photon with a frequency higher than kT. Hence, we may take $\omega_{max} \cong (kT/\hbar)$. Then we obtain

$$\left(\frac{d\mathscr{E}}{dt \, dV} \right) = \int_0^{(kT/\hbar)} d\omega \left(\frac{d\mathscr{E}}{d\omega \, dt \, dV} \right) \cong \left(\frac{q^6}{m^2 c^3} \right) \left(\frac{mkT}{\hbar^2} \right)^{1/2} n_e n_i. \tag{1.19}$$

(b) Consider a charged particle moving with velocity **v** in a magnetic field **B**. In the rest frame of the charge, there will be an electric field of magnitude $E' = (v/c)\gamma B \cong \gamma B$ (for $v \cong c$), inducing an acceleration $a' = (qE'/m)$. The energy radiated per second in the instantaneous rest frame will be

$$\left(\frac{d\mathscr{E}'}{dt'} \right) \cong \frac{q^2}{c^3} \left(a' \right)^2 \cong \frac{q^2}{c^3} \cdot \frac{q^2}{m^2} \gamma^2 B^2. \tag{1.20}$$

But since both $\mathscr{E}'$ and t' are the zeroth components of four-vectors, $\left(d\mathscr{E}'/dt' \right)_{rest} = \left(d\mathscr{E}/dt \right)_L$. Further, the energy density in the magnetic field is $U_B = \left(B^2/8\pi \right)$; hence we can write

$$\left(\frac{d\mathscr{E}}{dt} \right) = \left(\frac{d\mathscr{E}'}{dt'} \right) \cong 8\pi \left(\frac{q^2}{mc^2} \right)^2 \gamma^2 c U_B = 3 \left(\sigma_T c U_B \right) \gamma^2 \cong \left(\sigma_T c U_B \right) \gamma^2, \tag{1.21}$$

where $\sigma_T \equiv (8\pi/3) \left(q^2/mc^2 \right)^2$ is called the Thomson scattering cross-section for the charged particle (see problem 1.4).

1.4 Thomson scattering and Eddington limit

(a) Consider a charge q placed on an electromagnetic wave of amplitude E. The wave will induce an acceleration $a \cong (qE/m)$ causing the charge to radiate. The power radiated will be $P \cong (q^2 a^2/c^3) \cong (q^4/m^2 c^3) E^2$. Since the incident power in the electromagnetic wave was $S \cong (cE^2/4\pi)$, the scattering cross-section is $\sigma_T \equiv (P/S) \cong 4\pi (q^2/mc^2)^2$. The correct answer, taking into account the $(2/3)$ factor which arises due to angular averaging, is

$$\sigma_T = \frac{8\pi}{3}\left(\frac{q^2}{mc^2}\right)^2 \cong 6.65 \times 10^{-25}\,\text{cm}^2, \tag{1.22}$$

which is called the Thomson scattering cross-section.

(b) When a mass m falls from infinity to a radius R, in the gravitational field of a massive object with mass M, it gains the kinetic energy $E \cong (GMm/R)$. If this kinetic energy is converted into radiation with efficiency ϵ, then the luminosity of the accreting system will be $L = \epsilon (dE/dt) = \epsilon (GM/R)(dm/dt)$. It might seem that by increasing the 'accretion rate', $\dot{m}$, one can reach arbitrarily high L. This, however, is not possible for the following reason.

The photons which are emitted by this process will be continuously interacting with the infalling particles and will be exerting a force on the ionized gas. When this force is comparable to the gravitational force attracting the gas towards the central object, the accretion will effectively stop. The number density $n(r)$ of photons crossing a sphere of radius r, centred at the accreting object of luminosity L, is $(L/4\pi r^2)(\hbar\omega)^{-1}$, where ω is some average frequency. The rate of collision between photons and the electrons in the ionized matter will be $[n(r)\sigma_T]$ and each collision will transfer a momentum $(\hbar\omega/c)$. Since electrons and ions are strongly coupled in a plasma, this force will be transferred to the protons. Hence, the outward force on an infalling proton at a distance r will be

$$f_{\text{rad}} \cong (n\sigma_T)\left(\frac{\hbar\omega}{c}\right) = \left(\frac{L}{4\pi r^2}\right)\left(\frac{1}{\hbar\omega}\right)\sigma_T\left(\frac{\hbar\omega}{c}\right) = \left(\frac{L\sigma_T}{4\pi c r^2}\right). \tag{1.23}$$

This force will exceed the gravitational force attracting the proton, $f_g = (GMm_p/r^2)$, if $L > L_E$, where

$$L_E = \frac{4\pi G m_p c}{\sigma_T} M \cong 1.3 \times 10^{46}\left(\frac{M}{10^8\,\text{M}_\odot}\right)\text{erg s}^{-1}. \tag{1.24}$$

Note that the photons are strongly scattered by the *electrons* in the plasma with the scattering cross-section σ_T. The force exerted on the electrons is transmitted to the *protons* by normal electrostatic forces and balances the gravitational force on the *protons*.

(i) To estimate the minimum accretion rate needed to produce a given luminosity, we will assume that $\epsilon \cong 1$ and take $R \cong (GM/c^2)$. Then

$$\frac{dm}{dt} = \frac{L_E}{(GM/R)} \cong \left(\frac{4\pi G m_p}{\sigma_T c}\right)M \cong 0.2\,\text{M}_\odot\,\text{yr}^{-1}\left(\frac{M}{10^8\,\text{M}_\odot}\right). \tag{1.25}$$

(ii) The corresponding time-scale is $[M/(dm/dt)] \cong 5 \times 10^8$ yr. On this time-scale the mass of the central object increases by a factor e.

(iii) The energy acquired by the atom at the end of the accretion is $(GMm_p/R) \cong m_p c^2 \cong 1$ GeV. If this is converted into thermal radiation the peak frequency will be $\omega_{peak} \cong (1\,\mathrm{GeV}/\hbar) \cong 10^{14}$ GHz, which corresponds to the hard γ-ray band.

1.5 Inverse Compton scattering and CMBR

(a) The radiation field may be thought of as a random superposition of electromagnetic waves with $\langle E^2 \rangle = \langle B^2 \rangle = 8\pi U$, where U is the energy density. If a charged particle moves through this field with a velocity $\mathbf{v}$, then, in the rest frame of the charge, the electric field will have a magnitude $E' \cong \gamma E$. Accelerated by this field, the charged particle will radiate energy at the rate

$$\frac{d\mathscr{E}}{dt} = \frac{d\mathscr{E}'}{dt'} \cong \frac{q^2 a'^2}{c^3} \cong \left(\frac{q^2}{mc^2}\right)^2 c\gamma^2 E^2 \cong \sigma_T c U \gamma^2. \tag{1.26}$$

This result is similar in form to (1.21). In general, when a radiation field and a magnetic field coexist, the ratio between the power radiated by the inverse Compton process and the synchrotron process will be in the ratio (U/U_B).

(b) The lifetime for inverse Compton loss may be estimated to be

$$\tau = \frac{\mathscr{E}}{d\mathscr{E}/dt} \cong \frac{\mathscr{E}}{\sigma_T c \gamma^2 U_{CBR}} \cong \frac{10^{12}}{\gamma}\,\mathrm{yr}, \tag{1.27}$$

where we have taken $U_{CBR} = aT^4 \cong 4 \times 10^{-13}$ erg cm^{-3}. For a 100 GeV electron, $\tau \lesssim 10^7$ yr.

1.6 Interaction of matter and radiation

(a) Let the energy transferred from the photons to the electrons be ΔE on the average. (We shall omit the symbol $\langle \rangle$ for simplicity of notation.) Since $E \ll mc^2$ and $kT \ll mc^2$, we can expand $(\Delta E/mc^2)$ in a double Taylor series in (E/mc^2) and (kT/mc^2), retaining up to quadratic order:

$$\frac{\Delta E}{mc^2} = c_1 + c_2 \left(\frac{E}{mc^2}\right) + c_3 \left(\frac{kT}{mc^2}\right) + c_4 \left(\frac{E}{mc^2}\right)^2$$
$$+ c_5 \left(\frac{E}{mc^2}\right)\left(\frac{kT}{mc^2}\right) + c_6 \left(\frac{kT}{mc^2}\right)^2 + \cdots. \tag{1.28}$$

The coefficients $(c_1, \ldots, c_6)$ can be fixed by the following arguments. (i) Since $\Delta E = 0$ for $T = E = 0$, we must have $c_1 = 0$. (ii) Consider next the scattering of a photon with electrons at rest. This will correspond to $T = 0$ and $E \neq 0$. If the scattering angle is θ, the wavelength of the photon changes by

$$\Delta\lambda = \left(\frac{h}{mc}\right)(1 - \cos\theta). \tag{1.29}$$

Such a scattering of a photon, by an electron at rest, is symmetric in the forward–backward directions and the mean fractional change in the frequency is $(\Delta\omega/\omega) = -(\Delta\lambda/\lambda) = -(\hbar\omega/mc^2)$; so the average energy transfer to the electrons is $\Delta E = E^2/mc^2$. This implies that $c_2 = 0$ and $c_4 = 1$. (iii) If $E = 0$ and $T \neq 0$ the photon has zero energy and nothing should happen; hence $c_3 = c_6 = 0$. So our expression reduces to

$$\frac{\Delta E}{mc^2} = \left(\frac{E}{mc^2}\right)^2 + c_5 \left(\frac{E}{mc^2}\right)\left(\frac{kT}{mc^2}\right)^2. \tag{1.30}$$

(iv) To fix c_5, we can consider the following thought experiment. Suppose there is a very *dilute* gas of photons at the *same* temperature as the electrons. Then the number density $n(E)$ of photons is given by the Boltzmann limit of the Planck distribution:

$$n(E)\,dE \propto E^2 \left(e^{\beta E} - 1\right)^{-1} dE \propto E^2 \exp\left(-E/kT\right) dE. \tag{1.31}$$

In this case, since the temperatures are the same, we expect the net energy transfer between the electrons and photons to vanish. That is, we demand

$$0 = \int_0^\infty dE \, n(E)\,\Delta E. \tag{1.32}$$

Substituting for ΔE and $n(E)$ from (1.30) and (1.31), one can easily show that $(4kT + c_5kT) = 0$ or $c_5 = -4$. Hence, we obtain the final result:

$$\frac{\Delta E}{E} = \frac{(E - 4kT)}{mc^2}. \tag{1.33}$$

One may say that, in a typical collision between an electron and photon, the electron energy changes by (E^2/mc^2) and the photon energy changes by $(4kT/mc^2)E$.

(b) The mean-free-path of the photon due to Thomson scattering is $\lambda_\gamma = (n_e\sigma_T)^{-1}$. If the size of the region l is such that $(l/\lambda_\gamma) \gg 1$, then the photon will undergo several collisions in this region; but if $(l/\lambda_\gamma) \lesssim 1$ then there will be few collisions. It is convenient to define an 'optical depth' $\tau_e \equiv (l/\lambda_\gamma) = (n_e\sigma_T l)$ so that $\tau_e \gg 1$ implies strong scattering. photons

If $\tau_e \gg 1$, then the photon goes through $N_s(\gg 1)$ collisions in travelling a distance l. From standard random walk arguments, we have $N_s^{1/2}\lambda_\gamma \cong l$ so that $N_s = (l/\lambda_\gamma)^2 = \tau_e^2$. On the other hand, if $\tau_e \lesssim 1$, then $N_s \cong \tau_e$; therefore we can estimate the number of scatterings as $N_s \cong \max(\tau_e, \tau_e^2)$. The average fractional change in the photon energy, per collision, is given by the second term in (1.33), namely, $4(kT_e/m_ec^2)$. Hence, the condition for significant change of energy is

$$N_s \left(\frac{4kT_e}{m_ec^2}\right) = 4\left(\frac{kT_e}{m_ec^2}\right)\max\left(\tau_e, \tau_e^2\right) \cong 1. \tag{1.34}$$

We define a parameter y (called the Compton-y parameter) by

$$y = \left(\frac{kT_e}{m_ec^2}\right)\max\left(\tau_e, \tau_e^2\right). \tag{1.35}$$

Then we can write the condition for significant scattering as $y \cong 1/4$.

The optical depth of the region in which this process is significant is given by (1.34). Since $l = \tau_e/n_e\sigma_T$, we can estimate the size of the region in which this process will be important. The corresponding time-scale is $t_c \cong (l/c)$. We find that

$$t_c = \frac{l}{c} = \begin{cases} (n_e\sigma_T c)^{-1}(m_e c^2/4kT_e) & \text{(for } \tau_e \ll 1) \\ (2n_e\sigma_T c)^{-1}(m_e c^2/kT_e)^{1/2} & \text{(for } \tau_e \gg 1) . \end{cases} \qquad (1.36)$$

After a single scattering $(\mathscr{E}'/\mathscr{E}) = (1 + 4kT_e/m_e c^2) \cong \exp(4kT_e/m_e c^2)$ since $kT_e \ll m_e c^2$. After N_s scatterings

$$\left(\frac{\mathscr{E}'}{\mathscr{E}}\right) = \exp\left(\frac{4kT_e N_s}{m_e c^2}\right) = \exp(4y) , \qquad (1.37)$$

where we have used (1.34).

(c) Suppose that the initial mean frequency of the radiation field is ω_i with $\hbar\omega_i \ll kT_e$. The 'Comptonization' goes on till the mean energy of the photons rises to $(4kT_e)$. The optical depth needed for this is determined by the equation

$$\frac{\mathscr{E}'}{\mathscr{E}} = \left(\frac{4kT_e}{\hbar\omega_i}\right) = \exp\left[4\left(\frac{kT_e}{m_e c^2}\right)\tau_c^2\right] , \qquad (1.38)$$

giving,

$$\tau_{\text{crit}} = \left[\left(\frac{m_e c^2}{4kT_e}\right)\ln\left[\frac{4kT_e}{\hbar\omega_i}\right]\right]^{1/2} . \qquad (1.39)$$

The equilibrium spectrum has to be of Bose–Einstein form since photons are bosons. However, Compton scattering cannot change the number of photons. Hence, we cannot set chemical potential $\mu = 0$, and the final spectrum will, in general, be a Bose–Einstein spectrum with $\mu \neq 0$. The actual value of μ will be determined by the number density of the photons which does not change during Comptonization.

1.7 Diffusion in energy space

(a) The simplest procedure for deriving equations of this kind is based on the study of diffusion in a (abstract) space with energy as one of the coordinates. Figure 1.4 shows a region of this space in which the x-axis is the ordinary spatial x-direction but the y-axis is the energy. Particles move in and out of a small box in this space due to different processes. The motion along the x-direction is due to normal diffusion, while the motion along the E-direction is due to energy gain or loss. The number of particles inside the area shown in the diagram is $n(E, t, x)\, dE\, dx$. (For simplicity, we shall discuss the case with one spatial

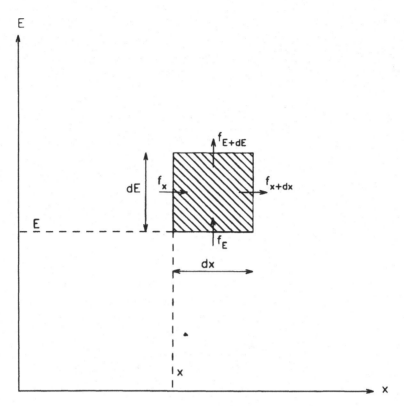

Fig. 1.4. The flux of particles in an abstract space, with energy along the vertical axis and one of the spatial directions along the horizontal axis. Gain or loss of energy makes the particle move in the vertical direction while normal diffusion makes the particle move in the horizontal direction.

dimension.) Therefore, the rate of change of particle density in this box is given by

$$\frac{\partial}{\partial t} n(E, x, t)\, dE\, dx = [f_x(E, x, t) - f_{x+dx}(E, x + dx, t)]\, dE$$
$$+ [f_E(E, x, t) - f_{E+dE}(E + dE, x, t)]\, dx$$
$$+ Q(E, x, t)\, dE\, dx, \tag{1.40}$$

where $f_x(E, x, t)$ denotes the flux of particles through the energy interval dE at the point x in space, etc. Expanding the various quantities in Taylor series and retaining the lowest order contributions we can write this equation as

$$\frac{\partial n}{\partial t} = -\frac{\partial f_x}{\partial x} - \frac{\partial f_E}{\partial E} + Q. \tag{1.41}$$

We know that the flux of particles f_x, at constant energy, is due to diffusion in space and is given by

$$f_x = -D \frac{\partial n}{\partial x}. \tag{1.42}$$

On the other hand, f_E is the flux of particles through dx which have energy E at some time in the interval dt. If $(dE/dt) = -b(E)$, then this flux can be written as

$$n(E) \frac{dE}{dt} = f_E = -b(E)n. \tag{1.43}$$

Substituting (1.42) and (1.43) into (1.41) and generalizing to three dimensions we obtain

$$\frac{\partial n}{\partial t} = D\nabla^2 n + \frac{\partial}{\partial E}(b(E)n) + Q. \tag{1.44}$$

(b) When the diffusion is ignored, the steady state ($\partial n/\partial t = 0$) solution to equation (1.44) can easily be found to be

$$n(E) = -\frac{1}{b(E)} \int Q(E)\, dE. \tag{1.45}$$

For a power law distribution with $Q(E) = kE^{-p}$ this solution reduces to

$$n(E) = \frac{kE^{-(p-1)}}{(p-1)b(E)} \qquad (p \neq 1). \tag{1.46}$$

(We have assumed that $n(E) \to 0$ as $E \to \infty$.) Using the form for $b(E)$ given in the question, the answer is

$$n(E) = \frac{k}{(p-1)} \frac{E^{-(p-1)}}{(A_1 + A_2 E + A_3 E^2)}. \tag{1.47}$$

Note that the final steady state spectrum is flatter by one power of E (compared to the injected spectrum) if ionization losses dominate. If adiabatic losses dominate, the final spectrum has the same index as the injected spectrum. If synchrotron or inverse Compton losses dominate, the spectrum is steeper by one power of E.

1.8 Line widths

(a) When the atom makes the transition from an excited state to the ground state it emits a photon of energy $\hbar\omega$. If the rate of transitions is A, then the net energy emitted per second is $(A\hbar\omega)$. By the correspondence principle, this should be equal to $(e^2/c^3)\, a^2$. The acceleration is about $a \cong \omega^2 x_0$, where x_0 is the size of the atom. Equating the two rates, we obtain $(A\hbar\omega) \cong (e^2/c^3)(\omega^2 x_0)^2$ or

$$A \cong \left(\frac{e^2}{\hbar c}\right)\left(\frac{x_0}{c/\omega}\right)^2 \omega. \tag{1.48}$$

For a hydrogen-like atom radiating in the optical band $x_0 \cong 0.5 \times 10^{-8}$ cm, $(c/\omega) \cong 10^{-5}$ cm and we obtain $A^{-1} \cong 4 \times 10^8 \omega^{-1}$. The quantity A^{-1} represents the lifetime of the excited state.

Using the energy–time uncertainty principle $\Delta E \Delta t \cong \hbar$, with $\Delta E = \hbar \Delta \omega$ and $\Delta t = A^{-1}$, we obtain the line width $\Delta \omega \cong A$ or $(\Delta \omega / \omega) = (A/\omega) \cong 2 \times 10^{-9}$. Note that, other things being equal, radio lines are about 10^{10} times sharper and have lifetimes which are 10^{15} (or more) larger.

The above analysis is correct only if the radiation is due to an electric dipole transition. Magnetic transitions are suppressed by a further factor of $(v/c)^2 \cong \alpha^2 \cong 10^4$ because magnetic moments are smaller than electric dipole moments by a factor (v/c).

For a hydrogen atom, $x_0 = a_0 = (\hbar^2/me^2)$ and $\omega \cong (me^4/\hbar^2)$. Using these expressions, we can rewrite the rate of transition as $A \cong (e^2/mc^3)\,\omega^2$.

(b) In thermal equilibrium, the probability distribution for the line-of-sight velocity [say, v_x] is a Gaussian proportional to $[\exp - (v_x^2/2\sigma^2)]$ with $\sigma^2 = (kT/m)$. When an atom moving with velocity v_x emits light, there will be a Doppler shift $(\Delta \omega / \omega) \cong (v_x/c)$. Hence, the line profile will be

$$I(\omega) \propto \exp - \left[\frac{(\omega - \omega_0)^2}{\omega_0^2} \left(\frac{c}{\sigma} \right)^2 \right]. \tag{1.49}$$

The fractional line width $(\Delta \omega / \omega)$ is (σ/c).

Note that $\sigma \cong 0.1\,\mathrm{km\,s^{-1}}\,(T/1\,\mathrm{K})^{1/2}$ for hydrogen. Since the natural line width is $A \cong 10^{-7}$, it is clear that Doppler width will dominate over the natural line width for $\sigma \gtrsim 10^{-7} c \cong 0.003\,\mathrm{km\,s^{-1}}$. It turns out, however, that the line profile due to natural width is Lorentzian; the Gaussian profile falls much more rapidly. Hence, the natural width will dominate over the Doppler width away from the resonance.

1.9 Photon: particle or wave?

For a system in thermodynamic equilibrium at temperature T, with $\beta \equiv (kT)^{-1}$, the mean energy $\bar{E}$ is given by

$$\bar{E} = \frac{\sum E e^{-\beta E}}{\sum e^{-\beta E}} = Z^{-1} \sum E e^{-\beta E}, \tag{1.50}$$

where $Z = \sum e^{-\beta E}$ is the partition function. It follows that

$$\bar{E} = -\frac{\partial}{\partial \beta} \ln Z = -\frac{1}{Z} \frac{\partial Z}{\partial \beta}. \tag{1.51}$$

Differentiating once again, we obtain an expression for mean square fluctuation in energy:

$$-\frac{\partial \bar{E}}{\partial \beta} = \frac{1}{Z} \frac{\partial^2 Z}{\partial \beta^2} - \left(\frac{1}{Z} \frac{\partial Z}{\partial \beta} \right)^2 = \langle E^2 \rangle - \bar{E}^2 = (\Delta E)^2. \tag{1.52}$$

In the case of a photon gas with $\bar{E} = \hbar\omega \left(e^{\beta\hbar\omega} - 1\right)^{-1}$, direct differentiation gives

$$(\Delta n)^2 \equiv \left(\frac{\Delta E}{\hbar\omega}\right)^2 = \left(\frac{\bar{E}}{\hbar\omega}\right)^2 + \left(\frac{\bar{E}}{\hbar\omega}\right) = \bar{n}^2 + \bar{n}, \qquad (1.53)$$

where $\bar{n}$ is the mean number of photons with frequency ω and Δn is the fluctuation in this number. If photons were to be interpreted as particles, then one would expect $(\Delta n)^2 \cong \bar{n}$, giving the usual Poisson fluctuations of $(\Delta n/n) \cong n^{-1/2}$. For this to occur we need $\bar{n} \gg \bar{n}^2$, that is, $\bar{n} \ll 1$, which happens for $\beta\hbar\omega \gg 1$. On the other hand, if $\beta\hbar\omega \ll 1$ we have $\bar{n} \gg 1$ and we obtain $(\Delta n)^2 \cong \bar{n}^2$, which characterizes the wave-like fluctuations.

For $\hbar\omega \ll kT$ and $\hbar\omega \gg kT$ the expression for $n(\omega)$ has simple asymptotic forms. When $\hbar\omega \ll kT$, we are in the long wavelength, classical regime of the radiation. Equipartition of energy suggests that each mode (having two polarization states) should have energy $\epsilon_\omega = 2 \times (kT/2) = (kT)$ or $n_\omega = \left(\epsilon_\omega/\hbar\omega\right) = (kT/\hbar\omega)$. This is what we obtain from the Planck spectrum. When $\hbar\omega \gg kT$ we are in the regime in which photons behave as particles. In that case we expect $n_\omega = \exp\left(-\hbar\omega/kT\right)$ based on Boltzmann statistics. (In this limit, $n_\omega \ll 1$ and quantum statistical effects can be ignored; hence we obtain Boltzmann statistics rather than Bose–Einstein statistics.) Again, this is what we obtain from the Planck spectrum. Thus, one may think of photons as particles when $\hbar\omega \gg kT$ and as waves when $\hbar\omega \ll kT$.

1.10 Sensitivities of telescopes

If a source with a flux density $N(v)$ is observed for a time t using a telescope with area A and waveband Δv, the total number of photons detected from the source will be about $S = N(v)A\Delta vt$. If the noise generates a total number of photons σ during the same time, then the signal-to-noise ratio will be (S/σ).

(a) The intrinsic uncertainty in the number of photons, due to Poisson fluctuations, is $\sigma = \sqrt{S}$. In this case, the signal-to-noise ratio is $(S/\sigma) = S^{1/2} = (NA\Delta vt)^{1/2}$. Clearly, to achieve a constant (S/σ) value, t must scale as A^{-1}. Since $A \propto D^2$ we obtain $t \propto D^{-2}$.

(b) Let the flux density of unwanted background radiation in the sky be $B(v)$ photons $cm^{-2}\,s^{-2}\,Hz^{-2}\,sr^{-1}$. If the effective solid angle subtended by the telescope to the sky is Ω, then the number of unwanted background photons received from the sky will be $(A\Omega B\Delta vt)$. The noise due to these photons will be $\sigma = (A\Omega B\Delta vt)^{1/2}$. Hence, we find that $(S/\sigma) = N(At/\Omega B)^{1/2}$. Therefore, we need $t \propto \Omega D^{-2}$ for constant (S/σ) ratio.

(c) If the dark current of electrons is generated at a constant rate, then the equivalent number of photons which will be counted due to the dark current will be Ct where C is a constant of proportionality that depends on the details of the instrument. This contributes a noise $\sigma = (Ct)^{1/2}$. A reasoning similar to the above leads to the conclusion that $t \propto A^{-2} \propto D^{-4}$.

1.11 Newtonian gravity

(a) To calculate $\nabla^2 \phi$, we need to know the Laplacian of $|\mathbf{x} - \mathbf{x}'|^{-1}$. Changing the origin to $\mathbf{x}'$, it is clear that we need to evaluate the function

$$f(\mathbf{r}) = \nabla^2 \left(\frac{1}{r} \right) = \frac{1}{r^2} \frac{d}{dr} \left(r^2 \frac{d}{dr} \right) \left(\frac{1}{r} \right). \tag{1.54}$$

Obviously, $f = 0$ for all $r \neq 0$. To study the behaviour of f near the origin, we integrate both sides of this equation over a spherical region $\mathscr{V}$ of radius R. Then

$$\int_{\mathscr{V}} f(\mathbf{r}) \, d^3 \mathbf{r} = \int_0^R 4\pi r^2 dr \left\{ \frac{1}{r^2} \frac{d}{dr} \left(r^2 \frac{d}{dr} \right) \left(\frac{1}{r} \right) \right\} = -4\pi. \tag{1.55}$$

Hence, we must have $f(\mathbf{x}) = -4\pi \delta_{\mathrm{Dirac}}(\mathbf{x})$, where δ_{Dirac} is the Dirac delta function. It follows that

$$\nabla^2 \phi = -G \int d^3 \mathbf{x}' \, \rho(t, \mathbf{x}') \nabla_{\mathbf{x}}^2 \left\{ \frac{1}{|\mathbf{x} - \mathbf{x}'|} \right\}$$

$$= 4\pi G \int d^3 \mathbf{x}' \, \rho(t, \mathbf{x}') \delta_{\mathrm{Dirac}}(\mathbf{x} - \mathbf{x}') = 4\pi G \rho(\mathbf{x}, t). \tag{1.56}$$

One solution to Poisson's equation is clearly that given in the question. However, we can add to this any other solution of the homogeneous equation $\nabla^2 \phi = 0$. Thus the solution to Poisson's equation is not unique and we need boundary conditions to decide on the correct solution. In the integral form, this boundary condition is already incorporated and the solution is unique.

(b) Consider a sphere of radius R and constant density ρ centred at the origin. Let there be a spherical cavity of radius L inside with centre at $\mathbf{l}$. Consider the force on a particle located inside the cavity at $(\mathbf{l} + \mathbf{r})$. This force can be calculated by first calculating the force due to a full sphere and subtracting the force due to the matter needed to fill the cavity. The force due to a constant density sphere is $\mathbf{F} = -(4/3)\pi G \rho \mathbf{x}$ (so that $-\nabla \cdot \mathbf{F} = 4\pi G \rho$). Hence, the force we want is

$$\mathbf{F} = \mathbf{F}_{\mathrm{sph}} - \mathbf{F}_{\mathrm{hole}} = -\frac{4}{3}\pi G \rho \, [\mathbf{l} + \mathbf{r}] + \frac{4}{3}\pi G \rho \mathbf{r} = -\frac{4\pi}{3} G \rho \mathbf{l}, \tag{1.57}$$

which is clearly a constant inside the hole. Thus a spherical hole inside a sphere is a region with constant gravitational force.

Consider now two such spheres – one with density ρ_1, radius R_1, hole radius L_1 and the centre of the hole located at $\mathbf{l}_1$ with respect to the centre of the sphere; the second one has density ρ_2, radius R_2, etc. We superpose the spheres such that: (i) $\mathbf{l}_1$ and $\mathbf{l}_2$ are in opposite directions; (ii) $\rho_1 l_1 = \rho_2 l_2$; and (iii) part of the spherical cavities overlap. The resulting density distribution is clearly not spherically symmetric. But in the cavity which is common to the holes of both spheres, the gravitational force is strictly zero. This is because each sphere produces an equal and opposite force in the cavity when $\rho_1 l_1 = \rho_2 l_2$.

One can now superpose several such *pairs* of spheres in such a way that a common cavity exists. Since any such unit of two spheres produces zero force in the cavity, the sum of them will also lead to zero force.

(c) This result can be verified by direct substitution. However, it is possible to provide a geometric interpretation of this result along the following lines.

We begin by noting that Poisson's equation for the gravitational field can be obtained from the action

$$A\left[g_{ik}, \phi, \rho\right] \equiv \frac{1}{2} \int \left(\frac{\partial \phi}{\partial x^i} \frac{\partial \phi}{\partial x_i}\right) \sqrt{g}\, d^3 x + 4\pi G \int \rho \phi \sqrt{g}\, d^3 x \qquad (1.58)$$

by varying ϕ. This action is invariant (except for terms which are expressible as total divergences) under the transformations

$$g_{ik} \rightarrow e^{-4\alpha(x)} g_{ik}, \quad \phi \rightarrow e^{\alpha(x)}\phi, \quad \rho \rightarrow e^{5\alpha(x)}\rho, \qquad (1.59)$$

provided $\alpha(\mathbf{x})$ satisfies the condition

$$(\nabla \alpha)^2 = \nabla^2 \alpha. \qquad (1.60)$$

So if $\phi(\rho, g; \mathbf{x})$ is a solution to $\nabla^2 \phi = 4\pi G \rho$, then $e^{\alpha}\phi$ is a solution with density distribution $e^{5\alpha}\rho$ in a space with the metric $e^{-4\alpha}g_{ik}$. In general, however, $e^{-4\alpha}g_{ik}$ will not be a flat three-dimensional space and we are only interested in the situations in which both g_{ik} and $e^{-4\alpha}g_{ik}$ represent flat spaces.

There is a simple choice of $\alpha(\mathbf{x})$ for which this happens. Consider the function $\alpha(\mathbf{x}) = \alpha(|\mathbf{x}|) = \ln(r/a)$ with some constant a. This function satisfies (1.60). Under the transformation (1.59) with this $\alpha(x)$, we find that

$$ds_{\text{NEW}}^2 = e^{-4\alpha}\, ds_{\text{OLD}}^2 = \left(\frac{a}{r}\right)^4 \left[dr^2 + r^2 \left(d\theta^2 + \sin^2\theta\, d\psi^2\right)\right]. \qquad (1.61)$$

By using a new radial coordinate $R = (a^2/r)$ we can re-express ds_{NEW}^2 as flat space:

$$\begin{aligned} ds_{\text{NEW}}^2 &= \left(\frac{a}{r}\right)^4 \left[dr^2 + r^2 \left(d\theta^2 + \sin^2\theta\, d\psi^2\right)\right] \\ &= \left[dR^2 + R^2 \left(d\theta^2 + \sin^2\theta\, d\psi^2\right)\right]. \end{aligned} \qquad (1.62)$$

Thus the two transformations – scaling $(g_{ik} \rightarrow e^{-4\alpha}g_{ik})$ and inversion $(r \rightarrow a^2/r)$ – bring the space back to the original form. But these two transformations leave a net change in ϕ:

$$\phi(r, \theta, \psi) \rightarrow \frac{r}{a}\phi(r, \theta, \psi) \rightarrow \frac{a}{R}\phi\left(\frac{a^2}{R}, \theta, \psi\right). \qquad (1.63)$$

During this process the density also changes from $\rho(r, \theta, \psi)$ to $(a/R)^5 \rho(a^2/R, \theta, \psi)$. Hence, we have the following result. If $[\phi(\mathbf{x}), \rho(\mathbf{x})]$ is a solution set to Poisson's equation, then $[(a/x)\phi(a^2\mathbf{x}/x^2), a^5/x^5 \rho(a^2\mathbf{x}/x^2)]$ is also a solution. The point $(a^2\mathbf{x}/x^2)$ is said to be related to $\mathbf{x}$ by 'inversion' by a sphere of radius a. Note

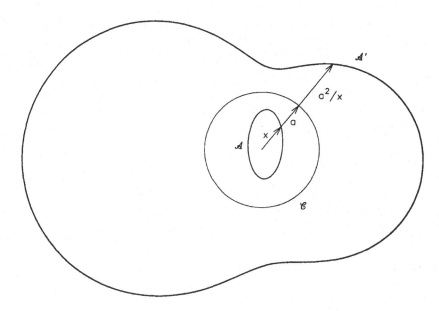

Fig. 1.5. Inversion of a region with vanishing gravitational force by a sphere which is placed asymmetrically. The surface obtained by inversion is shown by a thick line. The gravitational force everywhere outside this surface falls as the square of the distance.

that if the closed surface $\mathscr{S}$ is mapped to $\mathscr{S}'$ under inversion, then the points inside $\mathscr{S}$ will become mapped to regions outside $\mathscr{S}'$.

(d) The easiest way to construct such an example is to use the results of parts (b) and (c) above. In part (b), it was shown that one can have very asymmetric density distributions which will produce zero gravitational force in a compact region of space, $\mathscr{A}$, say. Let $\mathscr{C}$ be an imaginary spherical surface of radius a with centre inside $\mathscr{A}$ and placed asymmetrically with respect to $\mathscr{A}$ (see figure 1.5). Let us now 'invert' the surface $\mathscr{A}$ with respect to $\mathscr{C}$ and obtain a surface $\mathscr{A}'$. From the result of part (c), we know that the potential outside $\mathscr{A}'$ is the potential at the corresponding point inside $\mathscr{A}$, multiplied by (a/x). From the result of part (b) we know that the potential inside $\mathscr{A}$ is a constant. Hence, the potential outside $\mathscr{A}'$ falls as $|\mathbf{x}|^{-1}$. The matter distribution obtained by the inversion, of course, is not spherical.

1.12 Neutral gaseous system

(a) The distribution function for an ideal fermionic system is

$$f(\mathbf{p})\,d^3\mathbf{p} = \frac{g}{(2\pi\hbar)^3}\frac{d^3\mathbf{p}}{\exp\left[(E(\mathbf{p})-\mu)/kT\right]+1}, \tag{1.64}$$

where g is the spin degree of freedom, $E(\mathbf{p})$ is the energy, μ is the chemical potential and T is the temperature. The number density n of the particles, entropy S and the pressure P can be expressed as the integrals

$$n = \int f(\mathbf{p}) \, d^3\mathbf{p}, \quad S = -\int f(\mathbf{p}) \ln f(\mathbf{p}) \, d^3\mathbf{p}, \quad P = \int \frac{1}{3} \frac{p^2}{E} f(\mathbf{p}) \, d^3\mathbf{p}. \quad (1.65)$$

The first relation can be inverted to express μ in terms of n and T.

The Fermi–Dirac distribution can be approximated by a Boltzmann distribution when $f(\mathbf{p}) \ll 1$ for all relevant $\mathbf{p}$. In that case,

$$f(\mathbf{p}) \cong \frac{g}{(2\pi\hbar)^3} e^{+\mu/kT} e^{-E(\mathbf{p})/kT}. \quad (1.66)$$

We can take $E(\mathbf{p}) = (p^2/2m)$ for the non-relativistic case and $E(\mathbf{p}) = pc$ for relativistic fermions. Then we obtain

$$n = 2e^{\mu/kT} \lambda^{-3}, \quad S = -n \ln \left[n\lambda^3/2 \right], \quad P = nkT, \quad (1.67)$$

where

$$\lambda = \begin{cases} \lambda_{\text{Tnr}} = \left(2\pi\hbar^2/mkT \right)^{1/2} & \textit{(non-relativistic)} \\ \lambda_{\text{Tr}} = \pi^{2/3} \left(\hbar c/kT \right) & \textit{(relativistic)}. \end{cases} \quad (1.68)$$

In both the cases $\lambda \cong (\hbar/p)$, where p is the typical momentum of the particle, that is, λ is of the same order as the deBroglie wavelength. Hence, the classical description is valid for $n\lambda^3 \ll 1$.

In the quantum mechanical limit, $f(\mathbf{p}) \cong 1$ for all $p < p_{\max}$ and $f(\mathbf{p}) = 0$ for $p > p_{\max}$. From the distribution function it follows that $\mu = E(p_{\max})$. For both relativistic and non-relativistic cases,

$$n = \frac{2}{(2\pi\hbar)^3} \cdot \left(\frac{4\pi}{3} \right) p_{\max}^3. \quad (1.69)$$

The rest of the quantities can be worked out as in the classical case with $E = p^2/2m$ or $E = pc$. This will give

$$P = \begin{cases} [(3\pi^2)^{2/3}/5](\hbar^2/m)n^{5/3} & \textit{(non-relativisitic)} \\ [(3\pi^2)^{1/3}/4](\hbar c)n^{4/3} & \textit{(relativistic)}. \end{cases} \quad (1.70)$$

Let us now consider different regimes in the ρ–T plane. We start with the assumption that $kT \ll mc^2$ and $n\lambda_{\text{Tnr}}^3 \ll 1$, so that the gas is ideal and non-relativistic. At any temperature T, the ideal gas pressure is $P_{\text{g}} = \rho kT$, while the radiation pressure is $P_\gamma = (aT^4/3)$, where $a = (\pi^2/15)(k^4/\hbar^3 c^3) \cong 7.6 \times 10^{-15}\,\text{erg}\,\text{cm}^{-3}\,\text{K}^{-4}$. If $\rho \ll \rho_1$ with $\rho_1 = (aT^3/3k)$, then the radiation pressure will dominate over the gas pressure. This leads to the first entry in Table 1.1.

As we increase ρ (at the same T), we reach the ideal gas regime with $P = \rho kT$. This will hold until we have such high densities to give $n\lambda_{\text{Tnr}}^3 = 1$, that is, when

$n \cong n_Q$ with $n_Q = (mkT/2\pi\hbar^2)^{3/2}$. So we have the ideal gas behaviour for $\rho_1 \ll \rho \ll mn_Q$, which justifies the second entry in Table 1.1.

For $n > n_Q$, we are dealing with a quantum gas. As the density is increased further, p_F will increase (see equation (1.69)) and, at some stage, $p_F \cong mc$. When this happens, we have a *relativistic* degenerate gas. The critical density at which this happens is given by setting $p_F = mc$ in (1.69):

$$n_{rQ} = \frac{1}{3\pi^2} \left(\frac{mc}{\hbar} \right)^3. \tag{1.71}$$

So, for $n_Q \ll n \ll n_{rQ}$ we have a degenerate, non-relativistic system with pressure given by the first equation of (1.70). This explains the third entry in the table.

For $n_{rQ} \ll n$, we have a degenerate relativisitic gas with the pressure given by th second equation in (1.70). This is the last entry in the table.

(b) Figure 1.6 shows the various regimes of interest in a $\ln T$–$\ln \rho$ plane. Consider, for example, a temperature of $T \cong 10^3$ K (near the bottom of the diagram). We start with low densities (left bottom) and move to the right horizontally. At low densities, the radiation pressure dominates over the gas pressure. As we increase ρ, we cross the $\rho = (aT^3/3k)$ line (dot–dash line in the diagram) and enter the regime of neutral, ideal gas. As the density increases further, we cross the $n\lambda_{Tnr}^3 = 1$ line (triple dot–dash line in the diagram), to the right of which the gas is degenerate and non-relativistic. Finally, we cross the line $n = n_{rQ}$ beyond which the gas is degenerate and relativistic.

If the temperature is higher than about $T \cong 10^4$ K, the gas will show a significant amount of ionization and we will be dealing with a plasma rather than a neutral gas. (The dashed line separates plasma from the neutral gas. This aspect will be discussed in problem 1.21.) The diagram also shows various physical systems. For 'metals', the Fermi energy is used as an indicator of 'effective temperature'.

1.13 Transport coefficients for a gas

When the system is in strict thermodynamic equilibrium, it will have uniform temperature (T), bulk velocity (v) and density (n). If these quantities vary with space, then the system has not reached complete thermodynamic equilibrium. It is, however, possible to imagine a situation wherein various regions of the gas have reached local thermodynamic equilibrium characterized by the functions $T(\mathbf{x}, t)$, $n(\mathbf{x}, t)$ and $u(\mathbf{x}, t)$, which vary slowly in space and time. In such a context, one can derive an approximate expression for several transport coefficients.

Consider a fluid moving along the x-axis, with a *bulk* velocity $u_x(y)$ which varies along the y-direction. Let the root mean square *thermal* velocity of the particles be v so that $v^2 = (3kT/m)$. Consider an area ΔA in any plane perpendicular to the y-axis. The number of particles crossing this area ΔA, in unit time, either from above or from below, will be $(nv\Delta A/6)$. (The factor 6 arises because, on

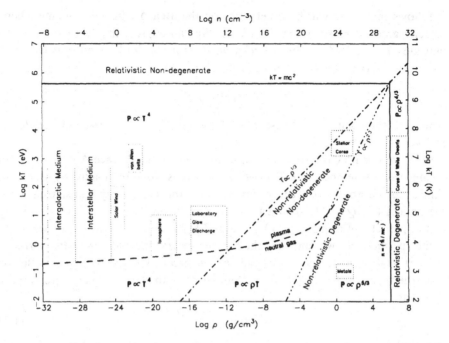

Fig. 1.6. Different regions in the ρ–T plane for the hydrogen gas. The dashed line distinguishes an ionized plasma (with 50% ionization) from a neutral gas. The dot–dash line separates the region where gas pressure dominates over radiation pressure. The triple dot–dash line separates the region where quantum mechanical consideration is important from the region in which a classical description is valid. The region above the solid horizontal line corresponds to a classical relativistic gas and the region to the right of the solid vertical line corresponds to a relativistic quantum gas.

the average, one-third of particles will be travelling along the y-axis, of which half will be travelling downwards.) The particles collide, on the average, within one mean-free-path $l = (n\sigma)^{-1}$, where σ is the collisional cross-section. Hence, the particles coming from above will transport properties of the gas at the point $(y + l)$, while the particles going from below will carry properties of the gas at the point $(y - l)$. If the properties of the gas vary along the y-axis, then this transport of molecules can lead to interesting physical effects.

(i) Consider first the net momentum transfer across the area ΔA. The x-component of the momentum of a typical particle coming from above will be $mu_x(y + l) \cong m\left[u_x + (\partial u_x/\partial y)\, l\right]$. Similarly, the typical momentum of a particle moving from below to above is $m\left[u_x - (\partial u_x/\partial y)\, l\right]$. Thus the net momentum transferred per second across the surface is

$$\frac{\Delta P}{\Delta t} \cong \frac{1}{6}\left(nv\Delta A\right)\left[2ml\left(\frac{\partial u_x}{\partial y}\right)\right]. \tag{1.72}$$

It follows that there will be a net force on the area ΔA in the xz-plane when a velocity gradient exists along the y-axis. We can express this 'viscous' force per unit area $F_x = \Delta P / \Delta t \Delta A$ by the relation $F_x \cong \eta \left(\partial u_x / \partial y \right)$, where η is called the coefficient of viscosity. Comparing the expressions we find

$$\eta = \frac{1}{3} mnlv \cong \frac{mv}{\sigma}. \tag{1.73}$$

Note that the coefficient of viscosity of the ideal gas is independent of density and varies as $\eta \propto v \propto T^{1/2}$.

(ii) Consider now the transport of energy across the surface ΔA. We will now assume that the temperature $T(y)$ of the gas varies along the y-axis. By arguments similar to the above, one can easily see that the net energy transferred across the area ΔA is $Q = (1/3) (nv\Delta A) (l) \left(\partial \epsilon / \partial y \right)$, where $\epsilon(y)$ is the average energy of the particle. Writing $\left(\partial \epsilon / \partial y \right) = \left(\partial \epsilon / \partial T \right) \left(\partial T / \partial y \right) = C_V \left(\partial T / \partial y \right)$ where C_V is the specific heat, we find that the net energy transported across the surface is proportional to the temperature gradient. We can define the thermal conductivity κ by the relation $Q = \kappa \left(\partial T / \partial y \right)$. Comparing with the previous expression, we find that

$$\kappa = \frac{1}{3} \left(nvC_V l \right) = \frac{\eta C_V}{m}. \tag{1.74}$$

(iii) The analysis for the coefficient of diffusion is identical. When there is a density gradient in the y-axis, a different amount of particles will flow from top to bottom and from bottom to top across any surface. Writing the particle current as $J_y = D \left(\partial n / \partial y \right)$, one can easily see that $D = (1/3) (vl)$.

The viscous force per unit *area* is $\bar{F}_{\text{vis}} \cong \left(\eta v / L \right)$, where L is the scale over which the velocity varies; so the force per unit *volume* will be $f_{\text{vis}} \cong \left(\eta v / L^2 \right)$. This will induce an acceleration $a_{\text{vis}} \cong \left(\eta v / L^2 \rho \right) \equiv \left(vv / L^2 \right)$, where $v \equiv \left(\eta / \rho \right)$ is called the 'kinematic viscosity'. This acceleration is to be compared with the regular 'convective term' for the acceleration of the fluid element, $a_{\text{con}} = (\mathbf{v} \cdot \nabla) \mathbf{v} \cong \left(v^2 / L \right)$. The ratio $\left(a_{\text{con}} / a_{\text{vis}} \right) \cong \left(v^2 / L \right) \left(L^2 / vv \right) \cong \left(vL / v \right) \equiv \mathcal{R}$ is called the Reynolds number. Viscous forces are important for scales at which $\mathcal{R}$ is small.

The collisions in the gas make any one particle execute a random walk in space with step length $l = (n\sigma)^{-1}$. After N collisions the particle could have diffused to a distance $q = \sqrt{N} l$. Since $N = \left(t / t_{\text{coll}} \right)$, we find that $q \cong \left(l^2 / t_{\text{coll}} \right)^{1/2} t^{1/2} \cong \left(v / n\sigma \right)^{1/2} t^{1/2}$. Diffusive effects are important when $q \cong R$; this happens over a time-scale $t_{\text{diff}} \cong \left(R^2 n\sigma / v \right)$. It is usual to write this as $t_{\text{diff}} \cong \left(R^2 / D \right)$, where $D \cong \left(v / n\sigma \right) \cong vl$ is the diffusion coefficient obtained above.

1.14 Time-scale for gravitational collisions

(a) Consider a gravitational encounter between two stars, each of mass m and relative velocity v. If the impact parameter is b, then the typical transverse velocity induced by the encounter is $\delta v_\perp \cong \left(Gm/b^2 \right) \left(2b/v \right) = \left(2Gm/bv \right)$. The deflection of the stars can be significant if $\left(\delta v_\perp / v \right) \gtrsim 1$, which occurs for collisions with impact parameter $b \lesssim b_c \cong (Gm/v^2)$. Thus the effective cross-section for collision with significant change of momentum is $\sigma \cong b_c^2 \cong (G^2m^2/v^4)$. The collisional relaxation time-scale will be

$$t_{gc} \cong \frac{1}{(n\sigma v)} \cong \frac{R^3 v^3}{N (G^2 m^2)} \cong \frac{N R^3 v^3}{G^2 M^2}. \tag{1.75}$$

If the system is in virial equilibrium, then $(GM/R) \cong v^2$ and we can write

$$t_{gc} \cong N \left(\frac{R}{v}\right) \left(\frac{R^2 v^4}{G^2 M^2}\right) \cong N \left(\frac{R}{v}\right). \tag{1.76}$$

Thus the collisional time-scale is N times the 'crossing time' (R/v) of the system.

There is, however, another effect which operates at a slightly shorter time-scale. When the impact parameter b is far larger than b_c, individual collisions impart a small transverse velocity $(\delta v_\perp) \cong (Gm/bv)$ to a given star. The effect of a large number of such collisions is to make the star perform a random walk in the velocity space. The net mean-square-velocity induced by collisions with impact parameters in the range $(b, b + db)$ in a time interval Δt is

$$\langle (\delta v_\perp)^2 \rangle = (2\pi b \, db) (v\Delta t) b \left(\frac{Gm}{bv}\right)^2. \tag{1.77}$$

The total mean-square transverse velocity due to all stars is found by integrating over b within a range (b_1, b_2):

$$\langle (\delta v_\perp)^2 \rangle_{\text{tot}} \cong \Delta t \int_{b_1}^{b_2} (2\pi b \, db) (vn) \left(\frac{G^2 m^2}{b^2 v^2}\right)$$
$$= \frac{2\pi n G^2 m^2}{v} \Delta t \ln \left(\frac{b_2}{b_1}\right). \tag{1.78}$$

It is reasonable to take $b_2 \cong R$, the size of the system; as regards b_1, notice that the effect discussed earlier becomes important for $b < b_c$. So we may take $b_1 \cong b_c \cong (Gm/v^2)$. Then $(b_2/b_1) \cong (Rv^2/Gm) = N (Rv^2/GM) \cong N$ in virial equilibrium. This effect is important over time-scales at which $\langle (\delta v_1)^2 \rangle_{\text{tot}} \cong v^2$, giving

$$(\Delta t)_{gc} \cong \frac{v^3}{2\pi G^2 m^2 n \ln N} \cong \left(\frac{N}{\ln N}\right) \left(\frac{R}{v}\right) \cong \left(\frac{t_{gc}}{\ln N}\right), \tag{1.79}$$

which is a shorter time-scale compared to t_{gc}.

(b) The analysis for this problem proceeds exactly as in part (a) above. Estimating the energy gained by 'field' particles (on the average), one can easily show that the time-scale should be

$$(\Delta t)_{\text{dyn fri}} \cong \frac{v^3}{2\pi G^2 M_t (m_f n_f) \ln N}$$
$$\cong \frac{10^{10} \text{yr}}{(\ln N)} \left(\frac{v}{10 \, \text{km s}^{-1}}\right)^3 \left(\frac{M_t}{M_\odot}\right)^{-1} \left(\frac{\rho_f}{10^3 \, M_\odot \, \text{pc}^{-3}}\right)^{-1}, \tag{1.80}$$

the only difference between this equation and (1.79) being the replacement of m^2 by $M_t m_f$. The numerical values correspond to those for a star cluster.

(c) The typical values are given in Table 1.1. The numbers clearly show the difference between gravitating systems and gaseous systems.

1.15 Collisionless evolution

Let the phase space density, that is, the number of particles per unit phase space volume, be $\mathcal{N}_0$ for the initial distribution and $\mathcal{N}$ for the final distribution. During a collisionless evolution, regions of high phase space density become mixed with regions of lower phase space density. *Hence, in this process the value of the maximum phase space density can only decrease.* Thus the present phase space density $\mathcal{N}$ must be less than the maximum value of $\mathcal{N}_0$.

The phase volume occupied by the halo particles today is $\mathcal{V} = (4\pi/3) R^3$ $(4\pi/3) (mv_{\text{max}})^3$, where $v_{\text{max}}^2 = (GM/R) \cong \sigma^2$. The present phase space density, therefore, is

$$\mathcal{N} = (N/\mathcal{V}) \cong \left(\frac{3}{4\pi}\right)^2 \left(\frac{M}{m}\right) \left(\frac{1}{R^3 m^3 \sigma^3}\right). \tag{1.81}$$

The maximum value of the initial phase space density for the Fermi–Driac distribution is $\mathcal{N}_{0,\text{max}} = f_{\text{FD}} (\mathbf{p} = 0) \cong (2\pi\hbar)^{-3}$. Using the condition $\mathcal{N} < \mathcal{N}_{0,\text{max}}$ we obtain the bound

$$m^4 > \frac{9\pi}{2} \left(\frac{c}{\sigma}\right) \left(\frac{c\hbar}{G}\right)^2 \left(\frac{G\hbar}{c^3 R^2}\right). \tag{1.82}$$

Substituting the numerical values we find that

$$m \gtrsim 30 \text{ eV} \left(\frac{\sigma}{220 \text{ km s}^{-1}}\right)^{-1/4} \left(\frac{R}{10 \text{ kpc}}\right)^{-1/2}. \tag{1.83}$$

1.16 When is gravity important?

We shall use units with $c = 1, \hbar = 1$ in this problem. Consider a spherical body with mass $M \cong NAm_p$ and radius $R \cong N^{1/3} a_0$. The gravitational potential energy of this body will be $E_g \cong (GM^2/R) \cong (Gm_p^2/a_0) A^2 N^{5/3}$. The atomic binding energy will be $E_{\text{atomic}} \cong NZ^2 \alpha^2 m_e$. The self-gravity will be important if $E_g \gtrsim E_{\text{atomic}}$, that is, when

$$A^2 N^{5/3} \left(\frac{Gm_p^2}{\alpha}\right) Z\alpha^2 m_e > N\alpha^2 m_e Z^2. \tag{1.84}$$

We define a dimensionless 'fine-structure constant' for gravity by $\alpha_G = (Gm_p^2/\hbar c)$ $= (Gm_p^2)$. Then we can write this condition as $N > N_{\text{crit}}$ with

$$N_{\text{crit}} = \left(\frac{Z}{A^2}\right)^{3/2} \alpha^{3/2} \alpha_G^{-3/2} \cong \left(\frac{Z}{A^2}\right)^{3/2} 10^{54}. \tag{1.85}$$

Table 1.1. Comparison of gravitating systems and ideal gas

	Gravitating Systems				'Ideal Gas'	
	General form	Star cluster	Galaxies	Clusters	General form	Value
Size of the particles	r_0	3×10^{-8} pc	3×10^{-8} pc	10 kpc	r_0	10^{-8} cm
Mass of the particles	m_0	2×10^{33} gm	2×10^{33} gm	$3\times10^{11}\,M_\odot$	m_0	10^{-24} gm
Size of system	R	10 pc	10 kpc	3 Mpc	R	30 cm
Number of particles	N	10^5	10^{11}	10^3	N	10^{24}
Velocity (v)	$\left\{\dfrac{Gm_0N}{R}\right\}^{1/2}$	10 km s^{-1}	300 km s^{-1}	10^3 km s^{-1}	$\{kT/m_0\}^{1/2}$	10^5 cm s^{-1}
Collisional range (b_c)	(R/N)	10^{-4} pc	10^{-7} pc	3 kpc	λ	$\cong 10^{-8}$ cm
Crossing time (t_{cross})	(R/v)	10^6 yr	3×10^7 yr	3×10^9 yr	(R/v)	3×10^{-4} s
Collisional relaxation time (t_{rel})	$\dfrac{R}{v}\dfrac{N}{\ln N}$	10^{10} yr	3×10^{17} yr	3×10^{11} yr	$\left\{\dfrac{R}{v}\right\}\left\{\dfrac{R}{\lambda}\right\}^2\dfrac{1}{N}$	3×10^{-9} s
Collisionality ($t_{\text{cross}}/t_{\text{rel}}$)	N^{-1}	10^{-5}	10^{-11}	10^{-3}	$N\left\{\dfrac{\lambda}{R}\right\}^2$	10^5
Mean-free-path ($l_{\text{mf}}=vt_{\text{rel}}$)	NR	10^6 pc	10^{12} kpc	3×10^3 Mpc	$\left\{\dfrac{R}{N}\right\}\left\{\dfrac{R}{\lambda}\right\}^2$	3×10^{-4} cm
'Structureness'	$\left\{\dfrac{r_0}{R}\right\}N^{1/3}$	10^{-7}	10^{-8}	3×10^{-2}	$\dfrac{r_0}{R}N^{1/3}$	3×10^{-2}

Taking $(Z/A^2) \cong 1$, such an object will have a radius $R \cong \alpha^{1/2}\alpha_G^{-1/2}a_0 \cong 10^{10}$ cm and mass of $M \cong N_{\max}m_p \cong 10^{30}$g. These correspond to the mass and radius of a large planet.

1.17 Making a star

(a) Consider a sphere of gaseous hydrogen with mass M and radius R. Classically, its potential energy is $U \cong -GM^2/R$ and its kinetic energy is $K \cong NkT = NT$, where T is the temperature. (In this problem, we use units with $k = 1, \hbar = 1$ and $c = 1$ to simplify the expressions.)   In virial equilibrium, $K \cong |U|$, giving $T \cong (GM^2/R)$. If this relation is exact, then the temperature can increase without bound as R decreases.   However, as the body contracts, each particle is confined to a smaller volume and, eventually, one has to take into account the quantum degeneracy effects. If the interparticle separation is $d \cong RN^{-1/3}$, then non-relativistic electrons will acquire a quantum kinetic energy of the order of $(p^2/2m_e) \cong (\hbar^2/2m_ed^2) \cong (m_ed^2)^{-1}$ in units with $\hbar = 1$. Writing the potential energy as $|U| \cong GM^2/R \cong N^{5/3}(Gm_p^2/d)$ and equating total kinetic and potential energies, $[NT + N/m_ed^2] \cong N^{5/3}(Gm_p^2/d)$, we find

$$T(d) \cong N^{2/3}\left(\frac{Gm_p^2}{d}\right) - \frac{1}{m_ed^2}.   \tag{1.86}$$

This function has a maximum at $d = d_{\max} = 2N^{-2/3}\alpha_G^{-1}m_e^{-1}$; the corresponding maximum temperature is

$$T_{\max} \cong N^{4/3}\alpha_G^2 m_e.   \tag{1.87}$$

(b) Consider two positively charged nuclei of mass m and charges q_1 and q_2 moving with relative velocity v. Classically, the Coulomb repulsion will prevent the particles from approaching each other closer than a distance $r_{\rm crit} \cong (2q_1q_2/mv^2)$. To reach the typical size of nuclear diameter $r_{\rm crit} \cong 10^{-13}$ cm, one needs an energy of $\frac{1}{2}mv^2 \cong 1\,{\rm MeV} \cong 10^{10}$ K. Fusion of nuclei can definitely take place at this temperature.

Nuclear reactions can, however, be triggered even at lower temperatures because of two effects: (i) at a given temperature T, most particles will have kinetic energy of the order of T; however, a fraction of particles at the tail of the Maxwell distribution can have significantly higher energies; (ii) quantum mechanics will permit the tunnelling of the particles through the Coulomb barrier (with a small probability), thereby allowing fusion. These two effects can be estimated as follows.

Let us begin with (ii). The tunnelling probability for particles to approach a separation $R \ll r_{crit}$ can be expressed as

$$P_1 = C \exp -\frac{2}{\hbar} \int_R^{r_c} dr \, (V(r) - E)^{1/2} \sqrt{2m}$$

$$= C \exp -\frac{2}{\hbar} \int_R^{r_c} \sqrt{2mE} \left(\frac{r_{crit}}{r} - 1\right)^{1/2} dr \equiv \exp(-I), \qquad (1.88)$$

where C is a constant. The integral I can be easily estimated by the substitution $(r/r_{crit}) = x^2$. Using the fact that $(R/r_{crit}) \ll 1$, we obtain

$$I \cong \frac{\pi q_1 q_2}{\hbar} \sqrt{\frac{2m}{E}} + \mathcal{O} \left(\frac{R^{1/2}}{r_{crit}^{1/2}}\right) \cong \frac{B}{\sqrt{E}}, \qquad (1.89)$$

with $B = \left(\pi q_1 q_2 \sqrt{2m}/\hbar\right)$. We see that the tunnelling probability varies with E as $\exp\left(-BE^{-1/2}\right)$.

Now consider the effect of (i). If the system is in thermal equilibrium at temperature T, then the probability that the particle will have an energy E is proportional to $\exp\left(-E/T\right)$. Thus the net probability for tunnelling will be proportional to

$$P_2 = \int_0^\infty dE \, \exp\left(-\frac{E}{T} - \frac{B}{\sqrt{E}}\right). \qquad (1.90)$$

Since $B^2 \gg T$, the argument of the exponential has a narrow maximum at $E = E_{max} = (BT/2)^{2/3}$. Near the maximum

$$-\frac{E}{T} - \frac{B}{\sqrt{E}} = -\frac{3E_{max}}{T} - \frac{3}{8} \frac{B}{E_{max}^{5/2}} (E - E_{max})^2 \cdots. \qquad (1.91)$$

Substituting (1.91) in (1.90) we find that

$$P_2 \propto \exp\left[-\left(\frac{T_{crit}}{T}\right)^{1/3}\right], \quad T_{crit} = \left(\frac{3}{2}\right)^3 \left(\frac{2\pi q_1 q_2}{\hbar}\right)^2 m_p. \qquad (1.92)$$

Numerically, $T_{crit} \cong 10 \, \text{MeV}$; however, even at $T \cong 0.01 \, \text{MeV}$, the factor $\exp\left[-(T_{crit}/T)^{1/3}\right]$ is only about $e^{-10} \cong 10^{-5}$. This means that a fraction of about 10^{-5} particles can undergo nuclear fusion even at temperatures of $T_{nuc} \equiv \eta \alpha^2 m_p$ with $\eta \cong 0.1$.

Combining with the result of part (a), we find that the condition $T_{max} > T_{nuc}$ implies $M > M_*$, with

$$M_* \cong \eta^{3/4} \left(\frac{m_p}{m_e}\right)^{3/4} \left(\frac{\alpha}{\alpha_G}\right)^{3/2} m_p \cong 10^{32} \, \text{gm} \qquad (1.93)$$

for $\eta \cong 0.1$.

(c)

(i) If a photon travels from the centre to the surface of the star after Q collisions, then $R \cong Q^{1/2}l$ or $Q \cong (R/l)^2$. Since the mean time between collisions is also l (for photons, with $c = 1$), a photon takes the time $t_{esc} \cong Ql \cong (R^2/l)$ to escape from the star. If the Thomson scattering dominates, then $t_{esc} \cong R^2\sigma_T n \cong 10^{10}$ s for the sun.

(ii) The luminosity of a star L is the ratio between the radiant energy content of the star, E_γ, and t_{esc}. Since $E_\gamma \cong (aT^4)R^3$, where $a = (\pi^2 k^4/15\hbar^3 c^3) = (\pi^2/15) \cong 1$ (in units with $k = \hbar = c = 1$), we find $L = (aR^3 T^4 l/R^2) \cong RT^4 l$. If Thomson scattering dominates, then we obtain

$$L \cong \frac{RT^4}{\sigma_T n} \cong \frac{T^4 R^4}{\sigma_T N} \cong \frac{G^4}{\alpha^2}m_p^5 m_e^2 M^3 \cong 10^{34} \text{ erg s}^{-1} \left(\frac{M}{M_*}\right)^3. \quad (1.94)$$

In arriving at the third equality we have also used the relation $T \cong (GMm_p/R)$.

(If, on the other hand, the plasma is only partially ionized, then $l \propto T^{7/2}n^{-2} \propto T^{7/2}R^6 M^{-2}$, and we have $L \propto RT^4 l \propto R^7 T^{15/2}M^{-2} \propto M^{11/2}R^{-1/2}$.)

(iii) If the average efficiency of the nuclear reaction is ϵ, then the nuclear reactions can support the star for a time-scale

$$t_* \cong \frac{\epsilon M}{L} \cong 3 \times 10^9 \text{ yr} \left(\frac{\epsilon}{0.01}\right)\left(\frac{M_*}{M}\right)^2 \quad (1.95)$$

for stars dominated by Thomson scattering. The gravitational collapse time-scale for a star like the sun will be

$$t_g = \left(\frac{GM}{R^3}\right)^{-1/2} \cong 26 \text{ min} \left(\frac{R}{R_\odot}\right)^{3/2}\left(\frac{M}{M_\odot}\right)^{-1/2}. \quad (1.96)$$

This is the time-scale over which the sun will collapse if the nuclear reactions are switched off.

The thermal time-scale for the sun will be $t_{th} \cong (aT^4)R^3/L_\odot \cong 2\times10^7$ yr. This is the time-scale in which the sun will radiate away its thermal energy.

(d) Consider a class of stars with almost same average temperature. For these stars $R \propto M$. Since the surface temperature T_s scales as $T_s \propto L^{1/4}R^{-1/2}$, we obtain

$$T_s \propto L^{1/4}R^{-1/2} \propto L^{1/4}M^{-1/2}. \quad (1.97)$$

From part (ii), we find that, if the stellar interior is only partially ionized, $L \propto M^{11/2}R^{-1/2} \propto M^5$ if $R \propto M$. Combining with (1.97) we obtain $T_s \propto L^{1/4}L^{-1/10} \propto L^{3/20}$. On the other hand, if Thomson scattering dominates with $L \propto M^3$, we obtain $T_s \propto L^{1/12}$. If the stars are plotted in a log T_s–log L plane, we expect them to lie within the lines with slopes $(3/20) = 0.15$ and $(1/12) \cong 0.08$. The observed slope is about 0.13.

(e) Since $E_\gamma \cong T^4 R^3$ and $E_{th} \cong NT$ we find that $(E_\gamma/E_{th}) \cong (T^3 R^3/N) \cong$ $(Gm_p^2)^3 N^2 = \alpha_G^3 (M/m_p)^2$. Setting this ratio to 10^2, we find that $M_{crit} \cong$ $10\alpha_G^{-3/2} m_p \cong 10\,M_\odot$. A more precise calculation gives $M_{crit} \cong 10^2\,M_\odot$.

1.18 Death of stars

The degeneracy momentum of a particle was worked out in problem 1.12 to be $p \cong \hbar n^{1/3}$. The energy of such a particle will be $\epsilon_{nr} \cong (p^2/m) \cong \hbar^2 n^{2/3} m^{-1}$ if the particle is non-relativistic and will be $\epsilon_r \cong p \cong \hbar n^{1/3}$ if it is relativistic. The pressure due to such particles, $P \cong n\epsilon$, will be $P_{nr} \cong (\hbar^2 m^{-1}) n^{5/3}$ and $P_r \cong \hbar n^{4/3}$ in the two cases. The pressure due to the gravity is $P_g \cong (GM^2/R^2)(1/R^2) \cong$ (GM^2/R^4). For a stellar remnant supported by the degeneracy pressure of *non-relativistic particles* we must have $P_{nr} = P_g$. This gives

$$M^{1/3} R = \frac{\hbar}{Gm_e^{8/3}} \left(\frac{m_e}{m_p}\right)^{5/3} = \text{constant.} \tag{1.98}$$

On the other hand, the degeneracy pressure of the *relativistic* particles, $P_r \cong$ $\hbar n^{4/3} \cong \hbar (M/m_p)^{4/3} R^{-4}$, has the same R-dependence as P_g. Hence, relativistic degeneracy pressure can support gravity only if $P_r > P_g$ or if $M < M_{Ch} \cong$ $\alpha_G^{-3/2} m_p \cong 1\,M_\odot$. This limit is known as the 'Chandrasekhar limit'.

1.19 Forming the galaxies

(a) Consider a region of size λ, containing gas with density ρ and temperature T. It has a thermal energy $E_{th} \cong (\rho\lambda^3/m_p) T$ and a gravitational potential energy $E_g \cong G (\rho\lambda^3)^2 \lambda^{-1}$. (We are using units with $c = \hbar = k = 1$.) For such a region to be stable against gravitational contraction, thermal energy should dominate over gravitational energy, that is, $T > (G\rho\lambda^2)$. If the body cools rapidly, leading to sudden decrease in T, this condition will tend to be violated. The system will become unstable and will tend to achieve stability at a reduced temperature by fragmenting to smaller objects (i.e. by reducing λ with the same ρ). The condition for this instability is that the time-scale for cooling, t_{cool}, should be less than the time-scale for gravitational collapse, t_g.

(b) For systems with temperature $T \cong (GMm_p/R)$, which is much higher than the ionization potential, $\alpha^2 m_e$, the dominant cooling mechanism is thermal Bremsstrahlung. The cooling time for this process can be computed from the results of problem 1.3(a). We found that $(d\mathscr{E}/dt\,dV) \cong (q^6/m^2)(mT)^{1/2} n^2$ (see equation (1.19) with $\hbar = c = k = 1$ and $n_i = n_e = n$). Since the energy density is (nT), we can estimate the cooling time to be

$$t_{cool} \cong \frac{nT}{(d\mathscr{E}/dt\,dV)} = \frac{m_e^2}{\alpha^3 n} \left(\frac{T}{m_e}\right)^{1/2}. \tag{1.99}$$

The time-scale for gravitational collapse is

$$t_g \cong \left(\frac{GM}{R^3} \right)^{-1/2}.$$

(1.100)

The condition for efficient cooling $t_{cool} < t_g$ leads to the constraint $R < R_g$ with

$$R_g \cong \alpha^3 \alpha_G^{-1} m_e^{-1} \left(\frac{m_p}{m_e} \right)^{1/2} \cong 74 \text{ kpc}.$$

(1.101)

The analysis assumed that $T > \alpha^2 m_e$; for $R \cong R_g$ this constraint is equivalent to the condition $M > M_g$ with

$$M_g \cong \alpha_G^{-2} \alpha^5 \left(\frac{m_p}{m_e} \right)^{1/2} m_p \cong 3 \times 10^{11} M_\bullet.$$

(1.102)

This result suggests that systems having a mass of about 10^{11} times the mass of a typical star could rapidly cool, fragment and form gravitationally bound structures. This is one possible scenario for forming galaxies.

1.20 Galaxy luminosity function

(i) Ignoring the effects of curvature (see chapter 6), one may relate the redshift z and observed luminosity f of a galaxy to its distance r and intrinsic luminosity L by $z \cong (H_0 r/c)$ and $f = (L/4\pi r^2)$. Then the joint distribution function in z and f for galaxies is

$$\frac{dN}{d\Omega\, dz\, df} = \frac{1}{4\pi} \int_0^\infty 4\pi r^2 dr\, \phi \left(\frac{L}{L_\bullet} \right) \delta_{\text{Dirac}} \left(z - \frac{H_0 r}{c} \right) \delta_{\text{Dirac}} \left(f - \frac{L}{4\pi r^2} \right).$$

(1.103)

The factor $(1/4\pi)$ in front converts the number density to density per solid angle and the Dirac delta functions select out the required redshift and flux. Using the known form of ϕ and evaluating the integral, we obtain

$$\frac{dN}{d\Omega\, dz\, df} = \frac{4\pi}{L_\bullet} \left(\frac{c}{H_0} \right)^5 z^4 \phi \left(z^2/z_{\text{crit}}^2 \right), \quad z_{\text{crit}}^2 = \frac{H_0^2 L_\bullet}{4\pi f c^2}.$$

(1.104)

(ii) The mean redshift of galaxies with a flux f can be obtained by multiplying the above expression by z, integrating over all z and normalizing the expression properly. This gives

$$\langle z \rangle = z_{\text{crit}} \frac{\int dy\, y^2 \phi(y)}{\int dy\, y^{3/2} \phi(y)} = z_{\text{crit}} \frac{\Gamma(3 + \alpha)}{\Gamma(5/2 + \alpha)}.$$

(1.105)

(iii) To find the number density of galaxies per unit flux interval we have to integrate (1.104) over all z. We obtain

$$\frac{dN}{d\ln f} = \frac{\phi_\bullet}{2} \left(\frac{L_\bullet}{4\pi f} \right)^{3/2} \Gamma \left(\frac{5}{2} + \alpha \right).$$

(1.106)

(iv) The mean luminosity per unit volume is given by

$$j = \int_0^\infty L\phi\left(\frac{L}{L_*}\right)\frac{dL}{L_*} = \Gamma(2+\alpha)\phi_* L_* = 1.0 \times 10^8 e^{\pm 0.26} h\, L_\odot\, \text{Mpc}^{-3}. \quad (1.107)$$

From this, one can estimate a typical number density of galaxies:

$$n_* \equiv \frac{j}{L_*} = \Gamma(2+\alpha)\phi_* = 0.01\, e^{\pm 0.4} h^3\, \text{Mpc}^{-3} \quad (1.108)$$

and a mean separation between the galaxies

$$d_* = n_*^{-1/3} = 4.7\, e^{\pm 0.13} h^{-1}\, \text{Mpc}. \quad (1.109)$$

(v) Multiplying the mass-to-light ratio $\mathcal{R} = (12 e^{\pm 0.2} h)\, \text{M}_\odot/\text{L}_\odot$ by the luminosity density j in (1.107) we obtain

$$\rho = j\mathcal{R} = 1.2 \times 10^9 e^{\pm 0.3} h^2\, \text{M}_\odot\, \text{Mpc}^{-3} = 8 \times 10^{-32} h^2\, \text{g cm}^{-3}. \quad (1.110)$$

1.21 Basic plasma physics

(a) Consider the reactions $H + \gamma \rightleftharpoons e + p$ in thermal equilibrium. If μ_e, μ_p and μ_H denote the chemical potentials of electrons, protons and neutral hydrogen atoms, then the conservation of chemical potential implies the relation $\mu_e + \mu_p = \mu_H$; further, $E_e + E_p - E_H = E_B$, where $E_B = 13.6$ eV is the binding energy of the hydrogen atom.

Using the expression $f(\mu, E)$ given in equation (1.66) we obtain

$$\frac{n_e n_p}{n_H} = \left(\frac{\lambda_H}{\lambda_e \lambda_p}\right)^3 \exp\left(-\frac{E_e + E_p - E_h}{kT}\right) \cong \lambda_e^{-3} \exp\left(-\frac{E_B}{kT}\right), \quad (1.111)$$

with $\lambda = (2\pi\hbar^2/mkT)^{1/2}$. We have assumed that $\lambda_H \cong \lambda_p$ in the prefactor and used the fact that the spin degeneracy factors for electron, proton and hydrogen atoms are 2, 2 and 4 respectively. It is conventional to express this result in terms of the number density, $n = n_p + n_H$, and the ionization fraction, $\alpha = (n_p/n_H)$ as

$$\frac{\alpha^2}{1-\alpha} = \frac{1}{n}\left(\frac{mkT}{2\pi\hbar^2}\right)^{3/2} \exp\left(-\frac{E_0}{kT}\right), \quad (1.112)$$

with $n = (N_H/L^3)$. The prefactor $n^{-1}(mkT/2\pi\hbar^2)^{3/2} \cong (N_{ph}/N_H)$ is the ratio between the total number of phase cells and the number of particles in the system. This factor should be large for the classical description to be valid. Because of this, significant ionization can occur even at $kT \ll E_0$. For example, 50% ionization ($\alpha = 0.5$) occurs at a temperature T_{crit} determined by

$$kT_{crit} = E_0 \left[\ln\frac{2}{n}\left(\frac{mkT}{2\pi\hbar^2}\right)^{3/2}\right]^{-1} \ll E_0. \quad (1.113)$$

Numerically, 50% ionization occurs at the values shown in figure 1.6 by the dashed line.

(b) The condition for 50% ionization was worked out in part (a) above. The condition for classical behaviour, $n\lambda_{\mathrm{Tnr}}^3 \ll 1$, was derived in problem 1.12. Finally, non-relativistic behaviour requires $kT \ll mc^2$. These constraints are also shown in figure 1.6. It is clear that most astrophysical plasmas satisfy these constraints.

(c) The Poisson equation for the plasma, with a charge Q added at the origin will be

$$\nabla^2\phi = 4\pi q\,(n_+ - n_-) + 4\pi Q\delta\,(\mathbf{r}) , \tag{1.114}$$

where $n_\pm(\mathbf{x})$ is the number density of ions (electrons) in the plasma. In thermal equilibrium we can assume that $n_\pm \propto \exp\,(\pm q\phi/kT)$. Then

$$n_+ - n_- = \bar{n}\left[\exp\,(q\phi/kT) - \exp\,(-q\phi/kT)\right] , \tag{1.115}$$

where $\bar{n}$ is the uniform density of particles in the absence of ϕ. (This was denoted by n in the question.) Now, far away from the charge Q, we will have $(q\phi/kT) \ll 1$ and

$$n_+ - n_- \cong \left(\frac{2\bar{n}q}{kT}\right)\phi. \tag{1.116}$$

Hence, equation (1.114) becomes

$$\nabla^2\phi = \frac{8\pi q^2\bar{n}}{kT}\phi + Q\,\delta\,(\mathbf{r}) , \tag{1.117}$$

which has the solution

$$\phi = \frac{Q}{r}\exp\left(-\frac{\sqrt{2}r}{\lambda_{\mathrm{D}}}\right) , \tag{1.118}$$

where $\lambda_{\mathrm{D}}^2 = \left(kT/4\pi q^2\bar{n}\right)$. Numerically,

$$\lambda_{\mathrm{D}} = 6.9\,\mathrm{cm}\left(\frac{T}{1\,\mathrm{K}}\right)^{1/2}\left(\frac{n}{1\,\mathrm{cm}^{-3}}\right)^{-1/2}. \tag{1.119}$$

(d) Consider a thermal fluctuation which moves the electrons (relative to ions) by a small distance δx along the x-axis. This deposits a charge $Q \cong q\,(nA\delta x)$ on a fictitious surface of area A perpendicular to the x-axis. This charge density leads to an electric field $E_x \cong 4\pi\,(Q/A) \cong 4\pi qn\delta x$ which acts on the electrons in this small volume, pulling it back. Thus the electrons will oscillate in accordance with the equation

$$m_{\mathrm{e}}\frac{d^2\delta x}{dt^2} = qE_x = -4\pi q^2 n\,\delta x. \tag{1.120}$$

The characteristic frequency of oscillation (called 'plasma frequency') is

$$\omega_{\mathrm{p}} = \left(\frac{4\pi q^2 n}{m}\right)^{1/2} = 5.64 \times 10^4\,\mathrm{Hz}\left(\frac{n}{1\,\mathrm{cm}^{-3}}\right)^{1/2}. \tag{1.121}$$

This analysis shows that to displace an electron by a distance λ, one requires an energy $(1/2)m_{\mathrm{e}}\omega_{\mathrm{p}}^2\lambda^2$. If the temperature of the plasma is T, then a typical thermal fluctuation has an energy of $(1/2)\,kT$. These spontaneous fluctuations can cause a displacement of the order of λ, where $(1/2)\,m_e\omega_p^2\lambda^2 \cong (1/2)\,kT$. Using (1.121) we obtain

$$\lambda^2 = \lambda_D^2 = \left(\frac{kT}{4\pi q^2 n}\right), \tag{1.122}$$

which is the same as Debye length. Such thermal fluctuations will lead to random electric fields of magnitude $E \cong 4\pi q n \lambda_D \cong (4\pi n k T)^{1/2}$. (The same result could have been obtained by equating the mean energy density in the electric field to the thermal energy density ('equipartition of energies'), $E^2/4\pi \cong nkT$.)

The corresponding time-scale is given by $t_D = (\lambda_D/v)$, where v is the typical velocity of the electron. This is the time it takes for the electrons to move over the distance λ_D neutralizing a random fluctuation. Using $v^2 \cong (kT/m)$ and equation (1.122) we obtain

$$t_D^2 = \left(\frac{kT}{4\pi q^2 n}\right)\left(\frac{m}{kT}\right) = \frac{m}{4\pi q^2 n} = \omega_p^{-2}. \tag{1.123}$$

For a plasma to behave collectively as a gas, it is necessary that the number of particles inside a Debye volume,

$$N_D \equiv \left(n\lambda_D^3\right) \cong 360 \left(\frac{T}{1\,\text{K}}\right)^{3/2}\left(\frac{n}{1\,\text{cm}^{-3}}\right)^{-1/2} \tag{1.124}$$

is sufficiently large. The quantity N_D can also be expressed as

$$N_D \cong \left(n\lambda_D^2\right)\lambda_D \cong \lambda_D\left(\frac{kT}{4\pi q^2}\right) \cong \left(\frac{\lambda_D}{r_{\text{crit}}}\right), \tag{1.125}$$

where $r_{\text{crit}} = (4\pi q^2/kT)$. For a plasma with number density n, the mean inter-particle distance is $n^{-1/3}$. The ratio $\mathscr{R}$ between the typical electrostatic energy of nearest neighbours, $q^2 n^{1/3}$, and the thermal energy, kT, is $\mathscr{R} = (q^2 n^{1/3}/kT) \cong r_{\text{crit}} n^{1/3} \cong \left[(r_{\text{crit}}^3/\lambda_D^3)\,N_D\right]^{1/3} \cong N_D^{-2/3}$. Hence, condition $N_D \gg 1$ implies that the typical electrostatic energy is small compared to kT.

(e)

(i) In a plasma the energy exchange takes place due to electron–electron, electron–ion and ion–ion collisions. The collisions between electrons will make the electron distribution reach an equilibrium in a time-scale τ_{ee}. The analysis to estimate τ_{ee} proceeds exactly as in problem 1.14(b). The time-scale for significant exchange of energy between electrons can be estimated by replacing (G) by (Zq^2/m^2) in that analysis. This gives

$$\tau_{ee} \cong \frac{v^3}{2\pi(Zq^2/m)^2 n \ln(b_2/b_1)}. \tag{1.126}$$

In the case of a plasma, b_2 should be of the order of the Debye length λ_D since the r^{-2} force is shielded at a larger distance. The value of b_1 is somewhat more difficult to determine and will be discussed in problem

4.5(a). The factor $\ln(b_2/b_1)$ is called the 'Coulomb logarithm' and is usually denoted by $\ln \Lambda$. Numerically,

$$\tau_{ee} \cong 10^9 \text{s} \left(\frac{\ln \Lambda}{10}\right)^{-1} \left(\frac{kT_e}{1 \text{ keV}}\right)^{3/2} \left(\frac{n}{1 \text{ cm}^{-3}}\right)^{-1}. \tag{1.127}$$

The corresponding mean-free-path is $\lambda \cong v\tau \cong 7 \times 10^5 (T^2 n^{-1})$ cm in c.g.s. units.

(ii) The analysis in part (i) shows that $\tau \propto m^2 v^3 \propto m^{1/2}$ at a given T. Therefore, the ion–ion collision time-scale τ_{pp} will be larger by the factor $(m_p/m_e)^{1/2} \cong 43$, giving $\tau_{pp} = (m_p/m_e)^{1/2} \tau_{ee} \cong 43\tau_{ee}$.

(iii) The time-scale for significant transfer of energy between electrons and ions will be considerably larger than τ_{ee} because of the following fact. When two particles (of unequal mass) scatter off each other, there is no energy exchange in the centre of mass frame. In the case of ions and electrons, the centre of mass frame differs from the lab frame only by a velocity $v_{CM} \cong (m_e/m_p)^{1/2} v_p \ll v_p$. Since there is no energy exchange in the centre of mass frame, the maximum energy transfer in the lab frame (which occurs for a head-on collision), is about

$$\Delta E = 2m_p v_{CM}^2 \cong 2m_e v_p^2, \tag{1.128}$$

giving

$$\frac{\Delta E}{(1/2)\, m_p v_p^2} \cong \frac{m_e}{m_p} \ll 1. \tag{1.129}$$

Therefore, it takes (m_p/m_e) times more collisions to produce equilibrium between electrons and ions. That is, the time-scale for electron–ion collisions is $\tau_{pe} = (m_p/m_e)\, \tau_{ee} \cong 1836\tau_{ee}$. The plasma will relax to Maxwellian distribution in this time-scale.

(iv) Suppose that the mean time between collisions of electrons in a plasma is τ_{ee}. The typical velocity achieved by a charged particle between collisions, when accelerated by an electric field E, will be $v_{drift} \cong (qE/m)\tau_{ee}$. This velocity leads to a current $J = nqv = (nq^2\tau_{ee}/m)\, E$. Using the definition of conductivity $\sigma = (J/E)$ we obtain $\sigma \cong (nq^2/m)\tau_{ee}$. Using equation (1.126) for τ_{ee}, and ignoring the Coulomb logarithm, we find

$$\sigma \cong \left(\frac{kT}{m}\right)^{3/2} \left(\frac{m}{q^2}\right) \cong 2.5 \times 10^{17} \text{ s}^{-1} \left(\frac{T}{10^6 \text{ K}}\right)^{3/2}. \tag{1.130}$$

Note that this result depends only on T and is independent of the density of the plasma.

(f) The analysis in this case proceeds exactly as in problem 1.13. Consider two points separated along the y-axis by the mean-free-path of the photon, $l = (n_e \sigma_T)^{-1}$. Photons moving from y to $(y+l)$ will carry an energy flux $caT^4(y)$.

Similarly, photons moving from $(y + l)$ will carry a flux $ca T^4(y + l)$. The net flux will be

$$q = ca \left[T^4 (y) - T^4 (y + l) \right] = -4 \, cal \, T^3 \frac{\partial T}{\partial y}$$

$$= -\frac{ca T^3}{n_e \sigma_T} \frac{\partial T}{\partial y} = -\kappa \frac{\partial T}{\partial y} . \tag{1.131}$$

This gives the necessary expression for the conductivity.

2
Dynamics and gravity

2.1 Dynamical systems

(a) Setting the variation of A to zero we obtain

$$0 = \delta A = \int_{\mathscr{P}_1}^{\mathscr{P}_2} dt \left[\delta p_i \, \dot{q}^i + p_i \, \delta \dot{q}^i - \frac{\partial H}{\partial q_i} \delta q^i - \frac{\partial H}{\partial p_i} \delta p^i \right]$$

$$= p_i \, \delta q^i \Big|_{\mathscr{P}_1}^{\mathscr{P}_2} + \int_{\mathscr{P}_1}^{\mathscr{P}_2} dt \left[\left(\dot{q}_i - \frac{\partial H}{\partial p_i} \right) \delta p^i - \left(\dot{p}_i + \frac{\partial H}{\partial q_i} \right) \delta q^i \right]. \qquad (2.1)$$

The first term vanishes since δq vanishes at the end points. The second term can vanish for arbitrary δp^i and δq^i only if

$$\dot{q}_i = \frac{\partial H}{\partial p_i}, \quad \dot{p}_i = -\frac{\partial H}{\partial q_i}. \qquad (2.2)$$

(b) The variation now gives

$$\delta A = \int_{\mathscr{P}_1}^{\mathscr{P}_2} dt \left[\frac{\partial L}{\partial q_i} \delta q^i + \frac{\partial L}{\partial \dot{q}_i} \delta \dot{q}^i \right]$$

$$= \int_{\mathscr{P}_1}^{\mathscr{P}_2} dt \left[\frac{\partial L}{\partial q_i} - \frac{d}{dt} \left(\frac{\partial L}{\partial \dot{q}_i} \right) \right] \delta q^i + \left(\frac{\partial L}{\partial \dot{q}_i} \right) \delta q^i \Big|_{\mathscr{P}_1}^{\mathscr{P}_2}. \qquad (2.3)$$

The second term vanishes since $\delta q = 0$ at the end points, giving

$$\frac{\partial L}{\partial q_i} = \frac{d}{dt} \left(\frac{\partial L}{\partial \dot{q}_i} \right). \qquad (2.4)$$

170

(c) In this case we have

$$
\delta A' = \int_{\mathscr{P}_1}^{\mathscr{P}_2} \delta L' dt = \int_{\mathscr{P}_1}^{\mathscr{P}_2} dt \left[\frac{\partial L}{\partial q} \delta q + \frac{\partial L}{\partial \dot{q}} \delta \dot{q} \right] - \delta \left(q \frac{\partial L}{\partial \dot{q}} \right) \Big|_{\mathscr{P}_1}^{\mathscr{P}_2}
$$

$$
= \int_{\mathscr{P}_1}^{\mathscr{P}_2} dt \left[\frac{\partial L}{\partial q} - \frac{d}{dt} \left(\frac{\partial L}{\partial \dot{q}} \right) \right] \delta q + \delta q \left(\frac{\partial L}{\partial \dot{q}} \right) \Big|_{\mathscr{P}_1}^{\mathscr{P}_2} - \delta q \left(\frac{\partial L}{\partial \dot{q}} \right) \Big|_{\mathscr{P}_1}^{\mathscr{P}_2} - q \delta \left(\frac{\partial L}{\partial \dot{q}} \right) \Big|_{\mathscr{P}_1}^{\mathscr{P}_2}
$$

$$
= \int_{\mathscr{P}_1}^{\mathscr{P}_2} dt \left[\frac{\partial L}{\partial q} - \frac{d}{dt} \left(\frac{\partial L}{\partial \dot{q}} \right) \right] \delta q - q \delta p \Big|_{\mathscr{P}_1}^{\mathscr{P}_2}. \tag{2.5}
$$

If we keep $\delta p = 0$ at the end points, then we get back the standard equations (2.4).

The crucial difference between L and L' is the following. Since $L = L(\dot{q}, q)$, the quantity $q(\partial L/\partial \dot{q})$ will also depend on $\dot{q}$. Hence, the term $d(q\partial L/\partial \dot{q})/dt$ will involve $\ddot{q}$. Thus L' contains second derivatives of q while L contains only up to first derivatives. In spite of the fact that L' contains second derivatives of q, the equations of motion arising from L' are only second order.

(In quantum theory, the probability amplitude for a particle to go from q_1 (at t_1) to q_2 (at t_2) is given by

$$
K(q_2, t_2; q_1, t_1) = \sum_{\text{paths}} \exp \frac{i}{\hbar} \int L\, dt, \tag{2.6}
$$

where the sum is over all paths connecting (q_1, t_1) and (q_2, t_2). The amplitude for the particle to have a momentum p_1 at t_1 and p_2 at t_2 is given by the Fourier transform

$$
G(p_2, t_2; p_1, t_1) \equiv \int dq_2\, dq_1\, K(q_2, t_2; q_1, t_1) \exp \left[-\frac{i}{\hbar}(p_2 q_2 - p_1 q_1) \right]. \tag{2.7}
$$

Using (2.6) in (2.7) we obtain

$$
G(p_2, t_2; p_1, t_1) = \sum_{\text{paths}} \int dq_1\, dq_2 \exp \frac{i}{\hbar} \left[\int L dt - (p_2 q_2 - p_1 q_1) \right]
$$

$$
= \sum_{\text{paths}} \int dq_1\, dq_2 \exp \frac{i}{\hbar} \left[\int dt \left\{ L - \frac{d}{dt}(pq) \right\} \right]
$$

$$
= \sideset{}{'}\sum_{\text{paths}} \exp \frac{i}{\hbar} \int dt\, L'. \tag{2.8}
$$

In arriving at the last expression, we have redefined the sum over paths to include integration over q_1 and q_2. It is now clear that our modified Lagrangian L' may be thought of as describing the transition amplitude between states with fixed momentum.)

2.2 Hamiltonian dynamics

(a) Since the same dynamical equations are to be obtained by varying the two different actions, made up of $(p\dot{q} - H)$ and $(P\dot{Q} - H')$, the two Lagrangians can only differ by a total time derivative (dS/dt). So

$$p\dot{q} - H - (P\dot{Q} - H') = \frac{dS}{dt} \qquad (2.9)$$

or

$$dS = p\,dq - P\,dQ + (H' - H)\,dt. \qquad (2.10)$$

It follows that, if $S = S(q, Q, t)$, then

$$p = \frac{\partial S}{\partial q}, \quad P = -\frac{\partial S}{\partial Q}, \quad H' = H + \frac{\partial S}{\partial t}. \qquad (2.11)$$

The function S is called the 'generating function' for the 'canonical transformation' from (q, p) to (Q, P).

   (b) The equation

$$\frac{\partial S}{\partial t} + H\left[\frac{\partial S}{\partial q}, q\right] = 0 \qquad (2.12)$$

can be solved – in principle – to obtain a particular solution $S = S(t, q^i, \alpha_i)$, where the α_i (with $i = 1, 2, ..., N$) are constants of integration. This generating function, of course, makes $H' = 0$ by construction. We now treat α_i as the new coordinates Q_i (which are constant in time since $H' = 0$). From equation (2.11) we see that the new momenta are $P_i = -(\partial S/\partial \alpha_i)$. But since $H' = 0$, these momenta should also be constant, say $P^i = \beta^i$. We thus obtain the equations $\beta_i = [\partial S(t, q_i, \alpha_i)/\partial \alpha_i]$ which implicitly determine the trajectories $q_i(t)$ in terms of the $2N$ constants (α_i, β_i).

2.3 Integrals of motion

(a) (i) Energy $E \equiv (1/2)\,\dot{q}_i\dot{q}^i + V(q)$ is the only obvious integral in this case. (ii) Energy E and the third component of the momentum $p_3 \equiv \dot{q}_3$ are now integrals of motion. (iii) Changing to (r, θ, z) coordinates with the z-axis along q_1, we find that the angular momentum along the z-axis, $J_z = r^2\dot{\theta}$, is conserved, in addition to the energy E. (iv) The angular momentum vector $\mathbf{J} = \mathbf{r} \times \mathbf{p}$ and the energy E are conserved.

   (b) In this particular case, we can easily integrate the equations of motion and obtain the solution

$$x(t) = A\cos(\omega_x t + \epsilon_x), \quad y(t) = B\cos(\omega_y t + \epsilon_y), \qquad (2.13)$$

where $(A, B, \epsilon_x, \epsilon_y)$ are the four constants which are to be determined in terms of $[x(0), y(0), p_x(0), p_y(0)]$.

(i) From the above solution we obtain

$$x = A\cos(\omega_x t + \epsilon_x), \quad p_x = -A\omega_x \sin(\omega_x t + \epsilon_x),$$
$$y = B\cos(\omega_y t + \epsilon_y), \quad p_y = -B\omega_y \sin(\omega_y t + \epsilon_y). \tag{2.14}$$

We can now express $(A, B, \epsilon_x, \epsilon_y)$ in terms of (x, p_x, y, p_y, t):

$$A^2 = x^2 + \frac{p_x^2}{\omega_x^2}, \qquad B^2 = y^2 + \frac{p_y^2}{\omega_y^2}, \tag{2.15}$$

$$\epsilon_x = -\omega_x t - \tan^{-1}\left(\frac{p_x}{x\omega_x}\right), \qquad \epsilon_y = -\omega_y t - \tan^{-1}\left(\frac{p_y}{y\omega_y}\right). \tag{2.16}$$

These are the constants of motion among which (ϵ_x, ϵ_y) depend on t explicitly. (Using the arbitrariness in the origin of t we can eliminate one of them.)

(ii) Since the Lagrangian depends on both x and y we expect only the energy to be a conserved quantity. This will define a three-dimensional region of the four-dimensional phase space in which the motion takes place. The analysis in (i) shows, however, that it is possible to write two independent integrals for the system since the x- and y-motions are uncoupled. We have the integrals (A, B) or, equivalently,

$$E_x \equiv \frac{1}{2}\omega_x^2 A^2 = \frac{p_x^2}{2} + \frac{1}{2}\omega_x^2 x^2, \quad E_y \equiv \frac{1}{2}\omega_y^2 B^2 = \frac{p_y^2}{2} + \frac{1}{2}\omega_y^2 y^2. \tag{2.17}$$

(The energy E is just $(E_x + E_y)$ and hence is not an independent integral.) These integrals clearly isolate the motion in the x–p_x and y–p_y planes. Since there are two integrals, the motion is confined to a $4-2 = 2$-dimensional surface on the phase space. The equation for this surface, given by (2.17), can be written parametrically as

$$p_x = \sqrt{2E_x}\sin\alpha, \quad x = \sqrt{\frac{2E_x}{\omega_x^2}}\cos\alpha \quad (0 \le \alpha < 2\pi), \tag{2.18}$$

$$p_y = \sqrt{2E_y}\sin\beta, \quad y = \sqrt{\frac{2E_y}{\omega_y^2}}\cos\beta \quad (0 \le \beta < 2\pi). \tag{2.19}$$

This is the two-dimensional surface of a torus on which motion takes place.

(iii) From the solution (2.13), it is clear that

$$\frac{\epsilon_x}{\omega_x} - \frac{1}{\omega_x}\cos^{-1}\left(\frac{x}{A}\right) = \frac{\epsilon_y}{\omega_y} - \frac{1}{\omega_y}\cos^{-1}\left(\frac{y}{B}\right) \tag{2.20}$$

or

$$\cos^{-1}\left(\frac{x}{A}\right) - \frac{\omega_x}{\omega_y}\cos^{-1}\left(\frac{y}{B}\right) = \text{constant} \equiv c. \tag{2.21}$$

This quantity c is clearly another integral of motion. But – in general – this does not isolate the motion further, because $\cos^{-1} z$ is a multiple-valued function. To see this clearly, let us write (2.21) as

$$x = A \cos \left[c + \frac{\omega_x}{\omega_y} \left(\mathrm{Cos}^{-1} \left(\frac{y}{B} \right) + 2\pi n \right) \right], \qquad (2.22)$$

where $\mathrm{Cos}^{-1} z$ denotes the principle value. For a given value of y we will obtain an infinite number of xs as we take $n = 0, \pm 1, \pm 2....$ Thus in general, the curve in (2.22) will fill a region in the (x, y)-plane.

A special case arises if (ω_x/ω_y) is a rational number. In that case, the curve closes on itself after a finite number of cycles. In this case, c is also an isolating integral and we have three isolating integrals: (E_x, E_y, c). The motion is confined to a closed (one-dimensional) curve on the surface of the torus.

(c) The four obvious constants of motion are energy E and three components of angular momentum **J**. They will restrict the motion to $6 - 4 = 2$ dimensions. This may be taken to be a region in a plane $\phi = \pi/2$ perpendicular to **J**, with coordinates $[r(t), \theta(t)]$. Given (r, θ), we can determine $\dot{r}$ and $\dot{\theta}$ via the integrals

$$E = \frac{1}{2}\dot{r}^2 + \frac{J^2}{2r^2} - \frac{\alpha}{r} - \frac{\beta}{r^2}, \quad J = r^2 \dot{\theta}. \qquad (2.23)$$

Of course, $\dot{\phi} = 0$.

Using the coordinate $u = r^{-1}$, it is easy to integrate the equations of motion and obtain

$$\frac{1}{r} = \frac{\alpha}{J^2} \left(1 - \frac{2\beta}{J^2} \right)^{-1} + A \cos \left[\left(1 - \frac{2\beta}{J^2} \right)^{1/2} (\theta - \theta_0) \right], \qquad (2.24)$$

where θ_0 is another constant. Solving for θ_0, we obtain

$$\theta_0 = \theta + \left(1 - \frac{2\beta}{J^2} \right)^{-1/2} \cos^{-1} \left[\frac{1}{A} \left\{ \frac{1}{r} - \frac{\alpha}{J^2} \left(1 - \frac{2\beta}{J^2} \right)^{-1} \right\} \right]. \qquad (2.25)$$

This quantity, θ_0, is definitely an integral of motion. If it is single-valued, we will have an extra *isolating* integral of motion reducing the motion from two to one degree of freedom, that is, a closed curve in the plane perpendicular to **J**.

In general, $\cos^{-1} z$ will introduce a $2\pi n$-factor and we obtain

$$\theta = \theta_0 - \left(1 - \frac{2\beta}{J^2} \right)^{-1/2} \left[\mathrm{Cos}^{-1} \frac{1}{A} \left\{ \frac{1}{r} - \frac{\alpha}{J^2} \left(1 - \frac{2\beta}{J^2} \right)^{-1} \right\} \pm 2\pi n \right]$$

$$= \theta_0 - \left(1 - \frac{2\beta}{J^2} \right)^{-1/2} \mathrm{Cos}^{-1} \frac{1}{A} \left\{ \frac{1}{r} - \frac{\alpha}{J^2} \left(1 - \frac{2\beta}{J^2} \right)^{-1} \right\} \pm 2\pi \left(1 - \frac{2\beta}{J^2} \right)^{-1/2} n.$$

$$(2.26)$$

This expression for θ will be single-valued if $\beta = 0$; hence the motion is in a closed curve if $\beta = 0$. If $\beta \neq 0$, then θ can be single-valued only if $(1 - 2\beta/J^2)^{1/2}$

is a rational number. This cannot happen for a general J. Hence, E and $\mathbf{J}$ are only isolating integrals if $\beta \neq 0$.

Note that if $\alpha = 0$ and $\beta \neq 0$, we still do not obtain an extra isolating integral. The $\alpha \neq 0, \beta = 0$ case is clearly special and corresponds to the Kepler problem in the inverse-square-law force.

2.4 Surfaces of section

For a system with $N = 2$, the phase space is four dimensional and any point in the phase space has the coordinates (x, y, p_x, p_y). Conservation of energy restricts the motion to a three-dimensional surface defined by

$$\frac{1}{2}\left(\dot{x}^2 + \dot{y}^2\right) + \phi(x, y) = E = \text{constant}. \tag{2.27}$$

(We shall take $m = 1$ and use the symbols $\dot{x}$ and $\dot{y}$ instead of p_x and p_y.) The intersection of this three-dimensional surface with the surface $y = 0$ defines the 'surface of section' for the dynamical problem, which has the coordinates $(x, \dot{x})$. Given x, $\dot{x}$ and $y = 0$, $\dot{y}$ is determined as

$$\dot{y} = + \left[2\left(E - \phi(x, 0)\right) - \dot{x}^2\right]^{1/2}, \tag{2.28}$$

where we have chosen the positive square root by convention.

Given the energy E and suitable initial conditions, the equations of motion can be integrated to give the orbit of the particle. This orbit will repeatedly cut the surface of section at different values of x and $\dot{x}$. If there are no other integrals of motion, we expect these points to be scattered in a two-dimensional region bounded by the curve

$$\frac{1}{2}\dot{x}^2 + \phi(x, 0) \leq E. \tag{2.29}$$

In case the system has some other hidden integrals of motion, the orbit will be confined to a lower-dimensional surface in phase space and hence will form a smooth curve in the x–$\dot{x}$ plane.

The surface of section for the potential given in the question is shown in figure 2.1(a). The following conclusions can be drawn from the diagram.

(i) The intersection of the orbit with the x–$\dot{x}$ plane occurs on reasonably smooth curves rather than on the entire allowed region in the plane. This suggests that there could be some other hidden integrals of motion.

(ii) The orbits can be classified as belonging to the two different classes, usually called 'box orbits' and 'loop orbits'. Orbits like 1 or 2 revolve around the points A or B, respectively, and have a definite sense of rotation ('loop orbits'). These orbits do not pass through the origin. In contrast, orbits like 3 enclose both A and B and have no definite sense of rotation ('box orbits'). These orbits do cross the origin. The behaviour of these orbits in physical space is shown in figure

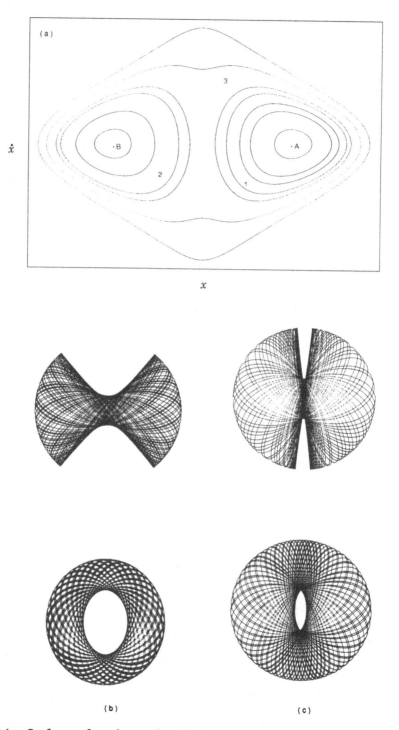

Fig. 2.1. Surfaces of section and orbits in an external potential; see text for discussion.

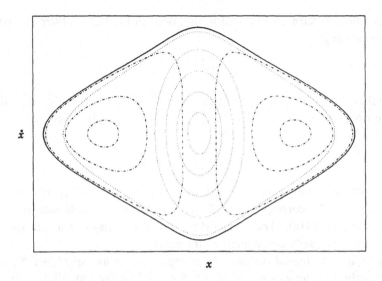

Fig. 2.2   Curves generated by constancy of $x\dot{y}$ and E_x in the surface of section.

2.1(b). The top frame shows the box orbits which oscillate through the origin with a superposed rotational motion. The bottom frame shows the loop orbits which are confined to an annular region.

(iii) As the initial conditions are changed, loop orbits can become more and more eccentric and the box orbits can become less and less elongated. Two such extreme cases are shown in figure 2.1(c).

(iv) The above discussion was based on a potential with $q = 0.9$. If $q = 1$ the potential depends only on $r^2 = x^2 + y^2$ and possesses an extra conserved quantity, namely, the z-component of the angular momentum, $J_z = x\dot{y} - y\dot{x} = x\dot{y}$ on the surface of section with $y = 0$. In this case, all generic orbits are loop orbits. For $q \neq 1$ the expression for J_z becomes

$$(J_z)_{y=0} = x\dot{y} = x \left[2E - v_0^2 \ln (x^2 + R_c^2) - \dot{x}^2\right]^{1/2}. \qquad (2.30)$$

Though this quantity is not strictly conserved when $q \neq 1$, it is useful to examine the nature of the curves in the x–$\dot{x}$ plane, generated by the constancy of J_z. These are shown in figure 2.2 by dot–dash lines. We see that these curves are similar to those corresponding to the loop orbits seen in figure 2.1(a). We may conclude that the loop orbits owe their existence to an approximate conservation of a quantity such as angular momentum.

(v) At the other extreme we may consider the motion in the potential when $r \ll R_c$ with $\phi(x, y)$ approximated as

$$\phi(x, y) \cong \frac{v_0^2}{2R_c^2} \left(x^2 + \frac{y^2}{q^2}\right) + \text{constant}. \qquad (2.31)$$

This is a double harmonic oscillator, discussed in the last problem, which has an additional integral

$$E_x = \frac{1}{2}\dot{x}^2 + \frac{v_0^2}{2R_c^2}x^2. \tag{2.32}$$

Constancy of E_x will lead to the elliptical orbits shown in the middle of figure 2.2. We do not see evidence of such orbits in figure 2.1(a). However, if we generalize E_x to the form

$$E_x = \frac{1}{2}\dot{x}^2 + \phi(x,0), \tag{2.33}$$

then we obtain better agreement with the orbits in figure 2.1(a). The outer dotted curve in figure 2.2 corresponds to $E_x = $ constant and is quite similar to the box orbits of figure 2.1(a). Thus box orbits owe their existence to an approximate conservation of the 'x-component of the energy'.

(vi) As q is decreased further, larger regions of phase space will be occupied by box orbits at the expense of loop orbits. When the potential is significantly non-axisymmetric, several 'island-like' regions will form in the x–$\dot{x}$ plane with loop orbits around each of them.

2.5 Fully integrable systems

Given the N integrals of motion, $F_i(q,p)$ with $i = 1, 2, ..., N$, it is convenient to make a canonical transformation to a new set of variables (Q, P) with $P_j = F_j(q,p)$. We have

$$\dot{p}_j = -\frac{\partial \mathcal{H}}{\partial Q_j} = \dot{F}_j = 0, \tag{2.34}$$

since the Fs are constants of motion. Hence, the new Hamiltonian $\mathcal{H}$ cannot depend on Q_i and is purely a function of the Ps. Let $(\partial \mathcal{H}/\partial P_j) = \Omega_j(P)$. Since the Ps are independent of t, the Ωs are also independent of t. So the equations of motion $\dot{Q}_j = (\partial \mathcal{H}/\partial P_j) = \Omega_j$ can be integrated to give

$$Q_j = \Omega_j t + \delta_j. \tag{2.35}$$

The generating function for this transformation can be obtained as follows. The constancy of F_j can be expressed by the set of equations $F_j(q,p) = f_j$, where the f_js are N constants. Inverting this equation, one can obtain p_i in terms of the qs and fs as $p_i = p_i(q_j, f_k)$. The generating function can be taken to be

$$S(q,P) = \int^q p_k(q,P)\,dq^k = S(q,f). \tag{2.36}$$

(Summation over k is assumed.) The new coordinates Q_i are given by the standard relation

$$Q_i = \frac{\partial S(q,f)}{\partial f_i}. \tag{2.37}$$

This equation relates the new and old coordinates and contains N arbitrary constants. Equation (2.35) gives the new coordinates as a function of time and another N constants. Combining, we can find the trajectories $q_i(t)$ in terms of $2N$ arbitrary constants of motion, which completely solves the problem.

Thus, in the case of integrable systems, the procedure for solving the dynamical problem can be summarized as follows. (i) Invert the relation $F_j(q, p) = f_j$ to obtain $p_i = p_i(q_j, f_k)$; here the F_js are the N given integrals of motion. (ii) Integrate $p_i(q_j, f_k)$ over dq_i to determine the function $S(Q, f)$ (see equation (2.36)). (iii) Use equation (2.37) to relate the new coordinates Q_j to the old coordinates q_i and N constants f_i. (iv) The new coordinates are linear functions of time. This allows one to determine the time evolution of the old coordinates $q_i(t)$.

2.6 Systems with many particles

(a) Consider two molecules which are moving towards each other. An electron of mass m, located at a distance of $D \cong 6000$ Mpc, will attract both these particles. The *relative* tidal acceleration will be of the order of $a \cong (Gml/D^3)$. This will render the point of impact of molecule-1 on molecule-2 inaccurate by an amount $\Delta s \cong at^2 \cong a(l/v)^2$. Hence, the angle of deflection will be uncertain by $\Delta \theta \cong (\Delta s/r) \cong (Gml^3/D^3 v^2 r)$. This, in turn, will cause the reflection angle after each succeeding collision to be uncertain by the factor $(l/r)\Delta\theta$, $(l/r)^2\Delta\theta$, $(l/r)^3\Delta\theta, \ldots$. The number of collisions N_{coll} after which the angle is uncertain by about one radian is given by

$$N_{coll} = \frac{\ln(1/\Delta\theta)}{\ln(l/r)} \cong \frac{\ln\left(D^3 v^2 r/Gml^3\right)}{\ln(l/r)}. \qquad (2.38)$$

If we take our system to be oxygen gas at room temperature and pressure, m to be the mass of the electron and $D \cong 6000$ Mpc, we find that $N_{coll} \cong 56$!

(b) Let $\rho(p, q, t)\, dp\, dq$ be the probability that the system is found in a small phase volume $d\Gamma \equiv dp\, dq$. When we look at a system after it has relaxed, physical quantities should become time-independent. Hence we expect the probability that the system is found in the phase volume $dp\, dq$ to also be time-independent, giving $\rho(p, q, t) = \rho(p, q)$. Then the mean value of any physical quantity $f(p, q)$ will be

$$\langle f \rangle \equiv \int f(p,q)\,\rho(p,q,t)\, dp\, dq = \int f(p,q)\,\rho(p,q)\, dp\, dq, \qquad (2.39)$$

which is clearly time-independent.

To determine the form of $\rho(p, q)$ we note that any probability density $\rho(p, q, t)$ satisfies a continuity equation:

$$\frac{\partial\rho}{\partial t} + \sum_{i=1}^{3N} \frac{\partial}{\partial q_i}(\rho\dot{q}_i) + \frac{\partial}{\partial p_i}(\rho\dot{p}_i) = 0. \qquad (2.40)$$

Using Hamilton's equations for $\dot{q}_i$ and $\dot{p}_i$, this equation can be written as

$$\frac{\partial \rho}{\partial t} + \dot{q}_i \frac{\partial \rho}{\partial q_i} + \dot{p}_i \frac{\partial \rho}{\partial p_i} = 0. \tag{2.41}$$

But the left-hand side of this equation is just the total time derivative $(d\rho/dt)$ of ρ. Hence,

$$\left(\frac{d\rho}{dt}\right) = 0. \tag{2.42}$$

A formal solution to (2.42) is provided by any function of the single-valued constants of motion. An equilibrium distribution function, however, cannot have any explicit time-dependence, that is, $(\partial \rho_{eq}/\partial t) = 0$. Therefore, ρ_{eq} must be a function of time-independent, single-valued constants of motion.

We argued earlier that if the system is confined by a box, linear and angular momenta are no longer conserved, leaving only the Hamiltonian function $\mathscr{H}(p, q)$ as a robust conserved quantity; any other single-valued constant of motion will be destroyed by small changes in the Hamiltonian. Therefore, we conclude that ρ_{eq} must be a function of $\mathscr{H}$ alone, that is, $\rho_{eq}(p, q) = \rho_{eq}(\mathscr{H}(p, q))$. Further, since we want to describe a system with fixed energy E, there should be no fluctuations in E. That is, $\langle \mathscr{H}^n \rangle = \langle \mathscr{H} \rangle^n$ for all n. This is possible only if

$$\rho_{eq} = C \delta_{\text{Dirac}}(E - \mathscr{H}(p, q)), \tag{2.43}$$

where δ_{Dirac} is the Dirac delta function and C is a normalization constant. Equation (2.43) shows that the system is equally likely to be found anywhere on its constant energy surface. From the normalization condition, it is clear that

$$C^{-1} = g(E) = \int \delta(E - \mathscr{H}(p, q)) \, dp \, dq. \tag{2.44}$$

To make the meaning of $g(E)$ more transparent we may write it as

$$g(E) = \frac{d}{dE} \int \theta(E - \mathscr{H}(p, q)) \, dp \, dq \equiv \frac{d\Gamma}{dE}, \tag{2.45}$$

where $\theta(z)$ is the Heaviside theta function and

$$\Gamma(E) = \int \theta(E - \mathscr{H}(p, q)) \, dp \, dq \tag{2.46}$$

is the volume of the region of the phase space where $\mathscr{H} < E$.

The basic problem of statistical mechanics is •o calculate $g(E)$. Once this is done, the average of any physical quantity $f(p, q)$ can be calculated as

$$\langle f \rangle = \frac{1}{g(E)} \int \delta(E - \mathscr{H}(p, q)) f(p, q) \, dp \, dq. \tag{2.47}$$

(c) For the purpose of computing $g(E)$, we may ignore the contribution to $\mathscr{H}$ from intermolecular forces; i.e. if we ignore $\mathscr{H}_{int}$. Then the Hamiltonian for a gas of N non-interacting particles (each of mass m) is

$$\mathscr{H}(p,q) = \sum_{i=1}^{3N} \frac{p_i^2}{2m}. \tag{2.48}$$

In this case,

$$\Gamma(E) = \int d^{3N}p\, d^{3N}q\; \theta\left(E - \sum_{i=1}^{3N} \frac{p_i^2}{2m}\right) = V^N \int d^{3N}p\; \theta\left(E - \sum_{i=1}^{3N} \frac{p_i^2}{2m}\right)$$

$$= V^N \mathscr{V}_{3N}\left(\sqrt{2mE}\right), \tag{2.49}$$

where $\mathscr{V}_M(R)$ is the volume of an M-dimensional sphere of radius R. It is obvious that $\mathscr{V}_M(R)$ must have the form $\mathscr{V}_M(R) = C_M R^M$. The factor C_M can be calculated by the following trick. Consider the integral

$$I = \int_{-\infty}^{\infty} \cdots \int_{-\infty}^{+\infty} dx_1 \ldots dx_M \exp\left(-(1/2)\left(x_1^2 + \ldots + x_M^2\right)\right) = (2\pi)^{M/2}, \tag{2.50}$$

which can also be written as

$$I = \int_0^\infty d\mathscr{V}_M \exp\left(-(1/2)R^2\right) = \int_0^\infty \left(M C_M R^{M-1} dR\right) \exp\left(-(1/2)R^2\right)$$

$$= M C_M \Gamma(M/2)\, 2^{(M/2)-1} \tag{2.51}$$

Comparing the two expressions, we obtain

$$C_M = \frac{\pi^{M/2}}{(M/2)\,\Gamma(M/2)}. \tag{2.52}$$

Hence, from (2.49),

$$\Gamma(E) = C_{3N}\,(2mE)^{3N/2}\, V^N, \quad g(E) = \frac{d\Gamma}{dE} \cong C_{3N}\left(\frac{3N}{2}\right)(2mE)^{3N/2}\, V^N, \tag{2.53}$$

where we have used the fact that, for realistic systems, $(3N/2) \gg 1$.

(d) The marginal probability distribution $\mathscr{P}_k$ for k particles to have momenta in a fixed range can be obtained by integrating out the remaining momenta:

$$\mathscr{P}_k(p_1,\ldots,p_{3k})\, dp_1 \ldots dp_{3k}\, dq_1 \ldots dq_{3k}$$

$$= g(E;N)^{-1} dp_1 \ldots dp_{3k}\, dq_1 \ldots dq_{3k} \int \delta\left(E - \epsilon - \sum_{i=3k+1}^{3N} \frac{p_i^2}{2m}\right) dp^{3N-3k}\, dq^{3N-3k}, \tag{2.54}$$

where $g(E;N)$ is the density of states for a system with N particles and energy E, and $\epsilon = \sum_{i=1}^{3k} p_i^2/2m$. The integral is just the density of states for a system with energy $(E - \epsilon)$ and $(N - k)$ particles. So, using (2.53), we obtain

$$\mathscr{P}_k(p_1,\ldots,p_{3k}) = \frac{g(E-\epsilon;N-k)}{g(E;N)} \propto (E-\epsilon)^{(3/2)(N-k)}, \tag{2.55}$$

where we have retained only the dependence on ϵ. The probability distribution for energy is

$$\mathscr{P}(\epsilon)\, d\epsilon \propto \left(1 - \frac{\epsilon}{E}\right)^{(3/2)N} g_k(\epsilon)\, d\epsilon \qquad \text{(for } N \gg k\text{)}, \qquad (2.56)$$

in which we have replaced the measure $\prod_{i=1}^{3k} dp_i\, dq_i$ by $g_k(\epsilon)\, d\epsilon$. Since $(\epsilon/E) \ll 1$ we can write this expression as

$$\mathscr{P}(\epsilon) \propto \exp\left(-(3N/2E)\epsilon\right) g_k(\epsilon). \qquad (2.57)$$

The factor $(3N/2E)$ in the exponent is purely a property of the full system of N particles. We shall write it as

$$\frac{3N}{2E} \equiv \beta \equiv \frac{\partial}{\partial E} \ln g_N(E). \qquad (2.58)$$

The expression for $\mathscr{P}(\epsilon)$ has the following interpretation. The probability that the subsystem is in some state with energy ϵ is proportional to $\exp(-\beta\epsilon)$, where β is some parameter characterizing the full system. The number of states in the range $(\epsilon, \epsilon + d\epsilon)$ is $g(\epsilon)\, d\epsilon$. Hence, the total probability for the subsystem to have an energy in the range $(\epsilon, \epsilon + d\epsilon)$ is

$$\left\{\begin{array}{l}\text{probability to have}\\[2pt]\text{an energy in the range}\\[2pt](\epsilon, \epsilon + d\epsilon) = \mathscr{P}(\epsilon)\, d\epsilon\end{array}\right\} = \left\{\begin{array}{l}\text{probability to be}\\[2pt]\text{in a state with}\\[2pt]\text{energy } \epsilon = e^{-\beta\epsilon}\end{array}\right\} \times \left\{\begin{array}{l}\text{number of states}\\[2pt]\text{in the range}\\[2pt](\epsilon, \epsilon + d\epsilon) = g(\epsilon)\, d\epsilon\end{array}\right\}.$$
$$(2.59)$$

Such a distribution is called 'canonical'.

The proportionality constant in (2.57) is easily fixed by normalizing $\mathscr{P}(\epsilon)$. We obtain, for the normalized probability distribution for k particles,

$$\mathscr{P}(\epsilon) = \frac{1}{Z(\beta)} e^{-\beta\epsilon} g_k(\epsilon), \qquad (2.60)$$

where

$$Z_k(\beta) = \int_0^\infty e^{-\beta\epsilon} g_k(\epsilon)\, d\epsilon = \int dp\, dq\, e^{-\beta H(p,q)}. \qquad (2.61)$$

(e) Suppose that we have two systems (1 and 2) in contact so that they can exchange energy. The full system $(1+2)$ is isolated and has a fixed energy E. The probability that system 1 has energy between E_1 and $E_1 + dE_1$ is

$$\mathscr{P}_1(E_1)\, dE_1 \propto g_1(E_1) g_2(E - E_1)\, dE_1. \qquad (2.62)$$

For realistic systems with large numbers of particles, $g(E)$ is a very sharply rising function of E. (It varies as E^N with $N \cong 10^{23}$.) Hence, the product $g_1(E_1) g_2(E - E_1)$ will have a very sharp maximum. In other words, the subsystem is most likely to be found in a state which maximizes this probability. This state (which is in the state of equilibrium) can be determined by maximizing $\mathscr{P}_1(E_1)$

or – equivalently – maximizing $\ln \mathscr{P}_1(E_1)$. Then we obtain the condition for equilibrium to be

$$\frac{\partial}{\partial E_1} \ln g_1(E_1) = \frac{\partial}{\partial E_2} \ln g_2(E_2)\Big|_{E_2=E-E_1}. \tag{2.63}$$

Or

$$\beta_1(E_1) = \beta_2(E - E_1), \tag{2.64}$$

showing that the two systems must have the same value for β.

The quantity $S(E) \equiv \ln g(E)$ is called the entropy and $\beta^{-1} = (\partial S/\partial E)^{-1}$ is called the temperature. For a subsystem, just as we regarded β formally as a function of ϵ, so can the entropy be regarded as a function of ϵ.

For two systems (1 and 2) in thermal contact the combined probability distribution is proportional to $g_1(E_1)g_2(E - E_1)$. Taking logarithms, the total entropy is $S_1(E_1) + S_2(E - E_1)$, that is, the entropy is additive. Maximizing the product $g_1 g_2$ is equivalent to maximizing the total entropy $S = S_1 + S_2$.

(f) We have seen that the probability of a system A, interacting with a much larger system B, having an energy ϵ is given by

$$\mathscr{P}(\epsilon)\,d\epsilon = \frac{d\epsilon}{Z} g(\epsilon)e^{-\beta\epsilon} = \frac{dp\,dq}{Z} e^{-\beta H(p,q)}, \tag{2.65}$$

where

$$Z = \int_0^\infty d\epsilon\, g(\epsilon)e^{-\beta\epsilon} = \int dp\,dq\, e^{-\beta H(p,q)}. \tag{2.66}$$

Since $g(\epsilon)$ is a rapidly increasing function of ϵ and $e^{-\beta\epsilon}$ is a rapidly decreasing function, their product will have a sharp peak at some $\epsilon = \bar{\epsilon}$. Consider the Taylor expansion of $F(\epsilon) = [\ln g(\epsilon) - \beta\epsilon]$ around $\bar{\epsilon}$:

$$\ln g - \beta\epsilon \cong \ln g(\bar{\epsilon}) - \beta\bar{\epsilon} + \frac{1}{2}\left[\frac{\partial^2}{\partial\epsilon^2}\ln g(\epsilon)\right]_{\bar{\epsilon}}(\epsilon - \bar{\epsilon})^2. \tag{2.67}$$

(The term linear in $(\epsilon - \bar{\epsilon})$ vanishes since $\epsilon = \bar{\epsilon}$ is also a maximum for $F(\epsilon)$.) Writing $(\partial^2 \ln g/\partial\epsilon^2) = (\partial\beta/\partial\epsilon) = -(\beta^2/C_v)$, where $C_v \equiv (\partial\epsilon/\partial T)$, we have the result

$$\mathscr{P}(\epsilon) = \frac{1}{Z}\exp[\ln g - \beta\epsilon] \cong Z^{-1}\exp\left[\ln g(\bar{\epsilon}) - \beta\bar{\epsilon} - \frac{\beta^2}{2C_v}(\epsilon - \bar{\epsilon})^2\right]$$

$$= P(\bar{\epsilon})\exp\left[-\frac{\beta^2}{2C_v}(\epsilon - \bar{\epsilon})^2\right], \tag{2.68}$$

where $\bar{\epsilon}$ is determined by the condition

$$\frac{\partial}{\partial\epsilon}\ln g(\epsilon)|_{\epsilon=\bar{\epsilon}} = \beta. \tag{2.69}$$

From the Gaussian nature of $\mathscr{P}(\epsilon)$ in (2.68) we see that the fluctuations in ϵ are given by $(\Delta\epsilon)^2 = C_v T^2$. This result shows that $C_v > 0$ for any system described by a canonical distribution. Since $C_v \propto N$, we have $(\Delta \epsilon)^2 / \bar{\epsilon}^2 \propto (N/N^2) \propto N^{-1}$, which is small for large N.

Since $g e^{-\beta \epsilon}$ is a sharply peaked function, we can approximate the integral for partition function Z in (2.66) as

$$Z = \int_0^\infty g(\epsilon) e^{-\beta \epsilon} \, d\epsilon \cong (T^2 C_v)^{1/2} g(\bar{\epsilon}) e^{-\beta \bar{\epsilon}} . \tag{2.70}$$

It follows that

$$\ln Z \cong \ln g(\bar{\epsilon}) - \beta \bar{\epsilon} + \frac{1}{2} \ln (T^2 C_v). \tag{2.71}$$

For normal systems $\bar{\epsilon} \propto N$, $\ln g \propto N$ and $C_v \propto N$. So the first two terms vary as N while the last term scales as $\ln N$, giving

$$\ln Z \cong \ln g(\bar{\epsilon}) - \beta \bar{\epsilon} + \mathcal{O} \left(\frac{\ln N}{N} \right). \tag{2.72}$$

The *most probable* value of ϵ is determined by (2.69). The *mean* value is decided by the relation

$$\langle \epsilon \rangle = \int\limits_0^\infty \frac{d\epsilon}{Z} g(\epsilon) \epsilon \, e^{-\beta \epsilon} = \frac{\partial \ln Z}{\partial (-\beta)} . \tag{2.73}$$

Differentiating expression (2.73) with respect to β and using $(\partial \ln g / \partial \epsilon) = \beta$, we obtain

$$-\langle \epsilon \rangle = \frac{\partial \ln Z}{\partial \beta} = \beta \frac{\partial \bar{\epsilon}}{\partial \beta} - \beta \frac{\partial \bar{\epsilon}}{\partial \beta} - \bar{\epsilon} = -\bar{\epsilon}. \tag{2.74}$$

Thus, to the accuracy of $(\ln N / N)$, the *mean* energy $\langle \epsilon \rangle$ is the same as the *most probable* energy $\bar{\epsilon}$.

(g) Consider a special class of variables, which can be expressed in the form

$$F_i = \frac{\partial H(p, q; \lambda_i)}{\partial \lambda_i} , \tag{2.75}$$

where the λ_is are some parameters which appear in the Hamiltonian. The mean value of any such quantity can easily be computed:

$$\langle F_i \rangle = \int dp \, dq \left(\frac{\partial H}{\partial \lambda_i} \right) \frac{e^{-\beta H}}{Z} = \frac{1}{\beta} \frac{\partial \ln Z}{\partial (-\lambda_i)} . \tag{2.76}$$

The pressure exerted by a gas can be expressed as the derivative of H with respect to the volume V occupied by the system:

$$P = -\frac{\partial H}{\partial V} . \tag{2.77}$$

To see this, note that the force exerted on the walls of a container, parametrized by coordinates $\mathbf{r}$, is

$$\mathbf{F} = -\frac{\partial H}{\partial \mathbf{r}} = -\frac{\partial H}{\partial V} \frac{\partial V}{\partial \mathbf{r}} = -\frac{\partial H}{\partial V} \cdot d\mathbf{s} = P \, d\mathbf{s}. \tag{2.78}$$

Here $d\mathbf{s} = (\partial V/\partial \mathbf{r})$ is a small element of area. Hence, the mean pressure exerted by the system will be

$$\langle P \rangle = \frac{1}{\beta} \frac{\partial \ln Z}{\partial V}. \tag{2.79}$$

2.7 Thermodynamics

From the expressions for $\bar{E} = \langle \epsilon \rangle$ and $\bar{P}$ (see (2.73), (2.79)), we have

$$d(\ln Z) = -\bar{E}\, d\beta + \beta \bar{P}\, dV = -d(\bar{E}\beta) + \beta\, d\bar{E} + \beta \bar{P}\, dV, \tag{2.80}$$

giving

$$d(\ln Z + \bar{E}\beta) \cong d(\ln g(\bar{E})) = \beta(d\bar{E} + \bar{P}\, dV), \tag{2.81}$$

which is accurate to $\mathcal{O}(\ln N)$. Writing this as

$$(d\bar{E} + \bar{P}\, dV) = \beta^{-1} d(\ln g), \tag{2.82}$$

we arrive at the central result of thermodynamics, namely, that the quantity $(d\bar{E} + \bar{P}\, dV)$ can be expressed as $\mathcal{T}\, d\mathcal{S}$, where $d\mathcal{S}$ is an exact differential. From the identification

$$\beta^{-1} d(\ln g) = T\, d(\ln g) = \mathcal{T}\, d\mathcal{S} \tag{2.83}$$

we obtain

$$\mathcal{T} = T = \left(\frac{\partial \ln g}{\partial E} \right)_V^{-1}, \quad d\mathcal{S} = d(\ln g) = \frac{dg}{g}, \tag{2.84}$$

which connects statistical mechanics and thermodynamics. The second relation gives

$$\mathcal{S} = \ln g(E) + f(N), \tag{2.85}$$

where $f(N)$ is some function independent of E and V and is possibly dependent on the number of particles N. These identifications, of course, allow a trivial change of units between (say) T and $\mathcal{T}$. We could have put $T = k\mathcal{T}$ and set $\mathcal{S} = k \ln g$. This scale change is of no fundamental importance and we will set $k = 1$. This means that we measure thermodynamic temperature in energy units.

Let us now see what these identifications imply for an ideal gas. We found earlier that, for an ideal gas (see equations (2.52) and (2.53)),

$$g(E) = \frac{3N}{2} \frac{\pi^{\frac{3N}{2}}}{\left(\frac{3N}{2}\right)\left(\frac{3N}{2}\right)!}(2m) \cdot V^N \cdot (2mE)^{\frac{3N}{2}}. \tag{2.86}$$

Taking logarithms and using $\ln N! \cong N \ln N - N$, we find

$$S = \ln g + f(N) = N \ln V + \frac{3N}{2} \ln \left(\frac{2E}{3N} \right) + q(N), \tag{2.87}$$

where $q(N)$ is an unknown function depending on N but independent of E and V. So we obtain

$$P = \frac{1}{\beta}\left(\frac{\partial S}{\partial V}\right)_{E,N} = \frac{N}{\beta V} = \frac{NT}{V}, \quad \frac{1}{T} = \left(\frac{\partial S}{\partial E}\right)_{V,N} = \frac{3N}{2E}, \qquad (2.88)$$

which are the usual ideal gas equations: $\overline{E} = \frac{3}{2}NT$ and $PV = NT$.

The expression (2.87) contains an arbitrary function $q(N)$. It is possible to determine this function by noting that the entropy S should be 'extensive'. In other words, we demand that $S(\lambda N, \lambda V, \lambda E) = \lambda S(N, V, E)$. Using this criterion, it is easy to see that $q(N)$ must satisfy the relation

$$q(N) - \frac{1}{\lambda}q(\lambda N) = N \ln \lambda. \qquad (2.89)$$

Differentiating this expression with respect to N and denoting the derivative (dq/dN) by the function $p(N)$ we find that

$$p(\lambda N) - p(N) = -\ln \lambda. \qquad (2.90)$$

Differentiating this further with respect to λ we obtain $p'(\lambda N) = -(\lambda N)^{-1}$; integrating we obtain $p(N) = -\ln N$ and $q(N) = -N \ln N + N \cong -\ln N!$. This suggests that the definition for entropy may be taken to be

$$S = \ln \left[g(E)/N! \right]. \qquad (2.91)$$

Or, equivalently, we may redefine the density of states by dividing it by $N!$.

2.8 Systems without extensivity of energy

(a) We are interested in the volume $g(E)$ of the constant energy surface $H = E$, which can be computed as

$$g(E) = \frac{d\Gamma(E)}{dE}, \qquad (2.92)$$

where

$$\Gamma(E) = \int d^3\mathbf{P}\, d^3\mathbf{p}\, d^3\mathbf{Q}\, d^3\mathbf{r}\Theta(E - H(\mathbf{P},\mathbf{Q},\mathbf{p},\mathbf{r})) \qquad (2.93)$$

is the volume of the region with $H < E$. Performing the momentum integrations and the integration over $\mathbf{Q}$ we obtain

$$\Gamma(E) = \frac{AR^3}{3} \int_a^{r_{\max}} r^2\, dr \left[E + \frac{Gm^2}{r} \right]^3 \qquad (2.94)$$

where $A = (64\pi^5 m^3/3)$. Differentiating with respect to E:

$$g(E) = AR^3 \int_a^{r_{\max}} r^2\, dr \left[E + \frac{Gm^2}{r} \right]^2. \qquad (2.95)$$

It is understood that the range of integration in (2.95) should be limited to the region in which the expression in the square brackets is positive. So we should use $r_{\max} = (Gm^2/|E|)$ if $(-Gm^2/a) < E < (-Gm^2/R)$, and use $r_{\max} = R$ if $(-Gm^2/R) < E < +\infty$. Since $H \geq (-Gm^2/a)$, we trivially have $g(E) = 0$ for $E < (-Gm^2/a)$. The constant A is unimportant for our discussions and hence will be omitted from the formulas hereafter.

The integration in (2.95) is straightforward and we obtain the following result:

$$
\frac{g(E)}{(Gm^2)^3} =
\begin{cases}
\dfrac{R^3}{3}(-E)^{-1}\left(1 + \dfrac{aE}{Gm^2}\right)^3, & (-Gm^2/a) < E < (-Gm^2/R), \\[4mm]
\dfrac{R^3}{3}(-E)^{-1}\left[\left(1 + \dfrac{RE}{Gm^2}\right)^3 - \left(1 + \dfrac{aE}{Gm^2}\right)^3\right], & (-Gm^2/R) < E < \infty.
\end{cases}
$$

$$(2.96)$$

This function $g(E)$ is continuous and smooth at $E = (-Gm^2/R)$. We define the entropy $S(E)$ and the temperature $T(E)$ of the system by the relations

$$S(E) = \ln g(E), \quad T^{-1}(E) = \beta(E) = \frac{\partial S(E)}{\partial E}. \tag{2.97}$$

All the interesting thermodynamic properties of the system can be understood from the $T(E)$ curve.

Consider first the case of very low energies, that is, $(-Gm^2/a) < E < (-Gm^2/R)$. Using (2.96) and (2.97) one can easily obtain $T(E)$ and write it in the dimensionless form as

$$t(\epsilon) = \left[\frac{3}{1+\epsilon} - \frac{1}{\epsilon}\right]^{-1}, \tag{2.98}$$

where we have defined $t = (aT/Gm^2)$ and $\epsilon = (aE/Gm^2)$.

This function exhibits the peculiarities characteristic of gravitating systems. At the lowest energy admissible for our system, which corresponds to $\epsilon = -1$, the temperature t vanishes. This describes a tightly bound low temperature phase of the system with negligible random motion. The $t(\epsilon)$ is clearly dominated by the first term of (2.98) for $\epsilon \cong -1$. As we increase the energy of the system, the temperature *increases*, which is the normal behaviour for a system. This trend continues up to

$$\epsilon = \epsilon_1 = -\frac{1}{2}(\sqrt{3} - 1) \cong -0.36, \tag{2.99}$$

at which point the $t(\epsilon)$ curve reaches a maximum and turns around. As we increase the energy further the temperature *decreases*. The system *exhibits negative specific heat in this range*.

Equation (2.98) is valid from the minimum energy $(-Gm^2/a)$ all the way up to the energy $(-Gm^2/R)$. For realistic systems, $R \gg a$ and hence this range is quite wide. For a small region in this range (from $(-Gm^2/a)$ to $(-0.36Gm^2/a)$), we have positive specific heat; for the rest of the region the specific heat is negative. *The positive specific heat region owes its existence to the non-zero short distance cut-off.* If we set $a = 0$, the first term in (2.98) will vanish; we will have $t \propto (-\epsilon^{-1})$ and negative specific heat in this entire domain.

For $E \geq (-Gm^2/R)$, we have to use the second expression in (2.96) for $g(E)$. In this case, we obtain

$$t(\epsilon) = \left[\frac{3 \left[(1+\epsilon)^2 - \frac{R}{a}(1 + \frac{R}{a}\epsilon)^2 \right]}{(1+\epsilon)^3 - (1 + \frac{R}{a}\epsilon)^3} - \frac{1}{\epsilon} \right]^{-1}. \qquad (2.100)$$

This function, of course, matches smoothly with (2.98) at $\epsilon = -(a/R)$. As we increase the energy, the temperature continues to decrease for a little while, exhibiting negative specific heat. However, this behaviour is soon halted at some $\epsilon = \epsilon_2$, say. The $t(\epsilon)$ curve reaches a minimum at this point, turns around and starts increasing with increasing ϵ. We thus enter another (high temperature) phase with positive specific heat. From (2.100) it is clear that $t \cong (1/2)\epsilon$ for large ϵ. (Since $E = (3/2)NkT$ for an ideal gas, we might have expected to find $t \cong (1/3)\epsilon$ for our system with $N = 2$ at high temperatures. This is indeed what we would have found if we had defined our entropy as $\ln \Gamma$. With our definition, the energy of the ideal gas is actually $E = ((3/2)N - 1)kT$; hence we obtain $t = (1/2)\epsilon$ when $N = 2$. Note that this is an effect arising from the small number of degrees of freedom and is not of any fundamental significance.)

The form of the $t(\epsilon)$ is shown in figure 2.3. The specific heat is positive along the portions AB and CD and is negative along BC.

The overall picture is now clear. Our system has two natural energy scales: $E_1 = (-Gm^2/a)$ and $E_2 = (-Gm^2/R)$. For $E \gg E_2$, gravity is not strong enough to keep $r < R$ and the system behaves like a gas confined by the container; we have a high temperature phase with positive specific heat. As we lower the energy to $E \cong E_2$, the effects of gravity begin to be felt. For $E_1 < E < E_2$, the system is unaffected by either the box or the short distance cut-off; this is the domain dominated entirely by gravity and we have negative specific heat. As we go to $E \cong E_1$, the hard core nature of the particles begins to be felt and the gravity is again resisted. This gives rise to a low temperature phase with positive specific heat.

We can also consider the effect of increasing R, keeping a and E fixed. (This is more in consonance with the spirit of microcanonical distribution, which describes a system with constant E.) Since we imagine the particles to be hard spheres of radius $(a/2)$, we should only consider $R > 2a$. It is amusing to note that, if $2 < (R/a) < (\sqrt{3} + 1)$, there is no region of negative specific heat. As we increase R, this negative specific heat region appears. It is easy to see that increasing R effectively increases the range over which the specific heat is negative. Suppose a system is originally prepared with some E- and R-values such that the specific heat is positive. If we now increase R, it can happen that the system finds itself in a region of negative specific heat. *This suggests the possibility that an instability*

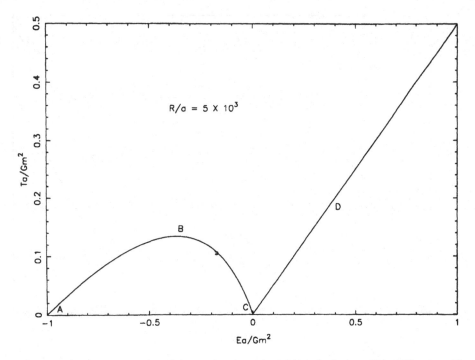

Fig. 2.3.   The behaviour of $T(E)$ in the microcanonical ensemble. The specific heat is positive along AB and CD but negative along BC.

may be triggered in a constant energy system if its radius increases beyond a critical value (see problem 2.10).

In the above discussion we have treated a as a fixed quantity and rescaled all variables using (Gm^2/a). To study the effect of a short distance cut-off we would like to vary a keeping R fixed. To do this, it is better to rescale variables using (Gm^2/R). This can easily be done and figure 2.4 shows the behaviour of the T–E curve as the lower cut-off a is changed. As we decrease the value of a the negative specific heat region becomes more and more pronounced. In fact, if a is zero, we have negative specific heat for all $E < (-Gm^2/R)$. We also see from figure 2.4 that the behaviour of the curve for large E is reasonably independent of the value of the cut-off a.

(b) In the partition function

$$Z(\beta) = \int d^3P \, d^3p \, d^3Q \, d^3r \exp(-\beta H), \qquad (2.101)$$

the integrations over P, p and Q can be performed trivially. Omitting an overall constant which is unimportant, we can write the answer as

$$Z(\beta) = R^3 \beta^{-3} \int_a^R dr \, r^2 \exp\left(\frac{\beta Gm^2}{r}\right), \qquad (2.102)$$

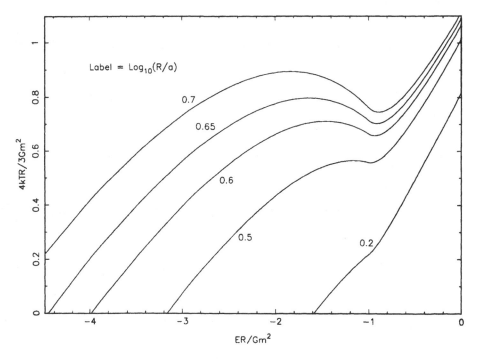

Fig. 2.4. The behaviour of $T(E)$ when the cut-off radius a is decreased. Note that the region of negative specific heat increases as R/a is increased.

which, in the dimensionless form, becomes

$$Z(t) = t^3 \left(\frac{R}{a}\right)^3 \int_1^{R/a} dx\, x^2 \exp\left(\frac{1}{xt}\right), \qquad (2.103)$$

where t is the dimensionless temperature defined in (2.98). Though this integral cannot be evaluated in closed form, all the limiting properties of $Z(\beta)$ can easily be obtained from (2.103). Note that the short distance cut-off is vital here for convergence of the integral, while we could have dispensed with it in the microcanonical description.

The integrand in (2.103) is large for both large and small x and reaches a minimum for $x = x_{\min} = (1/2t)$. The behaviour of the integral depends crucially as to whether this minimum falls within the limits of integration or not. At high temperatures, $x_{\min} < 1$ and hence the minimum falls outside the domain of integration. The exponential contributes very little to the integral and we can approximate Z adequately by

$$Z \cong t^3 \int_1^{R/a} dx\, x^2 \left[1 + \frac{2x_{\min}}{x}\right] = \frac{t^3}{3}\left(\frac{R}{a}\right)^6 \left(1 + \frac{3a}{2Rt}\right). \qquad (2.104)$$

On the other hand, if $x_{\min} > 1$ the minimum lies between the limits of the integration. Then the exponential part of the curve dominates the integral. We can easily evaluate this contribution by a saddle-point approach, and obtain

$$Z \cong \left(\frac{R}{a}\right)^3 t^4 (1-2t)^{-1} \exp\left(\frac{1}{t}\right) [1 - \exp -(R/ta)] \cong \left(\frac{R}{a}\right)^3 t^4 (1-2t)^{-1} \exp\left(\frac{1}{t}\right).$$

(2.105)

As we lower the temperature, making x_{min} cross 1 from below, the contribution switches over from (2.104) to (2.105). The transition is exponentially sharp. The critical temperature at which the transition occurs can be estimated by finding the temperature at which the two contributions are equal. This occurs at

$$t_{crit} = \frac{1}{3} \frac{1}{\ln(R/a)}.$$

(2.106)

For $t < t_{crit}$, we should use (2.105) and for $t > t_{crit}$ we should use (2.104).

Given $Z(\beta)$ all thermodynamic functions can be computed. In particular, the mean energy of the system is given by

$$E(\beta) = -\frac{\partial \ln Z}{\partial \beta}.$$

(2.107)

This relation can be inverted to give the $T(E)$, which can be compared with the $T(E)$ obtained earlier using the microcanonical distribution. (In making such a comparison we are identifying the constant energy at which the microcanonical distribution is defined, with the mean energy of the canonical distribution.) From (2.104) and (2.105) we obtain

$$\epsilon(t) = \frac{aE}{Gm^2} = 4t - 1$$

(2.108)

for $t < t_{crit}$ and

$$\epsilon(t) = 3t - \frac{3a}{2R}$$

(2.109)

for $t > t_{crit}$. Near $t \cong t_{crit}$, there is a rapid variation of the energy and we cannot use either asymptotic form. The system undergoes a phase transition at $t = t_{crit}$, absorbing a large amount of energy:

$$\Delta\epsilon \cong \left(1 - \frac{1}{3\ln(R/a)}\right).$$

(2.110)

The specific heat is, of course, positive throughout the range. This is to be expected because the canonical description cannot lead to negative specific heats.

The exact T–E curves obtained from the canonical and microcanonical distributions are shown in figure 2.5. (For convenience, we have rescaled the T–E curve of the microcanonical distribution so that $\epsilon \cong 3t$ asymptotically.) Note that the descriptions match very well in the regions of positive specific heat. The negative specific heat region of the microcanonical distribution is replaced by a phase transition (rapid change in E at almost constant T) in the canonical description.

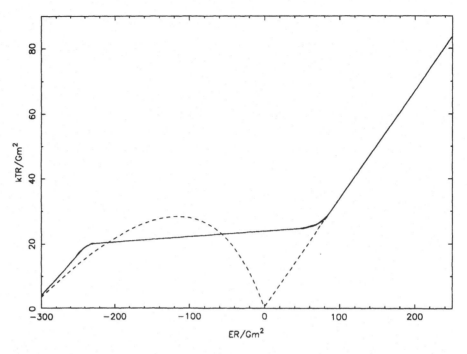

Fig. 2.5. Comparison of $T(E)$ curves in the canonical (unbroken line) and microcanonical (dashed line) ensembles. The negative specific heat region in the microcanonical ensemble is avoided by the occurrence of a phase transition in the canonical description.

The physics of canonical description is best understood by studying E as a function of T. As we increase the temperature from zero, the energy increases from the ground state value $(-Gm^2/a)$ in accordance with the relation (2.108). As the temperature approaches T_{crit} and crosses it, a phase transition occurs in the system and the energy increases rapidly. The latent heat in the system is large enough to push the system into the high temperature phase. At still higher temperatures, the energy increases steadily with the temperature in accordance with (2.109).

We can now compare the canonical and microcanonical descriptions of our system. At both very low and very high temperatures, the descriptions match. The crucial difference occurs at the intermediate energies and temperatures. Microcanonical description predicts negative specific heat and a reasonably slow variation of energy with temperature. Canonical description, on the other hand, predicts a phase transition with rapid variation of energy with temperature. Such phase transitions are accompanied by large fluctuations in the energy, which is the main reason for the disagreement between the two descriptions.

The two descriptions, therefore, disagree for a wide range of energies. This range of energies is of great practical importance since most astrophysical systems fall in this band.

2.9 Mean field equilibrium of gravitating systems

(a) Consider a system of N particles interacting with each other through the two-body potential $U(\mathbf{x}_i, \mathbf{x}_j)$. The entropy S of this system, in the microcanonical description, is defined through the relation

$$
e^S = g(E) = \frac{1}{N!} \int d^{3N}x\, d^{3N}p\, \delta(E - H) = \frac{A}{N!} \int d^{3N}x \left[E - \frac{1}{2} \sum_{i \neq j} U(\mathbf{x}_i, \mathbf{x}_j) \right]^{\frac{3N}{2}},
$$
(2.111)

wherein we have performed the momentum integrations and replaced $(3N/2 - 1)$ by $(3N/2)$. We shall approximate the expression in (2.111) in the following manner.

Let the spatial volume V be divided into M (with $M \ll N$) cells of equal size, large enough to contain many particles but small enough for the potential to be treated as a constant inside each cell. (It is assumed that such an intermediate scale exists.) Instead of integrating over the particle coordinates $(\mathbf{x}_1, \mathbf{x}_2, ..., \mathbf{x}_N)$ we shall sum over the number of particles n_a in the cell centred at $\mathbf{x}_a$ (where $a = 1, 2, ..., M$). Using the standard result that the integration over $(N!)^{-1} d^{3N}x$ can be replaced by

$$
\sum_{n_1=1}^{\infty} \left(\frac{1}{n_1!} \right) \sum_{n_2=1}^{\infty} \left(\frac{1}{n_2!} \right) \cdots \sum_{n_M=1}^{\infty} \left(\frac{1}{n_M!} \right) \delta\left(N - \sum_a n_a \right) \left(\frac{V}{M} \right)^N
$$
(2.112)

and ignoring the unimportant constant A, we can rewrite (2.111) as

$$
e^S = \sum_{n_1=1}^{\infty} \left(\frac{1}{n_1!} \right) \sum_{n_2=1}^{\infty} \left(\frac{1}{n_2!} \right) \cdots \sum_{n_M=1}^{\infty} \left(\frac{1}{n_M!} \right) \delta\left(N - \sum_a n_a \right) \left(\frac{V}{M} \right)^N
$$
$$
\times \left[E - \frac{1}{2} \sum_{a \neq b}^{M} n_a U_{ab} n_b \right]^{\frac{3N}{2}}
$$
$$
\cong \sum_{n_1=1}^{\infty} \sum_{n_2=1}^{\infty} \cdots \sum_{n_M=1}^{\infty} \delta\left(N - \sum_a n_a \right) \exp S[\{n_a\}],
$$
(2.113)

where

$$
S[\{n_a\}] = \frac{3N}{2} \ln \left[E - \frac{1}{2} \sum_{a \neq b}^{M} n_a U(\mathbf{x}_a, \mathbf{x}_b) n_b \right] - \sum_{a=1}^{M} n_a \ln \left(\frac{n_a M}{eV} \right).
$$
(2.114)

In arriving at the last expression we have used Sterling's approximation for the factorials. *The mean field limit is now obtained by retaining in the sum in* (2.113) only the term for which the summand reaches the maximum value, subject to the constraint on the total number. That is, we claim

$$
\sum_{\{n_a\}} e^{S[n_a]} \cong e^{S[n_{a,\max}]},
$$
(2.115)

where $n_{a,\text{max}}$ is the solution to the variational problem

$$\left(\frac{\delta S}{\delta n_a}\right)_{n_a = n_{a,\text{max}}} = 0 \quad \text{with} \quad \sum_{a=1}^{M} n_a = N. \tag{2.116}$$

Imposing this constraint by a Lagrange multiplier and using the expression (2.114) for S, we obtain the equation satisfied by $n_{a,\text{max}}$:

$$\frac{1}{T} \sum_{b=1}^{M} U(\mathbf{x}_a, \mathbf{x}_b) n_{b,\text{max}} + \ln\left(\frac{n_{a,\text{max}} M}{V}\right) = \text{constant}, \tag{2.117}$$

where we have defined the temperature T as

$$\frac{1}{T} = \frac{3N}{2} \left(E - \frac{1}{2} \sum_{a \neq b}^{M} n_a U(\mathbf{x}_a, \mathbf{x}_b) n_b\right)^{-1} = \beta. \tag{2.118}$$

We see from (2.114) that this expression is also equal to $(\partial S/\partial E)$; therefore T is indeed the correct thermodynamic temperature. We can now return to the continuum limit by the replacements

$$n_{a,\text{max}} \frac{M}{V} = \rho(\mathbf{x}_a), \qquad \sum_{a=1}^{M} \to \frac{M}{V} \int. \tag{2.119}$$

In this limit, the extremum solution (2.117) is given by

$$\rho(\mathbf{x}) = A \exp(-\beta\phi(\mathbf{x})), \quad \text{where} \quad \phi(\mathbf{x}) = \int d^3\mathbf{y}\, U(\mathbf{x}, \mathbf{y})\, \rho(\mathbf{y}), \tag{2.120}$$

which, in the case of gravitational interactions, becomes

$$\rho(\mathbf{x}) = A \exp(-\beta\phi(\mathbf{x})), \quad \phi(\mathbf{x}) = -G \int \frac{\rho(\mathbf{y})\, d^3\mathbf{y}}{|\mathbf{x} - \mathbf{y}|}. \tag{2.121}$$

Equation (2.121) represents the equilibrium configuration for a gravitating system in the mean field limit. The constant β is already determined through (2.118) in terms of the total energy of the system. The constant A has to be fixed in terms of the total number (or mass) of the particles in the system.

An important point needs to be noted about the mean field result we have obtained. The various manipulations in (2.111)–(2.117) tacitly assume that the expressions we are dealing with are finite. Unfortunately, for gravitational inter-actions *without* a short distance cut-off, the quantity e^S – and hence all the terms we have been handling – is divergent. We should, therefore, remember that a short distance cut-off is needed to justify the entire procedure and that (2.121) – which is based on a strict r^{-1} potential and does not incorporate any such cut-off – can only be approximately correct. We shall continue to work with (2.121) because of its mathematical convenience.

(b) Consider first the class of all $f(\mathbf{x}, \mathbf{v})$ with a fixed E, M and $\rho(\mathbf{x})$. Fixing $\rho(\mathbf{x})$ fixes M, $\phi(\mathbf{x})$ and U. Since E is also fixed $K = E - U$ is fixed. Thus extremizing

S subject to constant $\rho(\mathbf{x})$ and E is the same as extremizing S with a given $\rho(\mathbf{x})$ and K. Introducing two Lagrange multipliers $\lambda(\mathbf{x})$ and β, we have the variation

$$\delta S = - \int d^3\mathbf{x}\, d^3\mathbf{v} \left\{ \delta(f \ln f) + \lambda(\mathbf{x})\, \delta f + \frac{1}{2}\beta v^2\, \delta f \right\}$$

$$= - \int d^3\mathbf{x}\, d^3\mathbf{v} \left\{ (\ln ef)\, \delta f + \frac{(\delta f)^2}{2f} + \lambda\, \delta f + \frac{1}{2}\beta v^2\, \delta f \right\}. \qquad (2.122)$$

Equating the coefficient of f to zero, we obtain

$$f = \exp\left[-(\lambda(\mathbf{x}) + 1) - \frac{1}{2}\beta v^2 \right]. \qquad (2.123)$$

The multipliers $\lambda(\mathbf{x})$ and β have to be fixed in terms of $\rho(\mathbf{x})$ and K. Doing this, we can write f as

$$f(\mathbf{x}, \mathbf{v}) = (2\pi T)^{-3/2} \rho(\mathbf{x}) \exp\left(-\frac{v^2}{2T} \right), \qquad (2.124)$$

with $T = (2/3M)\, K$. From the coefficient of $(\delta f)^2$, we see that this is a true maximum for S. The maximum value of S can be obtained by substituting (2.124) into the expression for entropy. We find

$$S_{\max}(\rho(\mathbf{x}); K) = \frac{3M}{2} (\ln T) - \int d^3\mathbf{x}\, \rho \ln \rho + (\text{constant}). \qquad (2.125)$$

Equation (2.125) gives the maximum value of entropy possible for given $\rho(\mathbf{x})$ and K, or, equivalently, for a given $\rho(\mathbf{x})$ and E. This maximum is achieved when the velocity distribution is Maxwellian.

We next vary K and $\rho(\mathbf{x})$ keeping E and M fixed, so as to find the $\rho(\mathbf{x})$ for which (2.125) is extremum. We shall do this, retaining terms up to the second order. From the constraint

$$0 = \delta E = \delta K + \frac{1}{2} \int d^3\mathbf{x}\, (\delta\rho\, \phi + \rho\, \delta\phi + \delta\rho\, \delta\phi)$$

$$= \frac{3M}{2}\delta T + \int d^3\mathbf{x} \left(\phi\, \delta\rho + \frac{1}{2}\delta\rho\, \delta\phi \right), \qquad (2.126)$$

we find that

$$\delta T = -\frac{2}{3M} \int d^3\mathbf{x} \left(\phi\, \delta\rho + \frac{1}{2}\delta\rho\, \delta\phi \right). \qquad (2.127)$$

In arriving at (2.127) we have used the fact that $\int \phi\, \delta\rho\, d^3\mathbf{x} = \int \rho\, \delta\phi\, d^3\mathbf{x}$. From (2.125) we have

$$\delta S = \frac{3M}{2T}\delta T - \int d^3\mathbf{x}\, \delta\rho\, \{\ln \rho e\} - \frac{3M}{4T^2}(\delta T)^2 - \int d^3\mathbf{x}\, \frac{(\delta\rho)^2}{2\rho}; \qquad (2.128)$$

substituting for δT from (2.127) and imposing the $\delta M = 0$ constraint by a Lagrange multiplier α, we have the variation

$$\delta S + \alpha \, \delta M = -\int d^3\mathbf{x} \, \delta\rho \left\{ \ln(\rho e) - \alpha + \frac{\phi}{T} \right\} - \int d^3\mathbf{x} \left\{ \frac{\delta\rho \, \delta\phi}{2T} + \frac{(\delta\rho)^2}{2\rho} \right\}$$

$$- \frac{3M}{4T^2} \left\{ -\frac{2}{3M} \int d^3\mathbf{x} \left(\phi \, \delta\rho + \frac{1}{2}\delta\rho \, \delta\phi \right) \right\}^2$$

$$\equiv -\int d^3\mathbf{x} \, \delta\rho \left\{ \ln(\rho e) - \alpha + \frac{\phi}{T} \right\} + \delta^2 S,$$

where the second variation of S is

$$\delta^2 S = -\int d^3\mathbf{x} \left\{ \frac{\delta\rho \, \delta\phi}{2T} + \frac{(\delta\rho)^2}{2\rho} \right\} - \frac{1}{3MT^2} \left(\int d^3\mathbf{x} \, \phi \, \delta\rho \right)^2 + \mathcal{O}(\delta^3 \ldots)$$

$$= -\int d^3\mathbf{x} \left\{ \frac{\delta\rho \, \delta\phi}{2T} + \frac{(\delta\rho)^2}{2\rho} \right\} - \frac{1}{3MT^2} \left(\int d^3\mathbf{x} \, \phi \, \delta\rho \right)^2. \tag{2.129}$$

The extremum is decided by setting the coefficient of $\delta\rho$ to zero, which leads to equation (2.121). Whether the extremum is a true maximum is decided by the sign of $\delta^2 S$.

2.10 Isothermal sphere

(a) Using the results

$$\frac{d\phi}{dr} = \frac{GM(r)}{r^2}, \quad \frac{dM}{dr} = 4\pi r^2 \rho(r) \tag{2.130}$$

and the definitions of $m(x)$, $n(x)$ and $y(x)$, we obtain

$$y' = \frac{m}{x^2}, \quad m' = nx^2. \tag{2.131}$$

From the relation $\rho = \rho_c \exp{-\beta[\phi - \phi_c]}$ written in the form

$$\frac{1}{\rho}\frac{d\rho}{dr} = -\beta\frac{d\phi}{dr} = -\frac{G\beta M(r)}{r^2} \tag{2.132}$$

we find that

$$n' = -\frac{mn}{x^2}. \tag{2.133}$$

Thus the variables (y, n, m) satisfy the equations

$$y' = \frac{m}{x^2}, \quad m' = nx^2, \quad n' = -\frac{mn}{x^2}. \tag{2.134}$$

In terms of $y(x)$ the isothermal equation becomes

$$\frac{1}{x^2}\frac{d}{dx}\left(x^2\frac{dy}{dx} \right) = e^{-y}, \tag{2.135}$$

with the boundary condition $y(0) = y'(0) = 0$.

By direct substitution, we see that $n = (2/x^2), m = 2x, y = 2\ln x$ satisfies these equations. Consider now a change of variable, from x to $q = \ln x$ in equation (2.135). This leads to

$$\frac{d^2 y}{dq^2} + \frac{dy}{dq} = e^{2q-y}. \tag{2.136}$$

Changing to the variables $X \equiv 2q - y$, we obtain

$$\frac{d^2 X}{dq^2} + \frac{dX}{dq} = - \left(e^X - 2 \right). \tag{2.137}$$

This equation describes the motion of a particle with 'position' X at 'time' q in a potential $V(X) = (e^X - 2X)$, with a damping term (dX/dq). Because of the damping, the kinetic energy $(dX/dq)^2$ will keep decreasing and the particle will approach the minimum of the potential at $X = \ln 2$. This minimum point corresponds to $y = 2q - X = 2\ln x - \ln 2 = 2\ln x + \text{constant}$. Thus all solutions approach a $y = 2\ln x$ solution for large $q = \ln x$.

To obtain a more quantitative result, we can put $X = \ln 2 + \delta X$ and linearize (2.137) in δX. This gives

$$\left(\frac{d^2}{dq^2} + \frac{d}{dq} + 2 \right) \delta X = 0, \tag{2.138}$$

with the solution

$$\delta X = Ae^{-q/2} \cos \left[\frac{\sqrt{7}}{2} (q + \alpha) \right] = \frac{A}{x^{1/2}} \cos \left[\frac{\sqrt{7}}{2} (\ln x + \alpha) \right]. \tag{2.139}$$

This clearly shows that δX decreases to zero in an oscillatory manner.

(b) Equation (2.135) is invariant under the transformation $y \to y + a$; $x \to kx$, with $k^2 = e^a$. This invariance implies that, given a solution with some value of $y(0)$, we can obtain the solution with any other value of $y(0)$ by simple scaling. Therefore, only one of the two integration constants in (2.135) is really non-trivial. Hence, it must be possible to reduce the degree of the equation from two to one by a judicious choice of variables.

One such set of variables is the following:

$$v \equiv \frac{m}{x}, \quad u \equiv \frac{nx^3}{m} = \frac{nx^2}{v}. \tag{2.140}$$

In terms of v and u, we have

$$v' = \frac{m'}{x} - \frac{m}{x^2} = \frac{v}{x}(u-1), \quad u' = \frac{3x^2 n}{m} - nx - \frac{n^2 x^5}{m^2} = \frac{u}{x}(3 - v - u), \tag{2.141}$$

giving

$$\frac{u}{v} \frac{dv}{du} = - \frac{(u-1)}{(u+v-3)}. \tag{2.142}$$

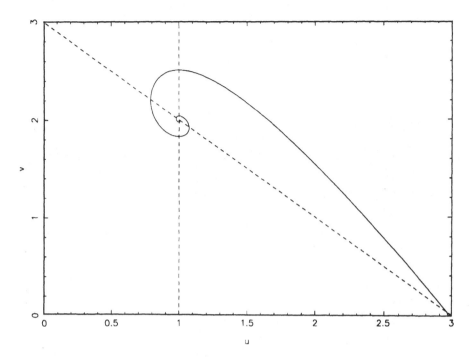

Fig. 2.6   The v–u curve for the isothermal sphere.

The boundary conditions $y(0) = y'(0) = 0$ translate into the following: v is zero at $u = 3$, and $(dv/du) = -5/3$ at $(3,0)$. The solution $v(u)$ is plotted in figure 2.6.

(c) The nature of the solution shown in figure 2.6 allows us to put interesting bounds on physical quantities, including energy. To see this, we shall compute the total energy E of the isothermal sphere. The potential and kinetic energies are

$$U = -\int_0^R \frac{GM(r)}{r} \frac{dM}{dr} dr = -\frac{GM_0^2}{L_0} \int_0^{x_0} mnx \, dx,$$

$$K = \frac{3}{2} \frac{M}{\beta} = \frac{3}{2} \frac{GM_0^2}{L_0} m(x_0) = \frac{GM_0^2}{L_0} \frac{3}{2} \int_0^{x_0} nx^2 \, dx, \tag{2.143}$$

where $x_0 = R/L_0$. The total energy is, therefore,

$$E = K + U = \frac{GM_0^2}{2L_0} \int_0^{x_0} dx \, (3nx^2 - 2mnx)$$

$$= \frac{GM_0^2}{2L_0} \int_0^{x_0} dx \frac{d}{dx} \{2nx^3 - 3m\} = \frac{GM_0^2}{L_0} \left\{ n_0 x_0^3 - \frac{3}{2} m_0 \right\}, \tag{2.144}$$

where $n_0 = n(x_0)$ and $m_0 = m(x_0)$. The dimensionless quantity (RE/GM^2) is given by

$$\lambda = \frac{RE}{GM^2} = \frac{L_0 x_0}{GM_0^2 m_0^2} \cdot \frac{GM_0^2}{L_0} \left\{ n_0 x_0^3 - \frac{3}{2} m_0 \right\} = \frac{1}{v_0} \left\{ u_0 - \frac{3}{2} \right\}. \tag{2.145}$$

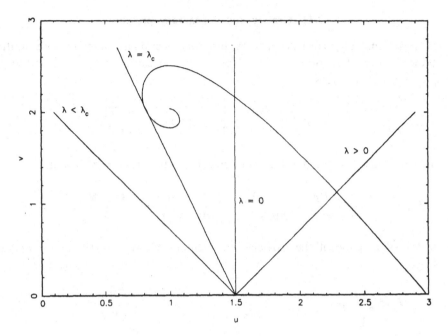

Fig. 2.7   Bound on RE/GM^2 for the isothermal sphere.

Note that the combination (RE/GM^2) *is a function of* (u,v) *alone.* Let us now consider the constraints on λ. Suppose we specify some value for λ by specifying R, E and M. Then such an isothermal sphere *must* lie on the curve

$$v = \frac{1}{\lambda}\left(u - \frac{3}{2}\right), \qquad (2.146)$$

which is a straight line through the point $(1.5, 0)$ with the slope λ^{-1}. Since, on the other hand, *all* isothermal spheres must lie on the u–v curve, *an isothermal sphere can exist only if the line in (2.146) intersects the u–v curve.*

For large positive λ (positive E) there is just one intersection (see figure 2.7). When $\lambda = 0$ (zero energy), we still have a unique isothermal sphere. (For $\lambda = 0$, (2.146) is a vertical line through $u = 3/2$.) When λ is negative (negative E), the line can cut the u–v curve at more than one point; thus more than one isothermal sphere can exist with a given value of λ. (Of course, specifying M, R, E individually will remove this non-uniqueness.) But as we decrease λ (more and more negative E) the line in (2.146) will slope more and more to the left; and when λ is smaller than a critical value λ_{crit}, the intersection will cease to exist. *Thus no isothermal sphere can exist if* (RE/GM^2) *is below a particular critical value* λ_{crit}. This fact follows immediately from the nature of the u–v curve and equation (2.146). The value of λ_{crit} can be found from the numerical solution in figure (2.6). It turns out to be about (-0.335).

2.11 Gravity and degeneracy pressure

(a) In the limit of perfect degeneracy, fermions occupy all states up to a limiting momentum p_F. The total number of particles that can be accommodated, if the spin degeneracy is g, is

$$N = g \int_0^{p_F} \frac{V d^3 p}{(2\pi\hbar)^3} = \frac{gV}{2\pi^2\hbar^3} \left(\frac{1}{3} p_F^3\right) = \frac{gV p_F^3}{6\pi^2\hbar^3}. \tag{2.147}$$

The total energy, if the particles are non-relativistic with $\epsilon = p^2/2m$, is

$$E_{nr} = g \int_0^{p_F} \frac{p^2}{2m} \frac{V d^3 p}{(2\pi\hbar)^3} = \frac{gV p_F^5}{20\pi^2 m\hbar^3} = \left(\frac{3}{10}\right) \left(\frac{6\pi^2}{g}\right)^{2/3} \frac{\hbar^2}{m} \left(\frac{N}{V}\right)^{2/3} N. \tag{2.148}$$

On the other hand, if the particles are extreme relativistic, the $\epsilon = pc$ and we obtain

$$E_r \cong g \int_0^{p_F} (pc) \frac{V d^3 p}{(2\pi\hbar)^3} = \frac{gV c p_F^4}{8\pi^2\hbar^3} = \frac{3}{4} \left(\frac{6\pi^2}{g}\right)^{1/3} \hbar c \left(\frac{N}{V}\right)^{1/3} N. \tag{2.149}$$

Consider now a gas with electrons and positively charged nuclei. For each electron, the Coulomb energy is about (Ze^2/a), where $a \cong (ZV/N)^{1/3}$ is the mean distance between the electrons and the nuclei. For the electrons to be treated as an ideal gas, this energy should be small compared to the kinetic energy per electron $\bar{\epsilon}_k \cong (E_{nr}/N)$. This is equivalent to the condition $(N/V) \gg (me^2/\hbar^2)^3 Z^2$ or, equivalently,

$$\rho \gg \left(\frac{me^2}{\hbar^2}\right)^3 \left(\frac{m_p A}{Z}\right) Z^2. \tag{2.150}$$

(The quantity $(m_p A/Z)$ gives the mass per electron of the substance.) Clearly, the system becomes more ideal at high densities.

(b) The pressure is $(2/3)(E/V)$ for non-relativistic systems and $(1/3)(E/V)$ for relativistic systems. So

$$P_{nr} = \frac{2}{3} \frac{E_{nr}}{V} = \frac{1}{5} \left(\frac{6\pi^2}{g}\right)^{2/3} \frac{\hbar^2}{m} \left(\frac{N}{V}\right)^{5/3} \tag{2.151}$$

and

$$P_r = \frac{1}{3} \frac{E_r}{V} = \frac{1}{4} \left(\frac{6\pi^2}{g}\right)^{2/3} \hbar c \left(\frac{N}{V}\right)^{4/3}. \tag{2.152}$$

Writing $(N/V) = (Z\rho/m_p A)$ and $g = 2$ we obtain

$$P_{nr} = \frac{(3\pi^2)^{2/3}}{5} \frac{\hbar^2}{m} \left(\frac{Z}{A}\right)^{5/3} \left(\frac{\rho}{m_p}\right)^{5/3} \equiv \lambda_{nr} \rho^{5/3} \tag{2.153}$$

and

$$P_r = \frac{1}{4}\left(3\pi^2\right)^{2/3}\hbar c\left(\frac{Z}{A}\right)^{4/3}\left(\frac{\rho}{m_p}\right)^{4/3} \equiv \lambda_r\rho^{4/3}. \tag{2.154}$$

The equations for a spherically symmetric system with gravity balancing the pressure will be

$$\frac{d\phi}{dr} = -\frac{1}{\rho}\frac{dP}{dr}, \quad \frac{1}{r^2}\frac{d}{dr}\left(r^2\frac{d\phi}{dr}\right) = 4\pi G\rho. \tag{2.155}$$

Combining, we obtain

$$\frac{1}{r^2}\frac{d}{dr}\left(\frac{r^2}{\rho}\frac{dP}{dr}\right) = -4\pi G\rho. \tag{2.156}$$

We now use equations (2.153), (2.154) to express P in terms of ρ. Then we obtain

$$\frac{1}{r^2}\frac{d}{dr}\left(\frac{r^2}{\rho^{1/3}}\frac{d\rho}{dr}\right) = -\frac{12\pi G}{5\lambda_{nr}}\rho \tag{2.157}$$

for the non-relativistic case and

$$\frac{1}{r^2}\frac{d}{dr}\left(\frac{r^2}{\rho^{2/3}}\frac{d\rho}{dr}\right) = -\frac{3\pi G}{\lambda_r}\rho \tag{2.158}$$

for the relativistic case.

(c) Let $r = Rx$ and $\rho = \rho_0 f(x)$, where R is the radius of the star (so that $f = 0$ at $x = 1$). Then equation (2.157) becomes

$$\frac{1}{x^2}\frac{d}{dx}\left(\frac{x^2}{f^{1/3}}\frac{df}{dx}\right) = -\rho_0^{1/3}R^2\frac{12\pi G}{5\lambda_{nr}}f. \tag{2.159}$$

We choose ρ_0 such that

$$\rho_0 = \left(\frac{5\lambda_{nr}}{8\pi G}\right)^3\frac{1}{R^6} \tag{2.160}$$

and set $f = y^{3/2}$. Then $y(x)$ satisfies the equation

$$\frac{1}{x^2}\frac{d}{dx^2}\left(x^2\frac{dy}{dx}\right) = -y^{3/2}. \tag{2.161}$$

Since $\rho'(0) = 0$ we must have $y'(0) = 0$; and, by definition, $y(1) = 0$. The density is $\rho(r) = \rho_0 f(r/R) = (\text{const})R^{-6}f(r/R)$ and the total mass is

$$M = \int_0^R 4\pi r^2\rho(r)\,dr = (\text{constant})\int_0^1 4\pi x^2\,dx\,f(x)\cdot\frac{R^3}{R^6} \propto R^{-3}. \tag{2.162}$$

So $R \propto M^{-1/3}$ and $\bar{\rho} \propto M/R^3 \propto R^{-6} \propto M^2$. To determine the constant of proportionality, we can use the relation $GM(R) = R^2\phi'(R)$. Using (2.155) and

(2.153) to express ϕ' in terms of y' and using the fact that $y'(1) \cong -132$, we obtain the result:

$$MR^3 \cong \left(\frac{92\hbar^6}{G^3 m^3 m_{\mathrm{p}}^5}\right)\left(\frac{Z}{A}\right)^5. \tag{2.163}$$

(d) The analysis is similar to that in part (c) above. Setting $\rho = \rho_0 f(x)$, $x = r/R$, $q = f^{1/3}$ and $\rho_0 = (\lambda_{\mathrm{r}}/\pi G)^{3/2} R^{-3}$ we can reduce the equation to

$$\frac{1}{x^2}\frac{d}{dx}\left(x^2\frac{dy}{dx}\right) = -y^3. \tag{2.164}$$

In this case $\rho \propto R^{-3} f(r/R)$ and M is independent of R. The total mass is estimated exactly as in the above case and we obtain

$$M \cong \frac{3.1}{m_{\mathrm{p}}^2}\left(\frac{Z}{A}\right)^2\left(\frac{\hbar c}{G}\right)^{3/2} \cong 5.8\left(\frac{Z}{A}\right)^2 M_\odot. \tag{2.165}$$

Systems with larger mass cannot be supported against gravity, even by the relativistic degeneracy pressure of the neutrons. This limiting mass is called the 'Chandrasekhar mass'.

2.12 Collisional evolution of gravitating systems

(a) Consider an infinitesimal area element centred at $\mathbf{p}_1$ and perpendicular to the p_x-direction in the momentum space. The flux of particles from left to right (i.e. the number of particles per unit area which cross this surface from left to right, in unit time) is given by

$$I_{\mathrm{L}} = \int_{q_x>0} d\mathbf{q} \int_{(p_1)_x - q_x}^{(p_1)_x} dk_x \int_{\mathrm{all}} d\mathbf{p}_2 \quad W\left(\mathbf{k} + \frac{1}{2}\mathbf{q}, \mathbf{p}_2 - \frac{1}{2}\mathbf{q}\right) f(\mathbf{p}_2) f(\mathbf{k}). \tag{2.166}$$

The integrand describes the scattering from $(\mathbf{k}, \mathbf{p}_2)$ to $(\mathbf{k} + \mathbf{q}, \mathbf{p}_2 - \mathbf{q})$ in which the particle with momentum $\mathbf{k}$ moves across the surface. To achieve this with a given momentum transfer $\mathbf{q}$ the value of k_x must be in the range $[(p_1)_x - q_x, (p_1)_x]$. We integrate over the momentum transfer $\mathbf{q}$, making sure it is a flow from left to right (i.e. $q_x > 0$).

By similar reasoning, the flux of particles moving from right to left across this surface is given by

$$I_{\mathrm{R}} = \int_{q_x<0} d\mathbf{q} \int_{(p_1)_x}^{(p_1)_x - q_x} dk_x \int_{\mathrm{all}} d\mathbf{p}_2 \quad W\left(\mathbf{k} + \frac{1}{2}\mathbf{q}, \mathbf{p}_2 - \frac{1}{2}\mathbf{q}; \mathbf{q}\right) f(\mathbf{p}_2) f(\mathbf{k}). \tag{2.167}$$

The integrand is the same as in (2.166) but the range is different. We are interested in the net flux $J_x \equiv I_{\mathrm{L}} - I_{\mathrm{R}}$. By a series of simple transformations, we can make the range of integration in both I_R and I_L the same. We have

$$
I_R = \int_{\text{all}} d\mathbf{p}_2 \int_{q_x>0} d\mathbf{q} \int_{(p_1)_x}^{(p_1)_x+q_x} dk_x\, W\left(\mathbf{k}-\frac{1}{2}\mathbf{q},\mathbf{p}_2+\frac{1}{2}\mathbf{q};-\mathbf{q}\right) f(\mathbf{k})\,f(\mathbf{p}_2)
$$

$$
= \int_{\text{all}} d\mathbf{p}_2 \int_{q_x>o} d\mathbf{q} \int_{(p_1)_x-q_x}^{(p_1)_x} dl_x\, W\left(\mathbf{l}+\frac{1}{2}\mathbf{q},\mathbf{p}_2+\frac{1}{2}\mathbf{q};-\mathbf{q}\right) f(\mathbf{l}+\mathbf{q})\,f(\mathbf{p}_2)
$$

$$
= \int_{\text{all}} d\mathbf{l}' \int_{q_x>0} d\mathbf{q} \int_{(p_1)_x-q_x}^{(p_1)_x} dl_x\, W\left(\mathbf{l}+\frac{1}{2}\mathbf{q},\mathbf{l}'-\frac{1}{2}\mathbf{q};-\mathbf{q}\right) f(\mathbf{l}+\mathbf{q})\,f(\mathbf{l}'-\mathbf{q}). \quad (2.168)
$$

In arriving at the first equality, we have changed $\mathbf{q}$ to $-\mathbf{q}$ in equation (2.167); in arriving at the second equality we have set $\mathbf{k}=\mathbf{l}+\mathbf{q}$; to arrive at the last equality, we have put $\mathbf{p}_2=(\mathbf{l}'-\mathbf{q})$. Subtracting I_R from I_L we can now express J_x as

$$
J_x = \int_{\text{all}} d\mathbf{l} \int_{q_x>0} d\mathbf{q} \int_{(p_1)_x-q_x}^{(p_1)_x} dl_x\ \ W\left(\mathbf{l}+\frac{1}{2}\mathbf{q},\mathbf{l}'-\frac{1}{2}\mathbf{q},\mathbf{q}\right) [f(\mathbf{l})f(\mathbf{l}')-f(\mathbf{l}+\mathbf{q})f(\mathbf{l}-\mathbf{q})].
$$
$$
(2.169)
$$

To proceed further we will make the assumption that the collisions are 'soft' so that most of the contribution to this integral comes from processes with small momentum transfer, that is, from small $\mathbf{q}$. In this case, we can (i) replace W in (2.169) by $W(\mathbf{l},\mathbf{l}',\mathbf{q})$; (ii) Taylor expand $f(\mathbf{l}+\mathbf{q})f(\mathbf{l}'-\mathbf{q})$ in $\mathbf{q}$ retaining only up to linear terms in $\mathbf{q}$; and (iii) replace the integral over l_x by multiplication by the range of integration, q_x. This brings $\mathbf{J}$ to the form

$$
J_\alpha = \frac{1}{2}\int d\mathbf{l}' \int d\mathbf{q}\, q_\alpha q_\beta W \left[f\frac{\partial f'}{\partial l'_\beta} - f'\frac{\partial f}{\partial l_\beta}\right]
$$

$$
= \int d\mathbf{l}'\, B_{\alpha\beta}(\mathbf{l},\mathbf{l}')\left[f\frac{\partial f'}{\partial l'_\beta} - f'\frac{\partial f}{\partial l_\beta}\right], \quad (2.170)
$$

where we have defined

$$
B_{\alpha,\beta}(\mathbf{l},\mathbf{l}') = \frac{1}{2}\int q_\alpha q_\beta W(\mathbf{l},\mathbf{l}';\mathbf{q})\,d\mathbf{q}. \quad (2.171)
$$

(b) Note that $B_{\alpha\beta}$ can only depend on $\mathbf{k}\equiv\mathbf{l}-\mathbf{l}'$; further, for soft collisions $B_{\alpha\beta}$ must be 'transverse' to k^α since the momentum transfer is in the transverse direction. Hence, we must have $B_{\alpha\beta}k^\alpha=0$. The most general second rank symmetric tensor, constructable from k_α and transverse to it, must have the form

$$
B_{\alpha\beta} = \frac{1}{2}B\left\{\delta_{\alpha\beta} - \frac{k_\alpha k_\beta}{k^2}\right\}, \quad k=|\mathbf{k}|. \quad (2.172)
$$

We therefore only need to compute

$$
B = B_\alpha^\alpha = \frac{1}{2}\int q^2 W(\mathbf{l},\mathbf{l}';\mathbf{q})\,d\mathbf{q}. \quad (2.173)
$$

Since W represents the rate of transitions and $\mathbf{q}$ is the momentum transfer per soft collision, the integral in (2.173) is essentially the same as the one in problem 1.14. Therefore,

$$B = \frac{1}{2} \int q^2 W \, d\mathbf{q} = \frac{m}{2} \int_{b_{\min}}^{b_{\max}} \left(\frac{2Gm^2}{bk} \right)^2 k \cdot 2\pi b \, db = 4\pi G^2 m^5 \cdot \frac{L}{k}, \qquad (2.174)$$

where

$$L = \int_{b_{\min}}^{b_{\max}} \frac{db}{b} = \ln \left(\frac{b_{\max}}{b_{\min}} \right) \cong \ln N \qquad (2.175)$$

is the logarithmic factor encountered before in problem 1.14. We will write B in momentum space as

$$B = \frac{B_0}{|\mathbf{k}|}, \quad B_0 = 4\pi G^2 m^5 L. \qquad (2.176)$$

Substituting (2.173) and (2.172) into (2.170) we obtain

$$J_\alpha = \frac{B_0}{2} \int d\mathbf{l}' \left\{ f \frac{\partial f'}{\partial l_\beta} - f' \frac{\partial f}{\partial l_\beta} \right\} \cdot \left\{ \frac{\delta_{\alpha\beta}}{k} - \frac{k_\alpha k_\beta}{k^3} \right\}, \qquad (2.177)$$

where $\mathbf{k}$ is the momentum change $(\mathbf{l} - \mathbf{l}')$.

(c) For any Maxwellian distribution, with $f(\mathbf{l}) \propto \exp(-\sigma l^2)$, we have

$$f \frac{\partial f'}{\partial l'_\beta} - f' \frac{\partial f}{\partial l_\beta} = f(\mathbf{l})(-2\sigma l'_\beta) f(\mathbf{l}') - f(\mathbf{l}')(-2\sigma l_\beta) f(\mathbf{l})$$

$$= 2\sigma f(\mathbf{l}) f(\mathbf{l}')(\mathbf{l} - \mathbf{l}')_\beta = 2\sigma f(\mathbf{l}) f(\mathbf{l}') k_\beta. \qquad (2.178)$$

Hence J_α vanishes because of the relation $k_\beta B^{\alpha\beta} = 0$. *Thus our collision term admits Maxwellian distribution as an equilibrium solution.*

(d) It is clear that J_α will have one term proportional to f and one term proportional to $(\partial f / \partial l_\beta)$. Using

$$\frac{\delta_{\alpha\beta}}{k} - \frac{k_\alpha k_\beta}{k^3} = \frac{\partial^2 |\mathbf{k}|}{\partial k_\alpha \, \partial k_\beta} = \frac{\partial^2 k}{\partial k_\alpha \, \partial k_\beta}, \qquad (2.179)$$

we can rewrite (2.177) as

$$J_\alpha = \frac{B_0}{2} f(\mathbf{l}) \int d\mathbf{l}' \frac{\partial f}{\partial l'_\beta} \cdot \frac{\partial^2 k}{\partial k_\alpha \, \partial k_\beta} - \frac{B_0}{2} \frac{\partial f}{\partial l_\beta} \cdot \int d\mathbf{l}' \frac{\partial^2 k}{\partial k_\alpha \, \partial k_\beta}. \qquad (2.180)$$

Further, since

$$\frac{\partial}{\partial k_\alpha} = \frac{\partial}{\partial l_\alpha} = -\frac{\partial}{\partial l'_\alpha}, \qquad (2.181)$$

we can write this expression in the form

$$J_\alpha = \frac{B_0}{2} f \int d\mathbf{l}' f'_\beta \frac{\partial^2 k}{\partial k_\alpha \, \partial k_\beta} - \frac{B_0}{2} f_\beta \cdot \frac{\partial^2 \psi}{\partial k_\alpha \, \partial k_\beta}, \qquad (2.182)$$

where

$$\psi(\mathbf{l}) = \int d\,\mathbf{l}' f(\mathbf{l}')|\mathbf{l} - \mathbf{l}'|, \quad f_\beta' \equiv \partial f(\mathbf{l}')/\partial l_\beta'. \tag{2.183}$$

The two terms in (2.182) can be transformed into a more conventional form by integrating them by parts and using (2.181) repeatedly:

$$\int d\,\mathbf{l}' \frac{\partial f}{\partial l_\beta'} \cdot \frac{\partial^2 k}{\partial k_\alpha \partial k_\beta} = -\int d\,\mathbf{l}' f(\mathbf{l}') \frac{\partial}{\partial l_\beta'} \frac{\partial^2 k}{\partial k_\alpha \partial k_\beta}$$

$$= +\int d\,\mathbf{l}' f(\mathbf{l}') \frac{\partial}{\partial k_\beta} \frac{\partial^2 k}{\partial k_\alpha \partial k_\beta} = \int d\,\mathbf{l}' f(\mathbf{l}') \frac{\partial}{\partial k_\alpha} \frac{\partial^2 k}{\partial k_\beta \partial k_\beta}$$

$$= \frac{\partial}{\partial l_\alpha} \frac{\partial}{\partial l^\beta \partial l_\beta} \int d\,\mathbf{l}' f(\mathbf{l}')|\mathbf{l} - \mathbf{l}'| = \frac{\partial}{\partial l_\alpha} \cdot \nabla_l^2 \psi(\mathbf{l}), \tag{2.184}$$

and

$$f_\beta \frac{\partial^2 \psi}{\partial k_\alpha \partial k_\beta} = \frac{\partial}{\partial k_\beta}\left(f\frac{\partial^2 \psi}{\partial k_\alpha \partial k_\beta}\right) - f\frac{\partial}{\partial k_\beta}\frac{\partial^2 \psi}{\partial k_\alpha \partial k_\beta}$$

$$= \frac{\partial}{\partial k_\beta}\left(f\frac{\partial^2 \psi}{\partial k_\alpha \partial k_\beta}\right) - f\frac{\partial}{\partial k_\alpha} \cdot \nabla^2 \psi, \tag{2.185}$$

where ∇^2 stands for the Laplacian in *momentum* space ($\partial^2/\partial k^\alpha \partial k_\alpha$). Defining

$$\eta(\mathbf{l}) \equiv \nabla^2 \psi(\mathbf{l}) = \frac{\partial^2}{\partial l^\alpha \partial l_\alpha} \int d\,\mathbf{l} f(\mathbf{l}')|\mathbf{l} - \mathbf{l}'| = 2\int d\,\mathbf{l}' \frac{f(\mathbf{l}')}{|\mathbf{l} - \mathbf{l}'|} \tag{2.186}$$

and substituting (2.184) and (2.185) in (2.182), we obtain

$$J_\alpha(\mathbf{l}) = \frac{B_0}{2} f(\mathbf{l}) \frac{\partial \eta}{\partial l_\alpha} - \frac{B_0}{2}\left\{\frac{\partial}{\partial l_\beta}\left(f\frac{\partial^2 \psi}{\partial l_\alpha \partial l_\beta}\right) - f\frac{\partial \eta}{\partial k_\alpha}\right\}$$

$$= B_0 f(\mathbf{l}) \frac{\partial \eta}{\partial l_\alpha} - \frac{B_0}{2}\frac{\partial}{\partial l_\beta}\left(f\frac{\partial^2 \psi}{\partial l_\alpha \partial l_\beta}\right)$$

$$\equiv a_\alpha(\mathbf{l})f(\mathbf{l}) - \frac{1}{2}\frac{\partial}{\partial l_\beta}\left\{\sigma_{\alpha\beta}^2 f\right\}, \tag{2.187}$$

where $a_\alpha = B_0\left(\partial \eta/\partial l_\alpha\right), \sigma_{\alpha\beta}^2 = B_0\left(\partial^2 \psi/\partial l_\alpha \partial l_\beta\right)$, with

$$\nabla^2 \psi = \eta, \quad \nabla_l^2 \eta(\mathbf{l}) = \nabla_l^2\left\{2\int d\mathbf{l}' \frac{f(\mathbf{l}')}{|\mathbf{l} - \mathbf{l}'|}\right\} = -8\pi f(\mathbf{l}). \tag{2.188}$$

(e) Consider the equation

$$\frac{\partial f(v, t)}{\partial t} = \frac{\partial}{\partial v}\left\{(\alpha v)f + \frac{\sigma^2}{2}\frac{\partial f}{\partial v}\right\} \equiv -\frac{\partial J}{\partial v}. \tag{2.189}$$

The J in this equation is similar in structure to the J_α in (2.187) if we confine our attention to one dimension and set $a = \alpha v$. Let us consider the effect of the two terms.

The second term $(\sigma^2/2)(\partial f/\partial v)$ has the standard form of a 'diffusion current' proportional to the gradient in the velocity space. As time goes on, this term will cause the mean square velocities of particles to increase in proportion to t, inducing the 'random walk' in the velocity space. Under the effect of this term, *all the particles in the system will have their $\langle v^2 \rangle$ increasing without bound. This unphysical situation is avoided by the presence of the first term ($\alpha v f$) in J.* This term acts as a friction term (called 'dynamical friction'). The combined effect of the two terms is to drive f to a Maxwellian distribution with a $\beta = (kT)^{-1} = (\alpha/\sigma^2)$. *In such a Maxwellian distribution the gain made in (Δv^2) due to diffusion is exactly balanced by the losses due to dynamical friction.* When two particles scatter, one gains the energy lost by the other; on the average, we may say that the one which has lost the energy has undergone 'dynamical friction' while the one which has gained energy has achieved 'diffusion' to higher v^2. The cumulative effect of such phenomena is described by the two terms in $J(v)$.

The above points can easily be illustrated by solving (2.189). Suppose we take an initial distribution $f(v,0) = \delta(v - v_0)$ peaked at velocity v_0. The solution of (2.189) with this initial condition is

$$f(v,t) = \left[\frac{\alpha}{\pi\sigma^2(1 - e^{-2\alpha t})} \right]^{1/2} \exp\left[-\frac{\alpha(v - v_0 e^{-\alpha t})^2}{\sigma^2(1 - e^{-2\alpha t})} \right], \qquad (2.190)$$

which is a Gaussian with mean

$$\langle v \rangle = v_0 e^{-\alpha t} \qquad (2.191)$$

and dispersion

$$\langle v^2 \rangle - \langle v \rangle^2 = \frac{\sigma^2}{\alpha}(1 - e^{-2\alpha t}). \qquad (2.192)$$

At late times ($t \to \infty$), the mean velocity $\langle v \rangle$ goes to zero while the velocity dispersion becomes (σ^2/α). Thus the equilibrium configuration is a Maxwellian distribution of velocities with this particular dispersion, for which $J = 0$. To see the effect of the two terms individually on initial distribution $f(v,0) = \delta(v - v_0)$, we can set α or σ to zero. When $\alpha = 0$, we obtain pure diffusion:

$$f_{\alpha=0}(v,t) = \left(\frac{1}{2\pi\sigma^2 t} \right)^{1/2} \exp\left\{ -\frac{(v - v_0)^2}{2\sigma^2 t} \right\}. \qquad (2.193)$$

Nothing happens to the steady velocity v_0; but the velocity dispersion increases in proportion to t, representing a random walk in the velocity space. On the other hand, if we set $\sigma = 0$, then we obtain

$$f_{\sigma=0}(v,t) = \delta(v - v_0 e^{-\alpha t}). \qquad (2.194)$$

Now there is no spreading in velocity space (no diffusion); instead the friction steadily decreases $< v >$. Note that the time-scale for the operation of the dynamical friction is $\alpha^{-1} = (a/v)^{-1}$.

Equation (2.187) allows the computation of the coefficient of dynamical friction for any given f. For a Maxwellian distribution of velocities:

$$f(v) = A e^{-v^2/q^2}, \qquad (2.195)$$

we obtain, for small v in one dimension,

$$a(v) \cong -\frac{32\sqrt{\pi}}{3} n \cdot (Gm)^2 L \frac{v}{q^3} \left(1 - \frac{3}{5}\frac{v^2}{q^2} + \cdots \right). \tag{2.196}$$

So that relaxation time-scale t_{dy} is determined by

$$t_{\mathrm{dy}}^{-1} \equiv \alpha(v) \equiv -\frac{a(v)}{v} \cong \frac{16}{3\sqrt{\pi}} \cdot \left(\frac{\ln N}{N}\right) \cdot \left(\frac{q}{R}\right). \tag{2.197}$$

This is the time-scale specified by $C(f)$. Our estimate of t_{dy} earlier in problem 1.14 is correct except for a numerical factor.

(f) We note that (2.188) is essentially Coulomb's law in the velocity space, with $f(\mathbf{l})$ as source. Suppose the velocity distribution is spherically symmetric, that is, $f(\mathbf{l}) = f(|\mathbf{l}|)$. In that case, the 'force' $(\partial \eta / \partial l_\alpha)$ at any given $\mathbf{l}$ is only due to the particles inside the (velocity) sphere of radius $|\mathbf{l}|$. In other words, dynamical friction force on a particle with momentum l is contributed *only* by the particles with lower momentum, if $f(\mathbf{l}) = f(|\mathbf{l}|)$.

2.13 Collisionless relaxation

(a) Let us assume that, at $t = 0$, the particles were located in some finite region $\mathscr{R}$ of the phase space. As the particles move, the shape of $\mathscr{R}$ will become distorted and we expect the evolution to spread the phase density all over the phase space. This arises due to two kinds of process. (a) Each particle is moving in the mean gravitational potential of the system. Since this potential is not harmonic (in general), the orbital period will depend on the amplitude and the occupied region in the phase space will become 'wound up' in the angular direction. (b) The gravitational potential is also changing with time, because of which the energies of individual particles will not be constant. In other words, particles will also move along the 'radial' direction in the phase space. Hence, we conclude that, as time goes on, particles become mixed in phase space significantly.

To study the effects of such mixing, we shall divide the phase space into 'micro-cells' of volume ω. These microcells are supposed to be so tiny that even a fine-grained distribution function f can be taken to be a constant over the cell. Initially, we assume that the phase density is constant ($f = \eta$) in some region of the phase space and zero elsewhere. So, initially, some microcells will contain $\eta\omega$ particles and others will have zero particles. The total number of particles is $N_{\mathrm{tot}} = N\eta\omega$, where N is the number of *occupied* microcells.

We shall examine the phase density after coarse graining it over cells of volume Ω ('macrocells'). Let $\nu = (\Omega/\omega)$ be the number of microcells contained in any one macrocell. Consider any one particular macrocell – say, the ith one – which has n_i microcells occupied and $(\nu - n_i)$ cells empty. This can be done in

$$W_i = \frac{\nu!}{(\nu - n_i)!} \tag{2.198}$$

ways. The coarse-grained distribution is completely specified by the set of numbers $\{n_i\}$. To compute the probability for this configuration, we have to multiply the product of the W_is by the number of ways of splitting N microcells into $\{n_i\}$. The latter factor is $(N!/\prod n_i!)$. Hence, the probability for the particular coarse-grained configuration is

$$p = N! \prod_i \frac{v!}{n_i! (v - n_i)!} . \tag{2.199}$$

Here we have assumed that particles are completely mixed up in phase space, giving equal a priori probability to all the configurations. The most probable coarse-grained distribution function can be obtained by maximizing this probability with respect to the n_is, subject to the constraints that the total number of particles in the system and the energy of the system remain constant. We note that the expression for P is the same as that of a Fermi–Dirac gas except for the (unimportant) $N!$ term. Therefore, the coarse-grained distribution with maximum probability will be the Fermi–Dirac distribution with

$$f_c(\mathbf{x}, \mathbf{v}) = \eta \left[\exp \beta (\epsilon - \mu) + 1 \right]^{-1} , \tag{2.200}$$

where

$$\epsilon = \epsilon(\mathbf{x}, \mathbf{v}) = \frac{1}{2} v^2 + \phi(\mathbf{x}) \tag{2.201}$$

and

$$\nabla^2 \phi = 4\pi G \int f_c(\mathbf{x}, \mathbf{v}) \, d^3 \mathbf{v} . \tag{2.202}$$

Equations (2.200), (2.201) and (2.202) represent a self-gravitating gas of fermions with two parameters μ and β. These parameters can be fixed in terms of the total energy and mass of the system.

In realistic situations the phase density will be significantly low compared to the maximum possible value. That is, we expect $f_c \ll \eta$. In this case one can approximate the Fermi–Dirac distribution by an equivalent Maxwell–Boltzmann distribution and obtain

$$f_c(\mathbf{x}, \mathbf{v}) \cong \eta \exp{-\beta (\epsilon - \mu)} \equiv A \exp{-\beta \epsilon} . \tag{2.203}$$

This equation, along with (2.202), implies that systems which have undergone collisionless relaxation are represented by isothermal spheres.

The mixing in phase space, described by the above process, can take place on a few orbital time-scales. Hence, this process operates at a far shorter time-scale compared to the collisional time-scale. Because of this reason, collisionless relaxation is also called 'violent relaxation'.

(b) Given a coarse-grained distribution function f_c we define the volume of phase space with phase density larger than q by the relation

$$V(q) = \int d\mathbf{x} \, d\mathbf{v} \, \theta(f_c(\mathbf{x}, \mathbf{v}) - q), \tag{2.204}$$

where $\theta(z)$ is unity for $z > 0$ and is zero otherwise. Similarly, the mass contained in the region with phase density larger than q is defined as

$$M(q) = \int dx\, dv\, f_c \theta(f_c - q). \qquad (2.205)$$

Clearly,

$$\frac{dM}{dq} = -\int dx\, dv\, f_c \delta(f_c - q) = -q\int dx\, dv\, \delta(f_c - q) = +q\frac{dV}{dq}. \qquad (2.206)$$

So that

$$M(V) = \int_0^V q(V')\, dV', \qquad (2.207)$$

where $q(V)$ is the inverse function of $V(q)$. Given the distribution function f_c we can obtain $M(V)$ by eliminating q between relations (2.204) and (2.205).

Now suppose we are given two distribution functions f_1 and f_2 with corresponding $M_1(V)$ and $M_2(V)$. We can then show that f_1 could have evolved into f_2 if and only if $M_2(V) \le M_1(V)$ for all V. To do this, we shall need a preliminary result, which we shall derive first.

Consider any functional H of the coarse-grained distribution function f_c defined by the integral

$$H[f_c] = -\int C(f_c)\, dx\, dv, \qquad (2.208)$$

where C is a convex function. Let the original ('fine-grained') distribution function corresponding to f_c be f; we shall also assume that at some initial moment $t = t_1$, $f_c(t_1) = f(t_1)$. Since the evolution is collisionless, $\dot{f} = 0$ and $\dot{C} = C'\dot{f} = 0$; C is conserved during evolution.

It can easily be shown that $H(t_2) \ge H(t_1)$ for all $t_2 > t_1$. We note that

$$H(t_2) - H(t_1) = \int dx\, dv\, [C(f_c(t_1)) - C(f_c(t_2))]$$

$$= \int dx\, dv\, [C(f(t_1)) - C(f_c(t_2))]$$

$$= \int dx\, dv\, [C(f(t_2)) - C(f_c(t_2))]. \qquad (2.209)$$

(The second equality follows from the fact that at $t = t_1, f_c(t_1) = f(t_1)$; the third from the conservation of $C(f)$ under time evolution.) But the coarse-grained distribution function f_c is obtained by averaging f over the phase cells. For any convex function $C(f)$, the quantity $C(\langle f \rangle)$ will be less than $\langle C(f) \rangle$, where $\langle \cdots \rangle$ denotes an averaging process with positive semidefinite weights. Therefore, we may conclude that

$$\int dx\, dv\, C(f_c) \le \int dx\, dv\, C(f) \qquad (2.210)$$

and hence

$$H(t_2) > H(t_1). \tag{2.211}$$

It should be stressed that $H(t)$, in general, is *not* a monotonically increasing function of t. The instant t_1 is special in the sense that we set $f_c = f$ at that instant. The above argument only shows that H-functions will always have values higher than the value taken at this special instant. For two arbitrary instants $(t_2, t_3) > t_1$, nothing can be said about the relative values of $H(t_2)$ and $H(t_3)$.

The above condition shows that collisionless evolution must proceed in such a manner that *all* convex H-functionals increase during the evolution. Suppose we are given two distribution functions $f_c(t_2)$ and $f_c(t_1)$ with $t_2 > t_1$. Collisionless evolution could have evolved $f_c(t_1)$ to $f_c(t_2)$ only if all H functionals satisfy the condition $H[f_c(t_2)] > H[f_c(t_1)]$.

We shall now show that if $M_2(V) \leq M_1(V)$, then $H(f_2) > H(f_1)$ for all H-functions. Now,

$$H(f_2) - H(f_1) = \int dx\, dv\, (C(f_1) - C(f_2)) = \int_0^\infty dV\, \{C(f_1) - C(f_2)\}. \tag{2.212}$$

We shall now use the fact that, for convex functions, $C(f_1) - C(f_2) \geq (f_1 - f_2)C'(f_2)$; then

$$H_2 - H_1 \geq \int_0^\infty dV(f_1 - f_2)\, C'(f_2) = \int_0^\infty dV\, \frac{d}{dV}(M_1 - M_2) \cdot C'(f_2). \tag{2.213}$$

Integrating by parts,

$$H_2 - H_1 \geq \{(M_1 - M_2)\, C'(f_2)\}_0^\infty - \int_0^\infty dV(M_1 - M_2)\, C'' \left(\frac{df_2}{dV}\right) dV. \tag{2.214}$$

The first term vanishes at both limits; the second term is non-negative because $(df_2/dV) \leq 0$ and (for convex functions) $C'' > 0$. Thus $H_2 > H_1$.

We can also prove that if f_1 can evolve into f_2 then $M_2(V) < M_1(V)$. This can easily be done by using the convex function

$$C(f) \doteq \begin{cases} 0 & f \leq \phi \\ f - \phi & f > \phi \end{cases}, \tag{2.215}$$

where $\phi = f(V_0)$ with some arbitrary V_0. The fact that the H-function constructed out of this particular $C(f)$ should be non-decreasing leads to the condition $M_2(V_0) \leq M_1(V_0)$.

One useful corollary of the above constraint is the following: the maximum value of the phase density can only decrease during collisionless evolution. Since $f_{max} = f(V = 0)$, the Taylor expansion of $M(V)$ gives $M(V) \cong f_{max} V$; the condition $M_2 < M_1$ immediately leads to the conclusion $f_{max}^{(2)} < f_{max}^{(1)}$.

2.14 Equilibria of collisionless gravitating systems

(a) For a collisionless system, we must have $(df/dt) = 0$. If f is a function of *constants* of motion, then this condition is automatically satisfied. Further, in steady state, f cannot explicitly depend on t. This means that f may be taken to be a function of single-valued *integrals* of motion.

(b) Let $E = v^2/2 + \phi < 0$ be the energy of a bound particle. We will define two positive quantities ψ and ϵ by $\epsilon = -E + \phi_0$, $\psi = -\phi + \phi_0$ and choose ϕ_0 so as to ensure positivity for ψ and ϵ. Then

$$\psi - \epsilon = E - \phi = \frac{1}{2}v^2, \quad v\,dv = -d\epsilon. \tag{2.216}$$

Any function $f(E) = f(\epsilon)$ clearly satisfies the collisionless Boltzmann equation. We now only need to satisfy Poisson's equation

$$\nabla^2\phi = -\nabla^2\psi = -\frac{1}{r^2}\frac{d}{dr}\left(r^2\frac{d\psi}{dr}\right) = 4\pi G\int f(\mathbf{x},\mathbf{v})\,d^3\mathbf{v}$$

$$= 16\pi^2 G\int_0^\psi d\epsilon\sqrt{2(\psi-\epsilon)}f(\epsilon). \tag{2.217}$$

Given any function $f(\epsilon)$, this equation determines a self-consistent $\phi(r)$.

(c) If $f(\epsilon) = A\epsilon^{n-3/2}$, the density becomes

$$\rho = 4\pi\int_0^\infty f\left(\psi - \frac{1}{2}v^2\right)v^2\,dv = 4\pi A\int_0^{\sqrt{2\psi}}\left(\psi - \frac{1}{2}v^2\right)^{n-\frac{3}{2}}v^2\,dv. \tag{2.218}$$

Evaluating the integral, we obtain

$$\rho = c(n)\psi^n, \quad c(n) = \frac{(2\pi)^{3/2}\,\Gamma\left(n-\frac{1}{2}\right)}{\Gamma(n+1)}A. \tag{2.219}$$

This expression is well defined only if $n > 1/2$. The model is completely determined by the solution to the equation

$$\frac{1}{r^2}\frac{d}{dr}\left(r^2\frac{d\psi}{dr}\right) = -4\pi Gc(n)\psi^n. \tag{2.220}$$

Changing variables to $s = (r/L)$, $L^2 = [4\pi Gc(n)]^{-1}$ we obtain

$$\frac{1}{s^2}\frac{d}{ds}\left(s^2\frac{d\psi}{ds}\right) = \begin{cases} -\psi^n & (\text{for } \psi \geq 0) \\ 0 & (\text{for } \psi \leq 0) \end{cases}. \tag{2.221}$$

When $n = 5$, this equation has the simple solution

$$\psi(s) = \frac{1}{\sqrt{1+\frac{1}{3}s^2}}, \tag{2.222}$$

as can be verified by direct substitution. The density scales as $\rho \propto \psi^5 \propto (1 + s^2/3)^{-5/2}$.

(d) Given $\rho(r)$ and $\psi(r)$ we can determine $\rho(\psi)$. We also know that ρ and f are related by

$$\rho(\psi) = 2\pi\sqrt{8}\int_0^\psi f(\epsilon)\sqrt{\psi - \epsilon}\; d\epsilon. \tag{2.223}$$

Since $\rho(\psi)$ is known and $f(\epsilon)$ is not, this is an integral equation for $f(\epsilon)$. Differentiating both sides with respect to ψ we obtain

$$\frac{1}{\pi\sqrt{8}}\frac{d\rho}{d\psi} = \int_0^\psi \frac{f(\epsilon)\, d\epsilon}{\sqrt{\psi - \epsilon}}. \tag{2.224}$$

Integral equations of this kind can be solved in a straightforward manner. Note that, if

$$A(x) = \int_0^x \frac{B(y)\, dy}{(y - x)^\alpha} \quad (0 < \alpha < 1), \tag{2.225}$$

then

$$B(y) = \frac{\sin \pi\alpha}{\pi}\frac{d}{dy}\int_0^y \frac{A(x)\, dx}{(y - x)^{1 - \alpha}}. \tag{2.226}$$

This result can be proved by substituting (2.225) into the right-hand side of (2.226) and using the relation

$$\int_0^1 \frac{du}{u^\alpha (1 - u)^{1 - \alpha}} = \frac{\pi}{\sin \pi\alpha}. \tag{2.227}$$

Applying (2.225) and (2.226) to (2.224) we obtain

$$f(\epsilon) = \frac{1}{\sqrt{8\pi^2}}\frac{d}{d\epsilon}\int_0^\epsilon \left(\frac{d\rho}{d\psi}\right)\frac{d\psi}{\sqrt{\epsilon - \psi}}, \tag{2.228}$$

which provides the solution we need.

To be physically acceptable we must have $f(\epsilon) > 0$ for all ϵ. There is no assurance that this condition will be satisfied in general. It has to be verified in each case separately.

(e) If we introduce the coordinates (v, α, β) in the velocity space with

$$v_r = v\cos\alpha, \quad v_\theta = v\sin\alpha\cos\beta, \quad v_\phi = v\sin\alpha\sin\beta, \tag{2.229}$$

then the density $\rho(r)$ due to a distribution function $f(\epsilon, J) = f\left(\psi - \frac{1}{2}v^2, |rv\sin\alpha|\right)$ is

$$\rho(r) = 2\pi \int_0^\pi \sin\alpha \, d\alpha \int_0^\infty v^2 dv f\left(\psi - \frac{1}{2}v^2, |rv\sin\alpha|\right). \tag{2.230}$$

Let us now assume that f depends on ϵ and J^2 only through the combination $Q = (\epsilon - J^2/2R^2)$, where R is a constant. At constant $r, dQ = -[1 + (r^2/R^2)\sin^2\alpha] v \, dv$ and (2.230) becomes

$$\rho(r) = 2\pi \int_0^\pi \sin\alpha \, d\alpha \int_0^\psi f(Q) \frac{\sqrt{2(\psi - Q)}}{[1 + (r^2/R^2)\sin^2\alpha]^{3/2}} dQ. \tag{2.231}$$

Interchanging the orders of integration and using

$$\int_0^\pi \frac{\sin\alpha \, d\alpha}{[1 + (r^2/R^2)\sin^2\alpha]^{3/2}} = \frac{2}{1 + (r^2/R^2)}, \tag{2.232}$$

we obtain

$$n(r) \equiv \left(1 + \frac{r^2}{R^2}\right)\rho(r) = 4\pi \int_0^\psi f(Q)\sqrt{2(\psi - Q)}\, dQ. \tag{2.233}$$

This is identical to (2.223) with $n(r)$ replacing $\rho(r)$ and Q replacing ϵ. Hence,

$$f(Q) = \frac{1}{\sqrt{8\pi^2}} \frac{d}{dQ} \int_0^Q \frac{dn}{d\psi} \frac{d\psi}{\sqrt{Q - \psi}}. \tag{2.234}$$

This is the required solution. Given a $\rho(r)$ we can construct either an $f(\epsilon)$ using (2.228) or an $f(Q)$ using (2.234); either of them will produce such a density distribution.

The main difference between $f(\epsilon)$ and $f(Q)$ is in the velocity dispersion. If $f = f(\epsilon)$, then it depends only on $(v_r^2 + v_\theta^2 + v_\phi^2)$; so, clearly,

$$\langle v_r^2 \rangle = \langle v_\theta^2 \rangle = \langle v_\phi^2 \rangle. \tag{2.235}$$

If $f = f(Q)$, then f depends on v^2 and $\sin\alpha$, that is, on v^2 and $(v_\theta^2 + v_\phi^2)$. Hence, we will have

$$\langle v_\theta^2 \rangle = \langle v_\phi^2 \rangle \neq \langle v_r^2 \rangle. \tag{2.236}$$

2.15 Axisymmetric systems and halos

(a) The density distribution is self-similar. This implies that the components of the force:

$$F_r = -\frac{\partial\Phi}{\partial r}, \quad F_\theta = -\frac{1}{r}\frac{\partial\Phi}{\partial\theta} \tag{2.237}$$

must both scale as r^{-1}. Hence, $(\partial\Phi/\partial\theta)$ should be independent of r, giving $(\partial F_r/\partial\theta) = -(\partial^2\Phi/\partial\theta\,\partial r) = 0$.

Since F_r cannot depend on θ, its value must be the same as that determined by the mass inside r. So

$$F_r(r) = -\frac{G}{r}\int\limits_0^r 2\pi r^2\rho(r)\,dr\int\limits_0^\pi S(\theta)\sin\theta\,d\theta = -\frac{4\pi\rho_0 r_0^2 G}{r} = -\frac{v_0^2}{r}. \tag{2.238}$$

(b) Integrating

$$\frac{\partial\Phi}{\partial r} = -F_r = \frac{v_0^2}{r}, \tag{2.239}$$

we obtain

$$\Phi = v_0^2\left[\ln\left(\frac{r}{r_0}\right) + P(\theta)\right]. \tag{2.240}$$

Using $\nabla^2\phi = 4\pi G\rho$ we can relate $P(\theta)$ to $S(\theta)$; we find that

$$\frac{1}{\sin\theta}\frac{d}{d\theta}\left(\sin\theta\frac{dP}{d\theta}\right) = S(\theta) - 1. \tag{2.241}$$

(c) The density distribution $\rho(r,\theta)$ is now a sum of a spherically symmetric part $\rho_1(r) = b\rho_0(r_0/r)^2$ and an axisymmetric part $\rho_2(r,\theta) = \rho_0(r_0/r)^2\,\delta_{\mathrm{Dirac}}(\theta - \pi/2)$. The potential can therefore be written as a sum $\phi_1(r) + \phi_2(r,\theta)$, where ϕ_1 is generated by ρ_1 and ϕ_2 is generated by ρ_2. Clearly,

$$\phi_1(r) = b\rho_0 r_0^2\ln\left(\frac{r}{r_0}\right). \tag{2.242}$$

To determine $\phi_2(r,\theta)$ we have to solve (2.241) with $S(\theta) = \delta_{\mathrm{Dirac}}(\theta - \pi/2)$. Changing variables to $\mu = \cos\theta$, we can write this function as

$$S_1(\theta) \equiv 2\delta_{\mathrm{Dirac}}\left(\theta - \pi/2\right) = 2\delta_{\mathrm{Dirac}}(\mu). \tag{2.243}$$

So we need to solve

$$\frac{d}{d\mu}\left((1-\mu^2)\frac{dP}{d\mu}\right) = 2\delta(\mu) - 1. \tag{2.244}$$

The solution to this equation is given by $P = \ln\left(1 + |\mu|\right)$; this can be verified by direct differentiation:

$$(1-\mu^2)\frac{dP}{d\mu} = \frac{(1-\mu^2)}{(1+|\mu|)}\frac{d|\mu|}{d\mu} = \frac{(1-\mu^2)}{(1+|\mu|)}Sg(\mu) = Sg(\mu) - \mu, \tag{2.245}$$

where $Sg(\mu) = (\mu/|\mu|) = \theta(\mu) - \theta(-\mu)$. Using $Sg'(\mu) = 2\delta(\mu)$, we find

$$\frac{d}{d\mu}\left[(1-\mu^2)\frac{dP}{d\mu}\right] = 2\delta(\mu) - 1. \tag{2.246}$$

So the density $S(\theta) = \delta_{\mathrm{Dirac}}\left(\theta - \pi/2\right)$ generates a $P(\theta)$ of the form

$$P(\theta) = \ln\left[1 + |\cos\theta|\right]. \tag{2.247}$$

3

Fluid mechanics

3.1 Basic equations of fluid mechanics

(a) In a small interval of time dt the distribution function changes due to two different processes. (i) The positions and velocities of the particles change due to the action of a smooth external field and the distribution function will change by the amount

$$df = \frac{\partial f}{\partial t} dt + \frac{\partial f}{\partial \mathbf{x}} \cdot d\mathbf{x} + \frac{\partial f}{\partial \mathbf{p}} \cdot d\mathbf{p} = \left[\frac{\partial f}{\partial t} + \mathbf{v} \cdot \frac{\partial f}{\partial \mathbf{x}} - \nabla U_{sm} \cdot \frac{\partial f}{\partial \mathbf{p}} \right] dt. \tag{3.1}$$

This change occurs even for a collisionless system. (ii) The distribution function will also change due to the collisions between molecules which change the number density of molecules in a phase cell. The net effect of collisions can be computed by calculating the rate at which particles are scattered *into* a small volume $d^3x \, d^3p$ and subtracting from it the rate at which the molecules are scattered *out* of this volume. These changes in the distribution function can be expressed in terms of the scattering cross-section for molecular collisions in the following manner.

Consider a collision between two molecules of momenta $\mathbf{p}$ and $\mathbf{p}_1$ leading to final momenta $\mathbf{p}'$ and $\mathbf{p}'_1$. Let the scattering cross-section for this process be $\sigma(\mathbf{p}, \mathbf{p}_1; \mathbf{p}', \mathbf{p}'_1)$. The rate at which such collisions will remove particles from the phase volume $d^3x \, d^3p$ is given by

$$\mathcal{R}_{out} = f(\mathbf{p}) \, d^3x \, d^3p \int |\mathbf{v} - \mathbf{v}_1| \sigma(\mathbf{p}, \mathbf{p}_1; \mathbf{p}', \mathbf{p}'_1) \, d\Omega \, f(\mathbf{p}_1) \, d^3\mathbf{p}_1. \tag{3.2}$$

The expression inside the integral gives the rate of collisions of a molecule having momentum $\mathbf{p}$ with molecules of all possible momenta $\mathbf{p}_1$. This is multiplied by the number of particles which are present in the small volume $d^3x \, d^3p$ under consideration. In order to simplify notation, we have suppressed the $\mathbf{x}$- and t-dependence of the distribution function. We have assumed that the probability of finding two particles at $(t, \mathbf{x})$ with momenta $\mathbf{p}$ and $\mathbf{p}_1$ simultaneously is proportional to the product $f(\mathbf{p})f(\mathbf{p}_1)$. This assumption tacitly ignores any correlation between particles.

One can similarly write an expression for the rate of scattering of molecules *into* the small element $d^3\mathbf{x}\,d^3\mathbf{p}$ as

$$\mathscr{R}_{\text{in}} = d^3\mathbf{x} \int |\mathbf{v}' - \mathbf{v}'_1| \sigma\left(\mathbf{p}', \mathbf{p}'_1; \mathbf{p}, \mathbf{p}_1\right) d\Omega\, f\left(\mathbf{p}'\right) d^3\mathbf{p}'\, f\left(\mathbf{p}'_1\right) d^3\mathbf{p}'_1. \tag{3.3}$$

The time reversability of molecular collision implies that the scattering cross-section is symmetric between the initial and final states. Hence, we must have

$$\sigma\left(\mathbf{p}', \mathbf{p}'_1; \mathbf{p}, \mathbf{p}_1\right) = \sigma\left(\mathbf{p}, \mathbf{p}_1; \mathbf{p}', \mathbf{p}'_1\right) \equiv \sigma\left(\Omega\right). \tag{3.4}$$

Further, for collisions conserving momentum and energy, $d^3\mathbf{p}\,d^3\mathbf{p}_1 = d^3\mathbf{p}'\,d^3\mathbf{p}'_1$. Using these results in (3.2) and (3.3), we can write

$$\mathscr{R}_{\text{in}} - \mathscr{R}_{\text{out}} = d^3\mathbf{x}\,d^3\mathbf{p} \int |\mathbf{v} - \mathbf{v}_1| \sigma\left(\Omega\right) d\Omega \left[f'f'_1 - ff_1\right] d^3\mathbf{p}_1, \tag{3.5}$$

where $f = f\left(\mathbf{x}, \mathbf{p}, t\right), f_1 = f\left(\mathbf{x}, \mathbf{p}_1, t\right), f' = f\left(\mathbf{x}, \mathbf{p}', t\right)$ and $f'_1 = f\left(\mathbf{x}, \mathbf{p}'_1, t\right)$. This integral represents the change in the distribution function due to the collisions and is usually expressed as $C[f]$. Combining with (3.1) we can write the evolution equation for the distribution function as an integro-differential equation

$$\frac{\partial f}{\partial t} + \mathbf{v} \cdot \frac{\partial f}{\partial \mathbf{x}} - \nabla U \cdot \frac{\partial f}{\partial \mathbf{p}} = C[f]. \tag{3.6}$$

In deriving this equation, we have assumed that a single collision can change the momentum of a particle by an arbitrary amount. Hence, $C[f]$ involves an integration over $\mathbf{p}_1$ and equation (3.6) is an integro-differential equation. In problem 2.12 we studied the case of gravitational 'soft collisions', based on the assumption that each collision changes the momentum of the particle only slightly. In that case, $C[f]$ can be represented as arising due to diffusion in velocity space, and the resulting equation is only a differential equation.

(b) We can write the evolution equation in the form

$$\frac{\partial f}{\partial t} + \frac{\partial}{\partial x^a}\left(v^a f\right) - \frac{\partial}{\partial p^a}\left(\frac{\partial U}{\partial x^a} f\right) = C[f]. \tag{3.7}$$

We multiply this equation by m, p^b or ϵ and integrate over $d^3\mathbf{p}$. Since the molecular collisions conserve the number of molecules involved in the collisions, the total momentum of the molecules and the total energy of the molecules, we must have

$$\int d^3\mathbf{p}\, C[f] = 0, \quad \int \epsilon C[f]\, d^3\mathbf{p} = 0, \quad \int \mathbf{p} C[f]\, d^3\mathbf{p} = 0. \tag{3.8}$$

So we obtain the equations

$$\frac{\partial \rho}{\partial t} + \frac{\partial}{\partial x^i}(\rho V^i) = 0, \tag{3.9}$$

$$\frac{\partial}{\partial t}(\rho V^i) + \frac{\partial}{\partial x^k}T^{ik} + \rho\frac{\partial U}{\partial x^i} = 0, \tag{3.10}$$

$$\frac{\partial}{\partial t}(N\bar{\epsilon}) + \frac{\partial}{\partial x^a}q^a + \rho\frac{\partial U}{\partial x^k}V^k = 0, \tag{3.11}$$

where

$$N = \frac{\rho}{m} = \int d^3\mathbf{p}\, f(\mathbf{x},\mathbf{p},t), \quad V^a \equiv \langle v^a \rangle = \frac{1}{N}\int v^a f\, d^3\mathbf{p}, \quad \bar{\epsilon} = \langle \epsilon \rangle = \frac{1}{N}\int \epsilon f\, d^3\mathbf{p}, \tag{3.12}$$

$$T^{ik} = mN\langle v^i v^k \rangle = \int m v^i v^k f\, d^3\mathbf{p}, \quad q^a = N\langle \epsilon v^a \rangle = \int \epsilon v^a f\, d^3\mathbf{p}. \tag{3.13}$$

Each of the terms in these equations has a simple interpretation. Equation (3.9) represents the conservation of mass in the smooth fluid limit. The quantity $\rho = Nm$ is the mass density of gas and the vector field $\mathbf{V}(\mathbf{x},t)$ represents the macroscopic flow velocity of the fluid at a location $\mathbf{x}$ at time t. This velocity field is the weighted average of the microscopic velocities of the molecules at the location $\mathbf{x}$ at time t.

Equation (3.10) is the force equation in the fluid limit. The quantity T^{ik} represents the stress-tensor of the fluid; the gradient of the stress-tensor gives rise to the acceleration of the fluid element. This equation could alternatively be thought of as a momentum conservation equation. The quantity $\rho\mathbf{V}$ denotes the momentum density in the fluid and the gradient of the stress-tensor determines how this quantity changes. The third term in this equation is due to the external forces acting on the fluid element.

Equation (3.11) ensures conservation of energy. The quantity $N\bar{\epsilon}$ is the energy density in the fluid and the vector field $\mathbf{q}$ represents the energy flux. The third term takes into account the change in the potential energy of the fluid element as it moves.

(c) We shall hereafter assume that $U_{\text{sm}} = 0$ and use units with $k = 1$. The collision term $C[f]$ vanishes if the distribution function has the form

$$f_0(\mathbf{p}) = \exp\left[\frac{\mu}{T} - \frac{m}{2T}(\mathbf{v} - \mathbf{V})^2\right], \tag{3.14}$$

with μ, T and $\mathbf{V}$ being constants. The left-hand side of the Boltzmann transport equation (BTE, hereafter) also vanishes for this choice of f_0 due to the absence of spatial or temporal gradients. It follows that f_0 represents the equilibrium distribution function characterized by a temperature T, chemical potential μ and a bulk velocity $\mathbf{V}$. Of these, the bulk velocity can be removed by transforming to another frame moving with a velocity $\mathbf{V}$; μ and T can be determined once the total energy and mass of the system are specified.

This distribution function describes a state of complete thermodynamic equilibrium. To obtain a more realistic situation, we may assume that the distribution function has the same form as in equation (3.14) but with μ, T and $\mathbf{V}$ varying in space and time. The collision term, being 'local' in space and time, still vanishes. But on the left-hand side we will get non-zero contributions due to spatial and temporal gradients in μ, T and $\mathbf{V}$. If we assume that these gradients are small and ignore them to the lowest order of approximation, we obtain a situation called the state of 'local thermodynamic equilibrium' (LTE, hereafter).

We can estimate T_{ik} and $\mathbf{q}$ in the context of LTE as follows. Let the symbol $\langle...\rangle$ denote the average value, computed using the distribution function f. Writing the microscopic velocity of the particles as $\mathbf{v} = \mathbf{v}' + \mathbf{V}$, where $\langle \mathbf{v} \rangle = \mathbf{V}$, we have

$$T_{ik} = mN\langle v_i v_k \rangle = mN\langle (v_i' + V_i)(v_k' + V_k) \rangle = mN V_i V_k + mN\langle v_i' v_k' \rangle, \quad (3.15)$$

since $\langle v_i' V_k \rangle = V_k \langle v_i' \rangle = 0$. Further, since there is no preferred direction for $\mathbf{v}'$, it follows that

$$\langle v_i' v_k' \rangle = \frac{1}{3}\langle v'^2 \rangle \delta_{ik} = \frac{T}{m}\delta_{ik} = \frac{P}{Nm}\delta_{ik}, \quad (3.16)$$

where $P = NT$ is the pressure. Hence, we obtain the stress-tensor for a gas in LTE to be

$$T_{ik} = \rho V_i V_k + P\delta_{ik}. \quad (3.17)$$

Similar analysis can be performed to compute $\mathbf{q}$. If the energy of a molecule is ϵ in the lab frame and ϵ' in the rest frame comoving with the fluid, then

$$\epsilon = \epsilon' + m\mathbf{V} \cdot \mathbf{v}' + \frac{1}{2}mV^2, \quad \mathbf{v} = \mathbf{v}' + \mathbf{V}. \quad (3.18)$$

Using these in the expression $\mathbf{q} = N\langle \epsilon \mathbf{v} \rangle$ we obtain

$$\mathbf{q} = N\mathbf{V}\langle \epsilon \rangle + Nm\langle \mathbf{v}'(\mathbf{V} \cdot \mathbf{v}') \rangle = N\mathbf{V}\left[\frac{1}{2}mV^2 + \langle \epsilon' \rangle + \frac{1}{3}m\langle v'^2 \rangle\right]$$

$$= \mathbf{V}\left[\frac{1}{2}\rho V^2 + P + N\langle \epsilon' \rangle\right] = \left[\frac{1}{2}V^2 + w\right]\rho\mathbf{V}, \quad (3.19)$$

where $w = (N\langle \epsilon' \rangle + P)/\rho$ is the heat function ('enthalpy') per unit mass of the fluid. Substitution of these expressions into (3.10) and (3.11) will lead to the equations of motion for an ideal gas in which all dissipative phenomena are ignored.

The fact that energy flux is $\rho\mathbf{V}(V^2/2 + w)$ rather than $\rho\mathbf{V}(V^2/2 + \epsilon)$ has a simple interpretation. Using $w = \epsilon + P/\rho$, we can write the flux of energy through a surface as

$$-\int \rho\mathbf{V}\left(\frac{1}{2}V^2 + \epsilon\right) \cdot \hat{\mathbf{n}}\,dS - \int P\mathbf{V} \cdot \hat{\mathbf{n}}\,dS. \quad (3.20)$$

The first term is the bulk kinetic energy and internal energy transported through the surface, while the second term is the work done by the pressure forces on the fluid.

3.2 Viscosity and heat conduction

(a) The description of gas in LTE given above ignores the effect of the spatial and temporal gradients of $T, \mathbf{V}$ and μ on the distribution function. To the next order in approximation, we may write $f = f_0 \left(1 + h/T\right)$, where h is linear in these gradients. This will allow us to obtain T_{ik} and q_i accurate to first order in the spatial gradient of $T, \mathbf{V}$ and μ. (As we shall see, time derivatives can be expressed in terms of spatial gradients.) The gradient in T will cause thermal conduction, the gradient in $\mathbf{V}$ will cause viscous dissipation and the gradient in the chemical potential μ will cause diffusion.

To do this we have to linearize BTE in h and in spatial gradients and solve the resulting equation. The solution h should, however, satisfy certain extra constraints. This is because the distribution function in LTE is fully specified only when the number, energy and momentum densities are specified, thereby allowing us to determine μ, T and $\mathbf{V}$. The small perturbation h to the equilibrium distribution f_0 should not change the values of these quantities. Hence, it follows that we must demand

$$\langle h \rangle = 0, \quad \langle h\epsilon \rangle = 0, \quad \langle h\mathbf{p} \rangle = 0. \tag{3.21}$$

We shall now linearize BTE with respect to h, starting from the right-hand side. The integrand on the right-hand side will contain h in the combinations $f_0' f_{01}' \left(h' + h_1'\right)$ and $f_0 f_{01} \left(h + h_1\right)$. Using the fact that $f_0 f_{01} = f_0' f_{01}'$ we can pull f_0 out of the integral and obtain $C[f] = f_0 I\left(h\right)/T$, where

$$I\left(h\right) = \int |\mathbf{v} - \mathbf{v}_1| f_{01} \left(h' + h_1' - h - h_1\right) \sigma\left(\Omega\right) d\Omega \, d^3\mathbf{p}_1. \tag{3.22}$$

Linearization of the left-hand side is somewhat more involved. To begin with, we note that we can evaluate the left-hand side most conveniently in a frame in which $\mathbf{V} = 0$. (Of course, this simplification should be made after calculating the derivatives.) Secondly, for a distribution function of the form (3.14) (with μ, T and $\mathbf{V}$ depending on $\mathbf{x}$) it is easier to evaluate the logarithmic derivatives. Differentiating (3.14) and then setting $\mathbf{V} = 0$, we obtain

$$\frac{T}{f_0} \frac{\partial f_0}{\partial t} = \left[\left(\frac{\partial \mu}{\partial T}\right)_P - \frac{\mu - \epsilon}{T}\right] \frac{\partial T}{\partial t} + \left(\frac{\partial \mu}{\partial P}\right)_T \frac{\partial P}{\partial t} + m\mathbf{v} \cdot \frac{\partial \mathbf{V}}{\partial t}. \tag{3.23}$$

We now use the thermodynamic relations

$$\left(\frac{\partial \mu}{\partial T}\right)_P = -s, \quad \left(\frac{\partial \mu}{\partial P}\right)_T = \frac{1}{N}, \quad \mu = w - Ts \tag{3.24}$$

to write the above equation in the form

$$\frac{T}{f_0} \frac{\partial f_0}{\partial t} = \left(\frac{\epsilon - w}{T}\right) \frac{\partial T}{\partial t} + \frac{1}{N} \frac{\partial P}{\partial t} + m\mathbf{v} \cdot \frac{\partial \mathbf{V}}{\partial t}. \tag{3.25}$$

Similarly, we find for the spatial gradient

$$\frac{T}{f_0} (\mathbf{v} \cdot \nabla) f_0 = \left(\frac{\epsilon - w}{T} \right) (\mathbf{v} \cdot \nabla) T + \frac{1}{N} (\mathbf{v} \cdot \nabla) P + m v_a v_b V_{ab} , \qquad (3.26)$$

where

$$V_{ab} \equiv \frac{1}{2} \left[\frac{\partial V_a}{\partial x_b} + \frac{\partial V_b}{\partial x_a} \right] . \qquad (3.27)$$

The left-hand side of BTE can now be found by adding the expressions for the spatial and temporal gradients of f_0. But before we do that, it is necessary to rewrite the time derivatives of macroscopic variables in terms of spatial derivatives. This can be done by using the equations for the ideal fluid, evaluated in a frame with $\mathbf{V} = 0$:

$$\frac{\partial \mathbf{V}}{\partial t} = -\frac{1}{Nm} \nabla P, \quad \frac{1}{N} \frac{\partial N}{\partial t} = -\nabla \cdot \mathbf{V}, \quad \frac{\partial s}{\partial t} = 0 . \qquad (3.28)$$

We now set $N = (P/T)$ in the second equation and use the thermodynamic relations $(\partial s/\partial T)_P = c_p/T, (\partial s/\partial P)_T = -1/P$ in the third equation. Simplifying the resulting set of equations we obtain

$$\frac{1}{T} \frac{\partial T}{\partial t} = -\frac{1}{c_v} \nabla \cdot \mathbf{V}, \quad \frac{1}{P} \frac{\partial P}{\partial t} = \frac{c_p}{c_v} \nabla \cdot \mathbf{V} . \qquad (3.29)$$

We have now expressed all the time derivatives in terms of spatial gradients. A straightforward calculation will now give the left-hand side of BTE to be

$$\frac{\partial f_0}{\partial t} + \mathbf{v} \cdot \nabla f_0 = \frac{f_0}{T} \left\{ \frac{\epsilon - w}{T} \mathbf{v} \cdot \nabla T + m v_a v_b V_{ab} + \frac{w - T c_p - \epsilon}{c_v} \nabla \cdot \mathbf{V} \right\} . \qquad (3.30)$$

For an ideal monotonic gas, $w = c_p T$; using this result and our linearized expression for $C[f]$ we obtain the final form of the linearized BTE:

$$\frac{\epsilon (\mathbf{p}) - c_p T}{T} \mathbf{v} \cdot \nabla T + \left[m v_i v_k - \delta_{ik} \frac{\epsilon (\mathbf{p})}{c_v} \right] V_{ik} = I (h) . \qquad (3.31)$$

This is an integral equation for h; once h is obtained by solving this equation, one can obtain T_{ik} and q_i accurate to first order in spatial gradients.

(b) (i) The vector g_1^i has to be made from v^i alone and hence should point in the same direction as $\mathbf{v}$. Any such vector can be written in the form $g_1^i = (v^i/v) g_1 (v)$, where g_1 is a scalar made from v.

By definition, g^{ik} is a traceless symmetric tensor made from v^i. The only tensor of that kind should have the form

$$g^{ik} = \left(v^i v^k - \frac{1}{3} v^2 \delta^{ik} \right) g_2 (v) , \qquad (3.32)$$

where g_2 is a scalar function of v.

(ii) From the definition for T_{ik} we have

$$T_{ik} = \int m v_i v_k f_0 \left(1 + \frac{h}{T} \right) d^3 p = \rho V_i V_k + P \delta_{ik} - \sigma_{ik} , \qquad (3.33)$$

with

$$\sigma_{ik} = -\frac{m}{T} \int v_i v_k f_0 h \, d^3 p$$

$$= -\frac{m}{T} \int f_0 v_i v_k \left[\frac{v_a}{v} g_1(v) \frac{\partial T}{\partial x_a} + \left(v^a v^b - \frac{1}{3} v^2 \delta^{ab} \right) g_2(v) V_{ab} + g_3 \nabla \cdot \mathbf{V} \right] d^3 p.$$

$$(3.34)$$

The form of these integrals can be determined by general symmetry considerations. The first integral has the form

$$A_{ik} = -\frac{m}{T} \left(\frac{\partial T}{\partial x_a} \right) \int d^3 p \, f_0 v_i v_k v_a \equiv -\frac{m}{T} \left(\frac{\partial T}{\partial x_a} S_{aik} \right). \qquad (3.35)$$

In evaluating this integral we may assume that we are working in a frame with $\langle v_i \rangle = \mathbf{V} = 0$, since the contribution from $\mathbf{V}$ is already taken into account. This implies that the completely symmetric tensor S_{aik} has to be constructed entirely from δ_{mn}, the only tensor available in the problem. No such S_{aik} can be constructed from δ_{mn}. Hence, $A_{ik} = 0$.

Consider now the second term; it has the form

$$B_{ik} = -\left(\frac{m}{T} \right) V^{ab} \int d^3 p \, v_i v_k \left(v_a v_b - \frac{1}{3} v^2 \delta_{ab} \right) g_2(v) = V^{ab} S_{ikab}. \qquad (3.36)$$

The tensor S_{abik} is symmetric in (a, b) and (i, k) and is traceless in (a, b). Again, it has to be constructed from δ_{mn}. The only tensor with these properties is

$$S_{ikab} = \eta \left[\delta_{ai} \delta_{bk} + \delta_{ak} \delta_{bi} - \frac{2}{3} \delta_{ab} \delta_{ik} \right], \qquad (3.37)$$

where η is some constant. On setting $i = a, k = b$, in this expression and in (3.36) we obtain

$$\eta \left[9 + 3 - \frac{2}{3} \times 3 \right] = 10\eta = -\frac{2}{3} \left(\frac{m}{T} \right) \int d^3 p \, v^4 g_2(v). \qquad (3.38)$$

Substituting for S_{ikab} in (3.36) from (3.37), B_{ik} can be written as

$$B_{ik} = \eta \left[V_{ik} + V_{ki} - \frac{2}{3} \delta_{ik} \nabla \cdot \mathbf{V} \right] = 2\eta \left[V_{ik} - \frac{1}{3} \delta_{ik} \nabla \cdot V \right]. \qquad (3.39)$$

Finally, the third term has the form

$$C_{ik} = -\frac{m}{T} \nabla \cdot \mathbf{V} \int f_0 v_i v_k \, d^3 p. \qquad (3.40)$$

By similar arguments we conclude that C_{ik} should be proportional to δ_{ik}. Hence,

$$C_{ik} = \zeta (\nabla \cdot \mathbf{V}) \delta_{ik}, \quad \zeta = -\frac{m}{3T} \langle v^2 g_3 \rangle. \qquad (3.41)$$

Adding the contributions B_{ik} and C_{ik} we obtain the final form of the stress-tensor correct to linear order in the gradient of the velocity:

$$T_{ik} = \rho V_i V_k + P\delta_{ik} - \sigma_{ik}, \quad \sigma_{ik} = 2\eta\left(V_{ik} - \frac{1}{3}\delta_{ik}\nabla\cdot\mathbf{V}\right) + \zeta\delta_{ik}\nabla\cdot\mathbf{V}, \quad (3.42)$$

with η and ζ determined by (3.38) and (3.41).

The analysis for determining the energy flux is similar. From the basic definition, we have

$$q_i = \int \epsilon v_i f_0 \left(1 + \frac{h}{T}\right) d^3p = \rho\left(\frac{1}{2}V^2 + w\right) V_i + Q_i, \quad (3.43)$$

where Q_i denotes the correction to the flux in the case of ideal gas:

$$Q_i = \frac{1}{T}\int d^3p\,\epsilon v_i \left[\frac{v_a}{v}g_1(v)\frac{\partial T}{\partial x_a} + g_{ab}V^{ab} + g_3\nabla\cdot\mathbf{V}\right]. \quad (3.44)$$

By arguments similar to the one given above, we can determine the form of the first term. It reduces to $-\kappa\left(\partial T/\partial x_i\right)$ with

$$\kappa = -\frac{1}{3T}\langle \epsilon v g_1(v)\rangle. \quad (3.45)$$

The second and third terms represent the contributions to the energy flux due to gradients in velocity. The values of these terms are most easily obtained by the following consideration. Since the energy density contributed by velocity gradients is already determined to be $-\sigma_{ik}$, the corresponding energy flux along the direction $-i$ must be $-V^k\sigma_{ik}$. Hence, the net energy flux must have the form

$$Q_i = -\kappa\frac{\partial T}{\partial x^i} - V^k\sigma_{ik}. \quad (3.46)$$

Note that the second term vanishes when computed in a frame of reference with $V_i = 0$. Hence, it needs to be obtained in a frame with $V_i \neq 0$ in which we cannot use the general symmetry arguments used earlier.

(iii) In an ideal fluid, the flux of energy is described by $\mathbf{q}$ obtained in (3.19). In the presence of temperature and velocity gradients we expect an additional flow of energy since the random energy of molecules at different locations will now be different. To the lowest order at which we are working, we can expand q_i as a Taylor series in the gradients and retain only the first non-trivial terms. This will contribute to $\mathbf{q}$ a term of the form $-\kappa\nabla T$ (where κ is called the thermal conductivity) in addition to terms arising from velocity gradients. This is what we found in the last part.

The contributions to T_{ik} arising from spatial variation of $\mathbf{V}$ are a bit more involved. To begin with, it is clear that we can again expand T_{ik} in a Taylor series and retain only terms linear in $\partial V_i/\partial x^k$. We can write this gradient as a sum of symmetric and antisymmetric tensors:

$$\frac{\partial V_i}{\partial x^k} = V_{ik} + \frac{1}{2}\left(\frac{\partial V_i}{\partial x^k} - \frac{\partial V_k}{\partial x^i}\right). \quad (3.47)$$

We do not, however, expect any viscous dissipation when the entire fluid is in a state of uniform rigid rotation, for which $\mathbf{V} = (\boldsymbol{\Omega} \times \mathbf{x})$. In this case, $V_{ik} = 0$ but the antisymmetric part does not vanish. It follows that T_{ik} should not have any contribution from the antisymmetric part if there should be no viscous dissipation in a state of uniform rotation. Hence, T_{ik} must be a linear combination of V_{ik} and $\delta_{ik} (\nabla \cdot \mathbf{V})$, which are the only two symmetric, independent, second rank tensors linear in the gradients of V. Such a linear combination can be written as $\alpha V_{ik} + \beta \delta_{ik} (\nabla \cdot \mathbf{V})$. It is, however, conventional to separate its trace out from V_{ik} and rewrite the expression in the form

$$\sigma_{ik} = 2\eta \left(V_{ik} - \frac{1}{3} \delta_{ik} \nabla \cdot \mathbf{V} \right) + \zeta \delta_{ik} \nabla \cdot \mathbf{V}. \tag{3.48}$$

This is the general form for the viscous energy tensor which we have obtained. Note that it is parametrized by two coefficients of viscosity, η and ζ. To linear order in the gradients of $\mathbf{v}$ and T no other contributions to T_{ik} can arise. Our system is now completely determined by the three constants η, ζ and κ.

3.3 Macroscopic flow of the fluid

(a) The mass conservation equation has the same form for ideal and non-ideal fluids:

$$0 = \frac{\partial \rho}{\partial t} + \frac{\partial}{\partial x^i} \left(\rho v^i \right) = \frac{\partial \rho}{\partial t} + v^i \frac{\partial \rho}{\partial x^i} + \rho \left(\nabla \cdot \mathbf{v} \right) \tag{3.49}$$

or

$$\frac{d\rho}{dt} + \rho \left(\nabla \cdot \mathbf{v} \right) = 0. \tag{3.50}$$

To obtain the Euler equation for a fluid with viscosity we have to substitute the expression for T_{ik}:

$$T_{ik} = \rho v_i v_k + p \delta_{ik} - 2\eta \left[v_{ik} - \frac{1}{3} \delta_{ik} \left(\nabla \cdot \mathbf{v} \right) \right] - \zeta \delta_{ik} \nabla \cdot \mathbf{v} \tag{3.51}$$

into the equation

$$\frac{\partial}{\partial t} \left(\rho v^i \right) + \frac{\partial T^{ik}}{\partial x^k} = 0. \tag{3.52}$$

This gives

$$\begin{aligned}
\frac{\partial}{\partial t} \left(\rho v_i \right) &+ \frac{\partial}{\partial x_k} \left(\rho v_i v_k \right) + \frac{\partial p}{\partial x^i} \\
&= \eta \left[\frac{\partial}{\partial x_k} \left(\frac{\partial v_i}{\partial x^k} \right) + \frac{\partial}{\partial x_k} \left(\frac{\partial v_k}{\partial x^i} \right) \right] - \frac{2}{3} \eta \frac{\partial}{\partial x^i} \left(\nabla \cdot \mathbf{v} \right) + \zeta \frac{\partial}{\partial x^i} \left(\nabla \cdot \mathbf{v} \right) \\
&= \eta \nabla^2 v_i + \left(\frac{1}{3} \eta + \zeta \right) \frac{\partial}{\partial x^i} \nabla \cdot \mathbf{v}. \tag{3.53}
\end{aligned}$$

On the left-hand side, we expand the first two terms and use the continuity equation to eliminate the $v_i \left(\partial \rho / \partial t \right)$ term. Then we find

$$\rho \frac{d\mathbf{v}}{dt} = -\nabla p + \eta \nabla^2 \mathbf{v} + \left(\zeta + \frac{1}{3}\eta \right) \nabla \left(\nabla \cdot \mathbf{v} \right). \tag{3.54}$$

This equation is called the Navier–Stokes equation. To derive the last equation, we start with the equation for energy flux:

$$\frac{\partial}{\partial t} \left(N\bar{\epsilon} \right) + \frac{\partial q^a}{\partial x^a} = 0 \tag{3.55}$$

and substitute

$$N\bar{\epsilon} = \langle \epsilon \rangle = \left\langle \frac{1}{2}m \left(\mathbf{v} + \mathbf{u} \right)^2 \right\rangle = \frac{1}{2}\rho v^2 + \rho \frac{3}{2}T = \rho \left(\frac{1}{2}v^2 + \mathscr{E} \right), \tag{3.56}$$

where $\left(\mathbf{v} + \mathbf{u} \right)$ is the microscopic velocity of the molecules with $\langle \mathbf{u} \rangle = 0$, $\langle u^2 \rangle = 3T/m$; $\mathscr{E} = \left(3T/2 \right)$ is the internal energy density and

$$q_a = \rho \left(\frac{1}{2}v^2 + w \right) v_a - v^b \sigma_{ab} - \kappa \frac{\partial T}{\partial x_a}. \tag{3.57}$$

(We are using units with $k = 1$.) Thus we obtain

$$\frac{\partial}{\partial t} \left[\rho \left(\frac{1}{2}v^2 + \mathscr{E} \right) \right] = -\frac{\partial}{\partial x^a} \left[\rho v^a \left(\frac{1}{2}v^2 + w \right) - v_b \sigma^{ab} - \kappa \frac{\partial T}{\partial x_a} \right]. \tag{3.58}$$

The left-hand side denotes the rate of change of energy density; the two contributions to the energy density are from steady motion (with velocity $\mathbf{v}$) and random motion (with an internal energy $(3/2)\rho T$). The energy flux on the right-hand side has three contributions: the first term represents the flow of kinetic energy and work done against pressure. The second term is the flux of energy due to internal friction. The last term is the heat flux due to conduction. We will now re-express this equation in a more convenient form.

Taking the dot product of the Euler equation with $\mathbf{v}$ and using the continuity equation, we obtain

$$\frac{\partial}{\partial t} \left(\frac{1}{2}\rho v^2 \right) + \frac{\partial}{\partial x^k} \left(\frac{1}{2}\rho v^2 v^k \right) = -v^i \frac{\partial p}{\partial x^i} + v^i \frac{\partial \sigma_{ki}}{\partial x_k}. \tag{3.59}$$

Subtraction of this equation from (3.58) gives, after some rearrangement,

$$\frac{\partial}{\partial t} \left(\rho \mathscr{E} \right) + \nabla \cdot \left(\rho \mathscr{E} \mathbf{v} \right) = -p \nabla \cdot \mathbf{v} + \nabla \cdot \left(\kappa \nabla T \right) + \sigma^{ik} \frac{\partial v_i}{\partial x^k}. \tag{3.60}$$

Using the continuity equation this can be rewritten as

$$\rho \frac{d\mathscr{E}}{dt} + p \nabla \cdot \mathbf{v} = \nabla \cdot \left(\kappa \nabla T \right) + \sigma^{ik} \frac{\partial v_i}{\partial x^k}. \tag{3.61}$$

But,

$$p \nabla \cdot \mathbf{v} = - \left(p/\rho \right) \left(d\rho/dt \right) = +p\rho \left(d\rho^{-1}/dt \right) = \rho p \frac{d\mathscr{V}}{dt}, \tag{3.62}$$

where $\mathscr{V} \equiv \rho^{-1}$ is the specific volume. Hence the left-hand side of (3.61) is

$$\rho \left(\frac{d\mathscr{E} + p\, d\mathscr{V}}{dt} \right) = \rho T \frac{ds}{dt} = \rho T \left(\frac{\partial s}{\partial t} + (\mathbf{v} \cdot \nabla)\, s \right), \qquad (3.63)$$

where s is the entropy density. Using this, we obtain

$$\rho T \frac{ds}{dt} = \rho T \left(\frac{\partial s}{\partial t} + (\mathbf{v} \cdot \nabla)\, s \right) = \nabla \cdot (\kappa \nabla T) + \sigma^{ik} \frac{\partial v_i}{\partial x^k}. \qquad (3.64)$$

In general, both the pressure and density (or specific volume) of the fluid element will change as it moves along. Among these, the change in pressure is negligible if the fluid velocity is small compared to the velocity of sound; the density variations will depend on temperature gradients as well and cannot, in general, be ignored. Hence, one may think of the derivatives on the left-hand side of (3.64) to be taken at constant pressure rather than at constant volume. In that case, we can write

$$\left(\frac{\partial s}{\partial t} \right) = \left(\frac{\partial s}{\partial T} \right)_P \left(\frac{\partial T}{\partial t} \right) = \frac{c_p}{T} \left(\frac{\partial T}{\partial t} \right), \quad \nabla s = \left(\frac{\partial s}{\partial T} \right)_P \nabla T = \frac{c_p}{T} \nabla T, \quad (3.65)$$

leading to the final equation

$$c_p \rho \left(\frac{dT}{dt} \right) = \sigma^{ik} \frac{\partial v_i}{\partial x^k} + \nabla \cdot (\kappa \nabla T). \qquad (3.66)$$

(b) The consistency of these equations with energy conservation is obvious from the way the last equation was derived.

The behaviour of entropy is slightly more tricky. To begin with, it can be verified by simple algebra that

$$\sigma^{ik} \frac{\partial v_i}{\partial x^k} = \frac{1}{2} \eta \left(2V_{ik} - \frac{2}{3} \delta_{ik} \nabla \cdot \mathbf{v} \right)^2 + \zeta \, (\nabla \cdot \mathbf{v})^2. \qquad (3.67)$$

Using this expression and the continuity equation, we can rewrite the energy conservation law in a different form. We note that

$$\frac{\partial (\rho s)}{\partial t} = \rho \frac{\partial s}{\partial t} + s \frac{\partial \rho}{\partial t} = -s \nabla \cdot (\rho \mathbf{v}) - \rho (\mathbf{v} \cdot \nabla)\, s + \frac{1}{T} \nabla \cdot (\kappa \nabla T)$$

$$+ \frac{\eta}{2T} \left(2V_{ik} - \frac{2}{3} \delta_{ik} \nabla \cdot \mathbf{v} \right)^2 + \frac{\zeta}{T} (\nabla \cdot \mathbf{v})^2. \qquad (3.68)$$

This equation represents how the entropy of a fluid element changes with time. On the right-hand side the first two terms can be combined to give $-\nabla \cdot (\rho s \mathbf{v})$. Further, we can write

$$\frac{1}{T} \nabla \cdot (\kappa \nabla T) = \nabla \cdot \left(\frac{\kappa \nabla T}{T} \right) + \kappa \left(\frac{\nabla T}{T} \right)^2. \qquad (3.69)$$

Using these results and integrating equation (3.68) over a large volume of the fluid we obtain

$$\frac{d}{dt} \int \rho s \, d^3x = \int \kappa \left(\frac{\nabla T}{T}\right)^2 d^3x + \int \frac{\eta}{2T} \left(2V_{ik} - \frac{2}{3}\delta_{ik}\nabla \cdot \mathbf{v}\right)^2 d^3x$$
$$+ \int \frac{\zeta}{T} (\nabla \cdot \mathbf{v})^2 d^3x. \tag{3.70}$$

(It is assumed that surface integrals vanish when taken over a surface at large distance.) The first term on the right gives the entropy increase due to thermal conduction and the other two terms describe the entropy increase due to viscous dissipation.

In the analysis above, we have been using standard results of equilibrium thermodynamics in a context in which the system is only in LTE and velocity and temperature gradients are present. Strictly speaking, the thermodynamic relations need to be modified in such a case. The quantities ρ, ϵ and $\mathbf{v}$, being defined through microscopic (primitive) variables, continue to retain their validity. The entropy $s = s(\rho, \epsilon)$, however, will not be the true thermodynamic entropy in the strict sense. But since the thermodynamic entropy should be a maximum in equilibrium, it cannot change when the variables deviate from their equilibrium values to first order. In other words, to first-order accuracy in the gradients of T and $\mathbf{v}$ the entropy used above may be taken to be the thermodynamic entropy.

Similar remarks apply in the case of pressure. If we take the pressure for a viscous fluid $p = p(\rho, \epsilon)$ to be of the same form as in thermal equilibrium, such a pressure will *not* give the correct normal component of force on a surface element. There is a correction to the pressure term proportional to $\nabla \cdot \mathbf{v}$ which arises when velocity gradients are present. This is most directly seen by noticing that the T_{ik} for a viscous fluid has two contributions proportional to δ_{ik}: one term is of the form $p\delta_{ik}$, where p is the thermodynamic pressure, and another term is $\zeta (\nabla \cdot \mathbf{v}) \delta_{ik}$, which arises from velocity gradients.

It should be noted that the form of corrections to the ideal fluid, to the lowest order in the gradients of macroscopic variables, is completely determined by η, ζ and κ. No other terms are permitted in this order of approximation in the description of the fluid.

3.4 Visualization of flow

(a) The quantity $\partial v_i/\partial x^k$ is a second rank tensor in three dimensions. Such a tensor can always be written as

$$\partial_k v_i = \frac{1}{2}(\partial_k v_i + \partial_i v_k) + \frac{1}{2}(\partial_k v_i - \partial_i v_k) \equiv \mathscr{S}_{ki} + A_{ki}$$
$$= \left(\mathscr{S}_{ki} - \frac{1}{3}\mathscr{S}\delta_{ki}\right) + \frac{1}{3}\mathscr{S}\delta_{ki} + \frac{1}{2}\epsilon_{kil}\Omega^l \equiv S_{ki} + \frac{1}{3}\theta\delta_{ki} + \frac{1}{2}\epsilon_{kil}\Omega^l, \tag{3.71}$$

where $\mathscr{S} = \text{Tr}\,\mathscr{S}_{ki} = \nabla \cdot \mathbf{v} = \theta$ and $\boldsymbol{\Omega} = \nabla \times \mathbf{v}$. This leads to the expression given in the problem.

(i) Consider a cubical fluid element with legs $L\mathbf{e}_x, L\mathbf{e}_y, L\mathbf{e}_z$ undergoing deformation as it moves along. After an infinitesimal time Δt, the vector $L\mathbf{e}_x$ would change to $[L\mathbf{e}_x + L\,(\partial\mathbf{v}/\partial x)\,\Delta t]$, and similarly for the y- and the z-axis. The change in the volume of the cube will be

$$\Delta V = \left(L\mathbf{e}_x + L\frac{\partial\mathbf{v}}{\partial x}\Delta t\right) \cdot \left[\left(L\mathbf{e}_y + L\frac{\partial\mathbf{v}}{\partial y}\Delta t\right) \times \left(L\mathbf{e}_z + L\frac{\partial\mathbf{v}}{\partial z}\Delta t\right)\right] - L^3$$

$$\cong \Delta t L^3\,(\nabla \cdot \mathbf{v}) = V\theta\Delta t. \tag{3.72}$$

Thus $\theta = \nabla \cdot \mathbf{v} = V^{-1}\,(dV/dt)$ represents the fractional rate of volume change of the fluid element.

(ii) The angle between the two legs of the cubical fluid element can be found from the dot product of the vectors representing the legs. The change in the angle α_{ij} between the legs i and j will be

$$\Delta\alpha_{ij} \cong \sin\,(\Delta\alpha_{ij}) \cong \cos\,(\alpha_{ij} + \Delta\alpha_{ij}) - \cos\alpha_{ij}$$

$$= (\mathbf{e}_i + \partial_i\mathbf{v}\Delta t) \cdot (\mathbf{e}_j + \partial_j\mathbf{v}\Delta t) - \delta_{ij}$$

$$= (\partial_j v_i + \partial_i v_j)\,\Delta t = S_{ji}\Delta t. \tag{3.73}$$

The last equality arises from the fact that we are only considering $i \neq j$ terms in defining the angles. This shows that S_{ij} changes the shape of the cubical element by shearing it.

(iii) Finally, consider a situation in which $\partial v_i/\partial x_k = -\frac{1}{2}\epsilon_{kil}\Omega^l$. In this case, the change in the fiducial vector separating two fluid elements can be written in the form

$$\frac{\partial l_j}{\partial t} = -\frac{1}{2}\epsilon_{jkl}\Omega^l l^k = \epsilon_{jlk}\left(\frac{1}{2}\Omega^l\right)l^k = \frac{1}{2}\,(\boldsymbol{\Omega} \times \mathbf{l})_j. \tag{3.74}$$

This shows that the effect of $\boldsymbol{\Omega}$ is to rotate the fluid element with angular velocity $(\boldsymbol{\Omega}/2)$.

(b) Consider a fluid element which possesses some vectorial property $\mathbf{A}\,(\mathbf{x}, t)$. When the fluid element moves from place to place, this quantity changes due to two different reasons. In a small time interval dt the vector changes by the amount $(\partial\mathbf{A}/\partial t)\,dt$ at any fixed location in space. Further, since $\mathbf{A}$ varies from place to place, it will change by the amount $(d\mathbf{x} \cdot \nabla)\mathbf{A}$ if $d\mathbf{x} = \mathbf{v}\,dt$ is the displacement of the fluid element in the time interval dt. Hence, the total change in $\mathbf{A}$ as the fluid element moves in space is

$$\frac{d\mathbf{A}}{dt} = \frac{\partial\mathbf{A}}{\partial t} + (\mathbf{v} \cdot \nabla)\mathbf{A}. \tag{3.75}$$

The vanishing of this expression implies that the quantity $\mathbf{A}$ does not undergo any intrinsic change as the fluid flows.

Consider now the integral curve to the vector field $\mathbf{A}$, and take two infinitesimally separated particles located on this curve. (The integral curve $\mathscr{C}$ to a vector field **A** is obtained by taking the tangent to $\mathscr{C}$ at each point to be **A** at that point. Such a curve is determined by the equations $dx/A_x = dy/A_y = dz/A_z$. The integral curves to the velocity field are called streamlines.) Let the separation be **l**. Since **l** is in the same direction as **A**, we can set $\mathbf{l} = \epsilon\mathbf{A}$ at the initial instant. As the fluid elements move, both **l** and **A** will change. We saw earlier that the infinitesimal vector separating two fluid particles **l** satisfies the equation

$$\frac{d\mathbf{l}}{dt} = (\mathbf{l}\cdot\nabla)\mathbf{v}. \tag{3.76}$$

This equation implies that the 'head' and 'tip' of the vector **l** are carried along adjacent streamlines as the fluid flows. On the other hand, we are told that **A** satisfies the equation $(d\mathbf{A}/dt) = (\mathbf{A}\cdot\nabla)\mathbf{v}$. Subtracting the two equations we find that the difference $(\mathbf{l} - \epsilon\mathbf{A})$ also satisfies the equation

$$\frac{d}{dt}[\mathbf{l} - \epsilon\mathbf{A}] = [(\mathbf{l} - \epsilon\mathbf{A})\cdot\nabla]\mathbf{v}. \tag{3.77}$$

This equation shows that if $\mathbf{l} = \epsilon\mathbf{A}$ initially, it will remain so for all times. In other words, if the 'tail' and 'tip' of the vector **A** were originally located on two infinitesimally separated streamlines, they will remain so as the fluid moves.

(c) Expanding $\nabla\times(\mathbf{v}\times\mathbf{A})$ and using the fact that $\nabla\cdot\mathbf{A} = 0$, we can write

$$0 = \frac{\partial\mathbf{A}}{\partial t} - \nabla\times(\mathbf{v}\times\mathbf{A}) = \frac{\partial\mathbf{A}}{\partial t} - (\mathbf{A}\cdot\nabla)\mathbf{v} + (\mathbf{v}\cdot\nabla)\mathbf{A} - \mathbf{A}(\nabla\cdot\mathbf{v}). \tag{3.78}$$

We substitute for $\nabla\cdot\mathbf{v}$ from the equation of continuity, written in the form

$$\nabla\cdot\mathbf{v} = -\frac{1}{\rho}\frac{\partial\rho}{\partial t} - \frac{(\mathbf{v}\cdot\nabla)\rho}{\rho} \tag{3.79}$$

and rearrange the terms to obtain

$$\left(\frac{\partial}{\partial t} + \mathbf{v}\cdot\nabla\right)\frac{\mathbf{A}}{\rho} - \left(\frac{\mathbf{A}}{\rho}\cdot\nabla\right)\mathbf{v} = 0 \tag{3.80}$$

or

$$\frac{d\mathbf{Q}}{dt} - (\mathbf{Q}\cdot\nabla)\mathbf{v} = 0, \tag{3.81}$$

where $\mathbf{Q} = (\mathbf{A}/\rho)$. This form of the equation was discussed in (b) above. It follows that the vector **Q** and an infinitesimal fluid line $\delta\mathbf{l}$ are transported by identical formulas. In other words, these two vectors, if they are initially in the same direction, will remain parallel and their lengths will remain in the same ratio. Hence, two infinitesimally close fluid particles which are on the field line of **A** will always remain on that line as the fluid moves. This shows that the field lines are frozen in the fluid.

Consider an imaginary closed curve $\mathscr{C}(t)$ in the fluid and a surface $\mathscr{S}(t)$ bounded by it (see figure 3.1). The flux of the vector field through this surface

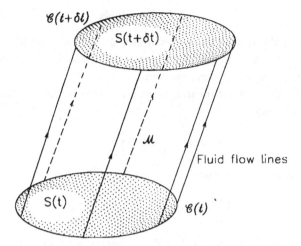

Fig. 3.1   Motion of an imaginary curve in the fluid. See text for discussion.

changes at the rate

$$\frac{dF}{dt} \equiv \dot{F} = \lim_{\Delta t \to 0} \left\{ \int_{\mathscr{S}(t+\Delta t)} \mathbf{A}(t+\Delta t) \cdot d\mathbf{s} - \int_{\mathscr{S}(t)} \mathbf{A}(t) \cdot d\mathbf{s} \right\} \frac{1}{\Delta t}. \tag{3.82}$$

Consider now the closed surface bounded by $\mathscr{S}(t)$, $\mathscr{S}(t+\Delta t)$ and the 'side' $\mathscr{M}$ of infinitesimal extent. Since $\nabla \cdot \mathbf{A} = 0$, the net flux through the closed surface at any given instant is zero. Hence, at $(t+\Delta t)$,

$$\int_{\mathscr{S}(t+\Delta t)} \mathbf{A}(t+\Delta t) \cdot d\mathbf{s} + \int_{\mathscr{M}} \mathbf{A}(t+\Delta t) \cdot d\mathbf{s} - \int_{\mathscr{S}(t)} \mathbf{A}(t+\Delta t) \cdot d\mathbf{s} = 0. \tag{3.83}$$

Consider a small element of length $d\mathbf{l}$ on the curve $\mathscr{C}$. Let the fluid element in this strip of the curve move a distance $d\mathbf{q} = \mathbf{v}\Delta t$ in a time interval Δt. In that case, the area element on $\mathscr{M}$ can be written as $d\mathbf{s} = d\mathbf{q} \times d\mathbf{l}$; using this result and (3.83), $\dot{F}$ becomes

$$\dot{F} = \lim_{\Delta t \to 0} \left\{ \int_{\mathscr{S}(t)} [\mathbf{A}(t+\Delta t) - \mathbf{A}(t)] \cdot d\mathbf{s} - \int_{\mathscr{M}} \mathbf{A}(t+\Delta t) \cdot (d\mathbf{l} \times \mathbf{v}) \Delta t \right\} \frac{1}{\Delta t}. \tag{3.84}$$

But

$$\int \mathbf{A} \cdot (d\mathbf{l} \times \mathbf{v}) = \int d\mathbf{l} \cdot (\mathbf{v} \times \mathbf{A}) = \int d\mathbf{s} \cdot [\nabla \times (\mathbf{v} \times \mathbf{A})]. \tag{3.85}$$

It follows that

$$\dot{F} = \int d\mathbf{s} \cdot \left[\frac{\partial \mathbf{A}}{\partial t} - \nabla \times (\mathbf{v} \times \mathbf{A}) \right]. \tag{3.86}$$

The flux F is clearly conserved when the integrand vanishes.

3.5 Ideal fluids

(a) Equation (3.64), applied to an ideal fluid, shows that $(ds/dt) = 0$. This implies that s is constant for a given fluid element but does *not* necessarily imply that all fluid elements have the same entropy.

To write the Euler equation in a convenient form we rewrite the $(\mathbf{v} \cdot \nabla)\mathbf{v}$ term by using the identity

$$(\mathbf{v} \cdot \nabla)\mathbf{v} = \frac{1}{2}\nabla v^2 - \mathbf{v} \times (\nabla \times \mathbf{v}). \tag{3.87}$$

Then Euler's equation becomes

$$\frac{\partial \mathbf{v}}{\partial t} - (\mathbf{v} \times \mathbf{\Omega}) = -\frac{1}{2}\nabla v^2 - \frac{1}{\rho}\nabla p, \tag{3.88}$$

where $\mathbf{\Omega} = \nabla \times \mathbf{v}$. Taking the curl of this equation we find

$$\frac{\partial \mathbf{\Omega}}{\partial t} = \nabla \times (\mathbf{v} \times \mathbf{\Omega}) - \nabla \times \left(\frac{1}{\rho}\nabla p\right) = \nabla \times (\mathbf{v} \times \mathbf{\Omega}) + \frac{1}{\rho^2}(\nabla\rho \times \nabla p). \tag{3.89}$$

Consider now the dot product of this equation with ∇s. Since $s = s(\rho, p)$, the gradient ∇s is a linear function of ∇p and $\nabla \rho$ and gives zero when multiplied by the last term. The rest of the terms can be transformed using a series of vector identities as follows:

$$\begin{aligned}
\nabla s \cdot \frac{\partial \mathbf{\Omega}}{\partial t} &= \nabla s \cdot \nabla \times (\mathbf{v} \times \mathbf{\Omega}) = -\nabla \cdot (\nabla s \times (\mathbf{v} \times \mathbf{\Omega})) \\
&= -\nabla \cdot (\mathbf{v}(\mathbf{\Omega} \cdot \nabla s)) + \nabla \cdot (\mathbf{\Omega}(\mathbf{v} \cdot \nabla s)) \\
&= -(\nabla \cdot \mathbf{v})(\mathbf{\Omega} \cdot \nabla)s - (\mathbf{v} \cdot \nabla)(\mathbf{\Omega} \cdot \nabla s) + (\mathbf{\Omega} \cdot \nabla)(\mathbf{v} \cdot \nabla s).
\end{aligned} \tag{3.90}$$

In the last term we use the relation $\mathbf{v} \cdot \nabla s = -\partial s/\partial t$ and combine it with the left-hand side, to obtain $(\partial/\partial t)(\mathbf{\Omega} \cdot \nabla s)$. Combining this with the second term we obtain $(d/dt)(\mathbf{\Omega} \cdot \nabla s)$. Finally, we substitute $\rho(\nabla \cdot \mathbf{v}) = -d\rho/dt$ to obtain

$$\frac{d}{dt}\left[\frac{(\mathbf{\Omega} \cdot \nabla s)}{\rho}\right] = 0. \tag{3.91}$$

This shows that each fluid particle carries a constant value of $(\mathbf{\Omega} \cdot \nabla s/\rho)$.

(b) Two of the most important examples of barotropic flow are: (i) incompressible flow with $\rho = $ constant in which ρ is a trivial, constant function of p; (ii) isentropic flow with $s(\rho, p) = $ constant, which can be solved to give $\rho = \rho(p)$.

For a barotropic fluid, ∇p and $\nabla \rho$ are parallel to each other and hence the last term in (3.89) vanishes, giving

$$\frac{\partial \mathbf{\Omega}}{\partial t} = \nabla \times (\mathbf{v} \times \mathbf{\Omega}). \tag{3.92}$$

From our general result in problem 3.4(c), it follows that the flux of $\mathbf{\Omega}$ is conserved.

That is,

$$F = \int \mathbf{\Omega} \cdot d\mathbf{s} = \int (\nabla \times \mathbf{v}) \cdot d\mathbf{s} = \oint \mathbf{v} \cdot d\mathbf{l} \qquad (3.93)$$

is conserved.

(c) In the case of a steady flow of an ideal fluid, both $(\partial s/\partial t)$ and (ds/dt) vanish, implying that $v^i (\partial s/\partial x^i) = 0$. In other words, the entropy is constant along each streamline. Further, we have

$$dw = d\epsilon + pd\left(\frac{1}{\rho}\right) + \frac{dp}{\rho} = T\,ds - pd\left(\frac{1}{\rho}\right) + pd\left(\frac{1}{\rho}\right) + \frac{dp}{\rho}$$

$$= T\,ds + \frac{dp}{\rho}\,. \qquad (3.94)$$

Since entropy is constant along streamlines, we have $dw = dp/\rho$ along any streamline. Now consider Euler's equation (3.88), which can be written in the form

$$\frac{1}{2}\nabla v^2 - (\mathbf{v} \times \mathbf{\Omega}) = -\frac{1}{\rho}\nabla p \qquad (3.95)$$

because $(\partial \mathbf{v}/\partial t) = 0$. Let $\mathbf{l}$ be a unit tangent vector to any streamline. Taking the dot product of the above equation with $\mathbf{l}$, the first term gives the rate of change of $(1/2)\,v^2$ along the streamline; the second term is perpendicular to $\mathbf{v}$ (and hence to $\mathbf{l}$) and thus gives no contribution; the last term becomes dw when evaluated along the streamline. Therefore, it follows that $B = \left(\frac{1}{2}v^2 + w\right)$ is a constant along the streamline. The constancy of B is called the Bernoulli theorem.

(d) Setting $\mathbf{\Omega} = 0$ and $\mathbf{v} = \nabla \psi$ in the Euler equation for an ideal fluid (3.88), we obtain

$$\nabla \left(\frac{\partial \psi}{\partial t} + \frac{1}{2}v^2\right) = -\frac{\nabla p}{\rho}\,. \qquad (3.96)$$

When the flow is isoentropic, $ds = 0$ and (3.94) gives $(\nabla p/\rho) = \nabla w$, where w is the enthalpy. This gives the relation

$$\frac{\partial \psi}{\partial t} + \frac{1}{2}v^2 + w = f(t)\,, \qquad (3.97)$$

where $f(t)$ is an arbitrary function of time. We can, however, set this function to zero without any loss of generality since one can always add to ψ an arbitrary function of time without changing the relation $\mathbf{v} = \nabla \psi$.

3.6 Viscous fluids

In problem 3.3(a) we derived the equation of motion governing the viscous fluid:

$$\left(\frac{\partial \mathbf{v}}{\partial t} + \mathbf{v} \cdot \nabla\right)\mathbf{v} = -\frac{\nabla p}{\rho} + \frac{\eta}{\rho}\nabla^2 \mathbf{v} + \left(\frac{\zeta}{\rho} + \frac{1}{3}\frac{\eta}{\rho}\right)\nabla(\nabla \cdot \mathbf{v})\,. \qquad (3.98)$$

Taking the curl of this equation and using the vector identity

$$\nabla \times (\mathbf{v} \cdot \nabla) \mathbf{v} = -\nabla \times (\mathbf{v} \times (\nabla \times \mathbf{v})) \,, \tag{3.99}$$

we obtain

$$
\begin{aligned}
\frac{\partial \mathbf{\Omega}}{\partial t} - \nabla \times (\mathbf{v} \times \mathbf{\Omega}) &= -\nabla \times \left(\frac{\nabla p}{\rho} \right) + \eta \nabla \times \left(\frac{\nabla^2 \mathbf{v}}{\rho} \right) + \nabla \times \left[\left(\frac{\zeta}{\rho} + \frac{1}{3} \frac{\eta}{\rho} \right) \nabla (\nabla \cdot \mathbf{v}) \right] \\
&= -\nabla \times \left(\frac{\nabla p}{\rho} \right) + \frac{\eta}{\rho} \nabla^2 \mathbf{\Omega} - \frac{\eta}{\rho^2} (\nabla \rho) \times \nabla^2 \mathbf{v} - \frac{1}{\rho^2} \left(\zeta + \frac{1}{3} \eta \right) (\nabla \rho \times \nabla \theta) \\
&= \frac{\eta}{\rho} \nabla^2 \mathbf{\Omega} + \frac{1}{\rho^2} (\nabla \rho) \times \left[\nabla p - \eta \nabla^2 \mathbf{v} - \left(\zeta + \frac{1}{3} \eta \right) \nabla \theta \right] .
\end{aligned}
\tag{3.100}
$$

On the right-hand side the first term represents diffusion of vorticity $\mathbf{\Omega}$, while the second term arises due to density gradients.

(b) We can estimate the fractional change in the density $(\Delta\rho/\rho)$ from the laws of thermodynamics and the equations of fluid motion. Demanding that $(\Delta\rho/\rho) \ll 1$ we can obtain the conditions for incompressible motion.

Given $\rho = \rho(p, s)$ it follows that $d\rho = (\partial\rho/\partial p)\, dp + (\partial\rho/\partial s)\, ds$. Now using the relations $c_s^2 = (\partial p/\partial\rho)_s$, $\beta = -(1/\rho)(\partial\rho/\partial T)_p$ and $c_p = T(\partial s/\partial T)_p$ we find that

$$d\rho = \frac{dp}{c_s^2} - \frac{\beta\rho T}{c_p} ds . \tag{3.101}$$

Here c_s is the speed of sound in the fluid, β is the coefficient of thermal expansion and c_p is the specific heat at constant pressure. Interpreting each of these small differentials as changes in physical quantities and treating $\Delta\rho, \Delta p$ and Δs as positive quantities we can rewrite the above equation as

$$\frac{\Delta\rho}{\rho} \lesssim \frac{1}{c_s^2} \frac{\Delta p}{\rho} + \frac{\beta T}{c_p} \Delta s . \tag{3.102}$$

In order to make an estimate of the right-hand side, we will evaluate the order of magnitude contribution from each of the terms in the equations for fluid motion.

To begin with, consider the equation for the viscous fluid in an external gravitational potential ϕ, written in the form

$$\rho \left[\frac{\partial v^i}{\partial t} + v^j \frac{\partial v^i}{\partial x^j} \right] = -\frac{\partial p}{\partial x^i} - \rho \frac{\partial \phi}{\partial x^i} + \frac{\partial}{\partial x^i} (\zeta\theta) + 2 \frac{\partial}{\partial x^j} (\eta S^{ij}) . \tag{3.103}$$

The last two terms arise from using (3.51) in (3.52) and using the notation of (3.71). If the length scale of variation is l and the time-scale of variation is τ, then the two terms on the left-hand side have the order of magnitude $(\rho v/\tau)$ and $(\rho v^2/l)$. On the right-hand side the four terms are of orders $(\Delta p/l)$, $(\rho\Delta\phi/l)$, $(\zeta v/l^2)$ and $(\eta v/l^2)$. Comparing the two sides we find that

$$\frac{\Delta p}{\rho} \lesssim \frac{vl}{\tau} + v^2 + \Delta\phi + \frac{(\eta + \zeta)}{\rho l} v . \tag{3.104}$$

Consider next the equation for energy conservation written in the form

$$T\left[\frac{\partial}{\partial T}(\rho s) + \nabla \cdot (\rho s v)\right] - T\nabla \cdot \left(\frac{\kappa \nabla T}{T}\right) = \zeta \theta^2 + 2\eta S_{jk}S^{jk} + \frac{\kappa}{T}(\nabla T)^2. \quad (3.105)$$

In this case the three terms on the left-hand side have the magnitudes $\Delta s \left(T\rho/\tau\right)$, $\Delta s \left(\rho v/l\right)$ and $\kappa \left(\Delta T/l^2 T\right)$; the three terms on the right-hand side are of order $\left(\zeta v^2/l^2\right)$, $\left(\eta v^2/l^2\right)$ and $\kappa(\Delta T)^2/Tl^2$. Note that for $\Delta T \lesssim T$ the last term is smaller than the corresponding κ-term on the left-hand side. By an analysis similar to the one performed above we can easily show that

$$T\Delta s \lesssim \frac{\kappa \Delta T}{\rho v l} + \frac{(\eta + \zeta)v}{\rho l}. \quad (3.106)$$

Using (3.104) and (3.106) in (3.102) and simplifying the terms we find

$$\frac{\Delta \rho}{\rho} \lesssim \frac{v}{c_s} \cdot \frac{l/\tau}{c_s} + \frac{v^2}{c_s^2} + \frac{\Delta \phi}{c_s^2} + \left(\frac{\eta + \zeta}{\rho l v}\right)\left(\frac{v}{c_s}\right)^2 + \frac{\beta}{c_p} \cdot \frac{\kappa}{\rho v} \cdot \frac{\Delta T}{l} + \left(\frac{\eta + \zeta}{\rho l v}\right)\left(\frac{v}{c_s}\right)^2 \frac{\beta c_s^2}{c_p}. \quad (3.107)$$

In order for the fluid to be considered incompressible we must demand that each of the terms on the right-hand side contributing to $(\Delta \rho/\rho)$ should be small. This implies that we must have

$$\frac{v}{c_s} \ll 1 \implies \text{flow is subsonic}$$

$$\frac{l/\tau}{c_s} \ll 1 \implies \text{no rapid variations in time}$$

$$\frac{\Delta \phi}{c_s^2} \ll 1 \implies \text{potential drops not too large} \quad (3.108)$$

$$\frac{\rho l v}{\eta} \gg \left(\frac{v}{c_s}\right)^2 \implies \text{viscosity is not too large}$$

$$\frac{\Delta T}{l} \ll \frac{\rho c_s^2 v}{\kappa} \implies \text{temperature gradient not too large.}$$

The condition for adiabacity can be obtained from equation (3.106) itself. This equation relates the dissipative heat generation $T\Delta s$ to other parameters in the problem. For adiabatic flow, we expect $T\Delta s$ to be small compared to ρc_s^2. This gives the conditions

$$\frac{\Delta T}{l} \ll \frac{\rho c_s^2 v}{\kappa}, \qquad \frac{\rho l v}{\eta} \gg \frac{v^2}{c_s^2}, \quad (3.109)$$

which are the same as two of the conditions derived above. In other words, incompressibility implies adiabaticity, though the converse is not necessarily true.

(c) In the case of incompressible fluids, $\rho = \text{constant}$ and $\nabla \cdot v = 0$. Hence, Euler's equation becomes

$$\frac{\partial v}{\partial t} + (v \cdot \nabla)v = -\nabla\left(\frac{p}{\rho}\right) + \frac{\eta}{\rho}\nabla^2 v. \quad (3.110)$$

Taking the curl of this equation we obtain

$$\frac{\partial \mathbf{\Omega}}{\partial t} - \nabla \times (\mathbf{v} \times \mathbf{\Omega}) = \frac{\eta}{\rho} \nabla^2 \mathbf{\Omega}. \tag{3.111}$$

(We can obtain the same result from (3.100) by setting $\rho = $ constant.) Expanding $\nabla \times (\mathbf{v} \times \mathbf{\Omega})$ and using $\nabla \cdot \mathbf{v} = 0$, we find that

$$\frac{\partial \mathbf{\Omega}}{\partial t} + (\mathbf{v} \cdot \nabla) \mathbf{\Omega} - (\mathbf{\Omega} \cdot \nabla) \mathbf{v} = \frac{\eta}{\rho} \nabla^2 \mathbf{\Omega}. \tag{3.112}$$

Or

$$\frac{d\mathbf{\Omega}}{dt} - (\mathbf{\Omega} \cdot \nabla) \mathbf{v} = \frac{\eta}{\rho} \nabla^2 \mathbf{\Omega}. \tag{3.113}$$

To relate the pressure to the velocity field, we take the divergence of (3.110). This gives, on using $\nabla \cdot \mathbf{v} = 0$,

$$\nabla \cdot [(\mathbf{v} \cdot \nabla) \mathbf{v}] = \frac{\partial}{\partial x^i} \left[v^k \frac{\partial v^i}{\partial x^k} \right] = \frac{\partial v^k}{\partial x^i} \frac{\partial v^i}{\partial x^k} = \frac{\partial^2}{\partial x^i \partial x^k} \left(v^i v^k \right) = -\frac{1}{\rho} \nabla^2 p. \tag{3.114}$$

Given the equations $\nabla \cdot \mathbf{v} = 0$ and (3.111), we can determine the velocity field. Then one can determine p by integrating (3.114). This provides the complete solution to the flow of incompressible viscous fluid.

(d) In the study of incompressible viscous fluid, the only dimensional parameter which appears in the equation is $v = (\eta/\rho)$, which has the dimension L^2/T. Consider now the motion of a body, characterized by size l and velocity u, through this fluid. The boundary conditions will introduce the variables l and u into the problem. The only dimensionless quantity which can be formed from l, v and u is $R = (ul/v)$. By re-expressing the equations of motion in terms of the ratios (v/u), (x/l) and (ut/l) we can reduce the equations to dimensionless forms involving only the parameter R. Further, in the steady state solution we will have no time dependence. It follows that the solutions to the equations of motion will only depend on dimensionless variables and R. In particular, the velocity and pressure must scale in proportion to u and ρu^2, which are the only two quantities with the correct dimensions. Hence, we must have

$$\mathbf{v} = u\mathbf{f}_1 \left(\frac{\mathbf{x}}{l}, R \right), \quad p = \rho u^2 f_2 \left(\frac{\mathbf{x}}{l}, R \right). \tag{3.115}$$

Similarly the only combination having the dimensions of force is $\rho u^2 l^2$. Hence, the force F acting on the solid body must have the form $F = \rho u^2 l^2 f_3(R)$. These relations express the following important fact. Bodies of geometrically similar shape differing only in overall size will induce flows which are self-similar if the Reynolds number R remains constant. That is, if we have solved the problem for a given Reynolds number for a body of given size, then the solution for a body of different size but the same shape can be found by rescaling the original solution, provided the Reynold number is same.

3.7 Instabilities and turbulence

We will work in a reference frame in which the fluid velocity is zero on one side of the separation and $\mathbf{v}$ on the other side. We take the direction of $\mathbf{v}$ as the x-axis and that normal to the surface along the z-axis. Let the surface of discontinuity receive a small perturbation, in which all quantities (the displacement of the surface in the z-direction, presssure and velocity) are periodic functions of the form $\exp\left[i\left(kx - \omega t\right)\right]$. We shall linearize the equations in the small perturbations and denote the deviations by primed quantities. If the dispersion relation between ω and k leads to an imaginary part for ω then the perturbations will lead to an instability.

The linearized equations in the absence of viscosity are

$$\nabla \cdot \mathbf{v}' = 0, \quad \frac{\partial \mathbf{v}'}{\partial t} + (\mathbf{v} \cdot \nabla)\mathbf{v}' = -\frac{\nabla p'}{\rho}. \tag{3.116}$$

Since the original velocity is along the x-axis, we can write $(\mathbf{v} \cdot \nabla)\mathbf{v}' = v\left(\partial \mathbf{v}'/\partial x\right)$. Taking the divergence of both sides of Euler's equation, we find that p' must satisfy Laplace's equation: $\nabla^2 p' = 0$. Since the variation along the x-axis is fixed, let us try an ansatz of the form $p' = f(z)\exp\left[i\left(kx - \omega t\right)\right]$. This leads us to the equation

$$\frac{d^2 f}{dz^2} = k^2 f, \tag{3.117}$$

with the solution $f \propto \exp(\pm kz)$. Assuming that the region under consideration (side 1, say) corresponds to positive values of z, the solution with proper behaviour for large z is

$$p_1' \propto e^{-kz} e^{i(kx - \omega t)}. \tag{3.118}$$

From (3.116) we see that this corresponds to the velocity distribution

$$v_z' = \frac{kp_1'}{i\rho_1 (kv - \omega)}. \tag{3.119}$$

Consider next the displacement of the surface $l(x, t)$. Since the velocity component normal to the surface is equal to the rate of displacement of the surface itself, we can write $\left(dl/dt\right) = v_z'$, correct to linear order. So,

$$\frac{\partial l}{\partial t} = v_z' - v\frac{\partial l}{\partial x}, \tag{3.120}$$

where the value of v_z' must be taken on the surface. Using $l \propto \exp[i(kx - \omega t)]$ in this equation, we find $v_z' = il(kv - \omega)$; substituting in (3.119), we obtain

$$p_1' = -\frac{l\rho_1}{k}(kv - \omega)^2. \tag{3.121}$$

On side 2, similar considerations apply except that we must set $v = 0$ and use the e^{kz} solution instead of the e^{-kz} solution. This gives

$$p_2' = \frac{l\rho_2}{k}\omega^2. \tag{3.122}$$

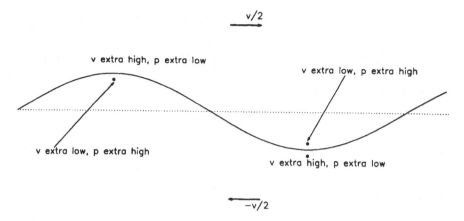

Fig. 3.2. Intuitive way of understanding Kelvin–Helmholtz instability. The perturbation in the diagram will grow because of the pressure difference.

If the liquids on both sides are the same, then $\rho_1 = \rho_2$; but the above analysis is applicable even when two different fluids are in contact with each other with $\rho_1 \neq \rho_2$. To obtain the dispersion relation we only need to use the fact that $p_1' = p_2'$ at the surface of separation. This leads to the quadratic equation $\rho_1 (kv - \omega)^2 = -\rho_2\omega^2$ with the solution

$$\omega = kv\frac{\rho_1 \pm i\sqrt{\rho_1\rho_2}}{\rho_1 + \rho_2}. \tag{3.123}$$

The frequency ω is complex with a positive imaginary part. Hence, the fluid is unstable to shearing flow of the kind described above. When $\rho_1 = \rho_2$ the imaginary part of ω is given by $\text{Im }\omega = (kv/2)$. The inverse of this quantity gives the time-scale for the growth of instability.

A simple, intuitive way of understanding this instability is illustrated in figure 3.2. This diagram shows the interface between the two liquids, at some moment of time, in a frame of reference moving with velocity $(\mathbf{v}/2)$. In this frame, the upper liquid moves with velocity $(\mathbf{v}/2)$ and the lower liquid with velocity $(-\mathbf{v}/2)$. Suppose the perturbation is such that the velocity is slightly higher on top of a crest and below a trough. Because of Bernoulli's theorem, the pressure will be lower in regions of higher velocity and vice versa. This will cause the perturbation to grow.

(b) The above analysis shows that instabilities will grow near any region with a strong gradient in a fluid. The growth of such an instability will further increase the velocity gradient and the flow pattern will proceed along the lines shown in figure 3.3. Such a motion will easily become very complex and turbulent, with the energy being transported to smaller and smaller scales until viscous effects become important.

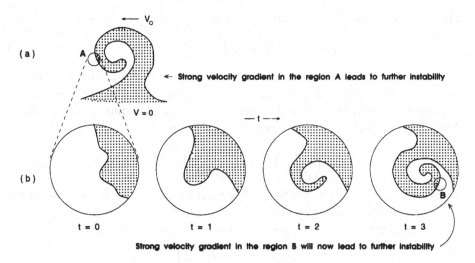

Fig. 3.3. Origin of turbulence due to instability. (a) The velocity gradient produces a Kelvin–Helmholtz instability at some region of the fluid. Such an instability will lead to strong velocity gradients in the region marked A. (b) The fluid configuration inside the region marked A is shown on a magnified scale. As time goes on, the strong velocity gradient in the local region leads to a secondary instability. The fluid configuration at $t = 3$ has strong velocity gradients in the region marked B. The situation is now similar to the initial configuration. The repeated occurrence of such instabilities leads to a highly turbulent flow of liquid with velocity gradients existing at all scales.

(c) The rate at which energy can cascade from larger scales to smaller scales is given by

$$\dot{\epsilon} \cong \left(\frac{1}{2} v_\lambda^2 \right) \left(\frac{v_\lambda}{\lambda} \right) \cong \frac{v_\lambda^3}{\lambda} . \qquad (3.124)$$

The first factor gives the energy per unit mass and unit volume in the eddies of scale λ and the second factor $(\lambda/v_\lambda)^{-1}$ gives the inverse time-scale at which the transfer of energy occurs at this scale. Assuming that the energy is transferred at a constant rate $\dot{\epsilon}$ all the way down to the smallest scales, it follows that (v_λ^3/λ) is a constant. Hence, we can write

$$v_\lambda \cong U \left(\frac{\lambda}{L} \right)^{1/3} , \qquad (3.125)$$

where U is the velocity at some macroscopically large scale L.

To estimate the smallest scale λ_s at which this process stops we have to calculate the energy dissipation due to viscosity and equate it to $\dot{\epsilon}$. The above analysis shows that $v_\lambda = (\dot{\epsilon}\lambda)^{1/3}$. The shear σ at some scale λ is of the order of (v_λ/λ).

Hence, viscous heating rate at any scale λ is

$$h_\lambda \cong \eta \sigma^2 \cong \eta \left(\frac{v_\lambda}{\lambda}\right)^2 \cong \eta \dot{\epsilon}^{2/3} \lambda^{-4/3}. \qquad (3.126)$$

At large scales h is negligible compared to $\dot{\epsilon}$ and hence the same amount of energy will be transmitted from scale to scale. As we reach smaller scales, viscous heating becomes important and when $h_\lambda \cong \rho\dot{\epsilon}$, the cascading of energy stops. Equating the expressions for h_λ and $\rho\dot{\epsilon} = \rho(u^3/L)$ we find

$$\lambda_s \cong \left(\frac{\eta}{\rho\dot{\epsilon}^{1/3}}\right)^{3/4} \cong L\left(\frac{\eta}{\rho U L}\right)^{3/4} \cong \frac{L}{R^{3/4}}, \qquad (3.127)$$

where R is the Reynolds number evaluated at the largest scale. As long as the Reynolds number is large, we have $\lambda_s \ll L$.

If the velocity field is $\mathbf{v}(\mathbf{x})$, then the correlation function of $\mathbf{v}$ can be defined as

$$\xi(\mathbf{x}) = \langle \mathbf{v}(\mathbf{x}+\mathbf{y}) \cdot \mathbf{v}(\mathbf{y}) \rangle \qquad (3.128)$$

if we are only interested in $|\mathbf{v}|$. The power spectrum $S(\mathbf{k})$ is the Fourier transform of $\xi(\mathbf{x})$. If the turbulence is spatially homogeneous and isotropic, then $S(\mathbf{k}) = S(k)$ and – within an order of magnitude – we will have $S(k)(dk/k) = v_\lambda^2\,(-d\lambda/\lambda)$, with $k \propto \lambda^{-1}$. That is, the power per logarithmic interval in k will be v_λ^2. This leads to the spectrum of velocity called the 'Kolmogorov spectrum':

$$S(k) \cong v_\lambda^2 k^{-1} \cong v_\lambda^2 \lambda \cong \dot{\epsilon}^{1/3}\lambda^{5/3}. \qquad (3.129)$$

3.8 Sound waves

(a) Consider a compressible fluid with small perturbations in pressure, density and velocity denoted by p', ρ' and $\mathbf{v}'$. Linearizing the equation of motion for the fluid around the equilibrium values we obtain

$$\frac{\partial\rho'}{\partial t} + \rho_0 \nabla \cdot \mathbf{v}' = 0, \qquad (3.130)$$

$$\frac{\partial\mathbf{v}'}{\partial t} + \frac{1}{\rho_0}\nabla p' = 0. \qquad (3.131)$$

To proceed further we need a relation between p' and ρ'. Since all motions in an ideal fluid can be taken to be adiabatic to the lowest order of approximation, we may set $p' = (\partial p/\partial\rho)_s\,\rho'$. Substituting for ρ' we can convert the equation of continuity to

$$\frac{\partial p'}{\partial t} + \rho_0\left(\frac{\partial p}{\partial\rho}\right)_s \nabla \cdot \mathbf{v}' = 0. \qquad (3.132)$$

To solve these equations we set $\mathbf{v}' = \nabla\phi$. Then equation (3.131) gives $p' = -\rho_0\,(\partial\phi/\partial t)$. Using this in (3.132) we find that ϕ satisfies the wave equation

$$\frac{\partial^2\phi}{\partial t^2} - c_s^2\nabla^2\phi = 0, \qquad (3.133)$$

where $c_s^2 = (\partial p/\partial \rho)$. (For an ideal gas in adiabatic flow, with $p \propto \rho^\gamma \propto \rho^{5/3}$, this gives $c_s = (5p/3\rho)^{1/2} \propto \rho^{1/3}$.) The solutions to the wave equation for waves propagating along the x-axis can be taken in the form $\phi(x, t) = f(x - c_s t)$. It is clear that $\mathbf{v}' = \nabla \phi$ is also along the x-axis. Hence, the waves are longitudinal, with the displacement along the direction of propagation.

The sound waves allow the propagation of perturbations in various physical quantities with the speed of sound along the fluid. However, the entropy and vorticity of the fluid are not propagated with the speed of sound. This follows from the fact that, in an ideal fluid, entropy and circulation are strictly conserved in each fluid element. They do not move relative to the fluid; relative to a fixed system of coordinates, they move with a velocity appropriate to each point in the fluid.

(i) For $\phi = f(x - c_s t)$ we have $v' = (\partial \phi/\partial x) = f'$ and $p' = -\rho(\partial \phi/\partial t) = \rho c_s f'$. Hence, $v' = (p'/\rho c_s)$. Using $p' = c_s^2 \rho$, this can be written as $v' = c_s(\rho'/\rho)$. (ii) For the temperature variation, we have $T' = (\partial T/\partial p)_s p'$; using the thermodynamic relation $(\partial T/\partial p)_s = (T/c_p)(\partial V/\partial T)_p$ and the relation $v' = (p'/\rho c_s)$ we find

$$T' = \frac{c_s \beta T}{c_p} v', \tag{3.134}$$

where $\beta = V^{-1}(\partial V/\partial T)_p$ is the coefficient of thermal expansion.

(b) In the presence of viscosity and thermal conduction, the energy in sound waves will be dissipated into heat. The easiest way to calculate the rate of dissipation of the mechanical energy $\dot{E}_{\mathrm{mech}}$ is as follows. The mechanical energy E_{mech} is the maximum amount of work that can be done in passing from a given non-equilibrium state to one of thermodynamic equilibrium. In other words, we can write $E_{\mathrm{mech}} = E_0 - E(s)$, where E_0 is the given, initial energy of the system, s is the initial entropy of the system and $E(s)$ denotes the energy of the system in thermodynamic equilibrium when the entropy is s. It follows that $\dot{E}_{\mathrm{mech}} = -(\partial E/\partial s)\dot{s} = -T_0 \dot{s}$, where T_0 is the temperature of the system in equilibrium. Since the maximum amount of energy which can be extracted from the system occurs when the transition from one state to another is reversible, it follows that $-T_0 \dot{s}$ will give us the energy dissipated in the system.

The entropy change $\dot{s}$ of a fluid with thermal conductivity and dissipation was computed in problem 3.3 (see equation (3.70)). From that expression, we see that

$$\dot{E} \cong -\frac{\kappa}{T_0} \int d^3x\, (\nabla T)^2 - \frac{1}{2}\eta \int d^3x \left(\frac{\partial v_i}{\partial x_k} + \frac{\partial v_k}{\partial x_i} - \frac{2}{3}\delta_{ik}\nabla \cdot \mathbf{v} \right)^2 - \zeta \int (\nabla \cdot \mathbf{v})^2\, d^3x. \tag{3.135}$$

In arriving at the above result, we have assumed the temperature gradients to be small and have pulled a factor T_0 out of the integration volume. If the wave is propagating along the x-axis, then $v_x = v_0 \cos(kx - \omega t)$ and $v_y = v_z = 0$. The last two terms in the above equation give

$$-\left(\frac{4}{3}\eta+\zeta\right)\int\left(\frac{\partial v_x}{\partial x}\right)^2 d^3x = -k^2 v_0^2\left(\frac{4}{3}\eta+\zeta\right)\int \sin^2(kx-\omega t)\,d^3x$$

$$= -\frac{1}{2}k^2 v_0^2 \mathscr{V}\left(\frac{4}{3}\eta+\zeta\right). \qquad (3.136)$$

To obtain the last expression, we have averaged the wave over one cycle and denoted the volume of the fluid by $\mathscr{V}$. To calculate the temperature gradient term, we use (3.134):

$$\frac{\partial T}{\partial x} = \frac{\beta c_s T}{c_p}\frac{\partial v}{\partial x} = -\frac{\beta c_s T}{c_p}v_0 k\sin(kx-\omega t). \qquad (3.137)$$

Substituting this expression and averaging over one cycle we obtain a contribution $\left(-\kappa c_s^2 T\beta v_0^2 k^2/2c_p^2\right)\mathscr{V}$ from the first term. This result can be rewritten using the thermodynamic formula

$$c_p - c_v = T\beta^2\left(\frac{\partial p}{\partial \rho}\right)_T = T\beta^2\left(\frac{c_v}{c_p}\right)\left(\frac{\partial p}{\partial \rho}\right)_S = T\beta^2\left(\frac{c_v}{c_p}\right)c_s^2. \qquad (3.138)$$

Collecting all the terms together, we find

$$\langle\dot{E}\rangle = -\frac{1}{2}k^2 v_0^2\mathscr{V}\left[\left(\frac{4}{3}\eta+\zeta\right)+\kappa\left(\frac{1}{c_v}-\frac{1}{c_p}\right)\right]. \qquad (3.139)$$

Since the total energy of the sound wave is $E_0 = (1/2)\rho v_0^2\mathscr{V}$, we may define the attenuation time-scale to be $\tau = (E_0/\langle\dot{E}\rangle)$. Alternatively, we can define an attenuation length $l = c_s\tau$ to be the distance travelled by the sound wave in time τ. The spatial attenuation coefficient is conventionally defined to be $\gamma = (1/l)$. Using the above expression we find that

$$\gamma = \frac{\omega^2}{\rho c_s^3}\left[\left(\frac{4}{3}\eta+\zeta\right)+\kappa\left(\frac{c_p-c_v}{c_p c_v}\right)\right]. \qquad (3.140)$$

(c) If a gas in steady motion, with velocity $\mathbf{v}$, is perturbed at some point O, then the perturbation is propagated in all directions with the speed of sound *relative to the gas*. With respect to a fixed coordinate system, the velocity of propagation of the disturbance has two components. The perturbations are 'carried along' with gas flow with velocity $\mathbf{v}$ and propagated in the rest frame of the gas with a velocity $c_s\hat{\mathbf{n}}$, where $\hat{\mathbf{n}}$ is a unit vector in any arbitrary direction. Thus the net velocity of propagation of the disturbance is $\mathbf{v} + c_s\hat{\mathbf{n}}$.

We can obtain all possible values of this vector by placing one end of the vector $\mathbf{v}$ at O and drawing a sphere with radius c_s from its tip (see figure 3.4). We see from the diagram that very different conclusions emerge depending on whether $v < c_s$ or $v > c_s$. If $v < c_s$ then $(\mathbf{v} + c_s\hat{\mathbf{n}})$ can take all possible directions and the perturbation can propagate in all possible directions. In other words, the

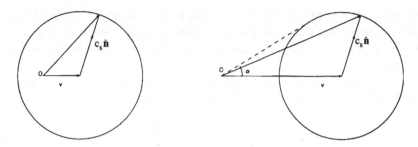

Fig. 3.4. Propagation of a disturbance in a fluid moving with a velocity v. When $v < c$ the disturbance can propagate in all directions with respect to O; when $v > c$, it can only propagate into a conical region in the forward direction.

disturbance which starts from any point O in a gas *which is in subsonic motion* can eventually reach any other point in the gas.

On the other hand, if $v > c_s$, the direction of the vector $\mathbf{v} + c_s \hat{\mathbf{n}}$ can only lie within a cone with the vertex at O with a semivertical angle α, where $\sin \alpha = (c_s/v)$. Thus a disturbance starting at O does not reach all the points downstream if the gas is flowing supersonically.

(d) We shall assume that the motion is along the x-axis with a velocity $v_x = v(\rho)$, which depends only on the density ρ. Further, since the flow is assumed to be isentropic, pressure can be expressed entirely in terms of density. In this case, the continuity and the Euler equations:

$$\frac{\partial \rho}{\partial t} + \frac{\partial (\rho v)}{\partial x} = 0, \quad \frac{\partial v}{\partial t} + \frac{v \partial v}{\partial x} + \frac{1}{\rho} \frac{\partial p}{\partial x} = 0 \tag{3.141}$$

can be transformed into the form

$$\frac{\partial \rho}{\partial t} + \frac{d (\rho v)}{d \rho} \frac{\partial \rho}{\partial x} = 0, \quad \frac{\partial v}{\partial t} + \left(v + \frac{1}{\rho} \frac{dp}{dv} \right) \frac{\partial v}{\partial x} = 0. \tag{3.142}$$

We now use the relation $(\partial x/\partial t)_\rho = - \left[(\partial \rho/\partial t) / (\partial \rho/\partial x) \right]$, etc., to obtain from these equations the partial derivatives:

$$\left(\frac{\partial x}{\partial t} \right)_\rho = \frac{d (\rho v)}{d \rho} = v + \rho \frac{dv}{d\rho}, \quad \left(\frac{\partial x}{\partial t} \right)_v = v + \frac{1}{\rho} \frac{dp}{dv}. \tag{3.143}$$

However, since $\rho = \rho(v)$, it follows that $(\partial x/\partial t)_\rho = (\partial x/\partial t)_v$. Equating the expressions for these two and simplifying, we easily find

$$v = \pm \int \frac{c_s}{\rho} d\rho = \pm \int \frac{dp}{\rho c_s}. \tag{3.144}$$

Substituting this result back in the expression for $(\partial x/\partial t)_v$ we obtain $(\partial x/\partial t)_v = v \pm c_s(v)$. Integrating,

$$x = t \left[v \pm c_s(v) \right] + f(v), \tag{3.145}$$

where $f(v)$ is an arbitrary function of v to be fixed by the initial conditions. This relation determines v as an implicit function of x and t. Since all other quantities such as ρ, p, etc., can be expressed in terms of v, we have completely solved the problem.

In the case of a polytropic gas, the adiabaticity condition is equivalent to $\rho T^{1/(1-\gamma)} = \text{constant}$. Since $c_s \propto \sqrt{T}$, it follows that

$$\rho = \rho_0 \left(\frac{c_s}{c_0}\right)^{2/(\gamma-1)}. \tag{3.146}$$

Using this relation in (3.144) we can determine $c_s(v)$:

$$c_s = c_0 \pm \frac{1}{2}(\gamma-1)v. \tag{3.147}$$

Hence,

$$\rho = \rho_0 \left(1 \pm \frac{1}{2}(\gamma-1)\frac{v}{c_0}\right)^{2/(\gamma-1)}, \quad p = p_0 \left(1 \pm \frac{1}{2}(\gamma-1)\frac{v}{c_0}\right)^{2\gamma/(\gamma-1)}. \tag{3.148}$$

Substituting these results in our general relation (3.145) we find that

$$x = t\left(\pm c_0 + \frac{1}{2}(\gamma+1)v\right) + f(v). \tag{3.149}$$

This relation can be rewritten in a more convenient form as

$$v = F\left[x - \left(\pm c_0 + \frac{1}{2}(\gamma+1)v\right)t\right], \tag{3.150}$$

where F is another arbitrary function.

The wave solution obtained here and described by (3.145) is qualitatively different from the one obtained in part (a) for small amplitudes. In equation (3.145) we may think of the disturbance as propagating due to two reasons: the fluid is moving with velocity v while the disturbance is moving *with respect to the fluid* with velocity $c_s(v)$. But since c_s depends on the density of gas, the net velocity of propagation, $v \pm c_s$, is different at different points. In general, the velocity of propagation increases with density. Thus high density regions propagate faster and overtake low density regions. This leads to the density profile changing progressively, as shown in figure 3.5. This suggests that, after a finite duration of time, the density is bound to become a multiple-valued function of x. Since this is physically impossible, a sharp discontinuity (shock) is formed in the fluid.

The location of this discontinuity can easily be ascertained. Let t_0 be the time of formation of discontinuity. At this instant the graph of velocity v as a function of x becomes vertical at some location x_0, that is, the derivative $(\partial x/\partial v)_t$ vanishes. It is also clear that this point should be a point of inflexion for the function $x(v)$,

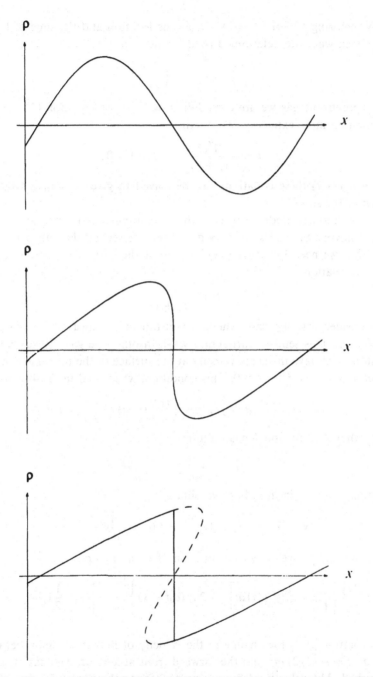

Fig. 3.5. The occurrence of discontinuities during the propagation of the sound wave. Since the speed of sound increases with density, high density regions move faster and overtake low density regions. This will eventually lead to a density configuration shown by the dashed lines in the last frame in which the density becomes multiple-valued. Such a situation is physically unacceptable and is avoided by the formation of strong discontinuity in the medium. The resulting density profile is shown by the unbroken line.

thereby requiring $(\partial^2 x/\partial v^2)_t = 0$. Hence the location and the time of formation of the shock wave are determined by the conditions

$$\left(\frac{\partial x}{\partial v}\right)_t = 0, \qquad \left(\frac{\partial^2 x}{\partial v^2}\right)_t = 0. \qquad (3.151)$$

For the polytropic gas we are considering, we can use equation (3.149) to write this condition more explicitly. We obtain

$$t = -\frac{2f'(v)}{(\gamma + 1)}, \qquad f''(v) = 0. \qquad (3.152)$$

Given $v = v(x, t)$ these equations can be solved to give $(x_{\text{crit}}, t_{\text{crit}})$, which is the location of the shock.

This condition is modified in case the discontinuity is formed at a boundary between moving gas and stationary gas. Then the second derivative f'' need not vanish; but we need to satisfy $(\partial x/\partial t)_v = 0$ at the surface where $v = 0$. So, we obtain the relation

$$t = -\frac{2f'(0)}{(\gamma + 1)}. \qquad (3.153)$$

(e) Consider first the case where the piston moves away from the gas with $v_{\text{piston}} = -at$. This sends a rarifaction wave through the gas. Since the gas and the piston must have the same velocity at the surface of the piston, we must have $v = -at$ at $x = -\frac{1}{2}at^2$ for $t > 0$. This condition, when used in (3.149), gives

$$f(-at) = -c_0 t + \frac{1}{2}\gamma a t^2 = \left(\frac{c_0}{a}\right)(-at) + \frac{1}{2}\frac{\gamma}{a}(-at)^2, \qquad (3.154)$$

showing that the function f has the form

$$f(z) = \left(\frac{c_0}{a}\right) z + \frac{1}{2}\frac{\gamma}{a}z^2. \qquad (3.155)$$

Using this result again in (3.149) we obtain

$$x - \left[c_0 + \frac{1}{2}(\gamma + 1)v\right]t = f(v) = \frac{c_0}{a}v + \frac{1}{2}\frac{\gamma}{a}v^2. \qquad (3.156)$$

Solving, we find the velocity as a function of x and t to be

$$v = \frac{1}{\gamma}\left[\left\{\left[c_0 + \frac{1}{2}(\gamma + 1)at\right]^2 - 2a\gamma(c_0 t - x)\right\}^{1/2} - \left[c_0 + \frac{1}{2}(\gamma + 1)at\right]\right].$$
$$(3.157)$$

This solution gives the change in the velocity of flow in a region between the piston at $x = -(1/2)at^2$ and the forward front at $x = c_s t$. The gas at $x > c_s t$ is undisturbed. The velocity of flow is everywhere to the negative x-axis (like that of the piston) and decreases monotonically to the right. The pressure and the density increase monotonically to the right.

If the piston moves to the right with a velocity $v_{\text{piston}} = +at$, the corresponding solution can be obtained by changing the sign of a in the expression. In this case, a compression wave propagates into the gas and a shock front is formed at the time determined by equation (3.153); using that equation, we find that $t = 2c_0/a\,(\gamma + 1)$.

3.9 Supersonic flow of gas

(a) We consider the steady adiabatic flow through a tube with a variable cross sectional area A. The conservation of mass implies that $\rho v A = $ constant or $d\,(\rho v)\,/\rho v = - \,(dA/A)$. The Euler equation for the one-dimensional steady flow becomes

$$\mathbf{v} \cdot \frac{d\mathbf{v}}{dt} = v\frac{dv}{dt} = -\frac{1}{\rho}\mathbf{v} \cdot \nabla p = -\frac{1}{\rho}\frac{dp}{dt} \qquad (3.158)$$

since $(\partial p/\partial t) = 0$. This implies that

$$v\,dv = -\frac{1}{\rho}dp = -\frac{c_s^2}{\rho}d\rho \qquad (3.159)$$

or, equivalently, $d\rho/\rho = -v\,(dv/c_s^2)$. We can use this to relate v and A:

$$-\frac{dA}{A} = \frac{d\,(\rho v)}{\rho v} = -v\frac{dv}{c_s^2} + \frac{dv}{v} = \left(1 - \frac{v^2}{c_s^2}\right)\frac{dv}{v}. \qquad (3.160)$$

Combining all these relations, we obtain

$$\frac{dA}{A} = \frac{c_s^2}{v^2}\left(1 - \frac{v^2}{c_s^2}\right)\frac{d\rho}{\rho} = \frac{p}{\rho v^2}\left(1 - \frac{v^2}{c_s^2}\right)\frac{dp}{p} = -\left(1 - \frac{v^2}{c_s^2}\right)\frac{dv}{v}. \qquad (3.161)$$

These equations show that the characteristics of the flow are very different depending on whether $v > c_s$ or $v < c_s$. Suppose that the flow is from left to right. When $v < c_s$ and the area A increases to the right ('subsonic diffuser'), we find that v decreases but ρ and p increase. If, on the other hand, A decreases to the right ('subsonic nozzle') then v increases and ρ and p decrease. This is what we 'normally' expect; in a constricted region v is larger and p is lower.

The behaviour is exactly the opposite in the case of supersonic flow with $v > c_s$. In this case, increase in area ('supersonic diffuser') leads to increase in v and decrease in ρ and p; similarly decrease in A ('supersonic nozzle') leads to decrease in v and increase in ρ and p. The transition from subsonic to supersonic flow can occur only at a point where $(dA/A) = 0$. Consider now the flow of gas through a tube in which A decreases, reaches a minimum and increases again. The flow begins subsonically and the speed of flow increases as the area decreases. From (3.160), we see that $(dv/dx) > 0$ for $(dA/dx) < 0$ as long as $v < c_s$; further, $v = c_s$ occurs only at the point where $(dA/dx) = 0$. Hence, two possibilities arise. (i) $v < c_s$ at the minimum area point ('throat'). In that case velocity increases, reaches a maximum and decreases again with $v < c_s$ everywhere. (ii) $v = c_s$ at the throat; then we have supersonic flow to the right with v increasing with increasing area.

(b) We will consider the flow of an ideal polytropic gas through a tube of variable cross-section, in which the cross-sectional area decreases to a minimum value and increases again. Let us assume that the flow is from left to right with the rate of mass flow being given by some constant value $\dot{M} = \rho v A$. Further, for an ideal polytropic gas in steady flow, we know that $(1/2)\,v^2 + w = w_0 = c_0^2/(\gamma - 1)$ remains a constant, where c_0 is the speed of sound at the initial point. The pressure and density for an ideal gas undergoing an adiabatic process can be written as $\rho = \rho_0 \left(T/T_0\right)^{1/(\gamma-1)}$, $p = p_0 \left(\rho/\rho_0\right)^{\gamma}$. Using these relations we can integrate the equations of flow (3.161) and obtain temperature, pressure and density in terms of the local velocity of flow:

$$T = T_0 \left[1 - \frac{1}{2}\,(\gamma - 1)\,\frac{v^2}{c_0^2}\right]$$

$$\rho = \rho_0 \left[1 - \frac{1}{2}\,(\gamma - 1)\,\frac{v^2}{c_0^2}\right]^{1/(\gamma-1)} \tag{3.162}$$

$$p = p_0 \left[1 - \frac{1}{2}\,(\gamma - 1)\,\frac{v^2}{c_0^2}\right]^{\gamma/(\gamma-1)} .$$

The velocity v at any location can be determined if the change in the cross-sectional area as a function of x is given.

A geometrical description of the above results can be provided along the following lines. For an ideal polytropic gas we have the relations

$$\rho = \rho_0 \left(\frac{c_s}{c_0}\right)^{2/(\gamma-1)} , \qquad w = \frac{c_s^2}{(\gamma - 1)} . \tag{3.163}$$

The conservation laws for the fluid flow through a region of varying cross-sectional area

$$\rho v A = \dot{M}, \quad \frac{1}{2}v^2 + w = w_0 , \tag{3.164}$$

can now be written as

$$v c_s^{2/(\gamma-1)} = k\frac{\dot{M}}{A}, \quad v^2 + \frac{2}{(\gamma - 1)}c_s^2 = \frac{2}{(\gamma - 1)}c_0^2 , \tag{3.165}$$

where k is an unimportant constant and c_0 is the speed of sound at the location $v = 0$. The second of these equations defines a quadratic curve in the v–c_s plane (see figure 3.6). The dashed curves are based on the first equation for different values of A. The flow begins around the point P and moves up along the quadratic curve as the area of cross-section decreases. If the area of cross-section decreases sufficiently to reach a value A_{min} before the flow reaches the line $v = c_s$, supersonic motion does *not* occur. On the other hand, if A_{min} is reached at $v = c_s$, the flow can become supersonic. In this case, we have to adjust $\dot{M}$ and A_{min} such

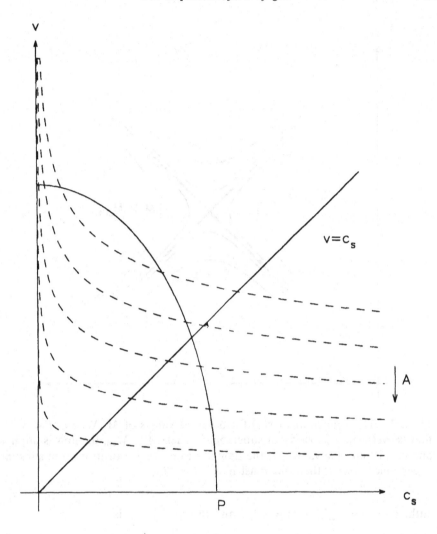

Fig. 3.6. Flow of a gas through a region of variable cross-section; see text for discussion.

that $A_{\min}$ is reached when $v = c_s$. The condition for this can easily be worked out to be

$$\dot{M} = \left(\frac{2}{\gamma + 1} \right)^{\frac{(\gamma + 1)}{2(\gamma - 1)}} \rho_0 c_0 A_{\min} \equiv \dot{M}_{\text{crit}}. \qquad (3.166)$$

Figure 3.7 gives $v(x)$ for different values of $\dot{M}$. We shall see a similar pattern of flow in spherical accretion later.

(c) We shall approximate the jet as a collimated one-dimensional flow with a variable area A, rate of mass flow $\dot{M}$ and speed v. For a gas with $\gamma = 5/3$ the

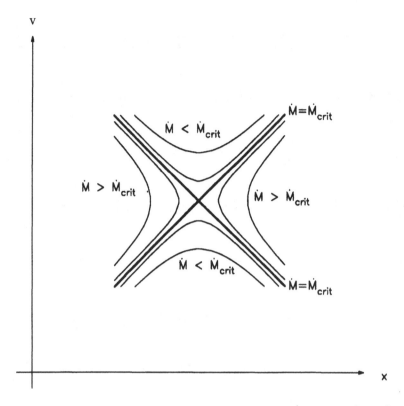

Fig. 3.7.   The behaviour of $v(x)$ for different values of $\dot{M}$. When $\dot{M} < \dot{M}_{\text{crit}}$ the flow is everywhere subsonic or supersonic. When $\dot{M} > \dot{M}_{\text{crit}}$ the flow is unphysical and $v(x)$ is multiple valued. If the flow has to make a transition from subsonic to supersonic velocity, then one must have $\dot{M} = \dot{M}_{\text{crit}}$.

enthalpy is $w = c_s^2/(\gamma - 1) = 3c_s^2/2$ and the energy flux is

$$\mathscr{E} = \dot{M}\left(\frac{1}{2}v^2 + w\right) = \frac{1}{2}\dot{M}v^2\left(1 + \frac{3c_s^2}{v^2}\right). \tag{3.167}$$

Further, conservation of mass implies that $\dot{M} = \rho A v$ must remain a constant, giving $\rho \propto (Av)^{-1}$; since for adiabatic flow $\rho \propto p^{3/5}$, $c_s \propto \rho^{1/3} \propto p^{1/5}$, we find that $v = \alpha p^{-3/5} A^{-1}$ and $(c_s/v) = \beta p^{4/5} A$, where α and β are constants. Substituting these relations in the expression for $\mathscr{E}$ we obatin

$$\frac{2\mathscr{E}}{\dot{M}} = \frac{\alpha^2}{p^{6/5}A^2}\left(1 + 3\beta^2 p^{8/5}A^2\right) = \text{constant}, \tag{3.168}$$

since $\mathscr{E}$ and $\dot{M}$ are constants. Or, equivalently,

$$A^2 = \alpha^2 p^{-6/5}\left(2\frac{\mathscr{E}}{\dot{M}} - 3\alpha^2\beta^2 p^{2/5}\right)^{-1}. \tag{3.169}$$

As p increases, the area first decreases, reaches a minimum and then increases again. The speed of the jet becomes transonic at the location where the area reaches a minimum. In the subsonic part of the flow, pressure and density are roughly constant and the area decreases inversely with the velocity. In the supersonic part, the velocity is mostly a constant and the area increases as $A \propto p^{-3/5}$ with decreasing pressure.

When $p \propto r^{-2}$ the angle subtended by the jet at the nucleus decreases as $A^{1/2} r^{-1} \propto r^{-2/5}$. Thus a supersonic jet can be well collimated even when its cross-sectional area increases.

3.10 Shock waves

(a) We shall consider the situation in a frame in which the shock front is at rest. The physical quantities on the two sides of the shock front are related by the conservation of mass, energy and momentum. The conservation of mass gives

$$\rho_1 v_1 = \rho_2 v_2 \equiv j. \tag{3.170}$$

The conservation of energy flux, taken in conjunction with the above equation, implies

$$w_1 + \frac{1}{2} v_1^2 = w_2 + \frac{1}{2} v_2^2. \tag{3.171}$$

Finally, the conservation of momentum flux gives

$$p_1 + \rho_1 v_1^2 = p_2 + \rho_2 v_2^2. \tag{3.172}$$

These three equations relate the set (p_2, ρ_2, v_2) to p_1, ρ_1 and v_1. The enthalpy w can be expressed in terms of other variables for a polytropic gas with index γ:

$$w = \frac{\gamma}{\gamma - 1} \frac{p}{\rho} = \frac{\gamma}{\gamma - 1} kT. \tag{3.173}$$

These equations can be solved by straightforward algebraic manipulations. It is most convenient to rewrite these equations expressing the ratios (ρ_2/ρ_1), (v_2/v_1), ..., etc., in terms of the Mach number $M_1 \equiv (v_1/c_s)$ as one side. Here $c_s^2 = \gamma p/\rho$. We find that

$$\frac{\rho_2}{\rho_1} = \frac{(\gamma + 1) M_1^2}{(\gamma + 1) + (\gamma - 1)\left(M_1^2 - 1\right)} = \frac{v_2}{v_1}, \tag{3.174}$$

$$\frac{p_2}{p_1} = \frac{(\gamma + 1) + 2\gamma\left(M_1^2 - 1\right)}{(\gamma + 1)}, \tag{3.175}$$

$$\frac{T_2}{T_1} = \frac{\left[(\gamma + 1) + 2\gamma\left(M_1^2 - 1\right)\right]\left[(\gamma + 1) + (\gamma - 1)\left(M_1^2 - 1\right)\right]}{(\gamma + 1)^2 M_1^2}. \tag{3.176}$$

(b) Since we have assumed the flow to be supersonic on side 1 we have $M_1 > 1$; it is clear from the above equations that $p_2 > p_1$, $\rho_2 > \rho_1$, $v_2 < v_1$ and $T_2 > T_1$. The strongest shock which one can have corresponds to the limit $M_1 \to \infty$. In this case we find that (ρ_2/ρ_1) tends to the finite value $(\gamma + 1)/(\gamma - 1)$. In the same limit, the pressure and temperature ratios diverge.

(c) In the above discussion, we have assumed the fluid to be ideal on either side of the shock front. In reality the shock front will have a small thickness over which the physical parameters will change rapidly. Since the gradients are large, dissipative processes will be important in the transition region which will determine the thickness of this region.

This thickness can be estimated as follows. Taking into account the viscous and thermal effects, the conservation laws for the flux of mass, momentum and energy can be written as

$$\rho v = \text{constant}, \tag{3.177}$$

$$\rho v^2 + p - \left(\frac{4}{3}\eta + \zeta\right)\frac{\partial v}{\partial x} = \text{constant}, \tag{3.178}$$

$$\rho v^2 \left(\frac{1}{2}v^2 + \epsilon + \frac{p}{\rho}\right) - \left(\frac{4}{3}\eta + \zeta\right)v\frac{\partial v}{\partial x} - \kappa\frac{\partial T}{\partial x} = \text{constant}. \tag{3.179}$$

In arriving at the above equations, we have used the fact that the component σ_{xx} of the viscous tensor is $\sigma_{xx} = \left((4/3)\eta + \zeta\right)(\partial v/\partial x)$. In the shock layer, η and ζ will be of comparable magnitude or η will dominate over ζ. Further, the momentum flux due to the friction in (3.178), given by the term $(4/3)\eta v\,(\partial v/\partial x)$, must have the same order of magnitude as the other two terms, in particular ρv^2. If the velocity changes by Δv over a region Δx, then $\eta\,(\partial v/\partial x) \cong \rho v\,(\Delta v/\Delta x)$. Equating this to ρv^2, we find that $\Delta x \cong v\,(\Delta v/v^2)$. For a strong shock, we may take $\Delta v \cong v$; using this and noting that $v \cong lv$, where l is the mean free path, we find that $\Delta x \cong l$. In other words, a strong shock has a thickness which is comparable to the mean free path.

The same result can be obtained from (3.179) by noting that $\kappa\,(\partial T/\partial x)$ must be of the same order as $\rho v\,(p/\rho) \cong nkTv$. This gives $\Delta x \cong (\kappa/nkv)\,(\Delta T/T)$. Using $\Delta T \cong T$ and $\kappa \cong nkvl$ we again find that $\Delta x \cong l$.

(d) As the piston begins to move, a shock wave is formed in front of the piston and moves to the positive x-direction. At $t = 0$, the shock and the piston are coincident but at $t > 0$ the shock is ahead of the piston. Thus, for $t > 0$, the gas is divided into two regions: region 1 ahead of the shock wave where the gas pressure is equal to its initial value p_1 and region 2, which lies between the shock front and the piston where the pressure is p_2. In region 2, the gas moves with constant velocity equal to that of the piston, U. Hence, the difference in velocity between regions 1 and 2 is also equal to U.

We shall now consider the motion in the frame in which the shock front is at rest and the gas has velocities v_1 and v_2 in the two regions. The velocity *difference* between the two regions $(v_1 - v_2)$ is, of course, the same in all frames and is given by U. The conservation laws (3.170), (3.171) and (3.172) can be rearranged to give the relations

$$\frac{V_2}{V_1} = \frac{p_1(\gamma+1) + p_2(\gamma-1)}{p_1(\gamma-1) + p_2(\gamma+1)},$$

$$j^2 = \frac{p_2 - p_1}{V_1 - V_2}, \tag{3.180}$$

$$v_1 - v_2 = j(V_1 - V_2) = [(p_2 - p_1)(V_1 - V_2)]^{1/2},$$

where $V = (1/\rho)$ is the specific volume. In our case we find that

$$v_1 - v_2 = U = [(p_2 - p_1)(V_1 - V_2)]^{1/2}$$

$$= (p_2 - p_1)\left[\frac{2V_1}{(\gamma-1)p_1 + (\gamma+1)p_2}\right]^{1/2}. \tag{3.181}$$

This equation can be rewritten as a quadratic equation for the pressure ratio (p_2/p_1), in terms of p_1, V_1 and U:

$$\left(\frac{p_2}{p_1}\right)^2 - \left(\frac{p_2}{p_1}\right)\left[2 + (\gamma+1)\frac{U^2}{2p_1 V_1}\right] + \left[1 - \frac{(\gamma-1)U^2}{2p_1 V_1}\right] = 0. \tag{3.182}$$

The solution can be expressed in terms of the velocity of sound, $c_1 = \sqrt{\gamma p_1 V_1}$, in the undisturbed gas:

$$\frac{p_2}{p_1} = 1 + \frac{\gamma(\gamma+1)U^2}{4c_1^2} + \frac{\gamma U}{c_1}\left[1 + \frac{(\gamma+1)^2 U^2}{16c_1^2}\right]^{1/2}. \tag{3.183}$$

Knowing this ratio, the conservation laws provide the expression for $|v_1|$ to be

$$|v_1| = \frac{\gamma+1}{4}U + \left[c_1^2 + \frac{(\gamma+1)^2 U^2}{16}\right]^{1/2}. \tag{3.184}$$

Note that these results are valid in the frame in which the shock front is at rest. In the frame in which the undisturbed gas is at rest, the shock front moves with a speed equal to v_1. Therefore, the above expression also gives the velocity of the shock front in the undisturbed medium. For $U \gg c_1$ the shock velocity reduces to

$$v_{\text{shock}} = \frac{1}{2}(\gamma+1)U. \tag{3.185}$$

The piston, on the other hand, moves with a velocity U; hence the ratio of the position of the shock front to the position of the piston is a constant given by $(v_{\text{shock}}/U) = (\gamma+1)/2$. For a monatomic perfect gas with $\gamma = 5/3$ this ratio has a value 4/3. Clearly, all the gas which was originally in the tube between $x = 0$ and the position of the shock wave is squeezed a smaller distance $(v_{\text{shock}} - U)t$. This will cause a density increase of

$$\frac{\rho_2}{\rho_1} = \frac{v_{\text{shock}}}{v_{\text{shock}} - U} = \frac{(\gamma+1)}{(\gamma-1)}, \tag{3.186}$$

which, of course, agrees with the result we found in part (b) above.

3.11 Particle acceleration mechanisms

(a) Let the rate of collisions be τ^{-1} so that the mean time between collisions is τ. Further let the number of particles which survive in the accelerating region for the time interval $(t, t + dt)$ be

$$n(t) \, dt = n_0 \, e^{-t/T} \left(\frac{dt}{T} \right). \tag{3.187}$$

Since the mean number of collisions in time t is (t/τ), the typical energy of these particles will be

$$E(t) = E_0 \beta^{t/\tau} = E_0 \exp \left[\left(\frac{\ln \beta}{\tau} \right) t \right]. \tag{3.188}$$

It follows that $t = (\tau/\ln \beta) \ln(E/E_0)$; so

$$n(E) \, dE = n(t) \left(\frac{dt}{dE} \right) dE \propto E^{-1} \exp \left[- \left(\frac{\tau}{T \ln \beta} \right) (\ln E) \right] dE \propto E^{-1 - \frac{\tau}{T \ln \beta}} \, dE. \tag{3.189}$$

The probability for a particle to remain in the accelerating region after one collision is $P = \exp(-\tau/T)$, giving $(\tau/T) = -\ln P$. Therefore, the energy spectrum can be written as

$$n(E) \propto E^{-p} \, dE, \quad p = 1 + \frac{\tau}{T \ln \beta} = 1 - \frac{\ln P}{\ln \beta}, \tag{3.190}$$

which is a power law.

(b) Consider a scatterer moving with a velocity V. Let the particle hit the scatterer with a momentum p, at an angle θ with respect to the normal. In the centre of mass frame, the energy of the particle will be conserved and its x-component of the momentum will be reversed. We can calculate the net gain in the energy by first transforming to the centre of mass frame and then transforming back (after the collision) to the lab frame.

Since the scatterer is infinitely massive, it will be moving with the same velocity V in the centre of mass frame. The energy and momentum of the particle in this frame, before the collision, are

$$E' = \gamma \, (E + V p_x), \qquad p'_x = \gamma \left(p_x + \frac{V E}{c^2} \right), \tag{3.191}$$

with $\gamma^2 = (1 - V^2/c^2)^{-1}$. After the collision E' remains as E' but p'_x changes to $-p'_x$. Transforming back to the lab frame, we find that the energy after the collision is

$$E'' = \gamma \, (E' + V p'_x). \tag{3.192}$$

Using (3.191) in (3.192) and recalling that $(p_x/E) = (v \cos \theta/c^2)$ we can express E'' in terms of E:

$$E'' = \gamma^2 E \left[1 + \frac{2 V v}{c^2} \cos \theta + \left(\frac{V}{c} \right)^2 \right]. \tag{3.193}$$

Expanding this expression in a Taylor series up to second order in (V/c), we obtain

$$\Delta E = E'' - E \cong \frac{2Vv}{c^2} \cos \theta + 2 \left(\frac{V}{c} \right)^2 . \tag{3.194}$$

We now have to average this expression over $\cos \theta$. The probability of a collision at an angle θ will be proportional to $\gamma[1 + (V/c)\cos \theta]$ for a highly relativistic particle with $v \cong c$. Averaging $(2V/c)\cos \theta$ over the angles we obtain

$$\left\langle \frac{2V}{c} \cos \theta \right\rangle = \left(\frac{2V}{c} \right) \frac{\int_{-1}^{1} \cos \theta \left[1 + (V/c) \cos \theta \right] d(\cos \theta)}{\int_{-1}^{1} \left[1 + (V/c) \cos \theta \right] d(\cos \theta)} = \left(\frac{2}{3} \right) \left(\frac{V}{c} \right)^2 . \tag{3.195}$$

Hence, $\langle \Delta E/E \rangle = (8/3)(V/c)^2$ is second order in (V/c). It follows that $\beta = [1 + (8/3)(V/c)^2]$. From our general expression we can determine p to be

$$p = 1 - \frac{\ln P}{\ln \beta} \cong 1 - \frac{3}{8} \frac{c^2}{V^2} \ln \tau . \tag{3.196}$$

(c) In the frame in which the shock front is at rest, the upstream gas flows with a velocity $v_1 = U$. In the case of a strong shock we know that $(\rho_2/\rho_1) = (\gamma+1)/(\gamma-1)$ and $(v_2/v_1) = (\rho_1/\rho_2)$. Taking $\gamma = 5/3$ for a monotonic fully ionized gas, we obtain $\rho_2 = 4\rho_1$ and $v_2 = (v_1/4) = (U/4)$.

Consider first the average increase in the energy of a particle which crosses from the upstream to the downstream side of the shock. In a frame at which the upstream gas is at rest, the gas on the downstream has a speed $V = (3/4) U$. Performing a Lorentz transformation, we find that the particle's energy when it passes to the downstream side is $E' = \gamma(E + p_x V)$. If the shock is non-relativistic but particles are ultrarelativistic, we can set $V \ll c$, $\gamma \cong 1$, $E \cong pc$ and $p_x \cong (E/c)\cos \theta$. In that case,

$$\frac{\Delta E}{E} = \frac{V}{c} \cos \theta . \tag{3.197}$$

We now need the probability that the particles that cross the shock arrive at an angle θ. The number of particles incident between the angles θ and $(\theta + d\theta)$ is proportional to $\sin \theta \, d\theta$; but the rate at which they approach the shock front is proportional to the x-component of their velocities, $c \cos \theta$. Thus the total probability should be proportional to $\sin \theta \cos \theta \, d\theta$. Averaging the expression $(\Delta E/E)$ using the properly normalized distribution, $P(\theta) = 2 \sin \theta \cos \theta \, d\theta$, we find that

$$\left\langle \frac{\Delta E}{E} \right\rangle = \frac{V}{c} \int_0^{\pi/2} 2 \cos^2 \theta \sin \theta \, d\theta = \frac{2}{3} \frac{V}{c} . \tag{3.198}$$

The velocity vector of the particle will be randomized by scattering in the downstream region without any loss of energy. Thus particles in the vicinity of a strong shock gain energy in each crossing of the shock front. It is easy to see that the same amount of fractional gain will occur when the particle crosses from downstream to upstream. Thus $\beta = [1 + (4V/3c)] = (1 + U/c)$ for this process.

To determine the escape probability we may proceed as follows. The number of particles crossing the shock is $(nc/4)$, where n is the number density of particles. This is the average number of particles crossing the shock in either direction. Downstream, however, the particles are swept away from the shock front by the bulk velocity of the flow at the rate $nV = (nU/4)$. Thus the fraction of particles lost per unit time away from the shock front is $[(1/4)nU/(1/4)nc] = (U/c)$. Since $U \ll c$ we may take the escape probability to be $P = (1 - U/c)$. We therefore find that

$$\ln P = \ln \left(1 - \frac{U}{c}\right) \cong -\frac{U}{c}, \qquad \ln \beta = \ln \left(1 + \frac{U}{c}\right) \cong \frac{U}{c}, \qquad (3.199)$$

giving $n(E)\, dE \propto E^{-2}\, dE$.

3.12 Spherical accretion

(a) For a steady spherically symmetric flow, the continuity equation

$$\frac{1}{r^2} \frac{\partial}{\partial r} \left(r^2 \rho v\right) = 0 \qquad (3.200)$$

shows that $\rho v r^2$ should be a constant, where v is the radial component of the velocity. We will set $\dot{M} = -4\pi r^2 \rho v$ to be the constant accretion rate. The Euler equation, with a force term arising from the gravitational field of the central object, is

$$v \frac{\partial v}{\partial r} + \frac{1}{\rho} \frac{\partial p}{\partial r} + \frac{GM}{r^2} = 0. \qquad (3.201)$$

In order to complete the system of equations we will add the polytropic relation $p \propto \rho^\gamma$. Writing $(\partial p/\partial r) = (\partial p/\partial \rho)(\partial \rho/\partial r) = c_s^2 (\partial \rho/\partial r)$ in the Euler equation (3.201) and using the result

$$\frac{1}{\rho} \frac{\partial \rho}{\partial r} = -\frac{1}{vr^2} \frac{\partial}{\partial r} \left(vr^2\right) \qquad (3.202)$$

(which can be obtained from (3.200)), we can rewrite Euler's equation as

$$v \frac{\partial v}{\partial r} - \frac{c_s^2}{vr^2} \frac{\partial}{\partial r} \left(vr^2\right) + \frac{GM}{r^2} = 0. \qquad (3.203)$$

This may be rearranged into a more convenient form

$$\frac{1}{2} \left(1 - \frac{c_s^2}{v^2}\right) \frac{\partial v^2}{\partial r} = -\frac{GM}{r^2} \left(1 - \frac{2c_s^2 r}{GM}\right), \qquad (3.204)$$

which allows one to understand several features of accretion.

To begin with, we note that, at very large distances from the central object, c_s^2 will tend to some constant value determined by the parameters of the gas at infinity. (In general, of course, c_s^2 depends on r.) Hence, the factor $(1 - 2c_s^2 r/GM)$

will be negative definite at large r, making the right-hand side of (3.204) positive at large r. On the left-hand side, we want $(\partial v^2/\partial r)$ to be negative because we expect the gas to be at rest at large distances and flow with an acceleration as it approaches the central body. Hence, we conclude that $v^2 < c_s^2$ at large r, that is, the flow is subsonic at large distances.

As the gas approaches the central object, the factor $(1 - 2c_s^2 r/GM)$ will tend to zero at some point. Using the fact that $c_s^2 = \gamma p/\rho = \gamma k T$ we can estimate the critical radius to be

$$r_{\text{crit}} = \frac{GM}{2c_s^2 (r_{\text{crit}})} \cong 7.5 \times 10^{13} \left(\frac{T}{10^4 \text{K}}\right)^{-1} \left(\frac{M}{M_\odot}\right) \text{cm}. \qquad (3.205)$$

This radius is much larger than the size of typical accreting objects encountered in astrophysics. For $r < r_{\text{crit}}$ we can repeat the above analysis and conclude that $v^2 > c_s^2$. In other words, the flow must pass through a transition from subsonic to supersonic speeds at $r = r_{\text{crit}}$. Further, at $r = r_{\text{crit}}$, we either must have $v^2 = c_s^2$ or $(\partial v^2/\partial r) = 0$.

We can now classify the solutions based on their behaviour near $r = r_{\text{crit}}$. (All these solutions are shown schematically in figure 3.8.) (1) $v^2 = c_s^2$ at $r = r_{\text{crit}}$; $v^2 \to 0$ as $r \to \infty$; and $v^2 < c_s^2$ for $r > r_{\text{crit}}$; $v^2 > c_s^2$ for $r < r_{\text{crit}}$. This is the most relevant solution for our problem and describes an accretion which starts out as subsonic at large distances and becomes supersonic near the compact object. (2) $v^2 = c_s^2$ at $r = r_{\text{crit}}$; $v^2 \to 0$ as $r \to 0$; $v^2 > c_s^2$ for $r > r_{\text{crit}}$; $v^2 < c_s^2$ for $r < r_{\text{crit}}$. This represents a time reversed situation to spherical accretion involving a wind flowing from the compact object. The flow is subsonic near the object and becomes supersonic at large distances. (3) $(\partial v^2/\partial r) = 0$ at $r = r_{\text{crit}}$; $v^2 < c_s^2$ for all r. (4) $(\partial v^2/\partial r) = 0$ at $r = r_{\text{crit}}$; $v^2 > c_s^2$ for all r. Solutions (3) and (4) are either subsonic or supersonic in the entire range and hence cannot describe accretion on to a compact object of sufficiently small radius. (5) $(\partial v^2/\partial r) = \infty$ when $v^2 = c_s^2$; $r > r_{\text{crit}}$ always. (6) $(\partial v^2/\partial r) = \infty$ when $v^2 = c_s^2$; $r < r_{\text{crit}}$ always. These two solutions are unphysical because they give two possible values for v^2 at any given r. Thus we conclude that only the type-1 solution is relevant to our problem.

(b) To proceed further, we shall integrate equation (3.201) using the fact that $(\partial p/\partial \rho) = \gamma (p/\rho) = c_s$ to obtain

$$\frac{v^2}{2} + \frac{c_s^2}{(\gamma - 1)} - \frac{GM}{r} = \text{constant}. \qquad (3.206)$$

Since we must have $v^2 \to 0$ as $r \to \infty$, the constant must be $c_s^2 (\infty) / (\gamma - 1)$. On the other hand, $v^2 (r_{\text{crit}}) = c_s^2 (r_{\text{crit}})$ and $(GM/r_{\text{crit}}) = 2c_s^2 (r_{\text{crit}})$. Using these relations we can relate $c_s (r_{\text{crit}})$ to $c_s (\infty)$, obtaining

$$c_s^2 (r_{\text{crit}}) \left[\frac{1}{2} + \frac{1}{\gamma - 1} - 2\right] = \frac{c_s^2 (\infty)}{(\gamma - 1)}. \qquad (3.207)$$

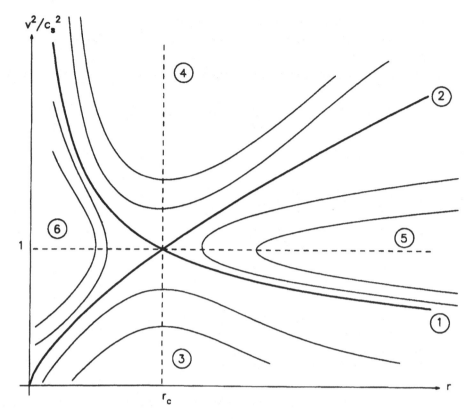

Fig. 3.8.   The fluid velocity in spherical accretion in different cases. See text for discussion.

Or

$$c_s\left(r_{\text{crit}}\right) = c_s\left(\infty\right) \left[\frac{2}{(5 - 3\gamma)}\right]^{1/2}. \tag{3.208}$$

Using $c_s^2 \propto \rho^{\gamma-1}$, we find that

$$\rho\left(r_{\text{crit}}\right) = \rho\left(\infty\right) \left[\frac{c_s\left(r_{\text{crit}}\right)}{c_s\left(\infty\right)}\right]^{2/(\gamma-1)}. \tag{3.209}$$

Finally, since $r^2 \rho v$ is a constant, we can write $\dot{M} = 4\pi r_{\text{crit}}^2 \rho\left(r_{\text{crit}}\right) c_s\left(r_{\text{crit}}\right)$, using which we can relate the accretion rate $\dot{M}$ to the conditions of the gas at infinity. Straightforward algebra gives

$$\dot{M} = \pi G^2 M^2 \frac{\rho\left(\infty\right)}{c_s^3\left(\infty\right)} \left[\frac{2}{5 - 3\gamma}\right]^{(5-3\gamma)/2(\gamma-1)}. \tag{3.210}$$

Numerically, this corresponds to the accretion rate:

$$\dot{M} = 1.4 \times 10^{11} \text{ g s}^{-1} \left(\frac{M}{M_\odot}\right)^2 \left(\frac{\rho(\infty)}{10^{-24} \text{ g cm}^{-3}}\right) \left(\frac{c_s(\infty)}{10 \text{ km s}^{-1}}\right)^{-3} \quad (3.211)$$

if $\gamma = 1.4$. To complete the solution we find $v(r)$ in terms of $c_s(r)$ from the conservation of $\dot{M}$. This gives

$$-v = \frac{\dot{M}}{4\pi r^2 \rho(r)} = \frac{\dot{M}}{4\pi r^2 \rho(\infty)} \left(\frac{c_s(\infty)}{c_s(r)}\right)^{2/(\gamma-1)}. \quad (3.212)$$

Substituting this relation into (3.206) gives an algebraic equation for $c_s^2(r)$. Since this equation has fractional exponents this has to be solved numerically. Given $c_s^2(r)$, equation (3.212) determines $v(r)$ and the relation $\dot{M} = r^2 \rho(r)v(r)$ fixes $\rho(r)$.

It may appear strange that the flow rate $\dot{M}$ is uniquely fixed at a value dependent only on M, $c_s(\infty)$, $\rho(\infty)$, implying that a gas can flow to a compact object only at a critical rate. This can be understood as follows. Using the relations $\dot{M} = -4\pi\rho v r^2$, $c_s^2(r) = c_s^2(\infty)(\rho/\rho(\infty))^{\gamma-1}$, and equation (3.206), we can estimate how $v(r)$ and $\rho(r)$ changes when $\dot{M}$ changes slightly. We find that, at fixed r,

$$\frac{\delta\dot{M}}{\dot{M}} = \left(1 - \frac{v^2}{c_s^2}\right)\frac{\delta v}{v}, \quad \frac{\delta\rho}{\rho} = -\frac{v^2}{c_s^2}\frac{\delta v}{v}. \quad (3.213)$$

As we increase v (at any r), $\dot{M}$ increases until $v = c_s$; then, as v increases, $\dot{M}$ decreases. This shows that, at any given r, there is a maximum permitted flow rate, which occurs when $v^2 = c_s^2$. The maximum value is

$$|\dot{M}_{\max}(r)| = 4\pi\rho c_s r^2 = 4\pi r^2 c_s(\infty)\rho(\infty) \cdot \left(\frac{c_s(r)}{c_\infty(\infty)}\right) \cdot \left(\frac{\rho}{\rho(\infty)}\right)$$

$$= 4\pi r^2 c_s(\infty)\rho(\infty) \left[\frac{c_s}{c(\infty)}\right]^{\frac{\gamma+1}{\gamma-1}}$$

$$= 4\pi r^2 c_s(\infty)\rho(\infty) \left[\frac{2(\gamma-1)}{(\gamma+1)}\left(\frac{GM}{r^2 c_s(\infty)} + \frac{1}{\gamma-1}\right)\right]^{\frac{\gamma+1}{2(\gamma-1)}}. \quad (3.214)$$

The last equality follows from (3.206) (with $v = c_s$ and the constant on the right-hand side set to $[c_s^2(\infty)/(\gamma-1)]$). For $1 < \gamma \le 5/3$, this function has the following behaviour: it decreases as $r^{-(5-3\gamma)/2(\gamma-1)}$ for small r, reaches a minimum value $\dot{M}_{\text{crit}}$ at $r = r_{\text{crit}}$ and increases as r^2 for large r. Since we must have $|\dot{M}| < |\dot{M}_{\max}|$ at all r, it follows that $\dot{M} > \dot{M}_{\text{crit}}$ is forbidden. Further, for $\dot{M} < \dot{M}_{\text{crit}}$, the flow is everywhere subsonic. In order to have a flow which makes the transition from subsonic to supersonic, we must have exactly $\dot{M} = \dot{M}_{\text{crit}}$.

3.13 Accretion disc

(a) Since gas flows in almost Keplerian circular orbits, we assume that $v_r \ll v_\phi \cong \sqrt{(GM/r)}$, where M is the mass of the central compact object, and drop all the terms which are quadratic or higher in v_r. The term linear in v_r gives the mass flux to be $\dot{M} = -2\pi r L v_r \rho$, where L is the thickness of the disc in the z-direction and ρ is the constant density of the disc. The dynamics of the flow is governed by the Navier–Stokes equation for an incompressible fluid in steady state:

$$(\mathbf{v} \cdot \nabla)\mathbf{v} = -\frac{1}{\rho}\nabla p - \nabla\psi + \nu\nabla^2\mathbf{v}, \qquad (3.215)$$

where ψ is the gravitational potential and $\nu = (\eta/\rho)$. Taking the ϕ-component of this equation and using $v_\phi^2 = GM/r$, we find that $v_r = -(3/2)(\nu/r)$, giving $\dot{M} = 3\pi L\eta$. Now taking the radial component of the equation and dropping the second-order terms in v_r we obtain

$$-\frac{v_\phi^2}{r} = -\frac{1}{\rho}\frac{\partial p}{\partial r} - \frac{\partial\psi}{\partial r}. \qquad (3.216)$$

Since $\psi = -GM/r$ and $v_\phi^2 = GM/r$, this gives $p = $ constant, at the level of approximation we are working with.

As the matter slowly falls into the compact object, angular momentum and energy flow outwards. By calculating the difference between the energy flowing outwards at a radius r and at a slightly larger radius $(r + dr)$ we can estimate the amount of energy which must have been dissipated as heat in an annular ring of radial width dr. The flux of energy through the sides of a cylinder of radius r and height L (with $L \ll r$) is given by

$$S(r) \cong 2\pi r L \left[\rho v_r\left(\frac{1}{2}v^2 + \psi\right) - \sigma_{ri}v^i\right]. \qquad (3.217)$$

The amount of energy dissipated in an annular ring of width dr is given by

$$\begin{aligned}
dE = [S(r+dr) - S(r)] &= \frac{\partial S}{\partial r}dr \\
&= \frac{\partial}{\partial r}\left[2\pi r L\left(\rho v_r\left(\frac{1}{2}v^2 + \psi\right) - \sigma_{ri}v^i\right)\right]dr \\
&= -\frac{\partial}{\partial r}\left[\dot{M}\left(\frac{1}{2}v^2 + \psi\right) + 2\pi r L v^i\sigma_{ri}\right]dr. \qquad (3.218)
\end{aligned}$$

In arriving at the last expression we have used the fact that $\dot{M} = -2\pi r L v_r \rho$. The quantities v^2 and $v^i\sigma_{ri}$ can be estimated easily:

$$\frac{1}{2}v^2 + \psi \cong \frac{1}{2}v_\phi^2 + \psi = \frac{1}{2}\frac{GM}{r} - \frac{GM}{r} = -\frac{GM}{2r} \qquad (3.219)$$

and

$$\begin{aligned}
v^i\sigma_{ri} = v^r\sigma_{rr} + v^\phi\sigma_{r\phi} &= 2\eta v^r\frac{\partial v_r}{\partial r} + \eta v^\phi\left(\frac{\partial v_\phi}{\partial r} - \frac{v_\phi}{r}\right) \\
&\cong \sqrt{\frac{GM}{r}}\,\eta\left(-\frac{1}{2}\frac{\sqrt{GM}}{r^{3/2}} - \frac{\sqrt{GM}}{r^{3/2}}\right) = -\frac{3}{2}\eta\frac{GM}{r^2}. \qquad (3.220)
\end{aligned}$$

Substituting this relation into the equation for dE and simplifying the terms we obtain

$$dE = -\frac{3}{2}\frac{GM\dot{M}}{r^2}dr = -\frac{3}{2}\dot{M}v_\phi^2\frac{dr}{r}. \tag{3.221}$$

In a Keplerian orbit, the kinetic energy is equal to half the potential energy in magnitude. Hence, the total energy in a Keplerian orbit is half the potential energy. When matter moves from a Keplerian orbit with radius $(r + dr)$ to an orbit at radius r, the release of gravitational binding energy will be $(GM\dot{M}/2r^2)$. Equation (3.221) shows that the energy dissipated is three times this quantity. The extra energy arises due to dissipative transport of angular momentum outwards.

The following fact is noteworthy regarding the above result. Even though the entire mechanism depends on the existence of viscous dissipation, the final expression for energy release can be stated entirely in terms of M, r and the accretion rate $\dot{M}$. The numerical value of the viscosity coefficient η (which is not known with any reliable accuracy) does not enter the final formula.

(b) The energy dissipated away per unit area from within a radius r of the disc is proportional to $(dE/\pi r^2) \propto (v_\phi^2/r^2) \propto r^{-3}$ if we ignore the logarithmic correction arising from dr/r integration. Since this flux is proportional to $T^4(r)$ it follows that $T(r) \propto r^{-3/4}$. Consider now the low frequency end of the Planckian distribution, where the intensity $I_\nu \propto \nu^2 T$. This form is valid for $h\nu \lesssim 3kT$; since $T \propto r^{-3/4}$ this result is valid up to a radius r_{max}, with $r_{max} \propto \nu^{-4/3}$. The total intensity emitted from the Rayleigh-end of the Planckian spectrum will be

$$I_\nu \propto \int^{r_{max}} \nu^2 T(r) r \, dr \propto \int^{r_{max}} \nu^2 r^{1/4} dr \propto \nu^2 r_{max}^{5/4}. \tag{3.222}$$

Using $r_{max} \propto \nu^{-4/3}$ we find that $I_\nu \propto \nu^{1/3}$.

3.14 The Sedov solution for a strong explosion

For a strong shock, $p_2 \gg p_1$ and we can ignore p_1. In the same limit, $(\rho_2/\rho_1) \cong (\gamma + 1)/(\gamma - 1)$ and hence ρ_2 is completely specified by ρ_1. Thus the strong explosion is entirely characterized by the total energy E and the density ρ_1 of the unperturbed gas.

Let the radius of the shock front at time t be $R(t)$. Using E, ρ_1 and t we can find only one quantity with the dimension of length, namely, $(Et^2/\rho_1)^{1/5}$. Hence, we *must* have

$$R(t) = R_0 \left(\frac{Et^2}{\rho_1}\right)^{1/5}, \tag{3.223}$$

where R_0 is a constant to be determined by solving the equations of motion.

All other physical quantities can now be determined in terms of the basic result for $R(t)$. The speed of the shock wave with respect to the undisturbed gas is

$$u_1 = \frac{dR}{dt} = \frac{2R}{5t} = \frac{2}{5}R_0 E^{1/5}\rho_1^{-1/5}t^{-3/5}. \tag{3.224}$$

The pressure p_2, density ρ_2 and the speed of shock propagation, $v_2 = u_2 - u_1$ (relative to a fixed coordinate system at the back of the shock), can be determined using our junction conditions across the shock front:

$$v_2 = \frac{2}{(\gamma + 1)} u_1, \quad \rho_2 = \rho_1 \frac{(\gamma + 1)}{(\gamma - 1)}, \quad p_2 = \frac{2}{(\gamma + 1)} \rho_1 u_1^2. \tag{3.225}$$

The density remains constant while v_2 and p_2 decrease as $t^{-3/5}$ and $t^{-6/5}$ respectively.

To proceed further we have to explicitly integrate the equations of motion and determine the velocity $v(r, t)$, pressure $p(r, t)$ and density $\rho(r, t)$ of the gas behind the shock. The relevant equations are

$$\frac{\partial v}{\partial t} + v \frac{\partial v}{\partial r} = -\frac{1}{\rho} \frac{\partial p}{\partial r}, \quad \frac{\partial \rho}{\partial t} + \frac{\partial (\rho v)}{\partial r} + \frac{2\rho v}{r} = 0, \tag{3.226}$$

$$\left(\frac{\partial}{\partial t} + v \frac{\partial}{\partial r} \right) \ln \left(\frac{p}{\rho^\gamma} \right) = 0. \tag{3.227}$$

The last equation – conservation of entropy – replaces the equation of energy conservation. Instead of p we can equivalently use the variable $c_s^2 = \gamma p / \rho$.

We shall look for a self-similar solution to these equations with all variables depending essentially on the quantity $\xi \equiv [r/R(t)]$ with suitable scaling. More precisely, we invoke the ansatz

$$v = \frac{2r}{5t} V(\xi), \quad \rho = \rho_1 G(\xi), \quad c_s^2 = \frac{4\gamma^2}{25 t^2} Z(\xi), \tag{3.228}$$

with the boundary conditions

$$V(1) = \frac{2}{(\gamma + 1)}, \quad G(1) = \frac{(\gamma + 1)}{(\gamma - 1)}, \quad Z(1) = \frac{2\gamma (\gamma - 1)}{(\gamma + 1)^2}. \tag{3.229}$$

It is possible to obtain a relation between Z and V which will simplify the integration of the equations. Since we have ignored (p_1/ρ_1) in comparison with E, it follows that the total energy of the gas contained within a sphere bounded by the shock is a constant equal to E. Further, since the flow is self-similar, the energy of the gas inside any sphere of smaller radius which follows the trajectory $\xi = $ constant must also remain constant. There is an amount of energy $4\pi r^2 \rho v \left(w + \frac{1}{2} v^2 \right) dt$ which flows out of a sphere of radius r in time dt. On the other hand, the volume of the sphere increases by $4\pi r^2 v_n \, dt$ during this time, where $v_n = 2r/5t$ is the radial velocity of this sphere. The energy of gas inside the extra volume is $4\pi r^2 \rho v_n (\epsilon + \frac{1}{2} v^2) \, dt$. Equating the two energies and simplifying the expression we obtain the result

$$Z = \frac{\gamma (\gamma - 1)(1 - V) V^2}{2 (\gamma V - 1)}. \tag{3.230}$$

Given this relation, it is fairly straightforward (though tedius) to integrate the equations (3.227). The second and third equations can be written as

$$\frac{dV}{d\ln \xi} - (1 - V)\frac{d\ln G}{d\ln \xi} = -3V, \tag{3.231}$$

$$\frac{d\ln Z}{d\ln \xi} - (\gamma - 1)\frac{d\ln G}{d\ln \xi} = -\frac{5 - 2V}{1 - V}. \tag{3.232}$$

Eliminating $[d(\ln G)/d\ln \xi]$ between these equations and using (3.230) to express $[dZ/d\ln \xi]$ in terms of $[dV/d\ln \xi]$, we can determine $(dV/d\ln \xi)$ as a function of V alone:

$$\frac{1}{V}\frac{dV}{d\ln \xi} = \frac{\gamma(1 - 3\gamma)V^2 + (8\gamma - 1)V - 5}{\gamma(\gamma + 1)V^2 - 2(\gamma + 1)V + 2}. \tag{3.233}$$

This equation allows integration in closed form. A straightforward integration with suitable boundary conditions gives

$$\xi^5 = \left[\frac{2}{(\gamma + 1)V}\right]^2 \left\{\frac{\gamma + 1}{7 - \gamma}\left[5 - (3\gamma - 1)V\right]\right\}^{n_1} \left[\frac{\gamma + 1}{\gamma - 1}(\gamma V - 1)\right]^{n_2}, \tag{3.234}$$

where

$$n_1 = -\frac{13\gamma^2 - 7\gamma + 12}{(3\gamma - 1)(2\gamma + 1)}, \quad n_2 = \frac{5(\gamma - 1)}{2\gamma + 1}. \tag{3.235}$$

We can also determine $G(V)$ in a similar way and obtain

$$G = \frac{\gamma + 1}{\gamma - 1}\left[\frac{\gamma + 1}{\gamma - 1}(\gamma V - 1)\right]^{n_3} \left\{\frac{\gamma + 1}{7 - \gamma}\left[5 - (3\gamma - 1)V\right]\right\}^{n_4} \left[\frac{\gamma + 1}{\gamma - 1}(1 - V)\right]^{n_5}, \tag{3.236}$$

with

$$n_3 = \frac{3}{2\gamma + 1}, \quad n_4 = -\frac{n_1}{2 - \gamma}, \quad n_5 = -\frac{2}{2 - \gamma}. \tag{3.237}$$

The solutions are shown in figure 3.9. This provides the complete solution to our problem. The constant R_0 can now be determined by the condition

$$E = \int_0^R 4\pi r^2 dr\rho \left[\frac{1}{2}v^2 + \frac{c_s^2}{\gamma(\gamma - 1)}\right], \tag{3.238}$$

which reduces to

$$R_0^5 \frac{16\pi}{25} \int_0^1 G\left[\frac{1}{2}V^2 + \frac{Z}{\gamma(\gamma - 1)}\right]\xi^4 d\xi = 1. \tag{3.239}$$

The integral needs to be evaluated numerically; for air with $\gamma = 7/5$ we obtain $R_0 = 1.033$.

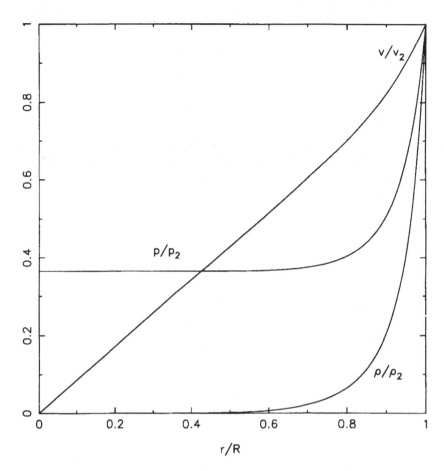

Fig. 3.9   The velocity, pressure and density for the Sedov solution.

3.15 Basics of magnetohydrodynamics

(a) In the outer zones of the Sun, $n_e \cong 10^{23}$ cm^{-3}, $L \cong 2 \times 10^{10}$ cm, $B \cong 10^3$ Gauss. From the equation $\nabla \times \mathbf{B} = \left(4\pi \mathbf{j}_e / c\right)$ and the expression for the current $\mathbf{j}_e = Z e n_i \mathbf{v}_i - e n_e \mathbf{v}_e = -e n_e \mathbf{v}_{\mathrm{diff}}$ we can make the order of magnitude estimate

$$\mathbf{v}_{\mathrm{diff}} \cong \frac{cB}{4\pi e n_e L} \cong 10^{-12} \text{ cm s}^{-1}. \tag{3.240}$$

(b) Since the fluid velocity is non-relativistic, the displacement current term $\left(\partial \mathbf{E}/c\partial t\right)$ is negligible compared to $\left(4\pi \mathbf{j}/c\right)$.

Consider the charge density and current in the lab frame, which can be obtained by a Lorentz transformation from the rest frame. For $v \ll c$, we have

$$\rho_e \cong \rho_e' + \mathbf{j}' \cdot \frac{\mathbf{v}}{c^2}, \quad \mathbf{j} \cong \rho_e' \mathbf{v} + \mathbf{j}'. \tag{3.241}$$

Since the charge density is negligible in the rest frame, that is, $\rho'_e \ll (j'/v)(v/c)^2$, it follows that $\rho_e \ll (|\mathbf{j}|/c)$. Hence, we can ignore ρ_e and retain j in the lab frame.

Further, using the relations $\nabla \cdot \mathbf{E} = 4\pi \rho_e$, $\nabla \times \mathbf{B} = (4\pi \mathbf{j}/c)$ and $c\rho_e \ll j$, we find that $|\mathbf{E}| \ll |\mathbf{B}|$. These are the conditions under which magnetohydrodynamics works.

(c) The equation of continuity, representing the conservation of mass, is unchanged in the presence of magnetic fields:

$$\frac{\partial \rho}{\partial t} + \nabla \cdot (\rho \mathbf{v}) = 0. \tag{3.242}$$

The Euler equation is modified due to the presence of electromagnetic force on matter. Since the electric field is negligible in the lab frame, the force on a charged particle is simply $q(\mathbf{v} \times \mathbf{B})/c$. Hence, the net force on a fluid element of unit volume can be written as $(\mathbf{j}_e \times \mathbf{B}/c)$. Substituting for $\mathbf{j}_e$ as $(c/4\pi)(\nabla \times \mathbf{B})$ we find the additional magnetic force to be $[(\nabla \times \mathbf{B}) \times \mathbf{B}]/4\pi$. Thus the Euler equation becomes

$$\rho \frac{d\mathbf{v}}{dt} = -\nabla p + \frac{1}{4\pi}(\nabla \times \mathbf{B}) \times \mathbf{B}. \tag{3.243}$$

The expression $(\nabla \times \mathbf{B}) \times \mathbf{B}$ has an elementary interpretation. Using a vector identity, the extra force term can be written as

$$\mathbf{f} = \frac{1}{4\pi}(\mathbf{B} \cdot \nabla)\mathbf{B} - \frac{1}{8\pi}\nabla B^2. \tag{3.244}$$

The first term may be thought of as a 'magnetic tension' and tends to straighten a magnetic field line which is curved. The second term can be thought of as originating due to an extra (magnetic) pressure $p_{mag} = B^2/8\pi$.

To find the equation satisfied by the magnetic field we start with Maxwell's equation $(\partial \mathbf{B}/\partial t) = -c(\nabla \times \mathbf{E})$ and substitute for $\mathbf{E}$ in terms of $\mathbf{j}$ using the relation

$$\mathbf{j} \cong \mathbf{j}' \cong \sigma \mathbf{E}' \cong \sigma \left[\mathbf{E} + \left(\frac{\mathbf{v}}{c} \times \mathbf{B} \right) \right]. \tag{3.245}$$

This gives

$$\frac{\partial \mathbf{B}}{\partial t} = -c\nabla \times \left[\frac{\mathbf{j}}{\sigma} - \left(\frac{\mathbf{v}}{c} \times \mathbf{B} \right) \right]. \tag{3.246}$$

We now substitute for $\mathbf{j}$ in terms of the magnetic field as $\mathbf{j} = (c/4\pi)(\nabla \times \mathbf{B})$. After some simple manipulation, we find that

$$\frac{\partial \mathbf{B}}{\partial t} = \nabla \times (\mathbf{v} \times \mathbf{B}) - \nabla \times \left(\frac{c^2}{4\pi\sigma} \nabla \times \mathbf{B} \right)$$

$$= \nabla \times (\mathbf{v} \times \mathbf{B}) + \frac{c^2}{4\pi\sigma} \nabla^2 \mathbf{B}. \tag{3.247}$$

In arriving at the last equation, we have used the fact that $\nabla \cdot \mathbf{B} = 0$. Note that, if $\nabla \cdot \mathbf{B} = 0$ initially, this equation will preserve that condition.

Equations (3.242), (3.243), (3.247), and an equation of state $p = p(\rho)$, together constitute a set of eight equations for the eight unknowns ρ, p, $\mathbf{v}$, and $\mathbf{B}$. Once $\mathbf{B}$ is determined, we can calculate $\mathbf{j}$ from the relation $\mathbf{j} = (c/4\pi)(\nabla \times \mathbf{B})$ and $\mathbf{E}$ using $\mathbf{E} = [\mathbf{j}/\rho - (\mathbf{v} \times \mathbf{B})/c]$. This completely solves the problem.

To obtain the equation for energy conservation, we proceed as follows. In the absence of any electromagnetic processes, we know that the rate of entropy increase is governed by an equation of the kind

$$\rho T \frac{ds}{dt} = \sigma_{ik} \frac{\partial v_i}{\partial x_k} + \nabla \cdot (\kappa \nabla T) . \tag{3.248}$$

The left-hand side of this equation is the amount of heat generated per unit time and volume in a moving fluid element. On the right-hand side, the first term gives the amount of heat dissipated due to viscosity and the second one is due to thermal conduction. In the presence of electromagnetic fields, we have to add the work done on the conduction current $\mathbf{j}'$ by the electric field $\mathbf{E}'$ in the rest frame of the fluid. This contributes an amount $\mathbf{j}' \cdot \mathbf{E}' = (j'^2/\sigma) \cong (j^2/\sigma)$. Adding this to the right-hand side we obtain the equation of energy conservation in magnetohydrodynamics:

$$\rho T \frac{ds}{dt} = \sigma_{ik} \frac{\partial v_i}{\partial x_k} + \nabla \cdot (\kappa \nabla T) + \frac{j^2}{\sigma} . \tag{3.249}$$

(d) The equation for magnetic field shows that the field $\mathbf{B}$ at any location changes due to two different processes. The term $\nabla \times (\mathbf{v} \times \mathbf{B})$, which is of the order of (vB/L), makes the magnetic field flow along with the fluid (a process usually called 'advection'). The second term $(c^2/4\pi\sigma) \nabla^2 B$ has the order of magnitude $(c^2/4\pi\sigma) (B/L^2)$ and makes the magnetic field diffuse through the fluid. The ratio between these two terms defines the magnetic Reynolds number

$$R_{\mathrm{M}} = \frac{4\pi\sigma vL}{c^2} . \tag{3.250}$$

Clearly, when $R_{\mathrm{M}} \gg 1$, the advection term dominates and when $R_{\mathrm{M}} \ll 1$ the diffusion term dominates. In the first case we can ignore the diffusion term and obtain the equation

$$\frac{\partial \mathbf{B}}{\partial t} - \nabla \times (\mathbf{v} \times \mathbf{B}) = 0 . \tag{3.251}$$

Based on the results of problem 3.4, we can interpret this situation as the magnetic field being frozen in the field and carried away with it.

On the other hand, when $R_{\mathrm{M}} \ll 1$, we ignore the advection term and obtain the diffusion equation

$$\frac{\partial \mathbf{B}}{\partial t} = \frac{c^2}{4\pi\sigma} \nabla^2 \mathbf{B} \cong \frac{c^2}{4\pi\sigma L^2} \mathbf{B} . \tag{3.252}$$

This shows that the diffusion time-scale is $t_{\mathrm{diff}} \cong 4\pi\sigma L^2/c^2$.

(i) For IGM, $\sigma \cong 3 \times 10^{14} \mathrm{s}^{-1}$, $v \cong 200 \, \mathrm{km \, s}^{-1}$, and $L \cong 10$ kpc; this gives $R_{\mathrm{M}} \cong 3 \times 10^{25} \gg 1$ and $t_{\mathrm{diff}} \cong 4 \times 10^{39}$ s.

(ii) For a plasma in the laboratory, $\sigma \cong 10^{16}\,\mathrm{s}^{-1}$, $v \cong 10\,\mathrm{m\,s}^{-1}$ and $L \cong 3$ cm. Then $R_M \cong 4$ and $t_{\mathrm{diff}} \cong 10^{-3}$ s.

This shows that the field is completely frozen to the fluid in the astrophysical context, while diffusion could be important for laboratory systems.

3.16 Landau damping

The z-component of the equation of motion for a charged particle is

$$m\left(\frac{dv_z}{dt}\right) = qE_0 \cos(kx - \omega t) = qE_0 \cos[(kv_x - \omega)t] \,, \qquad (3.253)$$

where we have used the fact that $x = v_x t$. Integrating this equation we find that

$$v_z = \frac{qE_0}{m(kv_x - \omega)} \sin[(kv_x - \omega)t]. \qquad (3.254)$$

Hence, the contribution from the z-component of the velocity to the kinetic energy, at any time t, is

$$\frac{1}{2}mv_z^2 = \frac{q^2 E_0^2}{2m(kv_x - \omega)^2} \sin^2[(kv_x - \omega)t]. \qquad (3.255)$$

Consider now a bunch of particles, described by a distribution function f, such that the number of particles with velocities between v_x and $(v_x + dv_x)$ is $f(v_x)\,dv_x$. The total kinetic energy imparted to these particles will be

$$K = \int_{-\infty}^{+\infty} dv_x f(v_x)\,(1/2)\,mv_z^2. \qquad (3.256)$$

From (3.255), we see that most of the contribution to this integral arises from particles with velocities near $v_x = (\omega/k)$. Assuming that f is a slowly varying function of v_x in this region we can approximate this expression by taking $f(\omega/k) \equiv f(v_x = \omega/k)$ outside the integral:

$$
\begin{aligned}
K &\cong \frac{q^2 E_0^2 f(\omega/k)}{2m} \int_{-\infty}^{+\infty} \frac{\sin^2[(kv_x - \omega)t]}{(kv_x - \omega)^2}\,dv_x \\
&\cong \frac{q^2 E_0^2 f(\omega/k)\,t}{2mk} \int_{-\infty}^{\infty} \frac{\sin^2 u}{u^2}\,du \cong \frac{\pi q^2 E_0^2 f(\omega/k)\,t}{2mk}.
\end{aligned} \qquad (3.257)
$$

This corresponds to a transfer of energy at the rate

$$P = \frac{dK}{dt} \cong \frac{\pi q^2 E_0^2 f(\omega/k)}{2mk}. \qquad (3.258)$$

4

Radiation processes

4.1 Feynman formula for classical radiation

To solve an equation of the type

$$\Box Q = P(x),\qquad(4.1)$$

one can use the method of the Green function. We define a retarded Green function D_{ret} to be the solution to the equation $\Box D_{\text{ret}} = \delta(x)$. The subscript ret implies that we choose the boundary conditions so as to ensure $D_{\text{ret}}(t, \mathbf{x}) = 0$ for $t < 0$. Given $D_{\text{ret}}(x)$, we can relate Q to P by

$$Q(x) = \int d^4 y \, D_{\text{ret}}(x - y) P(y).\qquad(4.2)$$

Thus we only need to find the retarded Green function $D_{\text{ret}}(x)$.

The conventional way of solving the equation $\Box D_{\text{ret}} = \delta(x)$ is by Fourier transforming both sides. There is, however, a simpler way of obtaining the result. Assume for a moment that we are working in four-dimensional *Euclidean* space (rather than *Minkowskian* space), so that the distance from the origin to x^i is $s^2 = (\tau^2 + |\mathbf{x}|^2)$ rather than $(-t^2 + \mathbf{x}^2)$. In this problem, we will temporarily use the signature $(-, +, +, +)$ and set $\tau = it$. In such a Euclidean space, it is trivial to verify that the spherically symmetric solution to $\Box D_{\text{ret}} = 0$ is proportional to s^{-2} except at the origin. Consider now the volume integral of $\Box D_{\text{ret}}$ over a region bounded by a sphere of radius R. We have

$$\int d^4 x \, \Box D_{\text{ret}} = \int d^3 x \, \hat{\mathbf{n}} \cdot \nabla D_{\text{ret}} = \left(2\pi^2 R^3\right)\left(-\frac{2}{R^3}\right) = -4\pi^2.\qquad(4.3)$$

In arriving at the last result, we have used the fact that the 'surface' area of a three-sphere of radius R is $2\pi^2 R^3$ and $\nabla D_{\text{ret}} = \left(-2/R^3\right)\hat{\mathbf{n}}$. It follows that

$$\Box\left(\frac{-1}{4\pi^2 s^2}\right) = \delta(x),\qquad(4.4)$$

giving $D_{\text{ret}} = (-4\pi^2 s^2)^{-1}$. Consider now the solution to the equation of the form $\Box Q = P$. Using (4.2), we can write

$$Q(x_i) = -\int \frac{d^4 y}{4\pi^2} \frac{P(y)}{(x-y)^2}. \tag{4.5}$$

This is in exact analogy with the solution to the Poisson equation in three dimensions and should be intuitively obvious. If we now continue analytically from the Euclidean to the Minkowskian space using $(d^4 y)_E = i(dt\, d^3 y)_M$ we obtain

$$Q(t,\mathbf{x}) = -i\int \frac{dt'\, d^3 \mathbf{y}}{4\pi^2} \frac{P(t',\mathbf{y})}{(t'-t-R)(t'-t+R)}, \qquad R = |\mathbf{x}-\mathbf{y}|. \tag{4.6}$$

This expression will provide the final answer to our problem, except for the fact that the integral has poles along the real axis. To evaluate the integral we need to specify the contour of integration which is equivalent to a choice of boundary condition. Taking the contour shown in figure 4.1 we see that only the pole at $t' = t - R < t$ contributes. This gives the result

$$Q(t,\mathbf{x}) = -\frac{1}{2\pi} \int d^3 \mathbf{y} \frac{P(t-R,\mathbf{y})}{(-2R)} = \frac{1}{4\pi} \int d^3 \mathbf{y} \frac{P(t-R,\mathbf{y})}{R}. \tag{4.7}$$

Using this result in the context of electromagnetic potentials we obtain

$$\phi(\mathbf{x},t) = \int d^3 \mathbf{y}\, dt' \frac{q}{R} \delta(\mathbf{y} - \mathbf{z}(t'))\, \delta\left(t' - t + \frac{R}{c}\right), \tag{4.8}$$

$$\mathbf{A}(\mathbf{x},t) = \frac{1}{c} \int d^3 \mathbf{y}\, dt' \frac{q\mathbf{v}(t')}{R} \delta(\mathbf{y} - \mathbf{z}(t'))\, \delta\left(t' - t + \frac{R}{c}\right), \tag{4.9}$$

where $R = |\mathbf{x} - \mathbf{y}|$.

(b) It is convenient for our purpose to find $\partial \mathbf{A}/\partial t$ and $\nabla \phi$, using ϕ and $\mathbf{A}$ in (4.8) and (4.9). We first evaluate $\nabla \phi$. Using the basic result,

$$\int f(\mathbf{x})\delta(\mathbf{x} - \mathbf{y})\, d^3 \mathbf{x} = f(\mathbf{y}), \tag{4.10}$$

we can carry out the spatial integrations in (4.8) and obtain

$$\phi(\mathbf{x},t) = \int dt' \frac{q}{R} \delta\left(t' - t + \frac{R}{c}\right), \tag{4.11}$$

where, now, $R(t') = |\mathbf{x} - \mathbf{z}(t')|$ and, hence, $\mathbf{v}(t') = -d\mathbf{R}/dt'$. So,

$$\nabla \phi \equiv \frac{\partial \phi}{\partial \mathbf{x}} = q \int dt' \left[-\frac{\mathbf{R}}{R^3} \delta\left(t' - t + \frac{R}{c}\right) + \frac{q\mathbf{R}}{cR^2} \frac{d\delta}{df} \right], \tag{4.12}$$

where $f(t') = t' - t + R/c$. The first term in the above integral can be simplified by using the formula

$$\delta[f(t')] = \sum_i \frac{\delta(t' - t_i)}{|df/dt'|_{t'=t_i}}, \tag{4.13}$$

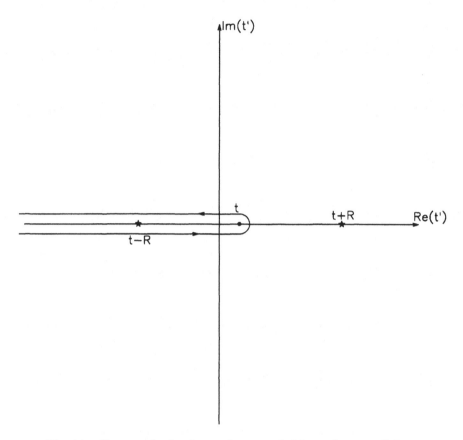

Fig. 4.1 Contour for implementing retarded boundary conditions.

where the t_i are the zeros of $f(t')$. To integrate the second term, we note that

$$\frac{d\delta}{df} = \left(\frac{df}{dt'}\right)^{-1} \frac{d\delta}{dt'} \tag{4.14}$$

and that

$$\frac{df}{dt'} = 1 - \frac{\mathbf{v} \cdot \mathbf{R}}{Rc} \equiv 1 - v_R. \tag{4.15}$$

Making the above substitutions, and carrying out a partial integration on the second term, finally gives

$$\nabla\phi = -\frac{q}{(1 - v_R)} \frac{\mathbf{R}}{R^3} - \frac{q}{(1 - v_R)} \frac{d}{c\, dt'} \left(\frac{\mathbf{R}}{R^2(1 - v_R)}\right), \tag{4.16}$$

where, now, t' is the retarded time and is determined by the equation

$$R(t') = c(t - t'). \tag{4.17}$$

Similarly, it can be shown that

$$\frac{\partial \mathbf{A}}{\partial t} = \frac{1}{1 - v_R} \frac{d}{c\, dt'} \left(\frac{q\mathbf{v}}{R(1 - v_R)} \right). \tag{4.18}$$

Thus we obtain

$$\mathbf{E}(\mathbf{x}, t) = \frac{\mathbf{n}}{R^2} \frac{q}{(1 - v_R)} + \frac{q}{(1 - v_R)} \frac{d}{c\, dt'} \left(\frac{\mathbf{R} - (\mathbf{v}/c)R}{R^2 (1 - v_R)} \right). \tag{4.19}$$

Equation (4.19) is our main result. From this point, one can follow two different routes. If we carry out the differentiations in (4.19) we will arrive at the standard results quoted in textbooks. To arrive at Feynman's formula, we should proceed differently. To begin with, we note from (4.17) that

$$\frac{dt'}{dt} = 1 - \frac{dR}{c\, dt} = (1 - v_R)^{-1} \equiv 1 - \frac{\dot{R}}{c}. \tag{4.20}$$

With this substitution, (4.19) becomes

$$\begin{aligned}
\mathbf{E} &= \frac{q\mathbf{n}}{R^2} \left(1 - \frac{\dot{R}}{c} \right) + q \frac{d}{c\, dt} \left[\frac{(1 - \dot{R}/c)}{R} \left(\mathbf{n} - \frac{\mathbf{v}}{c} \right) \right] \\
&= \frac{q\mathbf{n}}{R^2} - \frac{q\mathbf{n}}{R^2} \frac{\dot{R}}{c} + \frac{q}{c} \frac{d}{dt} \frac{\mathbf{n}}{R} - \frac{q}{c} \frac{d}{dt} \left(\frac{\mathbf{n}\dot{R}}{cR} \right) + \frac{q}{c^2} \frac{d}{dt} \left(\frac{1}{R} \frac{d}{dt} (R\mathbf{n}) \right) \\
&= q \frac{\mathbf{n}}{R^2} + q \frac{R}{c} \frac{d}{dt} \left(\frac{\mathbf{n}}{R^2} \right) + \frac{q}{c^2} \frac{d^2\mathbf{n}}{dt^2}.
\end{aligned} \tag{4.21}$$

We thus have the final result:

$$\mathbf{E} = -\nabla\phi - \dot{\mathbf{A}} = \left[\frac{q\mathbf{n}}{R^2} + q \frac{R}{c} \frac{d}{dt} \left(\frac{\mathbf{n}}{R^2} \right) + \frac{q}{c^2} \frac{d^2\mathbf{n}}{dt^2} \right]_{\text{ret}}. \tag{4.22}$$

This formula for the electric field has a curious interpretation. The first term represents the Coulomb field of the particle evaluated at the retarded time. The second term can be thought of as a 'first-order correction' to the retarded Coulomb field. This term is obtained by multiplying the time taken for the information to travel the distance R and the time rate of change of the Coulomb term $(\mathbf{n}/R^2)$. These two terms fall as R^{-2}. The third term depends on the acceleration of the charge and represents the radiation field of the charge.

For the sake of completeness, we also mention the derivation of the usual expression for the electromagnetic field due to a charged particle, starting from (4.19). All that one has to do is to carry out the time differentiation in (4.19), noting that

$$\frac{d}{dt'} v_R = \frac{1}{c} \frac{d}{dt'} (\mathbf{v} \cdot \mathbf{n}). \tag{4.23}$$

The final result can be written in the form

$$\mathbf{E} = \frac{q(1 - v^2/c^2)}{R^2 (1 - v_R)^3} \left(\mathbf{n} - \frac{\mathbf{v}}{c} \right) + \frac{q}{R^2 (1 - v_R)^3} \mathbf{R} \times \left[\left(\mathbf{n} - \frac{\mathbf{v}}{c} \right) \times \frac{d\mathbf{v}}{c^2\, dt'} \right], \tag{4.24}$$

where all quantities on the right-hand side refer to the retarded time t'. Since we have derived both (4.21) and (4.24) from (4.19), the equivalence of these formulas is apparent.

4.2 Radiation field in simple cases

(a) The field at any event $(t, \mathbf{r})$ depends on the behaviour of the charge at a time t', where $t' = t - |\mathbf{r} - \mathbf{x}(t')|$ and $\mathbf{x}(t)$ is the trajectory of the charge. At large distances from the system of charges, one can write

$$R \equiv |\mathbf{r} - \mathbf{x}(t')| \cong r - \mathbf{x} \cdot \mathbf{n}, \qquad (4.25)$$

where $\mathbf{n}$ is the unit vector $(\mathbf{r}/r)$. In calculating the field at large distances, we can replace the R^{-1} factor by r^{-1} and obtain, for example,

$$\mathbf{A}(t, \mathbf{r}) = \frac{1}{r} \int d^3x \quad \mathbf{J}(t - r + \mathbf{x} \cdot \mathbf{n}). \qquad (4.26)$$

Inside the integral we can ignore $\mathbf{x} \cdot \mathbf{n}$ compared to r if the charge distribution does not change too fast. If T is the characteristic time in which the charge distribution changes, then the typical wavelength of radiation emitted will be $\lambda \cong (cT)$. The quantity $(\mathbf{x} \cdot \mathbf{n}/c)$ is of the order of (a/c), where a is the size of the system. This term can be ignored if $(a/c) \ll T$ or when $a \ll \lambda$. This condition is clearly satisfied if $(a/T) \ll c$, that is, if $v \ll c$, where v is the velocity of the charges. Hence, we can use this approximation when the motion is non-relativistic. In that case, we obtain the simple formula

$$\mathbf{A}(t, \mathbf{r}) = \frac{1}{r} \int d^3x \, \mathbf{J}(t - r) = \frac{1}{r} \sum_i q_i \mathbf{v}_i(t - r) = \frac{\dot{\mathbf{d}}(t - r)}{r}, \qquad (4.27)$$

where $\mathbf{d} = \sum q_i \mathbf{x}_i$ is the dipole moment of the system and the sum is over all the charges in the system.

At large distances from the system of charges, the electromagnetic wave may be treated as a plane wave to the same degree of accuracy. Then $\mathbf{B} = \nabla \times \mathbf{A} \cong \dot{\mathbf{A}} \times \mathbf{n}$ and $\mathbf{E} = \mathbf{B} \times \mathbf{n}$ giving the energy flux $\mathbf{S} = (\mathbf{E} \times \mathbf{B})/4\pi = (B^2/4\pi)\mathbf{n}$. The amount of radiation propagating into a solid angle $d\Omega$ in unit time is $|\mathbf{S}|r^2 d\Omega$. So,

$$\frac{dE}{dt\, d\Omega} = |\mathbf{S}|r^2 = \frac{B^2 r^2}{4\pi}. \qquad (4.28)$$

In our case,

$$\mathbf{B} = \frac{1}{r}(\ddot{\mathbf{d}} \times \mathbf{n}), \qquad \mathbf{E} = \frac{1}{r}(\ddot{\mathbf{d}} \times \mathbf{n}) \times \mathbf{n}, \qquad (4.29)$$

leading to

$$\frac{dE}{dt\, d\Omega} = \frac{1}{4\pi}(\ddot{\mathbf{d}} \times \mathbf{n})^2 = \frac{|\ddot{\mathbf{d}}|^2}{4\pi c^3} \sin^2\theta. \qquad (4.30)$$

(In the last equation, we have reintroduced the c factor.) In equations (4.29) and (4.30) the right-hand sides should be evaluated at the retarded time.

The total energy radiated can be found by integrating over the solid angle. We then obtain

$$\frac{dE}{dt} = \frac{|\ddot{\mathbf{d}}|^2}{4\pi c^3} \int\limits_0^\pi 2\pi \left(\sin^2\theta\right)\left(\sin\theta \, d\theta\right) = \frac{2}{3c^3}|\ddot{\mathbf{d}}|^2. \qquad (4.31)$$

For a single charge, $\mathbf{d} = q\mathbf{x}$, and we obtain

$$\frac{dE}{dt} = \frac{2q^2}{3c^3}|\mathbf{a}|^2, \qquad (4.32)$$

where $\mathbf{a}$ is the acceleration. This is called the 'Larmor formula'.

(b) The Fourier transform of $\mathbf{B}(t)$ in (4.29) is given by $\mathbf{B}(\omega) = r^{-1}\left(\ddot{\mathbf{d}}(\omega) \times \mathbf{n}\right)$, where $\ddot{\mathbf{d}}(\omega) = -\omega^2 \mathbf{d}(\omega)$. Using the relation

$$\int\limits_{-\infty}^{+\infty} B^2(t)\,dt = \int\limits_{-\infty}^{+\infty} |\mathbf{B}(\omega)|^2 \frac{d\omega}{2\pi} = 2\int\limits_0^\infty |\mathbf{B}(\omega)|^2 \frac{d\omega}{2\pi}, \qquad (4.33)$$

we can write (4.28) as

$$\frac{dE}{d\Omega} = \frac{r^2}{4\pi} \int\limits_{-\infty}^{+\infty} B^2 dt = \frac{r^2}{2\pi} \int\limits_0^\infty |\mathbf{B}(\omega)|^2 \frac{d\omega}{2\pi}, \qquad (4.34)$$

giving

$$\frac{dE}{d\omega\,d\Omega} = \frac{r^2|\mathbf{B}(\omega)|^2}{4\pi^2} = \frac{\omega^4|\mathbf{d}(\omega)|^2}{4\pi^2 c^3} \cdot \sin^2\theta. \qquad (4.35)$$

Integrating over $d\Omega$ using the angular average $\langle \sin^2\theta \rangle = (2/3)$ we find that

$$\frac{dE}{d\omega} = \frac{\omega^4|\mathbf{d}(\omega)|^2}{4\pi^2 c^3} \int d\Omega\, \sin^2\theta = \frac{\omega^4|\mathbf{d}(\omega)|^2}{4\pi^2 c^3} 4\pi\langle\sin^2\theta\rangle = \frac{2}{3\pi}\frac{\omega^4}{c^3}|\mathbf{d}(\omega)|^2. \qquad (4.36)$$

(c) The Fourier transform of the magnetic field can be approximated as

$$\mathbf{B}(\omega) \equiv \int\limits_{-\infty}^{+\infty} \mathbf{B}(t)\,e^{-i\omega t}\,dt \cong \int\limits_{-\infty}^{+\infty} \mathbf{B}(t)\,dt \qquad (4.37)$$

if $\omega \ll \tau^{-1}$, where τ is the characteristic time-scale of variation of $\mathbf{B}(t)$. Using $\mathbf{B} = \dot{\mathbf{A}} \times \mathbf{n}$ we obtain

$$\mathbf{B}(\omega) = -\mathbf{n} \times \int\limits_{-\infty}^{+\infty} \dot{\mathbf{A}}\,dt = -\mathbf{n} \times [\mathbf{A}_2 - \mathbf{A}_1], \qquad (4.38)$$

where $\mathbf{A}_1$ and $\mathbf{A}_2$ are the initial and final values of the vector potential. The radiated energy is

$$\frac{dE}{d\omega\,d\Omega} = \frac{r^2}{4\pi^2}\left[(\mathbf{A}_2 - \mathbf{A}_1) \times \mathbf{n}\right]^2. \qquad (4.39)$$

The vector potentials can be related to the velocities v_1, v_2 of the charge by the Lienard–Wiechert formula:

$$\mathbf{A} = \frac{q\mathbf{v}}{r(1 - \mathbf{v} \cdot \mathbf{n})} \cong \frac{q\mathbf{v}}{r}, \tag{4.40}$$

where the second relation is valid for non-relativistic motion. Substituting for $\mathbf{A}$, we find that

$$\frac{dE}{d\omega \, d\Omega} = \frac{q^2}{4\pi^2} \left[\frac{\mathbf{v}_2 \times \mathbf{n}}{1 - \mathbf{n} \cdot \mathbf{v}_2} - \frac{\mathbf{v}_1 \times \mathbf{n}}{1 - \mathbf{n} \cdot \mathbf{v}_1} \right]^2 \qquad \text{(exact)}$$

$$\cong \frac{q^2}{4\pi^2} [(\mathbf{v}_2 - \mathbf{v}_1) \times \mathbf{n}]^2 = \frac{q^2}{4\pi^2} (\Delta v)^2 \sin^2 \theta \quad \text{(non-relativisitic)}. \tag{4.41}$$

(Note that $c = 1$ here.) The total energy emitted in all directions can be found by integrating over $d\Omega$, which is equivalent to replacing $\sin^2 \theta$ by $(4\pi)(2/3) = (8\pi/3)$; so

$$\frac{dE}{d\omega} \cong \frac{2}{3\pi} \frac{q^2}{c^3} (\Delta v)^2. \tag{4.42}$$

Notice that the energy emitted per unit frequency interval is independent of ω for $\omega \ll \tau^{-1}$. For $\omega \gtrsim \tau^{-1}$, there is very little radiation.

(d) Consider two Lorentz frames S and S', with S' moving along the z-axis with velocity v. Any given direction is represented by angles (θ, ϕ) and (θ', ϕ') in the two frames, with θ representing the direction with respect to the z-axis. Quite trivially, we have $\phi' = \phi$. The relation between θ and θ' was found in problem 1.2 (we take $c = 1$):

$$\cos \theta' = \frac{\cos \theta - v}{1 - v \cos \theta}. \tag{4.43}$$

To relate the intensities $(dE/dt \, d\Omega)$ in the two frames we use the transformations

$$d\Omega' = d(\cos \theta') \, d\phi' = \frac{1 - v^2}{(1 - v \cos \theta)^2} \, d(\cos \theta) \, d\phi = \frac{1}{\gamma^2} \frac{d\Omega}{(1 - v \cos \theta)^2},$$

$$dE' = \gamma (dE - \mathbf{v} \cdot d\mathbf{P}) = \gamma \, dE (1 - v \cos \theta), \tag{4.44}$$

$$dt' = dt\sqrt{1 - v^2} = \gamma^{-1} dt.$$

We find

$$\frac{dE'}{dt' \, d\Omega'} = \frac{\gamma (1 - v \cos \theta) \, dE}{(\gamma^{-1} dt) \, \gamma^{-2} (1 - v \cos \theta)^{-2} \, d\Omega} = \gamma^4 (1 - v \cos \theta)^3 \frac{dE}{dt \, d\Omega} \tag{4.45}$$

or

$$\left(\frac{dE}{dt \, d\Omega} \right)_L = \frac{(1 - v^2)^2}{(1 - v \cos \theta)^3} \left(\frac{dE'}{dt' \, d\Omega'} \right)_R. \tag{4.46}$$

As a check, consider the case in which $(dE'/dt' d\Omega')$ is independent of (θ', ϕ'), that is, the emission is isotropic in the rest frame. Then

$$\left(\frac{dE}{dt}\right)_L = \int d\Omega \left(\frac{dE}{dt\, d\Omega}\right)_L = \int d\Omega \frac{(1-v^2)^2}{(1-v\cos\theta)^3} \left(\frac{dE'}{dt'\, d\Omega'}\right)_R$$

$$= \left(\frac{dE'}{dt'\, d\Omega'}\right)_R \int_0^\infty 2\pi\, d\mu \frac{(1-v^2)^2}{(1-v\mu)^3}. \tag{4.47}$$

The integral is elementary and gives (4π). Hence,

$$\left(\frac{dE}{dt}\right)_L = 4\pi \left(\frac{dE'}{dt'\, d\Omega'}\right)_R = \left(\frac{dE'}{dt'}\right)_R, \tag{4.48}$$

as is to be expected.

4.3 Radiation reaction

(a) We can write the result obtained in (4.32) as

$$d\mathscr{E} = \frac{2}{3} \frac{q^2}{c^3} a^2(t')dt, \tag{4.49}$$

where $t' = t-(r/c)$ is the retarded time. Let us choose an instantaneous rest frame for the charge in which this non-relativistic formula is valid at $t = t'$. Because of symmetry, the net momentum radiated, $d\mathbf{P}$, will vanish in this instantaneous rest frame. Clearly, this result should be valid even for relativistic motion, if we can rewrite it in an invariant manner. If a^i is the four-acceleration, then $-a^2/c^4 = a^i a_i$ in the instantaneous rest frame of the charge. So we can express (4.49), as well as the condition $d\mathbf{P} = 0$, in the form

$$dP^k = -\frac{2}{3}\frac{q^2}{c}\left(a^i a_i\right) dx^k = -\frac{2}{3}\frac{q^2}{c}\left(a^i a_i\right) u^k\, ds, \tag{4.50}$$

where dP^k is the four-momentum radiated by the particle during the propertime interval ds. Being relativistically invariant this result is true for arbitrary velocities.

(b) Consider a frame S in which the particle has a velocity $\mathbf{v}$ and acceleration $\mathbf{a}$. We now make a Lorentz transformation to a frame S' at which the charge is instantaneously at rest. In this frame $\mathbf{E}'_\parallel = \mathbf{E}_\parallel$, $\mathbf{E}'_\perp = \gamma\left(\mathbf{E}_\perp + \mathbf{v}\times\mathbf{B}\right)$ and the acceleration is $\mathbf{a}' = (q/m)\, \mathbf{E}'$. (We have set $c = 1$ to simplify the expressions.) Hence, the instantaneous power radiated is

$$\frac{2}{3}q^2 a^2 = \frac{2}{3}\frac{q^4}{m^2}\left[\mathbf{E}_\parallel^2 + \gamma^2\left(\mathbf{E}_\perp + \mathbf{v}\times\mathbf{B}\right)^2\right] = \frac{2}{3}\frac{q^4}{m^2}\left[\mathbf{E}_\parallel^2 + \gamma^2\left(\mathbf{E} + \mathbf{v}\times\mathbf{B} - \mathbf{E}_\parallel\right)^2\right]$$

$$= \frac{2}{3}\frac{q^4}{m^2}\left[\gamma^2\left(\mathbf{E} + \mathbf{v}\times\mathbf{B}\right)^2 - \gamma^2 \mathbf{E}_\parallel^2 v^2\right]. \tag{4.51}$$

In arriving at the last equation we have used the relations $\mathbf{E}\cdot\mathbf{E}_\parallel = E_\parallel^2$ and $\mathbf{E}_\parallel\cdot(\mathbf{v}\times\mathbf{B}) = 0$. Writing $E_\parallel^2 v^2 = (\mathbf{E}\cdot\mathbf{v})^2$, we obtain

$$\Delta\mathscr{E} = \frac{2}{3}\left(\frac{q^4}{m^2}\right)\gamma^2\left[(\mathbf{E} + \mathbf{v}\times\mathbf{B})^2 - (\mathbf{E}\cdot\mathbf{v})^2\right]\Delta t. \tag{4.52}$$

The relativistic form of this result can be obtained by using $a^i = (q/m)\, F^i{}_k u^k$ in (4.50).

(c) Suppose the damping force is $\mathbf{f}$; then we expect the mean power radiated to be equal to the work done by the damping force, that is,

$$\left\langle \left(\frac{\Delta \mathscr{E}}{\Delta t} \right) \right\rangle = -\left\langle \left(\frac{2}{3} \right) q^2 a^2 \right\rangle = \langle \mathbf{f} \cdot \mathbf{v} \rangle, \tag{4.53}$$

when arranged over a period of time. Averaging a^2 over a time interval T, we obtain

$$\langle a^2 \rangle = \frac{1}{T} \int_0^T dt\, a^2 = \frac{1}{T} \int_0^T dt\, (\dot{\mathbf{v}} \cdot \dot{\mathbf{v}})$$

$$= \frac{1}{T} \int_0^T dt \left[\frac{d}{dt} (\mathbf{v} \cdot \dot{\mathbf{v}}) - \mathbf{v} \cdot \ddot{\mathbf{v}} \right] = \frac{1}{T} [\mathbf{v} \cdot \dot{\mathbf{v}}]_0^T - \langle \mathbf{v} \cdot \ddot{\mathbf{v}} \rangle. \tag{4.54}$$

The first term vanishes as $T \to \infty$ for any bounded motion, giving $\langle a^2 \rangle = -\langle \mathbf{v} \cdot \ddot{\mathbf{v}} \rangle$. Using this we see that $\mathbf{f}_{\text{damp}} = (2/3)\, q^2 \ddot{\mathbf{v}}$.

For non-relativistic motion, we have

$$\dot{\mathbf{v}} = (q/m)\, \mathbf{E} + (q/m)\, (\mathbf{v} \times \mathbf{B}), \qquad \ddot{\mathbf{v}} = (q/m) \left[\dot{\mathbf{E}} + \dot{\mathbf{v}} \times \mathbf{B} + \mathbf{v} \times \dot{\mathbf{B}} \right]. \tag{4.55}$$

We now transform to a frame at which the charge is instantaneously at rest so that $\dot{\mathbf{v}} = (q/m)\, \mathbf{E}$ and $\ddot{\mathbf{v}} = (q/m) \left[\dot{\mathbf{E}} + (q/m)\, (\mathbf{E} \times \mathbf{B}) \right]$ to the lowest order in (v/c). Then the damping force is

$$\mathbf{f}_{\text{damp}} = \frac{2}{3} \frac{q^3}{m} \dot{\mathbf{E}} + \frac{2}{3} \frac{q^4}{m^2} (\mathbf{E} \times \mathbf{B}). \tag{4.56}$$

(d) We have to find a four-vector g^i which reduces to $\left(0, (2/3)\, q^2 \ddot{\mathbf{v}} \right)$ in the rest frame of the charge. This condition is satisfied by any vector of the form $g^i = (2q^2/3) \left[(d^2 u^i / ds^2) - A u^i \right]$, where A is to be determined. To find A, we use the condition that $g^i u_i = 0$. This gives $A = u^k \left(d^2 u_k / ds^2 \right)$. Hence,

$$g^i = \left(\frac{2q^2}{3} \right) \left[\frac{d^2 u^i}{ds^2} - u^i u^k \frac{d^2 u_k}{ds^2} \right]. \tag{4.57}$$

The second term can be rewritten using

$$u^k \frac{da_k}{ds} = \frac{d}{ds} (u^k a_k) - a^k a_k = -a^k a_k, \tag{4.58}$$

since $u^k a_k = 0$. This gives another form for g^i:

$$g^i = \frac{2}{3} q^2 \left[\frac{d^2 u^i}{ds^2} + u^i \left(a^k a_k \right) \right]. \tag{4.59}$$

(e) When F^{ik} is a constant we have

$$a^i = \left(\frac{q}{m} \right) F^i{}_k u^k, \qquad \frac{da^i}{ds} = \left(\frac{q}{m} \right)^2 F^i{}_k F^k{}_j u^j. \tag{4.60}$$

Substituting these expressions in (4.57) and rearranging the terms we obtain

$$g^i = \frac{2}{3}\left(\frac{q^2}{m}\right)^2 \left[(F^{ka}F_{kj})\,u_a u^j u^i - F^{ki}F_{kj}u^j\right]. \tag{4.61}$$

Using the definition of T^{ab} we can write

$$F^{il}F_{kl} = F^{li}F_{lk} = -(4\pi)\,T_k^i + \frac{1}{4}\delta_k^i\left(F_{ab}F^{ab}\right). \tag{4.62}$$

Now we can express g^i in terms of T^{ab} alone. Note that, in the combination

$$\left(F^{ka}F_{kj}\right)u_a u^j u^i - F^{ki}F_{kj}u^j = u_a u^j u^i\left[-4\pi T_j^a + \frac{1}{4}\delta_j^a F^2\right] - u^j\left[-4\pi T_j^i + \frac{1}{4}\delta_j^i F^2\right]$$
$$= -4\pi\left(T^{aj}u_a u_j\right)u^i + 4\pi T^{ij}u_j, \tag{4.63}$$

the term involving $F^2 = F_{ab}F^{ab}$ cancels out. Therefore,

$$g^i = \frac{8\pi}{3}\left(\frac{q^2}{m}\right)^2\left[T^{ij}u_j - \left(T^{ab}u_a u_b\right)u^i\right]$$
$$= \left(\frac{\sigma_T}{c}\right)\left[T^{ij}u_j - \left(T^{ab}u_a u_b\right)u^i\right], \tag{4.64}$$

with $\sigma_T = (8\pi/3)\left(q^2/mc^2\right)^2$. This relation expresses the radiation reaction in terms of the energy density of the electromagnetic field.

4.4 Radiation in the ultrarelativistic case

The electromagnetic fields in this case are given by

$$\mathbf{E} = \frac{q}{r}\mu^3\left[\mathbf{n}\times(\mathbf{n}-\mathbf{v})\times\mathbf{a}\right],\quad \mathbf{B}=\mathbf{n}\times\mathbf{E},\quad \mu=(1-\mathbf{v}\cdot\mathbf{n})^{-1}, \tag{4.65}$$

where we have set $c=1$. The power emitted into a solid angle $d\Omega$ is

$$\frac{dE}{dt\,d\Omega} = \frac{1}{4\pi}(\mathbf{E}\times\mathbf{B})\cdot\mathbf{n}r^2 = \frac{E^2 r^2}{4\pi}. \tag{4.66}$$

To evaluate E^2, note that

$$\mathbf{n}\times(\mathbf{n}\times\mathbf{a}) = \mathbf{n}(\mathbf{a}\cdot\mathbf{n})-\mathbf{a},\quad \mathbf{n}\times(\mathbf{v}\times\mathbf{a})=\mathbf{v}(\mathbf{a}\cdot\mathbf{n})-\mathbf{a}(\mathbf{v}\cdot\mathbf{n}). \tag{4.67}$$

Squaring $\mathbf{E}$ and rearranging terms we obtain

$$\mu^6\left[\mathbf{n}\times(\mathbf{n}-\mathbf{v})\times\mathbf{a}\right]^2 = 2(\mathbf{n}\cdot\mathbf{v})(\mathbf{v}\cdot\mathbf{a})\mu^5 + a^2\mu^4 - (1-v^2)(\mathbf{n}\cdot\mathbf{a})^2\mu^6, \tag{4.68}$$

leading to

$$\frac{dE}{dt\,d\Omega} = \frac{q^2}{4\pi c^3}\left[2\mu^5\frac{(\mathbf{n}\cdot\mathbf{a})(\mathbf{v}\cdot\mathbf{a})}{c} + \mu^4 a^2 - \mu^6\gamma^{-2}(\mathbf{n}\cdot\mathbf{a})^2\right], \tag{4.69}$$

with c reintroduced appropriately. The radiation intensity is largest along the directions for which $\mu\gg1$ or $(1-\mathbf{v}\cdot\mathbf{n}/c)\ll1$. If θ is the angle between $\mathbf{v}$ and

n, then using $\beta = \left(1 - \gamma^{-2}\right)^{1/2} \cong 1 - \left(1/2\gamma^2\right)$ we obtain

$$(1 - \beta \cos\theta)^{-1} \cong \left[1 - \left(1 - \frac{1}{2\gamma^2}\right)\left(1 - \frac{1}{2}\theta^2\right)\right]^{-1} = \frac{2\gamma^2}{1 + \gamma^2\theta^2} \tag{4.70}$$

for $\theta \ll 1, \beta \cong 1$. For $\gamma \gg 1$, this expression is sharply peaked around $\theta = 0$ and has a width $\Delta\theta \cong \gamma^{-1} = \left(1 - v^2/c^2\right)^{1/2}$. Also note that the θ-dependence is *only* through the combination $\gamma\theta$.

The above expressions simplify if **v** and **a** are parallel or perpendicular. When **v** is perpendicular to **a** we obtain, from (4.69),

$$\frac{dE}{dt\,d\Omega} = \frac{q^2 a^2}{4\pi c^3} \left[\mu^4 - \mu^6 \gamma^{-2} \sin^2\theta \cos^2\Phi\right], \tag{4.71}$$

where θ is the angle between **n** and **v** and Φ is the azimuthal angle of **n** relative to the plane containing **a** and **v**. Similarly, when **a** and **v** are parallel we obtain

$$\frac{dE}{dt\,d\Omega} = \frac{q^2 a^2 \mu^6}{4\pi c^3} \left[\frac{2}{\mu} v \cos^2\theta + \frac{1}{\mu^2} - \gamma^{-2} \cos^2\theta\right]. \tag{4.72}$$

The expression in square brackets is actually

$$2v\cos^2\theta\,(1 - v\cos\theta) + (1 - v\cos\theta)^2 - (1 - v^2)\cos^2\theta = \sin^2\theta, \tag{4.73}$$

giving

$$\frac{dE}{dt\,d\Omega} = \frac{q^2 a^2 \mu^6}{4\pi c^3} \sin^2\theta. \tag{4.74}$$

4.5 Bremsstrahlung

(a) Consider an electron of velocity v which is scattered by an ion in a collision with impact parameter b. The acceleration experienced by the electron is about $a \cong \left(Ze^2/mb^2\right)$ and lasts for a time $(2b/v)$. From (4.42), we find the total energy emitted to be

$$\frac{dE}{d\omega} = \frac{2}{3\pi} \frac{e^2}{c^3} \left(\frac{Ze^2}{mb^2} \cdot \frac{2b}{v}\right)^2 = \frac{8}{3\pi} \frac{Z^2 e^6}{m^2 c^3} \left(\frac{1}{vb}\right)^2. \tag{4.75}$$

Since the acceleration lasted for a time $(2b/v)$, we expect very little power at frequencies $\omega > \omega_{max} \cong (v/2b)$.

This is the radiation emitted in a single collision. If the number density of ions (n_i) and electrons (n_e) is the same and is given by n, then the total amount of energy emitted per unit volume per second due to all collisions with impact parameter in the range $(b, b + db)$ will be $\left[(n_i n_e)(2\pi b\,db)(dE/d\omega)\right] = \left[n^2 v(2\pi b\,db) \cdot (dE/d\omega)\right]$. So

$$\left(\frac{dE}{dV\,d\omega\,dt}\right)_{tot} = n^2 v\,db \left(\frac{dE}{d\omega}\right)(2\pi b) = \frac{16Z^2 e^6 n^2}{3m^2 c^3 v} \cdot \frac{1}{b}\,db. \tag{4.76}$$

Integrating over b with the limits b_1 and b_2, we obtain

$$\left(\frac{dE}{dV\,d\omega\,dt}\right) = \frac{16Z^2e^6n^2}{3m^2c^3v}\ln\left(\frac{b_2}{b_1}\right). \tag{4.77}$$

The limits of integration b_2 and b_1 need to be determined by physical considerations.

The value of b_2 can be fixed by noticing that most of the radiation is at frequencies $\omega < (v/2b)$. So we must have $b < (v/2\omega)$ and we can set $b_2 = (v/2\omega)$. The lower limit is somewhat more difficult to determine. Our analysis, based on weak scattering, will break down below an impact parameter $b_{1,C}$, where

$$\frac{Ze^2}{b_{1,C}} = \frac{1}{2}m_e v^2 \tag{4.78}$$

giving $b_{1,C} = (2Ze^2/m_e v^2)$. On the other hand, quantum effects are important if $m_e v b_{1,Q} \cong \hbar$ or when $b_{1,Q} \cong (\hbar/m_e v)$. Which of these two expressions should be used for b_1 depends on the velocity of the particle. Clearly,

$$\frac{b_{1,Q}}{b_{1,C}} \cong \frac{v\hbar}{Ze^2} = \left(\frac{1}{Z\alpha}\right)\left(\frac{v}{c}\right), \tag{4.79}$$

where $\alpha = (e^2/\hbar c) \cong 10^{-2}$. Thus, for particles with $(v/c) \gtrsim 10^{-2}Z$ one should use $b_{1,Q}$; for particles with $(v/c) \lesssim 10^{-2}Z$ one should use $b_{1,C}$. Hence,

$$\left(\frac{b_2}{b_1}\right) \cong \begin{cases} (m_e v^3/Ze^2\omega) & \text{(for low velocities)} \\ (m_e v^2/\hbar\omega) & \text{(for high velocities)}. \end{cases}$$

Since this ratio appears only inside the logarithm, its exact numerical value is not very important. We shall ignore the logarithmic factor in the rest of the problem.

(i) For a plasma in thermal equilibrium, electrons will have Maxwellian distribution of velocities and we should average (4.77) over the velocities. Since an electron needs to have a minimum energy $\frac{1}{2}mv_{\min}^2 \cong \hbar\omega$ to emit a photon of energy $\hbar\omega$, the averaging of (4.77) will lead to a factor

$$\left\langle\frac{1}{v}\right\rangle = \left(\frac{m}{2\pi kT}\right)^{3/2}\int_{v_{\min}}^{\infty}\left(\frac{1}{v}\right)4\pi v^2 dv\left[\exp-\frac{mv^2}{2kT}\right]$$

$$= \sqrt{\frac{2m}{\pi kT}}\exp\left(-\frac{\hbar\omega}{kT}\right). \tag{4.80}$$

Hence, the specific emissivity of the plasma is

$$j(\omega) = \left(\frac{dE}{dV\,dt\,d\omega}\right) \cong \frac{16Z^2e^6n^2}{3m^2c^3}\left(\frac{2m}{\pi kT}\right)^{1/2}\exp\left(-\frac{\hbar\omega}{kT}\right) \propto n^2 T^{-1/2} \tag{4.81}$$

if $\hbar\omega \ll kT$.

(ii) To obtain the volume emissivity we have to integrate $j(\omega)$ over all the frequencies. Then,

$$J = \left(\frac{dE}{dVdt}\right) = \int_0^\infty d\omega \, J(\omega) = \frac{16Z^2 e^6 n^2}{3m^2 c^2 \hbar}\left(\frac{2mkT}{\pi}\right)^{1/2} \propto n^2 T^{1/2}. \qquad (4.82)$$

Numerically,

$$4\pi j_v = 6.8 \times 10^{-38} n^2 T^{-1/2} e^{-h\nu/kT} \text{ erg cm}^{-3} \text{ s}^{-1} \text{ Hz}^{-1} \qquad (4.83)$$

$$4\pi J \cong 1.4 \times 10^{-27} n^2 T^{1/2} \text{ erg cm}^{-3} \text{ s}^{-1} \qquad (4.84)$$

if all variables are expressed in c.g.s. units.

(iii) The cooling time can be estimated to be $t_{\text{cool}} \cong (E_{\text{th}}/J)$, where $E_{\text{th}} \cong nkT$ is the thermal energy per unit volume. So $t_{\text{cool}} \propto \left(nkT/n^2 T^{1/2}\right) \propto T^{1/2} n^{-1}$.

(b) For a plasma in equilibrium at temperature T, the free–free absorption rate R should match the Bremsstrahlung emission rate $j(\omega)$. The absorption rate is given by

$$R = n\sigma_{\text{ffa}}(\omega)I(\omega), \qquad (4.85)$$

where $I(\omega) \propto \omega^3 \left(e^{\hbar\omega/kT} - 1\right)^{-1}$ is the equilibrium distribution of photons at temperature T and σ_{ffa} is the cross-section for free–free absorption. (This relation, in fact, defines $\sigma_{\text{ffa}}(\omega)$.) Equating (4.85) and (4.81) we find

$$\sigma_{\text{ffa}}(\omega) = \frac{j(\omega)}{nI(\omega)} \propto \frac{nT^{-1/2}}{\omega^3}\left(1 - e^{-\hbar\omega/kT}\right). \qquad (4.86)$$

For $\left(\hbar\omega/kT\right) \lesssim 1$, this gives $\sigma_{\text{ffa}} \propto \left(nT^{-1/2}/\omega^3\right)\left(\omega/T\right) \propto \left(n/\omega^2 T^{3/2}\right)$. The corresponding time-scale for free–free absorption is $t_{\text{ff}} = (n\sigma_{\text{ffa}}c)^{-1}$. This is given by

$$t_{\text{ff}} \cong \frac{3}{8}\left(\frac{\pi}{2}\right)^{1/2}\frac{(mkT)^{1/2}}{Z^2 e^6 n^2} \cdot \frac{\omega^3}{\hbar^2 c^3 (2\pi)^3}\left(1 - e^{-\hbar\omega/kT}\right)^{-1}$$

$$\propto T^{1/2} n^{-2}\left(1 - e^{-\hbar\omega/kT}\right)^{-1}. \qquad (4.87)$$

In stellar interiors, most of the opacity arises due to photons being absorbed by fully ionized hydrogen and helium. In this case $\hbar\omega \cong kT$ and the effective free–free absorption coefficient scales with temperature as $nT^{-7/2}$. Hence, the mean-free-path scales as $l \propto T^{7/2} n^{-2}$. This was the result needed in problem 1.17(c).

The above result, in the limit $\hbar\omega \ll kT$, can be understood in a more intuitive way. Photons with $\hbar\omega \ll kT$ can be thought of as a classical electromagnetic wave with some electric field $E = E_0 \cos \omega t$. An electron subjected to this field will acquire a velocity $v \cong (eE_0/m\omega) \sin \omega T$ over and above the thermal velocity. Hence, the mean energy of the electron in the radiation field will be $\langle\epsilon\rangle = (3/2) kT + \left(e^2 E_0^2/4m\omega^2\right)$. The second term is constantly thermalized through collisions. The rate of collisions is $R \cong nv\sigma$, where $\sigma \cong \left(e^2/kT\right)^2$ and so the energy dissipated per unit time per unit volume is

$$\frac{dE}{dt\,dV} \cong nR\left(\frac{e^2 E_0^2}{4m\omega^2}\right). \tag{4.88}$$

This should be equal to $c\sigma_{\text{ffa}}n\left(E_0^2/8\pi\right)$, by definition, giving

$$\sigma_{\text{ffa}} \cong \frac{2\pi R}{c}\left(\frac{e^2}{m\omega^2}\right) \cong \frac{2\pi}{c}n\left(\frac{kT}{m}\right)^{1/2}\left(\frac{e^2}{kT}\right)^2\left(\frac{e^2}{m\omega^2}\right) \cong \frac{e^6 n}{(mkT)^{3/2}}\left(\frac{1}{\omega^2}\right). \tag{4.89}$$

(c) For a system of charges with the same charge-to-mass ratio, q/m, the second derivative of the dipole moment,

$$\ddot{\mathbf{d}} = \sum q_i \ddot{\mathbf{x}}_i = \sum \left(\frac{q_i}{m_i}\right)m_i \ddot{\mathbf{x}}_i = \frac{q}{m}\sum \dot{\mathbf{p}}_i, \tag{4.90}$$

vanishes due to momentum conservation. Hence, the dipole radiation from electron–electron collisions and ion–ion collisions will be subdominant to the effects from electron–ion collisions.

4.6 Magnetic fields and synchrotron radiation

(a) The motion of a charged particle in a magnetic field is governed by the equation

$$\frac{d\mathbf{p}}{dt} = \left(\frac{q}{c}\right)(\mathbf{v}\times\mathbf{B}) = \frac{E}{c^2}\frac{d\mathbf{v}}{dt}. \tag{4.91}$$

The second equation follows from $\mathbf{p} = \left(E\mathbf{v}/c^2\right)$ and the fact that energy E of the particle is conserved during motion in a magnetic field. Writing this equation as

$$\frac{d\mathbf{v}}{dt} = -\omega\hat{\mathbf{n}}\times\mathbf{v}, \quad \omega\hat{\mathbf{n}} = \frac{cq}{E}\mathbf{B}, \tag{4.92}$$

it is clear that the particle has a circular trajectory in the plane perpendicular to $\mathbf{B}$ with angular velocity

$$\omega = \frac{cqB}{E} = \frac{qB}{mc}\sqrt{1-\frac{v^2}{c^2}} = \frac{qB}{mc}\left(\frac{1}{\gamma}\right) = 1.76\times10^7\,\gamma^{-1}\,\text{Hz}\left(\frac{B}{1\,\text{Gauss}}\right). \tag{4.93}$$

If $\mathbf{v}\cdot\mathbf{B} = 0$, then the particle will move in a circular path of radius

$$r_B = \frac{v}{\omega} = \frac{mcv}{qB}\gamma. \tag{4.94}$$

If $\mathbf{v}\cdot\mathbf{B}/B \equiv v_\parallel \neq 0$, then the motion is a superposition of a circular motion perpendicular to $\mathbf{B}$ and a linear motion along $\mathbf{B}$. Table 4.1 gives the relevant numerical values for different systems.

Table 4.1. Plasmas with magnetic fields

System	n cm^{-3}	B Gauss	T Kelvin	ω_B Hz	ω_p Hz	r_B cm	l cm
Ionosphere	10^4	$1/3$	300	5×10^6	5×10^6	2.3	5.6×10^6
Solar chromosphere	10^{12}	10^3	10^4	2×10^{10}	5×10^{10}	3.4×10^{-3}	6.2×10
Solar corona	10^6	$10^{-5} - 1$	10^5	$10^2 - 10^7$	5×10^7	$2.1 - 2.1 \times 10^6$	6.2×10^9
Interstellar medium	10^{-2}	3×10^{-6}	10^4	50	5×10^3	1.3×10^6	6.2×10^{15}
Neutron-star magnetosphere	10^{12}	10^{12}	10^6	2×10^{19}	5×10^{10}	3.4×10^{-11}	6.2×10^5
Intergalactic medium	10^{-6}	10^{-9}	10^5	0.02	50	10^{10}	6.2×10^{21}

When $r_B \ll l$, the magnetic field can bring about some amount of cohesion among particles by making them move along the field lines. Thus fluid approximation might work even when $l \gtrsim R$ if $r_B \ll R$.

The plasma frequency is comparable to the gyrofrequency in the ionosphere and solar chromosphere. In the interstellar and intergalactic media, plasma frequency is considerably higher than the gyrofrequency, while in the neutron star magnetosphere gyrofrequency is higher than the plasma frequency. A plasma acts as a dispersive medium and allows propagation of electromagnetic radiation at $\omega > \omega_p$. Since synchrotron radiation is emitted at $\omega \cong \omega_B$ we see that the transmission will be seriously affected in interstellar and intergalactic media. Propagation will be distorted in the solar chromosphere and ionosphere but virtually unhindered in the neutron star magnetosphere.

(b) The energy radiated by a particle moving in a magnetic field can be found from our general formula (4.52). Using $\mathbf{E} = 0, (\mathbf{v} \times \mathbf{B})^2 = v^2 B^2 \sin^2 \alpha$, where α is the angle between $\mathbf{B}$ and $\mathbf{v}$, we can write

$$\frac{\Delta \mathscr{E}}{\Delta t} = \frac{2}{3} \left(\frac{q^2}{m} \right)^2 \gamma^2 v^2 B^2 \sin^2 \alpha = 2 \left(\frac{B^2}{8\pi} \right) \sigma_{\mathrm{T}} \gamma^2 v^2 \sin^2 \alpha . \qquad (4.95)$$

If we average over all possible angles and use $\langle \sin^2 \alpha \rangle = (2/3)$ we obtain

$$\frac{dE}{dt} = \frac{4}{3} \left(\sigma_T c \gamma^2 \beta^2 \right) \left(\frac{B^2}{8\pi} \right) = \frac{4}{3} \left(\sigma_T c \gamma^2 \beta^2 \right) U_B , \qquad (4.96)$$

where $U_B = (B^2/8\pi)$ is the energy density in the magnetic field. The time-scale for energy loss by this process can be found by dividing the energy of the electron (γmc^2) by (dE/dt) and obtaining

$$t_{\text{syn}} \equiv \frac{\gamma mc^2}{(dE/dt)} \cong 5 \times 10^8 \text{s} \left[\gamma^{-1} B_{\text{Gauss}}^{-2} \right] . \qquad (4.97)$$

(c) From our general result in problem 4.4, we know that most of the radiation is emitted into a cone with opening angle

$$\Delta\theta \cong \left(1 - \frac{v^2}{c^2} \right)^{1/2} \cong \gamma^{-1} = \frac{mc^2}{E} . \qquad (4.98)$$

This fact has two implications for the frequency spectrum of radiation emitted in the ultrarelativistic case. (i) An observer will receive the radiation only when the direction of observation is within a cone of angular width $\Delta\theta$. (ii) The electric field $E(t)$ at any time t (at the point of observation) can depend on the angle θ only through the combination $(\gamma\theta)$ (see equation (4.70)). We can relate the time of arrival of radiation, t, to the angle θ in the following way. Consider radiation *emitted* at two instances $t_1 = 0$ and $t_2 = t$ when the charged particle was at positions $\theta_1 = 0$ and $\theta_2 = \theta$. Clearly, $r_B \theta = v(t_2 - t_1) = vt$, giving $t = (r_B \theta/v)$. The arrival time of the pulses at the observer will be less by the amount of time $(r_B \theta/c)$ taken by the radiation to travel the extra distance $(r_B \theta)$. So $t_{\text{obs}} \cong t - r_B \theta/c = (r_B \theta/v)(1 - v/c)$. Hence,

$$\theta(t_{\text{obs}}) \cong \frac{v t_{\text{obs}}}{r_B} \frac{1}{(1 - v/c)} \cong \frac{2 v t_{\text{obs}}}{r_B} \gamma^2 , \qquad (4.99)$$

giving

$$\gamma\theta \cong t_{\text{obs}} \frac{2v}{r_B} \gamma^3 = t_{\text{obs}} \left(2\omega_B \gamma^3 \sin\alpha \right) \equiv \frac{4}{3} \omega_c t_{\text{obs}} . \qquad (4.100)$$

We have set $v = \omega_B r_B \sin\alpha$ with the $(\sin\alpha)$ factor, taking into account the fact that only the component of v perpendicular to B contributes in this analysis.

Since the electric field at the observed location is only a function of $\gamma\theta \propto \omega_c t_{\text{obs}}$, its Fourier transform:

$$E(\omega) = \int_{-\infty}^{\infty} E(t) e^{-i\omega t} \, dt = \int_{-\infty}^{+\infty} E(\omega_c t) e^{-i\omega t} \, dt$$

$$= \frac{1}{\omega_c} \int_{-\infty}^{+\infty} E(q) \exp\left[-i \left(\frac{\omega}{\omega_c} \right) q \right] dq = E(\omega/\omega_c) \qquad (4.101)$$

depends on ω only through the ratio (ω/ω_c). Hence, the energy radiated per orbital period, $(dE/d\omega)\,(1/T) \equiv (dE/dt\,d\omega)$ also depends on ω only through some function $F\,(\omega/\omega_c)$.

In fact, most of the energy will be emitted at frequencies close to $\omega_c \propto \omega_B\gamma^3 \propto B\gamma^2 \propto B\epsilon^2$, where ϵ is the energy of the charged particle. It follows that $\epsilon \propto \omega_c^{1/2}B^{-1/2}$. Numerically,

$$\epsilon \cong 4\nu_c^{1/2}B^{-1/2}\,\text{erg} \qquad (4.102)$$

in c.g.s. units.

Synchrotron radiation will be linearly polarized with a high degree of polarization. To see this, consider the radiation emitted by a single electron circling in a magnetic field. When the electron is viewed in the orbital plane, the radiation is 100% linearly polarized, with the electric vector oscillating perpendicular to the magnetic field. When viewed along the direction of the magnetic field, the radiation is circularly polarized. (If it is right-handed circular polarization when viewed from the top, it will be left-handed circular polarization when seen from the bottom.) Consider now the relativistic motion of the electron which beams the radiation within a narrow cone in the direction of motion. The radiation is now mostly confined to the plane of the orbit. So whenever significant radiation is received, it will be from an orbit which is a highly elongated ellipse. This alone will make the radiation almost linearly polarized. Further, in any realistic situation, there will be a bunch of charged particles with a distribution of pitch angles. In that case, the two components of the circular polarization will effectively cancel, while the linear polarization will survive to a significant extent. Hence, we expect the radiation to be linearly polarized to a high degree.

(d) The synchrotron power radiated by a particle of energy ϵ in a magnetic field B scales as $B^2\epsilon^2$ (see equation (4.95)). The power radiated by a set of particles with a spectrum $n(\epsilon)$ is

$$j_\nu \propto \begin{pmatrix} \text{power radiated} \\ \text{by a particle of} \\ \text{energy } \epsilon \end{pmatrix} \times \begin{pmatrix} \text{number of} \\ \text{particles with} \\ \text{energy } \epsilon \end{pmatrix} \times \begin{pmatrix} \text{Jacobian} \\ \text{from } \epsilon \text{ to } \nu \end{pmatrix}$$

$$\propto (B^2\epsilon^2)n(\epsilon)\frac{d\epsilon}{d\nu}\,. \qquad (4.103)$$

Since $\epsilon \propto \nu^{1/2}B^{-1/2}$ (see equation (4.102)), we have $(d\epsilon/d\nu) \propto B^{-1/2}\nu^{-1/2}$; writing $n(\epsilon) \propto \epsilon^{-p} \propto \nu^{-p/2}B^{p/2}$, we obtain

$$j_\nu \propto B^2 \cdot \frac{\nu}{B} \cdot \nu^{-p/2}B^{p/2} \cdot \nu^{-1/2}B^{-1/2} \propto \nu^{-\frac{1}{2}(p-1)}B^{\frac{1}{2}(p-1)} \propto \nu^{-\alpha}B^{1+\alpha}, \qquad (4.104)$$

with $p = 2\alpha + 1$.

4.7 Energy content of synchrotron sources

The luminosity of a synchrotron source of volume V, containing magnetic field B and a powerlaw distribution of relativistic electrons $n(E) = kE^{-p}$, can be expressed in the form

$$L = AVkB^{1+\alpha}v^{-\alpha}, \qquad (4.105)$$

where $\alpha = (p-1)/2$ (see (4.104)) and A is a constant. The magnetic energy of the source is $U_B = (B^2V/8\pi)$, while the kinetic energy of particles (protons and electrons) can be taken to be $(1+\beta)U_e$. Hence, the total energy density is

$$\frac{U_{tot}}{V} = (1+\beta) \int_{E_1}^{E_2} kEn(E)\,dE + \frac{B^2}{8\pi}. \qquad (4.106)$$

The typical frequency v at which an electron of energy E will be emitting radiation is given by $v = CBE^2$, where C is a constant (see equation (4.102)). Thus we can set $E_1 = (v_1/CB)^{1/2}$ and $E_2 = (v_2/CB)^{1/2}$ if the radiation is detected in the range of frequencies $v_1 < v < v_2$. The energy content in particles now becomes

$$\frac{U_{part}}{V} = (1+\beta) \int_{E_1}^{E_2} kE^{1-p}\,dE = \frac{(1+\beta)k}{(p-2)} (CB)^{(p-2)/2} \left[v_1^{(2-p)/2} - v_2^{(2-p)/2} \right]. \qquad (4.107)$$

Substituting for k, in terms of L and B, from equation (4.105), we obtain

$$\frac{U_{part}}{V} = \frac{(1+\beta)}{(p-2)} \left[\frac{L}{AVB^{1+\alpha}v^{-\alpha}} \right] (CB)^{(p-2)/2} \left[v_1^{(2-p)/2} - v_2^{(2-p)/2} \right] \equiv Q(1+\beta)LB^{-3/2}, \qquad (4.108)$$

where we have used the relation $2\alpha = p - 1$ and shown explicitly only the dependence on β, B, L. The total energy density now becomes

$$\frac{U_{tot}}{V} = \frac{B^2}{8\pi} + Q(1+\beta)LB^{-3/2}. \qquad (4.109)$$

This expression has a minimum when the magnetic field has the value

$$B_{min} = [6\pi Q (1+\beta) L]^{2/7}. \qquad (4.110)$$

Thus, in order to produce a particular amount of synchrotron luminosity using high energy particles, one requires a magnetic field of this minimum strength. It is easy to see that when $B = B_{min}$ the particle energy U_{part} and the magnetic energy U_B are comparable and $U_B = (3/4)U_{part}$.

4.8 Thomson scattering

(a) Consider an electromagnetic wave with $\mathbf{E} = \mathbf{E}_0 \cos(\mathbf{k} \cdot \mathbf{x} - \omega t)$ which is incident on a charged particle. We will assume that the motion of the charge is non-relativistic; in that case we can ignore the magnetic force on the charge and the $\mathbf{k} \cdot \mathbf{x}$ term in $\cos(\mathbf{k} \cdot \mathbf{x} - \omega t)$ and write the equation of motion for the charge as

$$m\ddot{\mathbf{x}} = qE_0 \cos \omega t = \left(\frac{m}{q}\right)\ddot{\mathbf{d}}, \tag{4.111}$$

where $\mathbf{d} = q\mathbf{x}$ is the dipole moment.

Such an oscillating charge will radiate energy at the same frequency as the incident wave. The intensity radiated in some direction $\hat{\mathbf{n}}'$ can be determined from our general formulas (see equation (4.30)):

$$\frac{dE}{dt\,d\Omega} = \frac{q^4}{4\pi m^2 c^3}\left(\mathbf{E}\times\hat{\mathbf{n}}'\right)^2 = \frac{q^4}{4\pi m^2 c^3}E^2 \sin^2\theta, \tag{4.112}$$

where θ is the direction between $\mathbf{E}$ and $\hat{\mathbf{n}}'$. Since the incident flux is $S = \left(cE^2/4\pi\right)$, the scattering cross-section is

$$\left(\frac{d\sigma}{d\Omega}\right) = \left(\frac{q^2}{mc^2}\right)^2 \sin^2\theta \equiv r_0^2 \sin^2\theta. \tag{4.113}$$

This formula is valid for radiation polarized along a specific direction $\hat{\mathbf{e}} = \left(\mathbf{E}/E\right)$. For unpolarized radiation we have to average $\sin^2\theta = 1 - \left(\hat{\mathbf{n}}'\cdot\hat{\mathbf{e}}\right)^2$ over all $\hat{\mathbf{e}}$ perpendicular to the direction of propagation $\hat{\mathbf{n}} = \left(\mathbf{k}/k\right)$. This can be done by noting that the average of $e_a e_b$ is

$$\langle e_a e_b \rangle = \frac{1}{2}\left(\delta_{ab} - \frac{k_a k_b}{k^2}\right) \tag{4.114}$$

and hence

$$\langle \sin^2\theta \rangle = 1 - n'_a n'_b \langle e_a e_b \rangle = \frac{1}{2}\left(1 + \left(\hat{\mathbf{n}}\cdot\hat{\mathbf{n}}'\right)^2\right). \tag{4.115}$$

Therefore, the scattering cross-section for unpolarized radiation is

$$\frac{d\sigma}{d\Omega} = \frac{1}{2}r_0^2\left[1 + \left(\hat{\mathbf{n}}\cdot\hat{\mathbf{n}}'\right)^2\right]. \tag{4.116}$$

The total scattering cross-section is

$$\sigma_{\mathrm{T}} = \frac{8\pi}{3}\left(\frac{e^2}{mc^2}\right)^2 = 6.7\times10^{-24}\,\mathrm{cm}^2 \tag{4.117}$$

and is called the Thomson scattering cross-section, where we have set $q = e$, the electric charge. Note that $\sigma_{\mathrm{T}} \ll \sigma_{\mathrm{m}}$ where $\sigma_{\mathrm{m}} \cong 10^{-15}\,\mathrm{cm}^2$ is the scattering cross-section of matter due to molecular collisions. Hence, the photon mean-free-path is usually much longer than the matter mean-free-path. In a plasma, Thomson scattering of photons by the charged particle will give the photons a mean-free-path of $l_{\mathrm{T}} = (n_e\sigma_{\mathrm{T}}c)^{-1}$ where n_e is the electron density. This will be the main source of scattering of photons in a fully ionized plasma. (We can ignore scattering by the protons since the Thomson cross-section for protons is lower by the factor $\left(m_e/m_p\right)^2 \cong 10^{-6}$.)

The validity of Thomson scattering depends on the condition that production of e^+e^- pairs is negligible, which requires $\hbar\omega \ll mc^2$. It turns out that the amount of energy transferred from the photon to the electron is $\Delta E \cong (\hbar\omega/mc^2)\,\hbar\omega \ll \hbar\omega$; see next problem. Hence, the photon transfers all its *momentum* to the electron but a negligible amount of *energy*.

(b) When an electromagnetic wave hits a charged particle, it makes the particle oscillate and radiate. The radiation will exert a damping force on the particle. This drag force can be obtained by averaging the force in equation (4.56) over one period of the wave. The first term with $\dot{\mathbf{E}}$ averages to zero and the second term gives

$$\langle \mathbf{f} \rangle = \frac{2}{3}\left(\frac{e^2}{mc^2}\right)^2 \langle E^2 \rangle \hat{\mathbf{n}} = \frac{8\pi}{3}\left(\frac{e^2}{mc^2}\right)^2 \frac{\langle E^2 \rangle}{4\pi}\hat{\mathbf{n}} = \sigma_{\mathrm{T}} U \hat{\mathbf{n}}, \qquad (4.118)$$

where $U\hat{\mathbf{n}}$ is the flux of radiation.

The same result can be obtained more elegantly from (4.64). In a frame in which the charge is at rest, $u^i = (1,0,0,0)$ and $g^i = (\gamma \mathbf{f}\cdot\mathbf{v}, \gamma\mathbf{f}) = (0,\mathbf{f})$. From (4.64) we obtain

$$g^i = \sigma_{\mathrm{T}}\left[T^{i0} - T^{00}u^i\right] = (0, \sigma_{\mathrm{T}} u\hat{\mathbf{n}}), \qquad (4.119)$$

which agrees with (4.118).

(c) Using $T^{ab} = U_{\mathrm{rad}}\,\mathrm{dia}\,(1, 1/3, 1/3, 1/3)$ for an isotropic radiation bath and $u^i = (\gamma, \gamma\mathbf{v})$ we obtain

$$T^{ab}u_a u_b = U_{\mathrm{rad}}\gamma^2\left(1 + \frac{1}{3}v^2\right), \qquad T^{ab}U_b = \left(U_{\mathrm{rad}}\gamma, -\frac{1}{3}U_{\mathrm{rad}}\gamma\mathbf{v}\right). \qquad (4.120)$$

This gives, on using (4.64),

$$g^i = \left(-\frac{4}{3}\sigma_{\mathrm{T}}U_{\mathrm{rad}}\gamma^3 v^2, -\frac{4}{3}\sigma_{\mathrm{T}}U_{\mathrm{rad}}\gamma^3\mathbf{v}\right) = (\gamma \mathbf{f}\cdot\mathbf{v}, \gamma\mathbf{f}). \qquad (4.121)$$

Comparing, we obtain

$$\mathbf{f} = -\frac{4}{3}\sigma_{\mathrm{T}}U_{\mathrm{rad}}\gamma^2\left(\frac{\mathbf{v}}{c}\right), \qquad -\mathbf{f}\cdot\mathbf{v} = \frac{4}{3}\sigma_{\mathrm{T}}U_{\mathrm{rad}}\gamma^2\left(\frac{v^2}{c^2}\right)c. \qquad (4.122)$$

This result is valid for any radiation field with energy density U_{rad}. The work done by this drag force is $\mathbf{f}_{\mathrm{drag}}\cdot\mathbf{v} = -(4/3)\,c\sigma_{\mathrm{T}}U_{\mathrm{rad}}\gamma^2\,(v/c)^2$. This expression should also be equal to the net power radiated by the electron. We shall see in the next problem that this is indeed the case.

An alternative way of deriving this result is as follows. A thermal bath of photons is equivalent to a random superposition of electromagnetic radiation with $\langle E^2/4\pi \rangle = \langle B^2/4\pi \rangle = aT^4$ at any location. If the charge is not moving, then there is no net flux hitting the charge and there is no drag force. Suppose the charge is moving with velocity $\mathbf{v}$, in a frame S in which radiation is isotropic. We will now transform to a frame S' in which the charge is at rest. The energy flux in S' along the x-axis is

$$T'^{0x} = \gamma^2 \left[\left(1 + \frac{v^2}{c^2} \right) T^{0x} - \frac{v_x}{c} \left(T^{00} + T^{xx} \right) \right]$$

$$= -\frac{v_x}{c} \gamma^2 \left(aT^4 \right) \left(1 + \frac{1}{3} \right) = -\frac{4}{3} \left(aT^4 \right) \left(\frac{v_x}{c} \right) \gamma^2. \qquad (4.123)$$

We have used the facts that $T^{0x} = 0$, $T^{00} = aT^4$, $T^{xx} = (1/3)\, aT^4$. From (4.118) we find that

$$\mathbf{f}_{\text{drag}} = -(4/3)\sigma_T U_{\text{rad}}\gamma^2 \left(\mathbf{v}/c \right) \cong -\frac{4}{3}\sigma_T \left(aT^4 \right) \left(\frac{\mathbf{v}}{c} \right). \qquad (4.124)$$

4.9 Compton and inverse Compton effects

(a) Let the initial and final four-momenta of the photon be $k_i^a = (\hbar\omega_i)\,[1, \mathbf{n}_i]$ and $k_f^a = (\hbar\omega_f)\,[1, \mathbf{n}_f]$, and those of the electron be $p_i^a = (m, 0)$ and $p_f^a = (E, \mathbf{p})$. The conservation of momentum and energy is expressed by the equation $p_i^a + k_i^a = p_f^a + k_f^a$. Squaring this equation and using the components will allow us to eliminate the final electron momentum. Rearranging the terms, we obtain

$$\frac{\omega_f}{\omega_i} = \left[1 + \left(\frac{\hbar\omega_i}{m_e c^2} \right) (1 - \cos\theta) \right]^{-1}, \qquad (4.125)$$

where $\cos\theta = (\mathbf{n}_i \cdot \mathbf{n}_f)$. When $\hbar\omega_i \ll m_e c^2$ we can expand the denominator in a Taylor series and obtain

$$\frac{\omega_f - \omega_i}{\omega_i} = \frac{\Delta\epsilon}{\epsilon} = - \left(\frac{\hbar\omega_i}{m_e c^2} \right) (1 - \cos\theta). \qquad (4.126)$$

To find the mean energy transfer, we have to average this expression over θ. In the rest frame of the electron, the scattering has front–back symmetry, making $\langle \cos\theta \rangle = 0$. Hence, the average energy lost by the photon per collision is

$$\langle \Delta\epsilon \rangle = - \left(\frac{\hbar\omega_i}{m_e c^2} \right) \hbar\omega_i . \qquad (4.127)$$

(b) We saw in the last problem that an electron moving through an isotropic radiation field will experience a drag and lose energy at the rate (see equation (4.122))

$$\frac{dE}{dt} = -\mathbf{f} \cdot \mathbf{v} = \frac{4}{3}\sigma_T U_{\text{rad}}\gamma^2 \left(\frac{v}{c} \right)^2 c. \qquad (4.128)$$

Clearly, this is the net power gained by the radiation field.

An alternative way of deriving the result (4.128) is as follows. We treat the radiation field as equivalent to an electromagnetic field with $\langle (E^2/8\pi) \rangle = \langle (B^2/8\pi) \rangle = (U_{\text{rad}}/2)$, with $\mathbf{E}$ and $\mathbf{B}$ randomly fluctuating around zero mean. In this case, we can again use (4.52) and average over $\mathbf{E}$ and $\mathbf{B}$ to obtain the net power. Now,

$$\mathscr{Q} \equiv \langle (\mathbf{E} + \mathbf{v} \times \mathbf{B})^2 - (\mathbf{E} \cdot \mathbf{v})^2 \rangle = \langle E^2 - (\mathbf{E} \cdot \mathbf{v})^2 \rangle + \langle (\mathbf{v} \times \mathbf{B})^2 \rangle \qquad (4.129)$$

since $\langle \mathbf{E} \cdot (\mathbf{v} \times \mathbf{B}) \rangle = \langle \mathbf{v} \cdot (\mathbf{B} \times \mathbf{E}) \rangle = 0$ due to random orientation of $\mathbf{v}$ with respect to $(\mathbf{E} \times \mathbf{B})$. Using the relation $\langle E^2 - (\mathbf{E} \cdot \mathbf{v})^2 \rangle = \langle E^2 \rangle - E^2 v^2 \langle \cos^2 \theta \rangle = E^2 \left(1 - v^2/3\right)$ and $\langle (\mathbf{v} \times \mathbf{B})^2 \rangle = \langle \mathbf{v} \cdot [\mathbf{v} B^2 - \mathbf{B}(\mathbf{v} \cdot \mathbf{B})] \rangle = v^2 B^2 - v^2 B^2/3 = (2/3)\beta^2 B^2$, we obtain $\mathcal{Q} = E^2 \left(1 - v^2/3\right) + (2/3)v^2 B^2$. Substituting these results in (4.52) we find that

$$\left(\frac{dE}{dt}\right)_{\text{scat}} = \frac{\sigma_T c}{4\pi} \gamma^2 \left(4\pi U_{\text{rad}}\right) \left(1 + \frac{1}{3}\frac{v^2}{c^2}\right) = \sigma_T c \gamma^2 \left(1 + \frac{1}{3}\frac{v^2}{c^2}\right) U_{\text{rad}}, \quad (4.130)$$

where we have used the relation $\langle E^2 \rangle = \langle B^2 \rangle = 4\pi U_{\text{rad}}$. The incident radiation energy has been absorbed by the electron. The rate at which this happens is

$$\left(\frac{dE}{dt}\right)_{\text{abs}} = \sigma_T c U_{\text{rad}}. \quad (4.131)$$

Hence the net addition of energy to the photon field is

$$\begin{aligned} P_{\text{in Comp}} = \left(\frac{dE}{dt}\right) &= \left(\frac{dE}{dt}\right)_{\text{scat}} - \left(\frac{dE}{dt}\right)_{\text{abs}} = \sigma_T c U_{\text{rad}} \left[\gamma^2 \left(1 + \frac{1}{3}\frac{v^2}{c^2}\right) - 1\right] \\ &= \frac{4}{3} \sigma_T c U_{\text{rad}} \gamma^2 \left(\frac{v}{c}\right)^2. \end{aligned}$$

In the non-relativistic limit, $\gamma \cong 1, v^2 = \left(3kT_e/m\right)$, giving

$$P_{\text{in Comp}}^{nr} \cong \frac{4}{3} \sigma_T c U_{\text{rad}} \left(\frac{3kT_e}{m_e c^2}\right). \quad (4.132)$$

The mean number of photons scattered per second is $N_{\text{Comp}} = \left(\sigma_T c n_{\text{rad}}\right) = \left(\sigma_T c U_{\text{rad}}/\hbar \omega_i\right)$, where $\hbar \omega_i$ is the average energy of the photon defined by $\hbar \omega_i = \left(U_{\text{rad}}/n_{\text{rad}}\right)$. Hence, the average energy gained by the photon in one collision is

$$\langle \Delta E \rangle = \frac{P_{\text{in Comp}}}{N_{\text{Comp}}} = \frac{4}{3} \gamma^2 \left(\frac{v}{c}\right)^2 \hbar \omega_i = \frac{4}{3} \gamma^2 \left(\frac{v}{c}\right)^2 \langle E \rangle. \quad (4.133)$$

In the relativistic limit $\langle \Delta E/E \rangle \cong (4/3) \gamma^2 \gg 1$ and this process can be a source of high energy photons. When $v \ll c$, the energy gain by photons per collision is $\langle \Delta E/E \rangle \cong \left(4kT_e/m_e c^2\right)$. Combining with (4.127) we find that the mean fractional energy change of photons, per collision, is

$$\left\langle \frac{\Delta \epsilon}{\epsilon} \right\rangle = -\frac{\langle \hbar \omega \rangle}{m_e c^2} + \frac{4kT_e}{m_e c^2}. \quad (4.134)$$

4.10 Comptonization

(a) The conservation of energy and momentum can be expressed by the equations

$$h\nu + \frac{p^2}{2m} = h\nu' + \frac{p'^2}{2m}, \quad \frac{h\nu}{c}\hat{\mathbf{n}} + \mathbf{p} = \frac{h\nu'}{c}\hat{\mathbf{n}}' + \mathbf{p}'. \quad (4.135)$$

Solving for $\mathbf{p}'$, squaring and substituting into the first equation, we will obtain a quadratic equation on $\Delta \equiv \nu' - \nu$. When $(\Delta/h\nu) \ll 1$ we can ignore the Δ^2 term in this equation and solve for Δ. This gives

$$h\left(v'-v\right) \equiv h\Delta \cong -\frac{hvc\mathbf{p}\cdot\left(\hat{\mathbf{n}}-\hat{\mathbf{n}}'\right)+h^2v^2\left(1-\hat{\mathbf{n}}\cdot\hat{\mathbf{n}}'\right)}{mc^2+hv\left(1-\hat{\mathbf{n}}\cdot\hat{\mathbf{n}}'\right)-c\mathbf{p}\cdot\hat{\mathbf{n}}'}. \qquad (4.136)$$

The second term in the numerator is a correction to the first of $\mathcal{O}\left(hv/mcv\right) \cong$ $\mathcal{O}\left(v/c\right)$; similarly the second and third terms in the denominator are small corrections to mc^2. Hence, to lowest order, we obtain

$$h\left(v'-v\right) \cong -\left(\frac{hv}{mc}\right)\mathbf{p}\cdot\left(\hat{\mathbf{n}}-\hat{\mathbf{n}}'\right). \qquad (4.137)$$

(b) For a homogeneous medium, we can ignore changes n due to the $\left(\partial n/\partial\mathbf{x}\right)$ term. The changes in the number $n\left(v\right)$ of photons of frequency v occur only due to scattering. The rate of scattering of photons from frequency v to frequency v' by electrons of energy E is described by the term

$$\int d^3p\, N\left(E\right)\int c\left(\frac{d\sigma}{d\Omega}\right)d\Omega\left[n\left(v\right)\left(1+n\left(v'\right)\right)\right]. \qquad (4.138)$$

The proportionality to $n\left(v\right)$ and $N\left(E\right)$ is obvious; the $\left[1+n\left(v'\right)\right]$ term takes into account the stimulated emission effects. (This will be discussed in problem 4.11.) Strictly speaking, one should also include a factor $\left[1-N\left(E\right)\right]$ to take into account the fermion nature of electrons. This is ignored because electrons are assumed to be non-degenerate with $N\left(E\right) \ll 1$.

Similarly, the scattering of photons from v' to v is described by the term

$$\int d^3p\, N\left(E'\right)\int c\left(\frac{d\sigma}{d\Omega}\right)d\Omega\left[n\left(v'\right)\left(1+n\left(v\right)\right)\right]. \qquad (4.139)$$

Note that the structure of the collision term is very similar to that encountered in problem 3.1, except for the $(1+n)$ factor. This is an integro-differential equation which is in general difficult to solve.

(c) Expanding $n\left(v'\right) = n(v+\Delta)$ and $N\left(E'\right) = N(E-h\Delta)$ in a Taylor series, we obtain

$$n\left(v'\right) = n\left(v\right) + \frac{h\Delta}{kT}\frac{\partial n}{\partial x} + \frac{1}{2}\left(\frac{h\Delta}{kT}\right)^2\frac{\partial^2 n}{\partial x^2} + \cdots. \qquad (4.140)$$

$$\begin{aligned}
N\left(E'\right) &= N\left(E\right) - h\Delta\frac{\partial N}{\partial E} + \frac{1}{2}h^2\Delta^2\frac{\partial^2 N}{\partial E^2} + \cdots \\
&= N\left(E\right) + \frac{h\Delta}{kT}N\left(E\right) + \frac{1}{2}\left(\frac{h\Delta}{kT}\right)^2 N\left(E\right)\ldots,
\end{aligned} \qquad (4.141)$$

where $x \equiv \left(hv/kT\right)$ and we have assumd that $N\left(E\right) \propto \exp\left(-E/kT\right)$. Substituting these expansions in the original equation we obtain

$$\frac{\partial n}{\partial t} = \frac{h}{kT} \left[\left(\frac{\partial n}{\partial x} \right) + n(n+1) \right] I_1$$

$$+ \frac{1}{2} \left(\frac{h}{kT} \right)^2 \left(\frac{\partial^2 n}{\partial x^2} + 2(1+n) \frac{\partial n}{\partial x} + n(n+1) \right) I_2 \qquad (4.142)$$

with

$$I_1 = \int d^3p \, d\Omega \left(\frac{d\sigma}{d\Omega} \right) cN(E) \Delta \qquad (4.143)$$

$$I_2 = \int d^3p \, d\Omega \left(\frac{d\sigma}{d\Omega} \right) cN(E) \Delta^2 . \qquad (4.144)$$

(d) Note that, in the lowest order, $\Delta \propto \mathbf{p} \cdot (\hat{\mathbf{n}} - \hat{\mathbf{n}}')$ and – at this order – I_1 will be zero when integrated over all $\mathbf{p}$. Thus to obtain non-zero contribution to I_1 we have to expand (4.136) to higher order in (v/c). We can, however, determine I_1 by an indirect procedure.

We note that, by definition, I_1 is the energy transfer rate divided by the mean energy (kT) of the electrons. If ΔE is the mean energy transfer per collision then, $I_1 = (\Delta E / kT)(\sigma_T n_e c)$ since the rate of collisions is $(\sigma_T n_e c)$. From our earlier discussion (see equation (4.134)) we know that $\Delta E = (hv/mc^2)(4kT - hv)$. This gives

$$I_1 = \left(\frac{kT}{mc^2} \right) \sigma_T n_e x (4 - x). \qquad (4.145)$$

(e) The integration for I_2 is straightforward. Using $\Delta = - (v/mc) \mathbf{p} \cdot (\hat{\mathbf{n}} - \hat{\mathbf{n}}')$, we obtain

$$I_2 = \left(\frac{v}{mc} \right)^2 \int c \, d\sigma \int d^3p \, N(E) \left[\mathbf{p} \cdot (\hat{\mathbf{n}} - \hat{\mathbf{n}}') \right]^2$$

$$= \left(\frac{v}{mc} \right)^2 \int c \, d\sigma \int_0^\infty 2\pi p^2 \, dp \, (\sin \psi \, d\psi) N(p) \, p^2 |\hat{\mathbf{n}} - \hat{\mathbf{n}}'|^2 \cos^2 \psi$$

$$= \frac{1}{3} \left(\frac{v}{mc} \right)^2 \int c \, d\sigma |\hat{\mathbf{n}} - \hat{\mathbf{n}}'|^2 \int_0^\infty 4\pi p^2 dp \, [N(p) p^2] . \qquad (4.146)$$

Since $N(p) \propto \exp(-p^2/2mkT)$, the p-integral gives $\langle p^2 \rangle n_e = 2mn_e \langle p^2/2m \rangle = 3kT m n_e$, where n_e is the number density of electrons. The angular integration gives

$$\int d\Omega \left(\frac{d\sigma}{d\Omega} \right) |\hat{\mathbf{n}} - \hat{\mathbf{n}}'|^2 = \int d\Omega \frac{1}{2} r_0^2 \left(1 + \cos^2 \theta \right) [2 - 2\cos\theta]$$

$$= r_0^2 \int d\Omega \left(1 + \cos^2 \theta \right) = r_0^2 4\pi \left(1 + \frac{1}{3} \right) = 2\sigma_T . \qquad (4.147)$$

Putting everything together:

$$I_2 = 2 \left(\frac{v}{mc}\right)^2 (kT)(mc)(n_e\sigma_T) = \frac{2n_e\sigma_T(kT)^3}{h^2 mc} x^2. \tag{4.148}$$

Using the form of I_2 and I_1, we obtain the final equation:

$$\left(\frac{mc^2}{kT}\right) \frac{1}{n_e\sigma_T c} \frac{\partial n}{\partial t} = \frac{1}{x^2} \frac{\partial}{\partial x}\left[x^4\left(\frac{\partial n}{\partial x}+n+n^2\right)\right] = \frac{\partial n}{\partial y}, \tag{4.149}$$

where the 'Compton-y parameter' is defined by the relation

$$y \equiv t\left(\frac{kT}{mc^2}\right) n_e\sigma_T c. \tag{4.150}$$

(f) The rate of change of the total energy of the radiation with respect to y is determined by the equation

$$\frac{h^3 c^3}{8\pi (kT)^4} \frac{dE_{\text{pho}}}{dy} = \frac{\partial}{\partial y} \int_0^\infty x^2 [n(x)x]\,dx$$

$$= \int_0^\infty dx\, x^3 \frac{\partial n}{\partial y} = \int_0^\infty dx\, x^3\left[\frac{1}{x^2}\frac{\partial}{\partial x}\left(x^4\left(\frac{\partial n}{\partial y}+n+n^2\right)\right)\right]. \tag{4.151}$$

If $n \ll 1$, we can ignore n^2 compared to n. Integrating the expressions by parts we obtain

$$\int_0^\infty dx\, x \frac{d}{dx}\left[x^4\left(\frac{\partial n}{\partial x}+n\right)\right] = -\int_0^\infty dx\, x^4\left[\frac{\partial n}{\partial x}+n\right]$$

$$= -\int_0^\infty dx\, n x^4 - \int_0^\infty dx\, x^4 \cdot \frac{\partial n}{\partial x} = -\int_0^\infty dx\, n x^4 + 4\int_0^\infty dx\, n x^3. \tag{4.152}$$

Therefore,

$$\frac{h^3 c^3}{8\pi (kT)^4} \frac{dE_{\text{pho}}}{dy} = 4\int_0^\infty n x^3\, dx - \int_0^\infty n x^4\, dx. \tag{4.153}$$

If $n(x)$ is significant mostly for $x \ll 1$, then the first integral dominates over the second and we obtain

$$\frac{dE_{\text{pho}}}{dy} = 4E_{\text{pho}} \tag{4.154}$$

or

$$E_{\text{pho}}(t) = E_{\text{pho}}(0)\exp 4y = E_{\text{pho}}(0)\exp\left(\frac{t}{t_{\text{Comp}}}\right), \tag{4.155}$$

with

$$t_{\text{Comp}} = \left(\frac{mc^2}{4kT}\right)\frac{1}{(n_e\sigma_T c)}. \tag{4.156}$$

In this limit, the optical depth is $\tau_e \cong (n_e \sigma_T)(ct)$ and the result (4.155) is the same as the one obtained in problem 1.6.

(g) When the $n(n+1)$ term is ignored we obtain the equation

$$\frac{\partial n}{\partial y} = \frac{1}{x^2} \frac{\partial}{\partial x} \left[x^4 \frac{\partial n}{\partial x} \right] = \frac{1}{x^3} \frac{\partial}{\partial (\ln x)} \left[x^3 \frac{\partial n}{\partial (\ln x)} \right]. \tag{4.157}$$

It is convenient to transform coordinates from $(y, q = \ln x)$ to (y, z), where $z = q + 3y$. Noting that

$$dn = \left(\frac{\partial n}{\partial y} \right)_q dy + \left(\frac{\partial n}{\partial q} \right)_y dq = \left(\frac{\partial n}{\partial y} \right)_q dy + \left(\frac{\partial n}{\partial q} \right)_y (dz - 3dy)$$

$$= \left[\left(\frac{\partial n}{\partial y} \right)_q - 3 \left(\frac{\partial n}{\partial q} \right)_y \right] dy + \left(\frac{\partial n}{\partial q} \right)_y dz = \left(\frac{\partial n}{\partial y} \right)_z dy + \left(\frac{\partial n}{\partial q} \right)_y dz, \tag{4.158}$$

we can transform our equation to the simple form

$$\left(\frac{\partial n}{\partial y} \right)_q = \left(\frac{\partial n}{\partial y} \right)_z + 3 \left(\frac{\partial n}{\partial q} \right)_y = \left(\frac{\partial^2 n}{\partial z^2} \right)_y + 3 \left(\frac{\partial n}{\partial q} \right)_y \tag{4.159}$$

or

$$\left(\frac{\partial n}{\partial y} \right)_z = \left(\frac{\partial^2 n}{\partial z^2} \right)_y. \tag{4.160}$$

This is a diffusion equation which has the solution

$$n(z, y) = \int_{-\infty}^{+\infty} dz' \left(\frac{1}{4\pi y} \right)^{1/2} \left[\exp -\frac{(z - z')^2}{4y} \right] n(z', 0). \tag{4.161}$$

Transforming back to x, y and writing $z' = \ln \mu$, we obtain

$$n(x, y) = \frac{1}{(4\pi y)^{1/2}} \int_0^{\infty} \frac{d\mu}{\mu} n(\mu, 0) \exp \left[-\frac{1}{4y} \left(3y + \ln \frac{x}{\mu} \right)^2 \right]. \tag{4.162}$$

For $n(x, 0) = x^{-1}$ the above equation reduces to

$$n(x, y) = \frac{1}{(4\pi y)^{1/2}} \int_0^{\infty} \frac{d\mu}{\mu^2} \exp \left[-\frac{1}{4y} \left(3y - \ln \frac{\mu}{x} \right)^2 \right]. \tag{4.163}$$

Substituting $p = \ln(\mu/x)$ the integral can be evaluated by elementary means and we find that

$$n(x, y) = x^{-1} e^{-2y}. \tag{4.164}$$

Since the effective temperature is proportional to the number density of photons at the Rayleigh–Jeans end, we find that the temperature is reduced by a factor e^{-2y}.

(h) For $n(x) = n_0 \exp\left(-x/\alpha\right) = n_0 \exp\left(-h\nu/\alpha k T\right)$, we have

$$4 \int_0^\infty nx^3 \, dx - \int_0^\infty nx^4 \, dx = -24\alpha^4 n_0 \left(\alpha - 1\right), \tag{4.165}$$

so that

$$\frac{dE_e}{dt} = -\frac{dE_{\text{pho}}}{dt} = \frac{8}{3} E_e \left(\frac{E_{\text{pho}}\sigma_{\text{T}}}{mc}\right) (\alpha - 1). \tag{4.166}$$

The parameter α measures the ratio of photon temperature to electron temperature. Taking $E_e = (3/2)k_{\text{B}} T_e$, $E_{\text{pho}} = a T^4$, this equation becomes

$$\frac{dT_e}{dt} = \frac{8}{3} \left(\frac{\sigma_{\text{T}}}{mc}\right) \left(a T^4\right) \left(T - T_e\right). \tag{4.167}$$

It often happens in astrophysical systems that the plasma is not fully ionized but has a fractional ionization of x_e. For a hydrogen plasma, we may define $x_e = n_e/(n_p + n_{\text{H}}) = n_p/(n_p + n_{\text{H}})$, where n_e, n_p, n_{H} denote the number densities of electrons, protons and hydrogen atoms. The collisions between e, p and H will maintain a common temperature for matter, T_{m}; the transfer of energy between photons and matter, however, is mainly due to Thomson scattering of electrons by photons. Then, in the energy balance equation (4.167), the left-hand side will be multiplied by $(n_e + n_p + n_{\text{H}}) = (n_p + n_{\text{H}})(1 + x_e)$, while the right-hand side will be multiplied by $n_e = (n_p + n_{\text{H}})x_e$. This changes the equation to

$$\frac{dT_{\text{m}}}{dt} = \left(\frac{x_e}{1 + x_e}\right) \frac{8}{3} \left(\frac{\sigma_{\text{T}}}{m_e c}\right) (a T^4)(T - T_{\text{m}}). \tag{4.168}$$

If the gas is not hydrogen, we need to multiply the right-hand side by a further factor of (μ/m_p) where μ is the molecular weight. The corresponding cooling time is

$$t_{\text{cool}} = \left(\frac{1}{T_{\text{m}}} \frac{dT_{\text{m}}}{dt}\right)^{-1} = \left(\frac{T}{T_{\text{m}}} - 1\right)^{-1} \left(\frac{1 + x_e}{x_e}\right) \frac{3}{8} \left(\frac{m_e m_p c}{\mu \sigma_{\text{T}}}\right) \left(\frac{1}{a T^4}\right)$$
$$\cong \left(\frac{mc^2}{kT}\right) \left(\frac{1}{n_\gamma \sigma_{\text{T}} c}\right). \tag{4.169}$$

The second equality is valid if $T_{\text{m}} \gg T$, $a T^4 \cong n_\gamma k T$, $x_e \cong 1$ and $m_p \cong \mu$.

4.11 Quantum theory of radiation

(a) Using the expansion for the vector potential

$$\mathbf{A}(t, \mathbf{x}) = \sum_{\mathbf{k}} \mathbf{q}_{\mathbf{k}} e^{i\mathbf{k}\cdot\mathbf{x}} = \sum_{\mathbf{k}} \left(\mathbf{a}_{\mathbf{k}} e^{i\mathbf{k}\cdot\mathbf{x}} + \mathbf{a}_{\mathbf{k}}^* e^{-i\mathbf{k}\cdot\mathbf{x}}\right) \tag{4.170}$$

and the condition $\nabla \cdot \mathbf{A} = 0$, we find that

$$\mathbf{k} \cdot \mathbf{q}_{\mathbf{k}} = \mathbf{k} \cdot \mathbf{a}_{\mathbf{k}} = \mathbf{k} \cdot \mathbf{a}_{\mathbf{k}}^* = 0. \tag{4.171}$$

For every value of $\mathbf{k}$, the vector $\mathbf{a}_{\mathbf{k}}$ is perpendicular to $\mathbf{k}$.

The electric and magnetic fields corresponding to this $A^i = (0, \mathbf{A})$ are

$$\mathbf{E} = -\dot{\mathbf{A}} = -\sum_{\mathbf{k}} \dot{\mathbf{q}}_{\mathbf{k}} e^{i\mathbf{k}\cdot\mathbf{x}}, \quad \mathbf{B} = \nabla \times \mathbf{A} = i \sum_{\mathbf{k}} (\mathbf{k} \times \mathbf{q}_{\mathbf{k}}) e^{i\mathbf{k}\cdot\mathbf{x}}. \tag{4.172}$$

To calculate the Hamiltonian for the system we will have to integrate E^2 and B^2 over the volume of the box. These integrals will reduce to expressions of the form

$$I = \int_0^L dx \, \exp i \left(k_x - k_x' \right) x = \int_0^L dx \, \exp \frac{2\pi i}{L} \left(n_x - n_x' \right) x. \tag{4.173}$$

Such an integral vanishes when $n_x \neq n_x'$ and gives $I = L$ for $n_x = n_x'$. Using this result and the identity

$$(\mathbf{k} \times \mathbf{q}_{\mathbf{k}}) \cdot (\mathbf{k} \times \mathbf{q}_{\mathbf{k}}^*) = \mathbf{q}_{\mathbf{k}}^* \cdot [(\mathbf{k} \times \mathbf{q}_{\mathbf{k}}) \times \mathbf{k}] = (\mathbf{q}_{\mathbf{k}}^* \cdot \mathbf{q}_{\mathbf{k}}) \, k^2, \tag{4.174}$$

we easily find that

$$H = \frac{1}{8\pi} \int_V d^3x \, \left(E^2 + B^2 \right) = \frac{V}{8\pi} \sum_{\mathbf{k}} \left(|\dot{\mathbf{q}}_{\mathbf{k}}|^2 + k^2 |\mathbf{q}_{\mathbf{k}}|^2 \right). \tag{4.175}$$

The wave equation satisfied by $\mathbf{A}(t, \mathbf{x})$ implies that $\mathbf{q}_{\mathbf{k}}(t)$ satisfies the harmonic oscillator equation

$$\ddot{\mathbf{q}}_{\mathbf{k}} + k^2 \mathbf{q}_{\mathbf{k}} = 0, \quad \mathbf{q}_{\mathbf{k}} \propto e^{-ikt}. \tag{4.176}$$

The Hamiltonian can be expressed in terms of $\mathbf{a}_{\mathbf{k}}$ by using the relation

$$\mathbf{q}_{\mathbf{k}} = \mathbf{a}_{\mathbf{k}} + \mathbf{a}_{-\mathbf{k}}^* \tag{4.177}$$

which follows from (4.170) and the fact $\mathbf{q}_{-\mathbf{k}} = \mathbf{q}_{\mathbf{k}}^*$. From (4.176) we find that $\mathbf{a}_{\mathbf{k}}$ also satisfies the harmonic oscillator equation. It is conventional to take $\mathbf{a}_{\mathbf{k}} \propto \exp(-ikt)$ so that $\mathbf{a}_{\mathbf{k}}^* \propto \exp(ikt)$ and

$$\dot{\mathbf{q}}_{\mathbf{k}} = -ik \left(\mathbf{a}_{\mathbf{k}} - \mathbf{a}_{-\mathbf{k}}^* \right). \tag{4.178}$$

When (4.177) and (4.178) are substituted into (4.175) terms like $\mathbf{a}_{\mathbf{k}} \cdot \mathbf{a}_{-\mathbf{k}}$ cancel out, while terms like $\mathbf{a}_{\mathbf{k}} \cdot \mathbf{a}_{\mathbf{k}}^*$ and $\mathbf{a}_{-\mathbf{k}} \cdot \mathbf{a}_{-\mathbf{k}}^*$ differ only in relabelling. Hence, we find

$$H = \sum_{\mathbf{k}} \frac{k^2 V}{2\pi} \mathbf{a}_{\mathbf{k}} \cdot \mathbf{a}_{\mathbf{k}}^*. \tag{4.179}$$

We now define two new variables $\mathbf{Q}_{\mathbf{k}}$ and $\mathbf{P}_{\mathbf{k}}$ by the relations

$$\mathbf{Q}_{\mathbf{k}} = \left(\frac{V}{4\pi} \right)^{1/2} (\mathbf{a}_{\mathbf{k}} + \mathbf{a}_{\mathbf{k}}^*), \quad \mathbf{P}_{\mathbf{k}} = -ik \left(\frac{V}{4\pi} \right)^{1/2} (\mathbf{a}_{\mathbf{k}} - \mathbf{a}_{\mathbf{k}}^*) = \dot{\mathbf{Q}}_{\mathbf{k}}. \tag{4.180}$$

Then H becomes

$$H = \sum_{\mathbf{k}} \frac{1}{2} \left(\mathbf{P}_{\mathbf{k}}^2 + k^2 \mathbf{Q}_{\mathbf{k}}^2 \right). \tag{4.181}$$

Finally, we note that $\mathbf{P_k}$ and $\mathbf{Q_k}$ are also orthogonal to $\mathbf{k}$ and hence have only two independent components in the plane perpendicular to $\mathbf{k}$. Denoting these components by $\{Q_{\mathbf{k}\alpha}, P_{\mathbf{k}\alpha}\}$ with $\alpha = 1, 2$, we obtain the final expression:

$$H = \sum_{\alpha=1}^{2} \sum_{\mathbf{k}} \frac{1}{2} \left(P_{\mathbf{k}\alpha}^2 + k^2 Q_{\mathbf{k}\alpha}^2 \right) = \sum_{\alpha} \sum_{\mathbf{k}} H_{\mathbf{k}\alpha}. \qquad (4.182)$$

This result shows that the Hamiltonian governing the free electromagnetic field can be expressed as a sum of Hamiltonians for harmonic oscillators, with each oscillator labelled by a wave vector $\mathbf{k}$ and polarization index α. To construct the quantum theory of the electromagnetic field, we only have to quantize each of these oscillators. This will make $P_{\mathbf{k}\alpha}$ and $Q_{\mathbf{k}\alpha}$ operators with the commutation rules

$$\left[Q_{\mathbf{j}\beta}, P_{\mathbf{k}\alpha} \right] = i\delta_{\mathbf{kj}}\delta_{\alpha\beta}. \qquad (4.183)$$

Using the relation between $a_{\mathbf{k}\alpha}$ and $Q_{\mathbf{k}\alpha}$ we can find the corresponding commutation rules for $a_{\mathbf{k}\alpha}$. We obtain

$$\left[a_{\mathbf{k}\alpha}, a_{\mathbf{p}\beta}^{\dagger} \right] = \left(\frac{2\pi}{V\omega_{\mathbf{k}}} \right) \delta_{\mathbf{kp}}\delta_{\alpha\beta}, \qquad (4.184)$$

with $\omega_{\mathbf{k}} = |\mathbf{k}|$.

To exhibit the operator nature of $\mathbf{A}(\mathbf{x}, t)$ and $\mathbf{a_k}$ it is best to separate out the time dependence in $\mathbf{a_k}(t)$ and write

$$\mathbf{a}_{\mathbf{k}\alpha} = c_{\mathbf{k}\alpha} \hat{\mathbf{e}}_{\alpha} \left(\frac{2\pi}{V\omega_{\mathbf{k}}} \right)^{1/2} \exp\left(-i\omega_{\mathbf{k}}t\right), \qquad (4.185)$$

where $\hat{\mathbf{e}}_1$ and $\hat{\mathbf{e}}_2$ are two unit vectors chosen such that $[\hat{\mathbf{e}}_1, \hat{\mathbf{e}}_2, \mathbf{k}/k]$ form an orthogonal system and $c_{\mathbf{k}\alpha} = \left(V\omega_{\mathbf{k}}/2\pi \right)^{1/2} a_{\mathbf{k}\alpha}$. From (4.184) it follows that

$$\left[c_{\mathbf{k}\alpha}, c_{\mathbf{p}\beta}^{\dagger} \right] = \delta_{\mathbf{kp}}\delta_{\alpha\beta}, \qquad (4.186)$$

which is the standard commutation rule for the creation and annihilation operators of a harmonic oscillator. We can, therefore, treat $c_{\mathbf{k}\alpha}$ and $c_{\mathbf{k}\alpha}^{\dagger}$ as annihilation and creation operators for photons with wave vectors $\mathbf{k}$ and polarization state α.

(b) The interaction Hamiltonian between the atom and the electromagnetic field is

$$H_{\mathrm{I}} = \int d^3\mathbf{x}\, \mathbf{J} \cdot \mathbf{A} = \int d^3\mathbf{x}\, \mathbf{J} \cdot \sum_{\mathbf{k}\alpha} \left(c_{\mathbf{k}\alpha} \mathbf{A}_{\mathbf{k}\alpha} + c_{\mathbf{k}\alpha}^{\dagger} \mathbf{A}_{\mathbf{k}\alpha}^{*} \right) \equiv H_{\mathrm{abs}} + H_{\mathrm{emis}}. \qquad (4.187)$$

Since the emission of a photon involves *creating* an extra photon, this process is governed by the term proportional to $c_{\mathbf{k}\alpha}^{\dagger}$. The emission process, in which the quantum system makes the transition from $|E_{\mathrm{i}}\rangle$ to $|E_{\mathrm{f}}\rangle$ and the electromagnetic field goes from a state $|n_{k\alpha}\rangle$ to $|n_{k\alpha}+1\rangle$, is governed by the amplitude

$$\mathscr{A} = \int dt \, \langle E_f | \langle n_{k\alpha}+1|H_I|n_{k\alpha}\rangle|E_i\rangle = \int dt \int d^3x \, \langle E_f|\mathbf{J}\cdot\mathbf{A}^*_{k\alpha}|E_i\rangle \, (n_{k\alpha}+1)^{1/2},$$

(4.188)

where we have used the fact that $\langle n_{k\alpha}+1|\mathbf{A}|n_{k\alpha}\rangle = \mathbf{A}^*_{k\alpha}\,(n_{k\alpha}+1)^{1/2}$. We will now use the expansion

$$\mathbf{A}^*_{k\alpha} = \left(\frac{2\pi}{V\omega_k}\right)^{1/2} \hat{\mathbf{e}}_\alpha \exp(-i\,(\omega t - \mathbf{k}\cdot\mathbf{x}))$$

(4.189)

and the fact that the energy eigenstate has the time dependence $\exp(-iEt)$, to rewrite $\mathscr{A}$ as

$$\mathscr{A} = \left(\frac{2\pi}{V\omega_k}\right)^{1/2} (n_{k\alpha}+1)^{1/2} \int dt \int d^3x \; \phi_f^*(\mathbf{x})\,(\mathbf{J}\cdot\hat{\mathbf{e}}_\alpha \exp(-i\mathbf{k}\cdot\mathbf{x}))$$
$$\times \phi_i(\mathbf{x}) \exp(-i\,(E_i - E_f - \omega)\,t),$$

(4.190)

where $\phi_i(\mathbf{x}), \phi_f(\mathbf{x})$ denote the wave functions of the two states. Denoting the matrix element

$$\langle E_f|e^{-i\mathbf{k}\cdot\mathbf{x}}\mathbf{J}|E_i\rangle = \left(\frac{q}{m}\right)\langle E_f|e^{-i\mathbf{k}\cdot\mathbf{x}}\mathbf{p}|E_i\rangle$$

(4.191)

by the symbol $\mathbf{M}_{fi}$, we can express the probability of transition $|\mathscr{A}|^2$ as

$$|\mathscr{A}|^2 = P = \int_{-\infty}^{+\infty} dt \int_{-\infty}^{+\infty} dt' |\hat{\mathbf{e}}_{k\alpha}\cdot\mathbf{M}_{fi}|^2 \cdot \left(\frac{2\pi}{V\omega_k}\right) (n_{k\alpha}+1) \exp(-i(E_i - E_f - \omega)(t - t'))$$

$$= \int_{-\infty}^{+\infty} dT \int_{-\infty}^{+\infty} d\tau \left(\frac{2\pi}{V\omega_k}\right)(n_{k\alpha}+1)|\hat{\mathbf{e}}_{k\alpha}\cdot\mathbf{M}_{fi}|^2 \exp(-i(E_i - E_f - \omega)\tau), \quad (4.192)$$

where we have introduced the variables $T = (t+t')$ and $\tau = (t - t')$.

The integration over τ gives a delta function expressing conservation of energy. We interpret the (divergent) integration over T as giving a finite rate of transition by the usual rule

$$\int_{-\infty}^{+\infty} dt = 2\pi\delta\,(0) = \lim_{T\to\infty}\,(2\pi T).$$

(4.193)

Then we get a finite rate for the emission of photons:

$$R \equiv \frac{dP}{dt} = 2\pi \left(\frac{2\pi}{V\omega_k}\right)(n_{k\alpha}+1)|\hat{\mathbf{e}}_{k\alpha}\cdot\mathbf{M}_{fi}|^2 2\pi\delta\,(E_i - E_f - \omega).$$

(4.194)

This expression gives the rate for the emission of a photon with a wave vector $\mathbf{k}$ and polarization α. It is conventional to multiply this rate by the density of states per unit energy interval and write

$$
\begin{aligned}
\left[\frac{dP}{dt\,d\omega\,d\Omega}\right]_{\text{emis}} &= \frac{dP}{dt}\cdot\frac{dN}{d\omega\,d\Omega} = \frac{dP}{dt}\cdot\frac{V}{(2\pi)^3}\cdot\frac{k^2dk\,d\Omega}{d\omega\,d\Omega} = \frac{dP}{dt}\cdot\frac{V\omega^2}{(2\pi)^3}\\
&= 2\pi\left(\frac{2\pi}{V\omega}\right)(n_{\mathbf{k}\alpha}+1)|\hat{\mathbf{e}}_{\mathbf{k}\alpha}\cdot\mathbf{M}_{\text{fi}}|^2 2\pi\,\delta(\omega-\omega_{\text{fi}})\cdot\frac{V\omega^2}{(2\pi)^3}\quad (4.195)\\
&= \left(\frac{\omega}{2\pi\hbar c^3}\right)(n_{\mathbf{k}\alpha}+1)|\hat{\mathbf{e}}_{\mathbf{k}\alpha}\cdot\mathbf{M}_{\text{fi}}|^2\,\delta(\omega-\omega_{\text{fi}}).
\end{aligned}
$$

In the last line we have introduced the $\hbar$ and c factors appropriately. (The correctness of this reinsertion can be verified as follows. The left-hand side is dimensionless. On the right-hand side, $\omega\delta(\omega-\omega_{\text{fi}})$ is dimensionless; the matrix element has dimensions of (q^2v^2). Since $(q^2v^2/\hbar c^3) = (q^2/\hbar c)(v/c)^2$ is dimensionless the $\hbar c^3$ factor in the last line is correct.)

Note that the final result is proportional to $(n_{\mathbf{k}\alpha}+1)$. The atom in an excited state has a certain rate of decay to lower states with the emission of photons, even when $n_{\mathbf{k}\alpha}=0$. This process is called 'spontaneous emission'. When $n_{\mathbf{k}\alpha}\neq 0$, this probability is enhanced by an extra term proportional to $n_{\mathbf{k}\alpha}$; this process is called 'induced emission'.

The analysis for the absorption rate of photons is identical except that only the annihilation operator $c_{\mathbf{k}\alpha}$ contributes. Since $\langle n_{\mathbf{k}\alpha}-1|c_{\mathbf{k}\alpha}|n_{\mathbf{k}\alpha}\rangle = n_{\mathbf{k}\alpha}^{1/2}$ we obtain $n_{\mathbf{k}\alpha}$ rather than $(n_{\mathbf{k}\alpha}+1)$ in the final result:

$$
\left[\frac{dP}{d\Omega\,dt\,d\omega}\right]_{\text{abs}} = \left(\frac{\omega}{2\pi\hbar c^3}\right)n_{\mathbf{k}\alpha}|\hat{\mathbf{e}}_{\mathbf{k}\alpha}\cdot\mathbf{M}_{\text{fi}}|^2\,\delta(\omega-\omega_{\text{fi}}).\quad (4.196)
$$

4.12 Classical limit and dipole approximation

(a) When $\lambda\gg a$ – where a is the typical atomic size – we can use the approximation $\exp(\pm i\mathbf{k}\cdot\mathbf{x})\cong 1$ in the matrix elements. Then $\mathbf{M}_{\text{fi}}$ can be expressed in terms of the matrix element $\mathbf{p}_{\text{fi}}$ of the momentum operator. But since $\mathbf{p}=(im/\hbar)[H_0,\mathbf{x}]$ it follows that $\mathbf{p}_{\text{fi}} = (im/\hbar)[H_0,\mathbf{x}]_{\text{fi}} = im\omega_{\text{fi}}\mathbf{x}_{\text{fi}}$. Therefore, the rate of spontaneous emission (corresponding to $n_{\mathbf{k}\alpha}=0$) becomes

$$
\begin{aligned}
\frac{dP}{d\Omega\,dt} &= \left(\frac{\omega}{2\pi\hbar c^3}\right)\frac{q^2}{m^2}|\hat{\mathbf{e}}_{\mathbf{k}\alpha}\cdot\mathbf{p}_{\text{fi}}|^2\,[\delta(\omega-\omega_{\text{fi}})\,d\omega]\\
&= \left(\frac{\omega^3}{2\pi\hbar c^3}\right)|\hat{\mathbf{e}}_{\mathbf{k}\alpha}\cdot q\mathbf{x}_{\text{fi}}|^2\,[\delta(\omega-\omega_{\text{fi}})\,d\omega]\\
&= \frac{\omega^3}{2\pi\hbar c^3}|\hat{\mathbf{e}}_{\mathbf{k}\alpha}\cdot\mathbf{d}_{\text{fi}}|^2\,[\delta(\omega-\omega_{\text{fi}})\,d\omega]\quad (4.197)
\end{aligned}
$$

where $\mathbf{d}_{\text{fi}} = \langle f|q\mathbf{x}|i\rangle$ is the dipole moment operator. Integration over ω will replace ω^3 by ω_{fi}^3. To average this expression over the polarizations we use the relation

$$
\langle(\mathbf{a}\cdot\mathbf{e})(\mathbf{b}\cdot\mathbf{e}^*)\rangle = \frac{1}{2}[\mathbf{a}\times\mathbf{n}]\cdot[\mathbf{b}\times\mathbf{n}],\quad (4.198)
$$

where $\mathbf{n} = (\mathbf{k}/k)$. Then,

$$\left\langle \frac{dP}{dt\,d\Omega} \right\rangle = \frac{\omega_{\text{fi}}^3}{4\pi\hbar c^3} |\mathbf{n} \times \mathbf{d}_{\text{fi}}|^2 = \frac{\omega_{\text{fi}}^3}{4\pi\hbar c^3} |\mathbf{d}_{\text{fi}}|^2 \sin^2\theta. \tag{4.199}$$

The average *energy* emitted per polarization state is found by multiplying by $\hbar\omega_{\text{fi}}$:

$$\left\langle \frac{dE}{dt\,d\Omega} \right\rangle = \hbar\omega_{\text{fi}} \left\langle \frac{dP}{dt\,d\Omega} \right\rangle = \frac{\omega_{\text{fi}}^4}{4\pi c^3} |\mathbf{d}_{\text{fi}}|^2 \sin^2\theta. \tag{4.200}$$

This agrees with the classical result in (4.30), provided we identify $|\ddot{\mathbf{d}}|^2$ with $\omega_{\text{fi}}^4 |\mathbf{d}_{\text{fi}}|^2$. Such an identification will arise automatically if we assume that the charged particle behaves as a harmonic oscillator with the equation of motion $\ddot{\mathbf{x}} = -\omega_{\text{fi}}^2 \mathbf{x}$.

(b) The absorption rate for unpolarized, isotropic radiation can be obtained from (4.199) by multiplying by $n_{\mathbf{k}\alpha} = [n(\omega)/2]$. To take two polarization states into account we multiply by 2 and finally integrate over the solid angle by replacing $\sin^2\theta$ by $4\pi\langle\sin^2\theta\rangle = (8\pi/3)$. This gives

$$\left(\frac{dP}{dt} \right)_{\text{abs}} = \frac{\omega_{\text{fi}}^3}{4\pi\hbar c^3} |\mathbf{d}_{\text{fi}}|^2 \cdot \left(\frac{n}{2} \right) \cdot 2 \cdot \frac{8\pi}{3} = \frac{2}{3} \frac{q^2}{\hbar c^3} n\omega_{\text{fi}}^3 |\mathbf{x}_{\text{fi}}|^2. \tag{4.201}$$

(c) Substituting the definition

$$\sigma_{\text{bb}}(\omega) = \frac{\pi q^2}{mc} f \left[2\pi\, \delta(\omega - \omega_{\text{fi}}) \right] \tag{4.202}$$

into

$$\left(\frac{dP_{\text{if}}}{dt} \right)_{\text{abs}} = \int_0^\infty \sigma_{\text{bb}}(\omega) \left[cn(\omega) \right] \frac{4\pi (\omega/c)^2 d(\omega/c)}{(2\pi)^3} \tag{4.203}$$

we obtain

$$\left(\frac{dP_{\text{if}}}{dt} \right)_{\text{abs}} = \left(\frac{q^2}{mc} \right) \cdot \frac{\omega_{\text{fi}}^2}{c^2} \cdot f n(\omega_{\text{fi}}). \tag{4.204}$$

Comparing this with (4.201) we find that

$$f = \frac{2}{3} \frac{m\omega_{\text{fi}}}{\hbar} |\mathbf{x}_{\text{fi}}|^2 = \frac{2}{3} \left[\frac{m\omega_{\text{fi}}^2 |\mathbf{x}_{\text{fi}}|^2}{\hbar\omega_{\text{fi}}} \right]. \tag{4.205}$$

If we treat the electron as a classical oscillator with frequency ω, then it will have an energy $m\omega^2 a^2$, where a is the amplitude of the oscillation. The energy of radiation, $\hbar\omega$, emitted or absorbed will be about the same order. Since we expect $a \cong |\mathbf{x}_{\text{fi}}|$, it is reasonable to expect f to be of order unity. This is indeed true for most cases. Inverting the argument, we may express $|\mathbf{x}_{\text{fi}}|^2$ in terms of f and set $f \cong 1$ for order of magnitude estimates.

Since

$$|\mathbf{x}_{\text{fi}}|^2 = \frac{3}{2} \left(\frac{\hbar}{m\omega} \right) f, \tag{4.206}$$

the absorption and emission rates become

$$\left(\frac{dP_{abs}}{dt}\right) = \left(\frac{q^2}{mc^3}\right) n\omega^2 f = \left(\frac{2q^2}{mc^3}\right) n_\alpha(\omega)\,\omega^2 f, \qquad (4.207)$$

$$\left(\frac{dP_{emis}}{dt}\right) = \left(\frac{2q^2}{mc^3}\right)[1 + n_\alpha(\omega)]\,\omega^2 f, \qquad (4.208)$$

with $n_\alpha(\omega) = [n(\omega)/2]$ being the number of photons per polarization state.

(d) To compare f we have to evaluate the matrix element x_{ij} between two atomic states. When the final state is degenerate, we have to sum over all possible final states with the same energy. In our case, the initial state has the wave function corresponding to $n = 1, l = 0, m = 0$:

$$\Psi_{1,0,0} = \frac{1}{\sqrt{\pi a^3}} \exp\left(-\frac{r}{a}\right), \quad a = \left(\frac{\hbar^2}{me^2}\right). \qquad (4.209)$$

The final states could be any of the following:

$$\Psi_{2,1,-1} = \frac{R_{21}(r)}{r} Y_{1,-1}(\theta,\phi), \quad \Psi_{2,1,1} = \frac{R_{21}(r)}{r} Y_{1,1}(\theta,\phi), \quad \Psi_{2,1,0} = \frac{R_{21}(r)}{r} Y_{1,0}(\theta,\phi). \qquad (4.210)$$

The matrix elements x, y and z between these states can be now computed by direct integration. However, it is possible to simplify the analysis by the following trick. Note that

$$x \pm iy = r(\sin\theta)\,e^{\pm i\phi} = r\left(\frac{8\pi}{3}\right)^{1/2} Y_{1,\pm 1},$$
$$z = r(\cos\theta) = r\left(\frac{4\pi}{3}\right)^{1/2} Y_{1,0} \qquad (4.211)$$

and

$$|x_{fi}|^2 = |z_{fi}|^2 + \frac{1}{2}|(x+iy)_{if}|^2 + \frac{1}{2}|(x-iy)_{if}|^2. \qquad (4.212)$$

The matrix element then becomes

$$|x_{fi}|^2 = \frac{1}{18}\frac{1}{a^8}\left(A^2 B^2\right), \qquad (4.213)$$

with

$$A = \int_0^\infty r^4 e^{-3r/2a}\,dr = \left(\frac{2}{3}\right)^5 24a^5 = \frac{2^8}{3^4}a^5 \qquad (4.214)$$

and

$$B = \left|\int Y_{1,m}^* Y_{1,1}\,d\Omega\right|^2 + \left|\int Y_{1,m}^* Y_{1,0}\,d\Omega\right|^2 + \left|\int Y_{1,m}^* Y_{1,-1}\,d\Omega\right|^2. \qquad (4.215)$$

In *B*, each integral is non-zero only for a particular value of *m* and – for that value – the integral is unity. Hence,

$$f = 2 \times 3 \times \left(\frac{2}{3} \frac{m\omega}{\hbar} \right) \times \frac{1}{18} \frac{1}{a^8} \times \frac{2^{16}}{3^8} a^{10}. \tag{4.216}$$

The factors 2 and 3 in front account for two spin states of the electron and the three possible *m*-values for the final state. So

$$f = \frac{2^{17}}{3^{10}} \left(\frac{a^2 m\omega}{\hbar} \right) = \frac{2^{17}}{3^{10}} \left(\frac{\hbar^3 \omega}{me^4} \right) = \frac{2^{17}}{3^{10}} \frac{(\hbar\omega)}{(me^4/\hbar^2)}. \tag{4.217}$$

The transition energy between $n = 2$ and $n = 1$ is

$$\hbar\omega = \frac{1}{2} \frac{me^4}{\hbar^2} \left(1 - \frac{1}{4} \right) = \frac{3}{8} \frac{me^4}{\hbar^2}, \tag{4.218}$$

giving

$$f = \frac{2^{14}}{3^9} \cong 0.83, \tag{4.219}$$

which is of order unity.

4.13 Photoionization

(a) The flux of incident photons is $(n/V) c$ times the density of states for photons: $[V d^3 p_\gamma / (2\pi\hbar)^3] = [4\pi (\omega/c)^2 d(\omega/c) / (2\pi)^3]$. This flux, multiplied by the cross-section σ_{bf}, should give the rate for a process – here the rate of ionization. Since the free electron is produced by the process, we should also include a density of state factor for the electron. Thus,

$$\left(\frac{dP}{dt} \right) \times \frac{V d^3 p_e}{(2\pi\hbar)^3} = \sigma_{bf}(\omega) \left[\frac{n}{V} c \right] \left[\frac{V d^3 p_\gamma}{(2\pi\hbar)^3} \right]. \tag{4.220}$$

This simplifies to the expression in the question.

(b) The initial state of the electron is

$$\phi_i(\mathbf{x}) = \left(\frac{1}{\pi a_0^3} \right)^{1/2} \exp\left(-\frac{|\mathbf{x}|}{a_0} \right), \qquad a_0 = \frac{\hbar^2}{me^2}. \tag{4.221}$$

while the final state can be approximated as a plane wave

$$\phi_f(\mathbf{x}) = \left(\frac{1}{V} \right)^{1/2} \exp(-i\mathbf{k}_e \cdot \mathbf{x}) \tag{4.222}$$

with momentum $\mathbf{p}_e = \hbar\mathbf{k}_e$. (We denote the electronic charge by q rather than by e in this problem.) The matrix element of $e^{i\mathbf{k}\cdot\mathbf{x}}\mathbf{p} = -i\hbar e^{i\mathbf{k}\cdot\mathbf{x}}\nabla$ between ϕ_i and ϕ_f

can be evaluated easily using the result

$$\int d^3x \exp\left[i\left(\mathbf{k} - \mathbf{k}_e\right) \cdot \mathbf{x} - |\mathbf{x}|/a_0\right]$$

$$= 2\pi \int_0^\infty dr\, r e^{-r/a_0} \int_{-1}^{+1} d\mu e^{ir\mu|\mathbf{k} - \mathbf{k}_e|} = \frac{8\pi a_0^3}{\left[1 + a_0^2|\mathbf{k} - \mathbf{k}_e|^2\right]^2}. \tag{4.223}$$

The matrix element becomes

$$|\langle \phi_f|e^{i\mathbf{k}\cdot\mathbf{x}}\hat{\mathbf{e}} \cdot \mathbf{p}|\phi_i\rangle|^2 = 64\pi \frac{\hbar^2 a_0^3}{V}\left(\mathbf{k}_e \cdot \hat{\mathbf{e}}_\alpha\right)^2 \left(1 + a_0^2|\mathbf{k} - \mathbf{k}_e|^2\right)^{-4}. \tag{4.224}$$

Substituting into our general formula for the rate of absorption

$$\left(\frac{dP_\alpha}{d\Omega\, dt\, d\omega}\right) = \left(\frac{\omega}{2\pi\hbar c^3}\right) n_{\mathbf{k}\alpha} \cdot \frac{q^2}{m^2} \cdot |\langle E_f|e^{i\mathbf{k}\cdot\mathbf{x}}\hat{\mathbf{e}}_\alpha \cdot \mathbf{p}|E_i\rangle|^2 \delta\left(\omega - \omega_{fi}\right) \tag{4.225}$$

we obtain the result

$$\left(\frac{dP_\alpha}{d\Omega\, dt\, d\omega}\right) = 32\left(\frac{q}{m}\right)^2 \left(\frac{\hbar a_0^3}{Vc^3}\right) \frac{\left(\omega n_{\mathbf{k}\alpha}\right)\left(\mathbf{k}_e \cdot \hat{\mathbf{e}}_\alpha\right)^2}{\left(1 + a_0^2|\mathbf{k} - \mathbf{k}_e|^2\right)^4} \delta\left(\omega - \omega_{fi}\right). \tag{4.226}$$

To proceed further we have to integrate over $d\omega, d\Omega$ and sum over the polarization states. Taking $\mathbf{k}$ along the z-axis, $\hat{\mathbf{e}}_\alpha$ along the x-axis and $\mathbf{k}_e$ along the direction specified by (θ, ϕ), one has the relations

$$\mathbf{k}_e \cdot \hat{\mathbf{e}}_\alpha = k_e \sin\theta \cos\phi; \quad |\mathbf{k} - \mathbf{k}_e|^2 = k^2 + k_e^2 - 2kk_e \cos\theta. \tag{4.227}$$

Further, since we expect $\hbar k \ll m_e c$ and $\hbar k_e \ll m_e c$ in the non-relativistic limit, we can write

$$1 + a_0^2|\mathbf{k} - \mathbf{k}_e|^2 \cong a_0^2 \frac{2m\omega}{\hbar}, \tag{4.228}$$

where we have used the fact that

$$k_e^2 = \frac{(2mE_f)}{\hbar^2} = \frac{2m}{\hbar^2}\left(\hbar\omega_{fi} - \frac{q^2}{2a_0}\right) = \frac{2mck}{\hbar} - \frac{1}{a_0^2}. \tag{4.229}$$

The angular integrations can now be performed using the result

$$\int_0^{2\pi} d\phi \int_0^\pi \sin^2\theta \cos^2\phi \sin\theta\, d\theta\, d\phi = \frac{4\pi}{3}, \tag{4.230}$$

giving

$$\left(\frac{dP}{dt}\right) = \left(\frac{8\pi}{3}\right)\left(\frac{a_0^3}{V}\right)\left(\frac{q^2}{\hbar c}\right)\left(\frac{\hbar k_e}{mc}\right)^2 \left(\frac{\hbar}{ma_0^2}\right)^4 \frac{n(\omega)}{\omega^3}. \tag{4.231}$$

The corresponding cross-section is

$$\sigma_{bf}(\omega) = \frac{8\pi}{3}\left(\frac{q^2}{mc}\right)\left(\frac{\hbar}{ma_0^2}\right)^4 \frac{1}{\omega^5}(a_0 k_e)^3 \tag{4.232}$$

with

$$k_e^2 = \frac{2m}{\hbar^2}\left(\hbar\omega - \frac{q^2}{2a_0}\right).\qquad(4.233)$$

We have assumed in our analysis that the final wave function of the electron is that of a free particle. This requires the final energy of the electron to be far greater than (q^2/a_0) or, equivalently, $\hbar\omega \gg q^2/a_0$. In that case $k_e^2 \cong (2m\omega/\hbar)$ and the final result becomes

$$\sigma_{bf} = \frac{2^8}{3}\left(\pi a_0^2\right)\left(\frac{q^2}{\hbar c}\right)\left(\frac{E_0}{\hbar\omega}\right)^{7/2}.\qquad(4.234)$$

(c) Strictly speaking, the final state of the electron is a scattering state in the $(-q^2/r)$ potential with an energy $E_p = (p^2/2m)$. Such a state is in general described as a superposition of partial waves with all values of l. However, since the photon has an angular momentum of one unit and the electron was initially in an s-state with zero angular momentum, the final state of the electron can only be a p-state. Hence, in the standard expansion of the scattering state,

$$\Psi_p = \frac{1}{p}\left(\frac{\pi}{2}\right)^{1/2}\sum_{l=0}^{\infty}i^l\,(2l+1)\,e^{-i\delta_l}R_{pl}\,(r)\,P_l\,(\hat{\mathbf{n}}\cdot\hat{\mathbf{n}}_1)\qquad(4.235)$$

where $R_{pl}(r)$ is the radial part of the scattering state with $\hat{\mathbf{n}} = (\mathbf{p}/p)$ and $\hat{\mathbf{n}}_1 = (\mathbf{r}/r)$, we only need to retain the $l = 1$ state. Omitting the unimportant phase factors, we can, therefore, write the final state as

$$\Psi_p = \frac{3}{p}\left(\frac{\pi}{2}\right)^{1/2}R_{p1}\,(r)\,(\hat{\mathbf{n}}\cdot\hat{\mathbf{n}}_1)\,.\qquad(4.236)$$

In this case, the relevant matrix element is

$$\hat{\mathbf{e}}_\alpha\cdot\mathbf{p}_{fi} = \frac{3}{\sqrt{2}}\left(\frac{1}{a_0}\right)^{5/2}\frac{1}{p}\int d\Omega_1\int r^2 dr\,(\hat{\mathbf{n}}_1\cdot\hat{\mathbf{n}})\,(\hat{\mathbf{n}}_1\cdot\hat{\mathbf{e}}_\alpha)\,e^{-r/a_0}R_{p1}\,(r)$$

$$= \frac{2^{3/2}}{pa_0^{5/2}}\,(\hat{\mathbf{n}}\cdot\hat{\mathbf{e}}_\alpha)\int_0^{\infty} r^2 dr e^{-r/a_0}R_{p1}\,(r)\,.\qquad(4.237)$$

The radial part of the wave function in a scattering state is

$$R_{p1} = \frac{2}{3a_0}\left[\frac{1+\mu^2}{\mu\left(1-e^{-2\pi\mu}\right)}\right]^{1/2}pre^{-ipr}F\left(2+i\mu,4,2ipr\right),\qquad(4.238)$$

where F is the hypergeometric function and $\mu = (q^2/\hbar c)\,(c/v)$. To evaluate the integral we need to use two results:

$$\int_0^{\infty} dz\,e^{-\lambda z}z^{\gamma-1}F\,(\alpha,\gamma,kz) = \Gamma\,(\gamma)\,\lambda^{\alpha-\gamma}\,(\lambda-k)^{-\alpha}\,,\qquad(4.239)$$

$$\left(\frac{\mu+i}{\mu-i}\right)^{i\mu} = e^{-2\mu\cot^{-1}\mu}\,.\qquad(4.240)$$

Using these, we find that

$$\hat{\mathbf{e}}_\alpha \cdot \mathbf{p}_{fi} = \frac{2^{7/2}\pi\mu^3\,(\hat{\mathbf{n}}\cdot\hat{\mathbf{e}}_\alpha)}{p^{1/2}\,\left(1+\mu^2\right)^{3/2}}\,\frac{\exp(-2\mu\cot^{-1}\mu)}{\left(1-\exp(-2\pi\mu)\right)^{1/2}}\,. \tag{4.241}$$

From the energy conservation we have

$$\hbar\omega = \frac{p^2}{2m}\left(1+\mu^2\right). \tag{4.242}$$

The rest of the analysis proceeds exactly as before and we obtain

$$\sigma_{bf} = \frac{2^9\pi^2}{3}\left(\frac{q^2}{\hbar c}\right)a_0^2\left(\frac{E_0}{\hbar\omega}\right)^4\frac{\exp(-4\mu\cot^{-1}\mu)}{1-\exp(-2\pi\mu)}\,. \tag{4.243}$$

This result is valid for both $\hbar\omega \cong E_0$ as well as $\hbar\omega \gg E_0$ provided $\hbar\omega \ll mc^2$. When $\hbar\omega \gg E_0$, that is, when $\mu \ll 1$,

$$\frac{\exp\left(-4\mu\cot^{-1}\mu\right)}{1-\exp(-2\pi\mu)} \cong \frac{1}{2\pi\mu} = \frac{1}{2\pi}\left(\frac{q^2}{\hbar c}\right)^{-1}\frac{(2m\hbar\omega)^{1/2}}{mc} = \frac{1}{2\pi}\left(\frac{\hbar\omega}{E_0}\right)^{1/2} \tag{4.244}$$

and the cross-section reduces to the expression derived before:

$$\sigma_{bf} = \frac{2^8\left(\pi a_0^2\right)}{3}\left(\frac{q^2}{\hbar c}\right)\left(\frac{E_0}{\hbar\omega}\right)^{7/2}. \tag{4.245}$$

On the other hand, when $\hbar\omega \gtrsim E_0$ (large μ) we obtain

$$\frac{\exp(-4\mu\cot^{-1}\mu)}{1-\exp(-2\pi\mu)} \cong e^{-4} \tag{4.246}$$

and

$$\sigma_{bf} = \left(\frac{2^9}{3e^4}\right)\left(\pi^2 a_0^2\right)\left(\frac{q^2}{\hbar c}\right). \tag{4.247}$$

The following point may be noted. Results (4.245) and (4.247) show that the exact ω dependence of σ_{bf} is complicated and has different limiting forms for $\hbar\omega \cong E_0$ and $\hbar\omega \gg E_0$. In general, it is conventional to write the result for σ_{bf} as

$$\sigma_{bf}(\omega) = \frac{8\pi}{3\sqrt{3}}\left(\frac{q^2}{\hbar c}\right)^5\left(\frac{mc^4}{\hbar\omega^3}\right)g_{bf}(\omega), \tag{4.248}$$

where $g_{bf}(\omega)$ is called a Gaunt factor. It varies slowly with ω and goes as $\omega^{-1/2}$ if $\hbar\omega \gg E_0$. If $g_{bf}(\omega)$ is nearly constant, $\sigma_{bf}(\omega) \propto \omega^{-3}$; such an approximation is often made in various astrophysical contexts.

(d) In equilibrium, the reaction $e^- + p \to H + \gamma$ should be balanced by the reaction $H + \gamma \to e^- + p$. This means that

$$p_i^2 g_i \sigma_{i\to f} = p_f^2 g_f \sigma_{f\to i} \tag{4.249}$$

where p_i, p_f are the momenta of relative motion of the particles and g_i and g_f take into account the spin degrees of the freedom. In our case, $g_i = 2 \times 2$ for electron and proton spins and $g_f = 2 \times 2 \times 2$ for the electron, proton and photon spins. So

$$\sigma_{rec} = \left(\frac{g_f}{g_i} \right) \left(\frac{p_f}{p_i} \right)^2 \sigma_{bf} = \frac{2 (\hbar k)^2}{(mv)^2} \sigma_{bf} = \frac{2 (\hbar \omega)^2}{(mv)^2} \left(\frac{\sigma_{bf}}{c^2} \right), \qquad (4.250)$$

where v is the velocity of the incident electron and ω is the frequency of the emitted photon.

(e) When the temperature of the electron gas is much lower than the ionization potential of the hydrogen atom, we can write

$$\hbar \omega = \frac{\hbar^2 k_e^2}{2m} + \frac{mq^2}{\hbar^2 a_0} \simeq \frac{mq^2}{\hbar^2 a_0} = \frac{m^2 q^4}{\hbar^4}. \qquad (4.251)$$

In this case, using (4.247) in (4.250), we find that.

$$\sigma_{rec} = \left(\frac{2^{10} \pi^2}{3 e^4} \right) \left(\frac{q^2}{\hbar c} \right) \left(\frac{a_0^2}{m^2 c^2 v_e^2} \right) \frac{1}{E_0^2}, \qquad (4.252)$$

where E_0 is the binding energy of hydrogen. Averaging $\sigma_{rec} v_e$ using the result that, for a Maxwellian distribution, $\langle v_e^{-1} \rangle = (2m/\pi T)^{1/2}$, we find

$$\alpha \equiv \langle \sigma_{rec} v_e \rangle = \left(\frac{2^{10} \pi^{3/2}}{3 e^4} \right) \left(\frac{q^2}{\hbar c} \right)^3 \left(\frac{a_0^3 E_0}{\hbar} \right) \left(\frac{E_0}{T} \right)^{1/2}$$

$$\simeq 35 \left(\frac{q^2}{\hbar c} \right)^3 \left(\frac{a_0^3 E_0}{\hbar} \right) \left(\frac{E_0}{T} \right)^{1/2}. \qquad (4.253)$$

Numerically,

$$\alpha = \langle \sigma_{rec} v \rangle \simeq 1.4 \times 10^{-13} \left(\frac{T}{1\,eV} \right)^{-1/2} cm^3 \, s^{-1}. \qquad (4.254)$$

4.14 Einstein's A, B coefficients

(a) Let $f_\alpha (t, \mathbf{x}, \mathbf{p}) \, d^3 x \, d^3 p$ denote the distribution function for photons at an event $(t, \mathbf{x})$ with momentum $\mathbf{p}$ and polarization α. A beam of photons travelling along the direction $\hat{\mathbf{p}}$ for a time dt through an area dA will span the volume:

$$d^3 x = (c \, dt) (\hat{\mathbf{p}} \cdot \hat{\mathbf{n}} \, dA). \qquad (4.255)$$

The momentum space volume element of these photons can be written as

$$d^3 p = p^2 dp \, d\Omega = \left(\frac{h\nu}{c} \right)^2 \left(\frac{h}{c} \right) d\nu \, d\Omega. \qquad (4.256)$$

Such a beam of photons carries the energy

$$dE = \sum_{\alpha=1}^{2} h\nu f_\alpha \, d^3 x \, d^3 p = \sum_{\alpha=1}^{2} h\nu f_\alpha \cdot \frac{h^3 \nu^2}{c^2} d\nu \, d\Omega \, dt \, dA \, (\hat{\mathbf{p}} \cdot \mathbf{n}), \qquad (4.257)$$

giving

$$I_v \equiv \frac{dE}{dv\, dt\, (\mathbf{n}\, dA) \cdot (\hat{\mathbf{p}}\, d\Omega)} = \sum_{\alpha=1}^{2} \left(h^3 f_\alpha \right) \frac{hv^3}{c^2}. \tag{4.258}$$

The occupation number n_α of photons is defined to be the dimensionless quantity $\left(h^3 f_\alpha \right)$, so that $n_\alpha \left(d^3\mathbf{x}\, d^3\mathbf{p}/h^3 \right) = f_\alpha d^3\mathbf{x}\, d^3\mathbf{p}$. Hence,

$$I_v = \sum_{\alpha=1}^{2} I_{v\alpha} = \sum_{\alpha=1}^{2} n_\alpha \frac{hv^3}{c^2} = n\left(\frac{2hv^3}{c^2} \right), \tag{4.259}$$

where the last expression is valid for an unpolarized beam.

(b) Using $n_\alpha = \left(c^2/2hv^3 \right) I_v$ in (4.207) and (4.208) we obtain

$$\left(\frac{dP}{dt} \right)_{abs} = \left(\frac{2q^2}{mc^3} \right) \omega^2 f \cdot \frac{c^2}{2hv^3} I_v = 4\pi^2 \left(\frac{q^2}{mc} \right) \left(\frac{f}{hv} \right) I_v \equiv B I_v \tag{4.260}$$

and

$$\left(\frac{dP}{dt} \right)_{emis} = \left(\frac{2q^2}{mc^3} \right) \left[1 + \frac{c^2}{2hv^3} I_v \right] \omega^2 f = A + B I_v, \tag{4.261}$$

with

$$A = 4\pi^2 \left(\frac{q^2}{mc} \right) \left(\frac{2v^2}{c^2} \right) f, \quad B = 4\pi^2 \left(\frac{q^2}{mc} \right) \left(\frac{f}{hv} \right). \tag{4.262}$$

Using (4.202), we find

$$\sigma_{bb} = \frac{\pi q^2}{mc} f \delta\left(v - v_{12} \right) = \left(\frac{B}{4\pi} \right) hv\, \delta\left(v - v_{12} \right) \equiv \frac{Bhv}{4\pi} \phi_v. \tag{4.263}$$

(c) In the steady state N_1, N_2 and $n(v)$ cannot change with time. The number of atoms making a downward transition per second, $N_2 \left(dP_{emis}/dt \right)$ should be balanced by the number of atoms making the upward transition per second, $N_1 \left(dP_{abs}/dt \right)$. It follows that

$$\frac{N_2}{N_1} = \frac{\left(dP_{emis}/dt \right)}{\left(dP_{abs}/dt \right)} = \frac{1 + n(v)}{n(v)} = \frac{1}{n} + 1. \tag{4.264}$$

Hence,

$$n(v) = \frac{1}{\left(N_2/N_1 \right) - 1}. \tag{4.265}$$

This is a general condition for the existence of steady state between radiation at frequency v and matter distribution with level populations N_2 and N_1. Note that, in the steady state, $n(v)$ is completely determined by the population of matter at different energy levels. It is conventional to define a temperature T_{12} for two energy levels E_1 and E_2 by the relation

$$\frac{N_2}{N_1} \equiv \exp-\frac{(E_2 - E_1)}{kT_{12}} = \exp\left(-\frac{hv}{kT_{12}} \right). \tag{4.266}$$

If matter is in thermodynamic equilibrium, T_{12} will be the thermodynamic temperature of the system and will be the same for all levels E_1 and E_2. Equation (4.265) shows that

$$n(v) = \frac{1}{e^{hv/kT} - 1} \tag{4.267}$$

when the radiation is in equilibrium with matter. The intensity I_v is

$$I_v = \frac{2hv^3}{c^2} n = \frac{2hv^3}{c^2 \left(e^{\beta hv} - 1 \right)}, \tag{4.268}$$

with $\beta = (kT)^{-1}$. To find the corresponding energy density U_v we proceed as follows. From the definitions

$$U_v = \frac{dE}{d^3x \, dv} = \frac{dE}{dv \, dA \, (c \, dt)}, \quad I_v = \frac{dE}{dv \, dt \, (\mathbf{n} \, dA) \cdot (\mathbf{k} \, d\Omega)}, \tag{4.269}$$

it follows that

$$cU_v = \frac{dE}{dv \, dA \, dt} = 4\pi I_v \tag{4.270}$$

for isotropic radiation. Therefore,

$$U_v = \frac{4\pi}{c} I_v = \frac{8\pi h}{c^3} \frac{v^3}{e^{\beta hv} - 1} \tag{4.271}$$

and the total energy density is

$$U = \int_0^\infty dv \, U_v = \frac{8\pi h}{c^3} \left(\frac{1}{\beta h} \right)^4 \int_0^\infty \frac{dx \, x^3}{e^x - 1} = \frac{\pi^2}{15 \, (\hbar c)^3} \frac{1}{\beta^4}. \tag{4.272}$$

(The integral has the value of $\pi^4/15$.)

(d) Each atom decays spontaneously with a probability of A per second. Hence, the number of decays per second per unit volume of material is $(n_2 A)$, where n_2 is the number of excited atoms per unit volume. Each decay produces the energy $hv\delta \, (v - v_{12}) \, dv$ in the frequency range dv. To obtain the rate of emission of energy per unit solid angle we have to divide by (4π). Thus we obtain

$$\left(\frac{dE}{d^3x \, dt \, dv \, d\Omega} \right)_{\text{spon emis}} \equiv j_v = \frac{n_2 A_{21}}{4\pi} hv \, \delta \, (v - v_{12}). \tag{4.273}$$

The emissivity per atom is

$$J_v = \left(\frac{dE}{dt \, dv \, d\Omega} \right)_{\text{atom}} = \frac{A_{21} hv}{4\pi} \delta \, (v - v_{12}). \tag{4.274}$$

(e) Let the number of atoms at the energy level E_1 in a small volume d^3x be $(n_1 \, d^3x)$. The fraction of those which will make a transition to the upper level absorbing the energy $h\nu\phi_\nu\, d\nu$ in a small time interval dt is $B_{12}I_\nu\, dt\, (d\Omega/4\pi)$. So the net energy absorbed is

$$dE_{\mathrm{abs}} = \left(n_1 d^3x\right)\left(B_{12}I_\nu\, dt\right)\left(d\Omega/4\pi\right)\left(h\nu\phi_\nu\, d\nu\right)$$
$$= \alpha_\nu I_\nu d^3x\, dt\, d\Omega\, d\nu\,. \tag{4.275}$$

The second equation follows from the definition of α_ν. Comparing, we obtain

$$\alpha_\nu = n_1 B_{12} h\nu\phi_\nu\, (4\pi)^{-1} = n_1 \sigma_{\mathrm{bb}}\, (\nu)\,. \tag{4.276}$$

The second equality follows from (4.263). The analysis for $\alpha_\nu^{\mathrm{ind}}$ is identical except for the replacement of $n_1 B_{12}$ by $n_2 B_{21}$, which is equal to $n_2 B_{12}$. Hence,

$$\alpha_\nu^{\mathrm{net}} = \frac{h\nu}{4\pi}\phi_\nu\, (n_1 - n_2)\, B_{12} = \sigma_\nu\, (n_1 - n_2)\,. \tag{4.277}$$

The ratio $\left(j_\nu/\alpha_\nu^{\mathrm{net}}\right)$ can be expressed in terms of the ratios (A/B) and (n_2/n_1). We obtain

$$\frac{j_\nu}{\alpha_\nu^{\mathrm{net}}} = \frac{A_{21}n_2}{(n_1 - n_2)\, B_{21}} = \frac{(A/B)}{(n_1/n_2) - 1} = \frac{2h\nu^3}{c^2}\frac{1}{(n_1/n_2) - 1}\,. \tag{4.278}$$

The ratio is entirely fixed by the ratio of atoms in the two levels (n_1/n_2) and can be calculated from the state of the *matter* distibution. If matter is in thermal equilibrium, then $(n_1/n_2) = \exp\beta h\nu$ and

$$\alpha_\nu^{\mathrm{net}} = \frac{c^2}{2h\nu^3}\left(e^{h\nu/kT} - 1\right) j_\nu\,. \tag{4.279}$$

(f) From the definition of intensity I_ν in terms of the photon occupation number, it is clear that $I_\nu \propto \nu^3 n$. Since n is Lorentz invariant it follows that (I_ν/ν^3) is invariant.

The change of intensity dI_ν due to emission can be written as $dI_\nu \propto \rho J_\nu\, dt$; so $(dI_\nu/\nu^3) \propto (\rho J_\nu/\nu^2)\, (dt/\nu)$. Both dt and ν are zeroth components of four-vectors and hence the ratio (dt/ν) is Lorentz invariant. It follows that $(\rho J_\nu/\nu^2)$ is Lorentz invariant.

Similarly, $dI_\nu \propto \rho\kappa_\nu I_\nu\, dt$, giving $(dI_\nu/\nu^3) \propto \rho\kappa_\nu\nu\, (dt/\nu)\, (I_\nu/\nu^3)$. Since (dt/ν), (dI_ν/ν^3), etc., are Lorentz invariant it follows that $\rho\kappa_\nu\nu$ is Lorentz invariant.

4.15 Absorption and emission in continuum case

From the distribution function $F(\mathbf{x},\mathbf{p},t)$ for the electrons, one can compute the number of electrons per unit volume with energy between E and $E + dE$:

$$n(E)\, dE = \int d^3x\, F(\mathbf{x},\mathbf{p},t)\left(\frac{d^3p}{dE}\right)dE \equiv f(\mathbf{p},t)\frac{d^3p}{dE}\, dE\,. \tag{4.280}$$

For an isotropic, steady-state, distribution $f(\mathbf{p},t) = f(p)$ and $(d^3p/dE) = 4\pi p^2\, (dp/dE)$. Then

$$n(E) = 4\pi p^2\, (E)\left(\frac{dp}{dE}\right)f(p)\,. \tag{4.281}$$

Since each electron emits the energy $d\epsilon = P(v, E) \, dt \, dv$, the total emissivity over all solid angles is

$$4\pi j_v = \int d^3 p \, f(p) \, P(v, E) = \int dE \, 4\pi p^2(E) \left(\frac{dp}{dE}\right) f(E) \, P(v, E) \qquad (4.282)$$

or

$$j_v = \frac{d\epsilon}{d\Omega \, dv \, dt \, d^3 x} = \int_0^\infty dE \, n(E) \, P(v, E). \qquad (4.283)$$

To determine the absorption and induced emission, we will relate $P(v, E)$ to the A, B coeffiiicients. We know that

$$P(v, E_2) = \begin{bmatrix} \text{energy radiated per second,} \\ \text{at frequency } v, \text{ by an electron} \\ \text{of energy } E_2 \end{bmatrix}$$

$$= hv \sum_{E_1} A_{21} \phi_{21}(v) = \left(\frac{2hv^3}{c^2}\right) hv \sum_{E_1} B_{21} \phi_{21}(v). \qquad (4.284)$$

The first relation follows from the definition of A_{21} and the second relation arises from the fact that $A = \left(2hv^3/c^2\right) B$. The absorption coefficient α_v, on the other hand, is defined to be

$$\rho \kappa_v = \alpha_v = \frac{hv}{4\pi} \sum_{E_1} \sum_{E_2} \left[n(E_1) B_{12} - n(E_2) B_{21}\right] \phi_{21}(v) \qquad (4.285)$$

(see equation (4.277)). The sum is over all states E_1 and E_2 but $\phi_{21}(v)$ will ensure that only states with $v = v_{12} = (E_2 - E_1)/h$ contribute significantly. The first term represents true absorption while the second one represents stimulated emission. They can be re-expressed in terms of P by using (4.284). The second term is

$$-\frac{hv}{4\pi} \sum_{E_2} n(E_2) \sum_{E_1} B_{21} \phi_{21}(v) = -\frac{hv}{4\pi} \sum_{E_2} n(E_2) \left(\frac{c^2}{2hv^3}\right) \frac{P(v, E_2)}{hv}$$

$$= -\frac{c^2}{8\pi hv^3} \sum_{E_2} n(E_2) P(v, E_2). \qquad (4.286)$$

The first term is

$$\frac{hv}{4\pi} \sum_{E_1} \sum_{E_1} n(E_1) B_{12} \phi_{21} = \frac{hv}{4\pi} \sum_{E_2} n(E_2 - hv) \sum_{E_1} B_{21} \phi_{21}$$

$$= \frac{c^2}{8\pi hv^3} \sum_{E_2} n(E_2 - hv) P(v, E_2). \qquad (4.287)$$

Hence,

$$\alpha_v = \frac{c^2}{8\pi h v^3} \sum_{E_2} \left\{ n(E_2 - hv) - n(E_2) \right\} P(v, E_2)$$

$$= \frac{c^2}{8\pi h v^3} \int d^3 p_2 \left[f(p_2') - f(p_2) \right] P(v, E_2) \qquad (4.288)$$

where p_2' is the momentum corresponding to energy $(E_2 - hv)$. In the second line we have converted the sum to an integral.

In thermal equilibrium, $f(p) \propto \exp(-\beta E(p))$ and

$$f(p_2') - f(p_2) = f(p_2) \left(e^{\beta h v} - 1 \right), \qquad (4.289)$$

so that

$$\alpha_v = \frac{c^2}{8\pi h v^3} \left(e^{\beta h v} - 1 \right) \int d^3 p_2 \, f(p_2) \, P(v, E_2)$$

$$= \frac{c^2}{8\pi h v^3} \left(e^{\beta h v} - 1 \right) \cdot 4\pi j_v = \frac{c^2}{2 h v^3} \left(e^{\beta h v} - 1 \right) j_v . \qquad (4.290)$$

Clearly, $j_v = I_v^{eq} \alpha_v = I_v^{eq} \rho \kappa_v$, as it should.

4.16 Self-absorption of continuum radiation

(a) Consider the propagation of radiation through a medium which is both emitting and absorbing the photons. When the radiation travels a distance dl the intensity I_v will increase by the amount $\rho j_v \, dl$ due to the emission of photons; the absorption of the photons will cause the intensity to decrease by the amount $(-\alpha_v I_v \, dl)$ during the same time. Hence, the rate of change of intensity can be described by a differential equation of the form

$$\frac{dI_v}{dl} = \rho j_v - \alpha_v I_v \qquad (4.291)$$

if scattering is ignored. This can be rewritten as

$$\frac{dI_v}{ds} = S_v - I_v, \qquad S_v \equiv \frac{\rho j_v}{\alpha_v}, \qquad s = \alpha_v l, \qquad (4.292)$$

which has the solution

$$I_v = S_v \left[1 - e^{-\alpha_v l} \right] = \frac{\rho j_v}{\alpha_v} \left[1 - e^{-\alpha_v l} \right] . \qquad (4.293)$$

This solution satisfies the boundary condition $I_v = 0$ at $l = 0$. If the source has a size R, then

$$I_v(R) = \frac{\rho j_v}{\alpha_v} \left[1 - e^{-\alpha_v R} \right] = \begin{cases} \rho j_v R & \text{(if } \alpha_v R \ll 1) \\ (\rho j_v / \alpha_v) & \text{(if } \alpha_v R \gg 1). \end{cases} \qquad (4.294)$$

(b) For a system emitting thermal Bremsstrahlung radiation, $\alpha_v \propto T^{-3/2} v^{-2}$ (see equation (4.86)) and $(\rho j_v / \alpha_v) = I_v^{eq}$. Thus, at low frequencies, $\alpha_v R \gg 1$ and we will observe the intensity to be $I_\nu^{eq} \propto \nu^3 \left[e^{h\nu/kT} - 1 \right]^{-1} \propto T\nu^2$. At intermediate and high frequencies we have $I_\nu \propto j_\nu$, which is independent of ν at intermediate scales and decreases as $\exp(-h\nu/kT)$ at high frequencies.

(c) To work out the corresponding results for synchrotron radiation, we have to compute α_ν for a powerlaw distribution of electrons. This can be done using (4.288). We will assume that the electrons are ultrarelativistic with $E \cong pc$, so that

$$f(p)\, d^3 p = 4\pi p^2 f(p)\, dp = 4\pi (E/c)^2 (dE/c) f(E/c) = N(E)\, dE, \qquad (4.295)$$

that is, $N(E) = f(E) E^2$. Then (4.288) becomes

$$\alpha_\nu = \frac{c^2}{8\pi h\nu^3} \int dE\, P(\nu, E) E^2 \left[\frac{N(E-h\nu)}{(E-h\nu)^2} - \frac{N(E)}{E^2} \right]. \qquad (4.296)$$

When $h\nu \ll \bar{E}$, where $\bar{E}$ is the typical energy of the electron, we can expand $N(E-h\nu)$ in a Taylor series in $h\nu$ and obtain

$$\alpha_\nu = -\frac{c^2}{8\pi\nu^2} \int dE\, P(\nu, E) E^2 \frac{\partial}{\partial E} \left[\frac{N(E)}{E^2} \right]. \qquad (4.297)$$

For a powerlaw distribution $N(E) = CE^{-p}$:

$$-E^2 \frac{d}{dE} \left(\frac{N}{E^2} \right) = (p+2) CE^{-(p+1)} = \frac{(p+2) N(E)}{E} \qquad (4.298)$$

and

$$\alpha_\nu = \frac{(p+2) c^2}{8\pi\nu^2} \int dE\, P(\nu, E) \frac{N(E)}{E}. \qquad (4.299)$$

From problem 4.6, we know that $P(\nu, E) \propto F(\nu/\nu_c) \propto F(\nu/E^2)$; using this result, α_ν becomes

$$\alpha_\nu \propto \frac{1}{\nu^2} \int \left[\nu^{1/2} \frac{dx}{x^{3/2}} \right] F(x) \left(\frac{\nu}{x} \right)^{-(p+1)/2} \propto \nu^{-\frac{1}{2}(p+4)}. \qquad (4.300)$$

Thus $\alpha_\nu R \gg 1$ at low frequencies and $\alpha_\nu R \ll 1$ at high frequencies. When $\alpha_\nu R \gg 1$, the observed intensity is

$$I_\nu \propto \left(\frac{j_\nu}{\alpha_\nu} \right) \propto \frac{\nu^{-(p-1)/2}}{\nu^{-(p+4)/2}} \propto \nu^{5/2} \qquad (4.301)$$

at low frequencies. When $\alpha_\nu R \ll 1$, the observed intensity is $I_\nu \propto j_\nu \propto \nu^{-(p-1)/2}$.

Note that low frequency synchrotron intensity has a $\nu^{2.5}$ behaviour in contrast to thermal spectrum which has a ν^2 behaviour. It is also clear from the preceding discussion that synchrotron radiation will peak at some intermediate frequency.

5

General relativity

5.1 Accelerated frames, special relativity and gravity

(a) Consider an observer A travelling along the X-axis in a trajectory $X = f(\tau), T = h(\tau)$, where f and h are specified functions and τ is the proper time on the clock carried by the observer. We would like to assign a suitable coordinate system to this observer.

Let $\mathscr{P}$ be some event with Minkowski coordinates (T, X). The observer sends a light signal from the event $\mathscr{A}$ (at $\tau = t_A$) to the event $\mathscr{P}$. The signal is reflected back to the observer at $\mathscr{P}$ and reaches him (her) at event $\mathscr{B}$ (at $\tau = t_B$). Since the light has travelled for a time interval $(t_B - t_A)$, it is reasonable to attribute the coordinates

$$t = \frac{1}{2}(t_B + t_A), \quad x = \frac{1}{2}(t_B - t_A) \tag{5.1}$$

to the event $\mathscr{P}$. To relate (t, x) to (T, X) we have to proceed as follows. Since the events $\mathscr{P}(T, X)$, $\mathscr{A}(T_A, X_A)$ and $\mathscr{B}(T_B, X_B)$ are connected by light signals travelling in forward and backward directions, it follows that

$$X - X_A = T - T_A, \quad X - X_B = -(T - T_B). \tag{5.2}$$

Or

$$X - T = X_A - T_A = f(t_A) - h(t_A) = f(t - x) - h(t - x), \tag{5.3}$$

$$X + T = X_B + T_B = f(t_B) + h(t_B) = f(t + x) + h(t + x). \tag{5.4}$$

Given f and h, these equations can be solved to find (X, T) in terms of (x, t).

Let us now apply this to an observer travelling along the X-axis with a uniform acceleration g. The equation of motion

$$\frac{d}{dT}\left(\frac{v}{\sqrt{1 - v^2}}\right) = g \tag{5.5}$$

has the solution

$$v = gT\left(1 + g^2 T^2\right)^{-1/2} = \frac{dX}{dT} \tag{5.6}$$

310

if we take the initial condition to be $v = 0$ at $T = 0$. Integrating this equation again and setting $X = 0$ at $T = 0$, we find

$$X = \frac{1}{g}\left(\sqrt{1 + g^2 T^2} - 1\right).$$ (5.7)

The proper time τ on the clock carried by the observer is

$$\tau(T) = \int_0^T dT\sqrt{1 - V^2} = \frac{1}{g}\sinh^{-1}(gT).$$ (5.8)

Using this, we can express the trajectory as

$$T = \frac{1}{g}\sinh g\tau, \quad X = \frac{1}{g}(\cosh g\tau - 1).$$ (5.9)

By shifting the spatial origin by g^{-1} (i.e. by using $X + g^{-1}$ rather than X) we can write the trajectory in a more symmetric form as

$$gT = \sinh g\tau \equiv g\, h(\tau), \quad gX = \cosh g\tau \equiv g\, f(\tau).$$ (5.10)

Equations (5.3) and (5.4) now become

$$X - T = g^{-1}\exp\left[-g(t - x)\right], \quad X + T = g^{-1}\exp\left[g(t + x)\right].$$ (5.11)

Or

$$X = g^{-1}e^{gx}\cosh gt, \quad T = g^{-1}e^{gx}\sinh gt.$$ (5.12)

This provides the transformation between the inertial coordinate system and that of a uniformly accelerated observer. Using

$$dT^2 - dX^2 = d(T - X)\, d(T + X) = e^{2gx}\left(dt^2 - dx^2\right),$$ (5.13)

we obtain the metric for the accelerated observer:

$$ds^2 = dT^2 - dX^2 - dY^2 - dZ^2 = e^{2gx}\left(dt^2 - dx^2\right) - dy^2 - dz^2.$$ (5.14)

If we change to a new space coordinate $\bar{x}$ with $(1 + g\bar{x}) = e^{gx}$ and $e^{gx}dx = d\bar{x}$ we obtain

$$ds^2 = (1 + g\bar{x})^2\, dt^2 - d\bar{x}^2 - dy^2 - dz^2.$$ (5.15)

(b)

(i) Since the same parameter m multiplies the kinetic and potential energy terms of the action, it has no influence on the equations of motion. In other words, the trajectories of material particles (with the same initial conditions) will be independent of the properties of the particle and will depend only on the gravitational potential $\phi(x^\alpha)$. All particles at an event $\mathscr{P}$ will experience an acceleration $g^\alpha \equiv -(\partial\phi/\partial x^\alpha)|_{\mathscr{P}}$, where $\alpha = 1, 2, 3$. If we now choose a new set of spatial coordinates:

$$\xi^\alpha = x^\alpha - \frac{1}{2}g^\alpha t^2 \quad (\alpha = 1, 2, 3),$$ (5.16)

near $\mathscr{P}$, then in the new frame the particles will experience no acceleration. It follows that the trajectories of particles in a given gravitational field are indistinguishable locally from the trajectories of free particles viewed from an accelerated frame. Hence, a gravitational field is locally indistinguishable from a suitably chosen non-inertial frame as far as the laws of mechanics are concerned. (Note that we could arrive at this result only because the trajectories are independent of any property of the particle. This is *not* the case, for example, in electromagnetism, in which the trajectories depend on the (q/m) ratio for the particles.)

(ii) Consider two frames of reference S and S' with S' moving along the positive x-axis with acceleration g. An observer in S' will feel all particles in his (her) vicinity to be moving in the $(-\hat{x})$ direction with an acceleration g. Hence, he (she) will attribute a gravitational potential $\phi = gx$ in the vicinity.

Let there be a series of clocks along the x-axis in S and let the observer in S' compare the readings of his (her) clock with these stationary clocks as he (she) passes each one of them. Since $t' = g^{-1}\sinh^{-1}(gt/c^2)$, we have

$$\Delta t' = \frac{\Delta t}{\sqrt{1 + g^2 t^2 / c^2}} \cong \Delta t \left(1 - \frac{1}{2}\frac{g^2 t^2}{c^2}\right). \tag{5.17}$$

But $x = (1/2)gt^2$ is the location of the observer at this instant. So we can also write

$$\Delta t' \cong \Delta t \left(1 - \frac{gx}{c^2}\right) = \Delta t \left(1 - \frac{\phi}{c^2}\right). \tag{5.18}$$

In the frame S', the observer will attribute the difference between $\Delta t'$ and Δt to the presence of a gravitational potential ϕ and conclude that the rate of flow of the clocks is affected by the gravitational field.

(c) For the purpose of the argument given below, we may assume that the gravitational potential near the surface of earth varies linearly with height; then the potential difference between two points A and B, separated by height L, will be gL, where g is the acceleration due to gravity. Let us assume that a pair of sufficiently high energy photons, each of energy $\hbar\omega$, was converted into an e^+e^- pair at the point A. These particles move from point A to point B, losing the gravitational potential energy mgL per particle. (We have assumed that B is above A.) At B we annihilate the two particles and produce two photons which are sent back to A. The sequence of events described above has restored the original statusquo. Since the amount of energy available in the form of particles was lower at B than at A, it follows that the energy of the photons must change in going from B to A. Since the energy of the photon is proportional to the frequency, the frequency of the photons at A and B must be related by $\omega_A = \omega_B(1 + gL/c^2)$. If this is not the case, we could use the above sequence of events to increase the energy of the system repeatedly.

Since the frequency shift derived above is independent of $\hbar$ (even though the original argument was quantum mechanical), it follows that the result should be true even in the classical limit. But in the classical limit, one can consider photons as electromagnetic radiation with some frequency. This frequency can be determined by counting the number of crests N of a wave train which crosses an observer in a time interval Δt. In a static gravitational field, the head of the wave train and the tail of the wave train will propagate in an identical manner. Hence, two observers located at A and B will attribute different frequencies to the wave *only if* the reference clocks which they have at A and B run at different rates. To reproduce the frequency shift derived above, the time intervals of the clocks have to obey the relation obtained in equation (5.18).

The conclusion has far-reaching implications. Consider the line interval ds between two infinitesimally separated events in spacetime. When the spatial interval dx between the two points vanishes, ds measures the lapse of proper time at a given spatial location. If a global Lorentz time exists, then $ds = dt$, which is the same for all observers at rest. The conclusion we arrived at in the last problem shows that ds should vary from point to point even when $dx = 0$. This is possible only if $ds^2 = g_{00}(\mathbf{x}) \, dt^2$, where g_{00} is a non-trivial function of the spatial coordinates. Using the above discussion we can even fix the form of g_{00}; to arrive at the correct flow of time, we need $g_{00}(x) = (1 + 2\phi/c^2)$. It follows that the line interval between two events in the spacetime cannot have the form $ds^2 = (dt^2 - d\mathbf{x}^2)$ but should *at least* be modified to a form

$$ds^2 = \left(1 + \frac{2\phi}{c^2}\right) c^2 \, dt^2 - d\mathbf{x}^2 \qquad (5.19)$$

in the presence of a gravitational field.

(d) The action for a particle in a gravitational field is

$$A = -mc^2 \int dt \sqrt{1 - \frac{v^2}{c^2}} - \int dt \, m\phi. \qquad (5.20)$$

This can be rewritten as

$$A = -(mc) \int (c \, dt) \left[\sqrt{1 - \frac{v^2}{c^2}} + \frac{\phi}{c^2} \right] \cong -(mc) \int \left[\left(1 + \frac{2\phi}{c^2}\right) c^2 \, dt^2 - d\mathbf{x}^2 \right]^{1/2}, \qquad (5.21)$$

provided $\phi \ll c^2$. The last form suggests that, if we introduce a metric g_{ik} with

$$ds^2 = g_{ik} \, dx^i \, dx^k = \left(1 + \frac{2\phi}{c^2}\right) c^2 \, dt^2 - d\mathbf{x}^2, \qquad (5.22)$$

then the action can be expressed as

$$A \cong -mc \int \sqrt{g_{ik} \, dx^i \, dx^k} = -mc \int ds. \qquad (5.23)$$

In the case of a uniformly accelerated observer (corresponding to a constant gravitational field) we found that

$$ds^2 = (1 + g\bar{x})^2\, dt^2 - d\mathbf{x}^2 \cong (1 + 2g\bar{x})\, dt^2 - d\mathbf{x}^2, \qquad (5.24)$$

which agrees with the result derived above in (a). The result is also, of course, consistent with that obtained in (c).

5.2 Gravity and the metric tensor

(a) Consider a flat spacetime with inertial coordinates X^i and metric $\eta_{ik} =$ dia $(1, -1, -1, -1)$. We make a coordinate transformation from X^i to a new set of coordinates x^k which are arbitrary functions of the original coordinates. The line interval in the new coordinates now becomes

$$ds^2 = \eta_{ik}\, dX^i\, dX^k = \eta_{ik} \frac{\partial X^i}{\partial x^a} \frac{\partial X^k}{\partial x^b}\, dx^a\, dx^b \equiv g_{ab}(x)\, dx^a\, dx^b. \qquad (5.25)$$

(i) Note that the transformation from X^i to x^k involves only four arbitrary functions. Therefore, the g_{ab} in the line interval can contain only four independent functions. In general, $g_{ab}(x)$ contains ten independent functions. (This can be seen as follows. When a and b range through 0,1,2,3 the pair (a, b) can take $4 \times 4 = 16$ different values. Since g_{ab} is symmetric, only the diagonal values (four in number) and half of the non-diagonal values $(16 - 4)/2 = 6$ in number, are independent.) It is, therefore, impossible to obtain an arbitrary set of $g_{ab}(x)$ by transforming the coordinates from the inertial frame to the non-inertial frame. Conversely, it is not possible to reduce an arbitrary metric $g_{ab}(x)$ to η_{ab} by a coordinate transformation.

(ii) Consider a coordinate transformation from the coordinates x^i to $\bar{x}^i$ around an event $\mathscr{P}$. We expand $\bar{x}^i$ in terms of x^i in a Taylor series as

$$\bar{x}^i = A^i + B^i_k x^k + C^i_{jk} x^j x^k + D^i_{jkl} x^j x^k x^l + \cdots. \qquad (5.26)$$

We will now attempt the following (i) Choose the values of B^i_k such that the transformed metric tensor $\bar{g}_{ik}$ at the event $\mathscr{P}$ is the same as η_{ik} and (ii) choose C^i_{jk} such that the first derivatives of $\bar{g}_{ik}$ at $\mathscr{P}$ vanish. The first requirement amounts to ten independent conditions, while the second requirement has $4 \times 10 = 40$ conditions. In B^i_k we have 16 independent parameters and in C^i_{jk} we have $4 \times 10 = 40$ independent parameters (since $C^i_{jk} = C^i_{kj}$ by definition). Using these 16 parameters of B^i_k we can certainly satisfy the ten conditions $g_{ik} = \eta_{ik}$. The six parameters which are left free correspond to the Lorentz transformations and rotations (three components of the velocity vector and three Euler angles) which are still allowed.

The 40 parameters in C^i_{jk} can be used to impose the 40 conditions $(\partial g_{ik}/\partial x^a) = 0$. Hence, one can always choose the coordinates in such a way that the metric reduces to the inertial form in an infinitesimal *region* around an event $\mathscr{P}$.

There are $10 \times 10 = 100$ independent components in the second derivatives $(\partial^2 g_{ik}/\partial x^a \partial x^b)$. But D^i_{jkl} has only $4 \times 20 = 80$ independent components. Hence, we cannot make all the second derivatives of g_{ik} vanish at $\mathscr{P}$. We can set 80 of them to preassigned values, leaving 20 independent components.

(b) When the coordinate system is changed from x^i to x'^i, the components of the tangent vector v^i to a curve $x^i(\lambda)$ change to

$$v'^i = \frac{dx'^i}{d\lambda} = \frac{\partial x'^i}{\partial x^a}\frac{dx^a}{d\lambda} = \frac{\partial x'^i}{\partial x^a}v^a . \qquad (5.27)$$

This *defines* the transformation law for any vector v^a under the coordinate transformation. That is, we define the four-vector $v^a(x)$ to be a set of four functions which change according to the law

$$v'^a(x'^i) = \left(\frac{\partial x'^a}{\partial x^b}\right)v^b(x^i) \qquad (5.28)$$

under a coordinate transformation $x^i \to x'^i$. Note that, in general, the coefficients $(\partial x'^a/\partial x^b)$ depend on the event x^i. Hence *vectors located at different points in the spacetime transform differently under coordinate transformation.*

Consider a quantity $T^{ab} = v^a v^b$. From the known transformation law for v^a we can determine the transformation law for T^{ab}. We find that

$$T'^{ab} \equiv v'^a v'^b = \frac{\partial x'^a}{\partial x^c}\frac{\partial x'^b}{\partial x^d}v^c v^d = \frac{\partial x'^a}{\partial x^c}\frac{\partial x'^b}{\partial x^d}T^{cd} . \qquad (5.29)$$

The quantity T^{ab} is clearly a second rank tensor, though it is a special kind of tensor. The transformation law arrived at above, however, is entirely in terms of the components T^{ab} and does not depend on the vector v^a. It is therefore natural to define a transformation law for a tensor T^{ab} based on the above equation.

The definition can be generalized easily for higher rank tensors. To every index we introduce a factor $(\partial x'^a/\partial x^b)$ in such a way as to relate the component a in the new coordinates to a component b in the old coordinates.

(c) The transformation law for g^{ik} can be obtained by noting that the relation $g^{ik}g_{kl} = \delta^i_l$ should hold in all coordinate frames. Since

$$g'_{ik} = \frac{\partial x^a}{\partial x'^i}\frac{\partial x^b}{\partial x'^k}g_{ab} , \qquad (5.30)$$

it follows that we must have

$$g'^{ik} = \frac{\partial x'^i}{\partial x^a}\frac{\partial x'^k}{\partial x^b}g^{ab} . \qquad (5.31)$$

The differential dg of the determinant g can be obtained by taking the differential of each component of the tensor g_{ik} and multiplying it by its cofactor in the determinant, that is, by the corresponding minor. On the other hand, the tensor g^{ik} is obtained by taking the minors of g_{ik} and dividing by determinant g. Hence, the minor of g_{ik} is gg^{ik}, giving $dg = gg^{ik} \, dg_{ik}$. It follows that

$$\frac{\partial g}{\partial x^a} = gg^{ik} \frac{\partial g_{ik}}{\partial x^a} = -gg_{ik} \frac{\partial g^{ik}}{\partial x^a} . \tag{5.32}$$

The second equality arises from the fact that $\partial(g_{ik}g^{ik})/\partial x^a = 0$.

(d) When we transform from x^i to x'^i the quantity T_k^i transforms as

$$
\begin{aligned}
T_k'^i = g_{ka}' T'^{ia} &= \frac{\partial x^b}{\partial x'^k} \frac{\partial x^c}{\partial x'^a} g_{bc} \cdot \frac{\partial x'^i}{\partial x^m} \frac{\partial x'^a}{\partial x^n} T^{mn} \\
&= g_{bc} T^{mn} \frac{\partial x^b}{\partial x'^k} \frac{\partial x'^i}{\partial x^m} \delta_n^c = g_{bn} T^{mn} \frac{\partial x^b}{\partial x'^k} \frac{\partial x'^i}{\partial x^m} \\
&= T_b^m \frac{\partial x^b}{\partial x'^k} \frac{\partial x'^i}{\partial x^m} .
\end{aligned}
\tag{5.33}
$$

This relation shows that T_k^i transforms correctly when the coordinates are changed. Similar results can be obtained for T_{ik}. Thus raising and lowering indices using g_{ik} is a legitimate tensorial operation.

5.3 Particle trajectories in a gravitational field

(a) Since u^b transforms as a vector under coordinate transformation, this equation will retain its form provided the derivative $u^a{}_{,b} \equiv (\partial u^a / \partial x^b)$ transforms as a tensor. The transformation law for this quantity can be obtained directly along the following lines. We have

$$
\begin{aligned}
u'^a{}_{,b} = \frac{\partial u'^a}{\partial x'^b} &= \frac{\partial x^c}{\partial x'^b} \frac{\partial}{\partial x^c} \left\{ \frac{\partial x'^a}{\partial x^d} u^d \right\} \\
&= \left(\frac{\partial x^c}{\partial x'^b} \frac{\partial x'^a}{\partial x^d} \right) u^d{}_{,c} + \left[\left(\frac{\partial x^c}{\partial x'^b} \right) \frac{\partial^2 x'^a}{\partial x^c \partial x^d} \right] u^d .
\end{aligned}
\tag{5.34}
$$

If $u^a{}_{,b}$ has to transform as a tensor, the second term has to vanish. This term vanishes for Lorentz transformations for which x'^i are linear functions of x^i. But for arbitrary coordinate transformations, this term will not vanish. Hence $u^a{}_{,b}$ is not a tensor under general coordinate transformation.

Another way of understanding this result is as follows. The derivative $u^a{}_{,b}$ is defined by subtracting $u^a(x^b)$ from $u^a(x^b + \Delta x^b)$. Since the transformation law for vectors at x^b and $x^b + \Delta x^b$ is different, the subtraction of vectors located at two different events is not a tensorial operation.

(b) The action for a particle in a spacetime with metric g_{ik} is taken to be

$$A = -m \int ds = -m \int \sqrt{g_{ik} \, dx^i \, dx^k} = -m \int \left(g_{ik} \frac{dx^i}{d\lambda} \frac{dx^k}{d\lambda} \right)^{1/2} d\lambda . \tag{5.35}$$

The trajectory $x^i(\lambda)$ can be found by varying this action. Since we can invoke special relativity at a local region around any event, it follows that u^i is timelike for massive particles and zero for massless particles. Also, since the Lagrangian multiplying $d\lambda$ in (5.35) is independent of λ, it follows that $(g_{ik}u^iu^k)$ is a constant along the trajectory. For massive particles we will choose this constant to be unity, which is equivalent to choosing $\lambda = s$, the proper time. The Euler–Lagrange equations then give

$$\frac{d}{ds}\left[\frac{1}{2}(g_{ik}u^i u^k)^{-1/2} 2g_{ab}u^b\right] = \frac{1}{2}(g_{ik}u^i u^k)^{-1/2}\frac{\partial g_{ik}}{\partial x^a}u^i u^k. \tag{5.36}$$

Setting $g_{ik}u^i u^k = 1$ and writing $(dg_{ab}/ds) = (\partial g_{ab}/\partial x^i)u^i$ we find that

$$g_{ab}\frac{du^b}{ds} = \frac{1}{2}\frac{\partial g_{ik}}{\partial x^a}u^i u^k - \frac{\partial g_{ab}}{\partial x^i}u^i u^b = \frac{1}{2}\left(\frac{\partial g_{ik}}{\partial x^a} - \frac{\partial g_{ak}}{\partial x^i} - \frac{\partial g_{ai}}{\partial x^k}\right)u^i u^k. \tag{5.37}$$

In arriving at the last equality, we have used the symmetry of $u^i u^k$ to write the second term in a symmetric manner. This equation may be written as

$$\frac{du^i}{ds} = -\frac{1}{2}g^{ik}\left(-\frac{\partial g_{mn}}{\partial x^k} + \frac{\partial g_{mk}}{\partial x^n} + \frac{\partial g_{kn}}{\partial x^m}\right)u^m u^n \equiv -\Gamma^i_{mn}u^m u^n, \tag{5.38}$$

where the second equality defines the quantities Γ^i_{km} (called the 'affine connection' or 'Christoffel symbols') in terms of the derivatives of the metric. Notice that Γ^i_{km} is symmetric in the lower indices; $\Gamma^i_{km} = \Gamma^i_{mk}$. Equation (5.38) (called the 'geodesic equation') governs the motion of material particles in a given gravitational field. We can rewrite (5.38) as

$$\frac{du^i}{ds} = \left(\frac{dx^a}{ds}\right)\left(\frac{\partial u^i}{\partial x^a}\right) = u^a\left(\frac{\partial u^i}{\partial x^a}\right) = -\Gamma^i_{ja}u^j u^a \tag{5.39}$$

or,

$$u^a\left[\frac{\partial u^i}{\partial x^a} + \Gamma^i_{ja}u^j\right] \equiv u^a u^i_{;a} = 0, \tag{5.40}$$

where the first equality defines the symbol $u^i_{;a}$. Since our action principle is covariant, this equation must be valid in any frame of reference. It follows that the quantity

$$u^a_{;b} \equiv u^a_{,b} + \Gamma^a_{ib}u^i \tag{5.41}$$

must transform as a tensor.

From the transformation law for g_{ik}, we can directly verify that Γ^a_{bc} transforms as

$$\Gamma'^a_{bc} = \frac{\partial x'^a}{\partial x^i}\frac{\partial x^k}{\partial x'^b}\frac{\partial x^m}{\partial x'^c}\Gamma^i_{km} + \frac{\partial x'^a}{\partial x^i}\frac{\partial^2 x^i}{\partial x'^b \partial x'^c}. \tag{5.42}$$

Comparing the second term in (5.42) with the second term in (5.34), we see that $u^a_{,b}$ does transform as a tensor, even though neither Γ^a_{bc} nor $u^a_{,b}$ is a tensor.

(c)

(i) Since the ordinary derivative of a scalar $\phi_{,a} = (\partial\phi/\partial x^a)$ does transform like a covariant vector, we define $\phi_{;a} = \phi_{,a}$ for scalars. Since $(u^i u_i)$ is scalar, $(u^i u_i)_{,b} = (u^i u_i)_{;b}$. Using the chain rule on both sides and the known form of $u^i_{;b}$, we find that

$$u_{i;b} = u_{i,b} - \Gamma^k_{ib} u_k. \tag{5.43}$$

This defines the covariant derivative of a covariant vector.

(ii) The covariant derivatives of higher rank tensors are defined by using these results and the chain rule. Consider, for example, the tensor $T^a_b = u^a u_b$. Using $T^a_{b;i} = (u^a u_b)_{;i} = (u^a_{;i})u_b + u^a(u_{b;i})$, and the known forms for the covariant derivatives of vectors, one obtains

$$T^a_{b;i} = T^a_{b,i} + \Gamma^a_{ki} T^k_b - \Gamma^k_{bi} T^a_k. \tag{5.44}$$

This will be the definition for $T^a_{b;i}$ for an arbitrary T^a_b. A similar procedure is adopted for other tensors.

It can be verified by direct computation that $g_{ik;b} = 0$. This fact allows us to interchange the order of covariant differentiation and the process of raising and lowering the indices. For example,

$$u_{a;b} = (g_{ai} u^i)_{;b} = (g_{ai;b})u^i + g_{ai}(u^i_{;b}) = g_{ai}(u^i_{;b}). \tag{5.45}$$

(d) The covariant derivative $A^a_{;b}$ of a vector field is

$$A^a_{;b} = A^a_{,b} + \Gamma^a_{mb} A^m. \tag{5.46}$$

Differentiating once again, treating $A^a_{;b}$ as a second rank tensor, we obtain

$$A^a_{;b;c} = \left(A^a_{;b}\right)_{,c} + \Gamma^a_{nc} A^n_{;b} - \Gamma^m_{bc} A^a_{;m} \tag{5.47}$$

which reduces to

$$A^a_{;b;c} = A^a_{,b,c} + \left(\Gamma^a_{bm} A^m\right)_{,c} + \Gamma^a_{cn}\left(A^n_{,b} + \Gamma^n_{mb} A^m\right) - \Gamma^m_{bc}\left(A^a_{,m} + \Gamma^a_{nm} A^n\right). \tag{5.48}$$

Computing the difference $(A^a_{;b;c} - A^a_{;c;b})$ we find that the terms with first derivatives of A^a vanish, giving

$$A^a_{;b;c} - A^a_{;c;b} = -\left(\Gamma^a_{ic,b} - \Gamma^a_{ib,c} + \Gamma^a_{kb}\Gamma^k_{ic} - \Gamma^a_{kc}\Gamma^k_{ib}\right) A^i \equiv -R^a_{ibc} A^i. \tag{5.49}$$

Since the left-hand side is a tensor and A^i is a vector, it follows that R^a_{ibc} is a tensor.

5.4 Parallel transport

Let $v^a(x)$ be a vector field and let $x^a(\lambda)$ be some curve. Then the quantity

$$v^a_{;b}\left(\frac{dx^b}{d\lambda}\right) = v^a_{,b}\left(\frac{dx^b}{d\lambda}\right) + \Gamma^a_{ib}v^i\left(\frac{dx^b}{d\lambda}\right) \tag{5.50}$$

can be thought of as the generalization of the 'directional derivative' of the vector field along a particular direction specified by the tangent vector $(dx^b/d\lambda)$. Suppose we are given a vector k^a at one event $\mathscr{P}$ and some curve $x^a(\lambda)$ passing through $\mathscr{P} = x^a(0)$. We can then solve the differential equation

$$\frac{dv^a}{d\lambda} + \Gamma^a_{ik}(\lambda)\frac{dx^i}{d\lambda}v^k = 0 \tag{5.51}$$

(where $\Gamma^a_{ik}(\lambda) = \Gamma^a_{ik}[x(\lambda)]$, etc.), with the boundary condition $v^a(\lambda = 0) = k^a$. The solution $v^a(\lambda)$ allows us to *define* a vector *field* along the curve $x^a(\lambda)$, given the vector k^a at only one point. If the spacetime is actually flat, then this construction is equivalent to 'moving' the vector from event to event, maintaining the same Cartesian components, that is, the vector is moved 'parallel' to itself. The above construction generalizes the notion of 'parallel transport' to curved spacetime.

5.5 Formulas for covariant derivative

(a) From the definition of Γ^i_{km} it follows that

$$\Gamma^a_{ba} = \frac{1}{2}g^{ad}\frac{\partial g_{ad}}{\partial x^b} = \frac{1}{2g}\frac{\partial g}{\partial x^b} = \frac{\partial}{\partial x^b}(\ln\sqrt{-g}), \tag{5.52}$$

where we have used (5.32). Similarly,

$$g^{bc}\Gamma^a_{bc} = g^{bc}g^{ad}\left(\frac{\partial g_{db}}{\partial x^c} - \frac{1}{2}\frac{\partial g_{bc}}{\partial x^d}\right) = -\frac{1}{\sqrt{-g}}\frac{\partial(\sqrt{-g}g^{ab})}{\partial x^b}. \tag{5.53}$$

(b) Consider now the covariant divergence of a vector, $A^a_{;a}$. From the definition of the covariant derivative we obtain

$$A^a_{;a} = \frac{\partial A^a}{\partial x^a} + \Gamma^a_{ca}A^c = \frac{\partial A^a}{\partial x^a} + A^c\frac{\partial(\ln\sqrt{-g})}{\partial x^c} = \frac{1}{\sqrt{-g}}\frac{\partial(\sqrt{-g}A^a)}{\partial x^a}, \tag{5.54}$$

where we have used (5.52). Further, if A^a was a gradient of some function ϕ, $A^a = \partial^a\phi$, then $A^a_{;a}$ will represent the covariant Laplacian $\phi^{;a}_{;a}$ of the scalar ϕ. We see that

$$\phi^{;i}_{;i} = \frac{1}{\sqrt{-g}}\frac{\partial}{\partial x^i}\left(\sqrt{-g}g^{ik}\frac{\partial\phi}{\partial x^k}\right). \tag{5.55}$$

In particular, note that $\phi^{;i}_{;i}$ is *not* given by $g^{ik}\partial_i\partial_k\phi$. The above considerations remain valid in any dimensions.

(c) A similar result can be derived for the covariant derivative of an antisymmetric tensor A^{ab}. We have

$$A^{ab}_{\;\;;b} = \frac{\partial A^{ab}}{\partial x^b} + \Gamma^a_{db}A^{db} + \Gamma^b_{db}A^{ad} = \frac{\partial A^{ab}}{\partial x^b} + \Gamma^b_{db}A^{ad} \qquad (5.56)$$

since $\Gamma^a_{db}A^{db} = 0$ for an antisymmetric A^{ab}. Using (5.52), we obtain

$$A^{ab}_{\;\;;b} = \frac{1}{\sqrt{-g}}\frac{\partial(\sqrt{-g}A^{ab})}{\partial x^b}. \qquad (5.57)$$

Also notice that

$$V_{a;b} - V_{b;a} = \frac{\partial V_a}{\partial x^b} - \frac{\partial V_b}{\partial x^a} \qquad (5.58)$$

for any vector field V_a since the term involving the Γs cancels out in this expression.

(d) From our formula for the covariant derivative, we obtain

$$T^k_{\;i;k} = \frac{\partial T^k_i}{\partial x^k} + \Gamma^k_{lk}T^l_i - \Gamma^l_{ik}T^k_l = \frac{1}{\sqrt{-g}}\frac{\partial\left(\sqrt{-g}\,T^k_i\right)}{\partial x^k} - \Gamma^l_{ki}T^k_l. \qquad (5.59)$$

Expanding out Γ^l_{ki}, we find that the last term is equal to

$$-\frac{1}{2}\left(-\frac{\partial g_{ki}}{\partial x^l} + \frac{\partial g_{kl}}{\partial x^i} + \frac{\partial g_{il}}{\partial x^k}\right)T^{kl}. \qquad (5.60)$$

Because of the symmetry of T^{kl}, the first and third terms in the brackets cancel, giving

$$T^k_{\;i;k} = \frac{1}{\sqrt{-g}}\frac{\partial\left(\sqrt{-g}\,T^k_i\right)}{\partial x^k} - \frac{1}{2}\frac{\partial g_{kl}}{\partial x^i}T^{kl}. \qquad (5.61)$$

5.6 Physics in curved spacetime

(a) Consider the action for the electromagnetic field in flat spacetime (Cartesian coordinates) which is Lorentz invariant:

$$\mathscr{A}_{em} = -\frac{1}{16\pi}\int F_{ab}F^{ab}\,d^4X, \quad F_{ab} = A_{b,a} - A_{a,b}. \qquad (5.62)$$

This action can be rewritten in a form which is generally covariant by the following procedure. We replace the volume element d^4X by the covariant volume element in arbitrary coordinates $\sqrt{-g}\,d^4x$; we also change the ordinary derivative $A_{b,a}$ to the covariant derivative $A_{b;a}$, etc. This leads to the action

$$\mathscr{A}_{em} = -\frac{1}{16\pi}\int F_{ab}F^{ab}\sqrt{-g}\,d^4x, \quad F_{ab} = A_{b;a} - A_{a;b}. \qquad (5.63)$$

The electromagnetic field equation in a curved spacetime can be obtained by varying this action; we find that $F^{ab}_{\;\;;b} = 0$. Using the identities (5.57) and (5.58) derived in the last problem, we can write these equations as

$$\frac{1}{\sqrt{-g}}\partial_a(\sqrt{-g}F^{ab}) = 0, \quad F_{ab} = \frac{\partial A_b}{\partial x^a} - \frac{\partial A_a}{\partial x^b}. \qquad (5.64)$$

These equations describe the influence of the gravitational field on the electro-magnetic phenomenon. The second of these equations can also be rewritten in terms of F_{ik} as

$$\frac{\partial F_{ik}}{\partial x^l} + \frac{\partial F_{li}}{\partial x^k} + \frac{\partial F_{kl}}{\partial x^i} = 0. \tag{5.65}$$

Thus the first pair of Maxwell's equations has the same form in curved spacetime as in flat spacetime.

(b) Consider a spacetime in which the metric is separated out in the form

$$ds^2 = g_{ik}\, dx^i\, dx^k = g_{00}\, dt^2 + 2g_{0\alpha}\, dt\, dx^\alpha + g_{\alpha\beta}\, dx^\alpha\, dx^\beta. \tag{5.66}$$

For the sake of convenience, we will introduce the notation $h \equiv g_{00}$ and define a vector $\mathbf{g}$ with components $g_\alpha \equiv -(g_{0\alpha}/g_{00})$. The three-dimensional metric $\gamma_{\alpha\beta}$ is defined to be $[-g_{\alpha\beta} + h g_\alpha g_\beta]$. Given these definitions, it is straightforward to work out various components of Maxwell's equations in the $(3+1)$ form. The equation

$$\frac{\partial F_{ik}}{\partial x^l} + \frac{\partial F_{li}}{\partial x^k} + \frac{\partial F_{kl}}{\partial x^i} = 0 \tag{5.67}$$

becomes

$$\nabla \cdot \mathbf{B} = 0, \qquad \nabla \times \mathbf{E} = -\frac{1}{\sqrt{\gamma}} \frac{\partial}{\partial t} \left(\sqrt{\gamma} \mathbf{B} \right). \tag{5.68}$$

The equation

$$\frac{1}{\sqrt{-g}} \partial_a \left(\sqrt{-g} F^{ab} \right) = 0 \tag{5.69}$$

can be written as

$$\nabla \cdot \mathbf{D} = 0, \qquad \nabla \times \mathbf{H} = \frac{1}{\sqrt{\gamma}} \frac{\partial}{\partial t} \left(\sqrt{\gamma} \mathbf{D} \right), \tag{5.70}$$

where

$$\mathbf{D} = \frac{\mathbf{E}}{\sqrt{h}} + \mathbf{H} \times \mathbf{g}, \qquad \mathbf{B} = \frac{\mathbf{H}}{\sqrt{h}} + \mathbf{g} \times \mathbf{E}. \tag{5.71}$$

In static gravitational fields with $\mathbf{g} = 0$ and other metric components independent of time, these equations become

$$\nabla \cdot \mathbf{B} = 0, \quad \nabla \times \mathbf{E} = -\frac{\partial \mathbf{B}}{\partial t}, \quad \nabla \cdot \left(\frac{\mathbf{E}}{\sqrt{h}} \right) = 0, \quad \nabla \times \left(\sqrt{h} \mathbf{B} \right) = \frac{1}{\sqrt{h}} \frac{\partial \mathbf{E}}{\partial t}. \tag{5.72}$$

These equations are identical to Maxwell's equations in a material media with electric and magnetic permeability $\epsilon = \mu = (1/\sqrt{h})$.

5.7 Curvature

(a) The parallel transport of a vector along a curve is equivalent to solving the partial differential equation

$$\frac{\partial v^a}{\partial x^c} = -\Gamma^a_{bc}(x)v^b. \tag{5.73}$$

In general, such an equation cannot be integrated to give a unique vector field $v^a(x)$. This equation will have an acceptable, unique solution only if the integrability conditions

$$\frac{\partial^2 v^a}{\partial x^b \, \partial x^c} = \frac{\partial^2 v^a}{\partial x^c \, \partial x^b} \tag{5.74}$$

are satisfied. (Note that we are essentially demanding $dv^a = -\Gamma^a_{bc}(x)v^b(x)dx^c$ to be an exact differential.) Differentiating (5.73) with respect to x^d and using (5.73) again, we find

$$\frac{\partial^2 v^a}{\partial x^d \, \partial x^c} = -\left[\frac{\partial \Gamma^a_{lc}}{\partial x^d} - \Gamma^a_{bc}\Gamma^b_{dl}\right]v^l. \tag{5.75}$$

Hence, the integrability condition is equivalent to $R^a_{ldc}v^l = 0$, where

$$R^a_{ldc} = \frac{\partial \Gamma^a_{lc}}{\partial x^d} - \frac{\partial \Gamma^a_{ld}}{\partial x^c} + \Gamma^a_{bd}\Gamma^b_{cl} - \Gamma^a_{bc}\Gamma^b_{dl}. \tag{5.76}$$

If this result should hold for an arbitrary v^a, then R^a_{ldc} has to vanish. In general, this quantity will *not* vanish and hence parallel transport will not give a unique vector.

This condition has a simple physical meaning. To begin with, we note that R^i_{klm} is a tensor since it is the same object encountered earlier in problem 5.3(d). In a flat spacetime one can always choose a system of coordinates such that $\Gamma^i_{kl} = 0$. In such a coordinate system R^i_{klm} will vanish. Since R^i_{klm} is a tensor, it follows that it will vanish in any coordinate system in flat spacetime.

The converse is also true. Suppose we are given that R^i_{klm} vanishes identically in a given spacetime. We then choose some event in the spacetime and construct a locally inertial Cartesian coordinate system around it. Since parallel transport leads to a unique vector field when $R^i_{klm} = 0$, we can parallel transport this Cartesian coordinate system to any other event in the spacetime in a unique manner. This permits us to construct a global inertial coordinate system in the spacetime, proving that the spacetime is flat. Thus the vanishing of R^i_{klm} is both necessary and sufficient for the spacetime to be flat.

(b) This identity is most easily proved in a locally inertial frame around some event. In such a frame, $\Gamma^i_{kl} = 0$ and we have

$$R^n_{ikl;m} = \frac{\partial R^n_{ikl}}{\partial x^m} = \frac{\partial^2 \Gamma^n_{il}}{\partial x^m \, \partial x^k} - \frac{\partial^2 \Gamma^n_{ik}}{\partial x^m \, \partial x^l}. \tag{5.77}$$

For this expression it is straightforward to verify that

$$R^n_{ikl;m} + R^n_{imk;l} + R^n_{ilm;k} = 0. \tag{5.78}$$

Since this is a tensor equation, it will remain valid in any coordinate system if it is valid in the locally inertial frame.

Contracting this equation on ik and ln (i.e. by setting $i = k$ and $l = n$), we obtain the relation $R^l_{m;l} = (1/2)(\partial R/\partial x^m)$. This can be rewritten as

$$\left(R^l_m - \frac{1}{2}\delta^l_m R \right)_{;l} = 0. \tag{5.79}$$

(c) We begin by noting that R_{iklm} is antisymmetric in ik and lm but symmetric under pair exchange:

$$R_{iklm} = -R_{kilm} = -R_{ikml} = R_{lmik}. \tag{5.80}$$

It is also easy to verify that the cyclic sum of any three indices vanishes, that is,

$$R_{iklm} + R_{imkl} + R_{ilmk} = 0. \tag{5.81}$$

Thus the pairs of indices ik and lm can take six different sets of values which are independent. Hence, there are six components of R_{iklm} with identical values for ik and lm and $(6 \times 5)/2 = 15$ components of R_{iklm} with different values for ik and lm, giving a total of 21 independent components which respect the symmetries of (5.80). The components in which all the indices are different are, however, not independent of one another because of the identity in (5.81). There is one independent constraint which these components have to satisfy, namely,

$$R_{0123} + R_{0312} + R_{0231} = 0. \tag{5.82}$$

Hence, the number of independent components of R_{iklm} is $(21 - 1) = 20$. Note that this result agrees with the counting argument given in problem 5.2 as regards the second derivatives of g_{ik}.

5.8 Practice with metrics

(a) For a metric of the form

$$ds^2 = e^\nu \, dt^2 - e^\lambda \, dr^2 - r^2 \left(d\theta^2 + \sin^2 \theta \, d\phi^2 \right), \tag{5.83}$$

the non-vanishing Christoffel symbols are

$$\begin{gathered}
\Gamma^1_{11} = \frac{\lambda'}{2}, \quad \Gamma^0_{10} = \frac{\nu'}{2}, \quad \Gamma^2_{33} = -\sin\theta\cos\theta, \\[2mm]
\Gamma^0_{11} = \frac{\lambda}{2}e^{\lambda-\nu}, \quad \Gamma^1_{22} = -re^{-\lambda}, \quad \Gamma^1_{00} = \frac{\nu'}{2}e^{\nu-\lambda}, \\[2mm]
\Gamma^2_{12} = \Gamma^3_{13} = \frac{1}{r}, \quad \Gamma^3_{23} = \cot\theta, \quad \Gamma^0_{00} = \frac{\dot\nu}{2}, \\[2mm]
\Gamma^1_{10} = \frac{\dot\lambda}{2}, \quad \Gamma^1_{33} = -r\sin^2\theta \, e^{-\lambda}.
\end{gathered} \tag{5.84}$$

Here the prime denotes the derivative with respect to r and the dot denotes differentiation with respect to t. The components of G^i_k can be computed in a straightforward manner and we find

$$G^1_1 = -e^{-\lambda} \left(\frac{v'}{r} + \frac{1}{r^2} \right) + \frac{1}{r^2},$$

$$G^2_2 = G^3_3 = -\frac{1}{2} e^{-\lambda} \left(v'' + \frac{v'^2}{2} + \frac{v' - \lambda'}{r} - \frac{v'\lambda'}{2} \right) + \frac{1}{2} e^{-v} \left(\ddot{\lambda} + \frac{\dot{\lambda}^2}{2} - \frac{\dot{\lambda}\dot{v}}{2} \right),$$

$$G^0_0 = -e^{-\lambda} \left(\frac{1}{r^2} - \frac{\lambda'}{r} \right) + \frac{1}{r^2},$$

$$G^1_0 = -e^{-\lambda} \frac{\dot{\lambda}}{r}.$$

(5.85)

All other components vanish identically.

(b) In this case, the spacetime components G^i_0 vanish identically and the space–space components G^i_j are diagonal. The non-zero components are

$$G^0_0 = \frac{3}{a^2} \left(\dot{a}^2 + k \right), \qquad G^i_j = \frac{1}{a^2} \left(2a\ddot{a} + \dot{a}^2 + k \right) \delta^i_j, \qquad (5.86)$$

where the dot denotes differentiation with respect to time.

(c) In this line element, the spatial components of the metric tensor are constants and hence their derivatives vanish. The Christoffel symbols Γ^i_{kl} are made up entirely of the derivatives of g_{00}. Among these derivatives, the time derivatives are one order lower compared to spatial derivatives since time derivatives involve a factor ct. Therefore, to the lowest order, the non-trivial contribution to the Γs arises only from the spatial derivatives g_{00}. Hence, we obtain

$$\Gamma^\alpha_{00} \cong -\frac{1}{2} g^{\alpha\beta} \frac{\partial g_{00}}{\partial x^\beta} = \frac{1}{c^2} \frac{\partial \phi}{\partial x^\alpha}. \qquad (5.87)$$

By the same argument, the only non-zero component (to leading order) in R_{ik} is

$$R_{00} = R^0_0 = \frac{\partial \Gamma^\alpha_{00}}{\partial x^\alpha} = \frac{1}{c^2} \nabla^2 \phi. \qquad (5.88)$$

5.9 Dynamics of gravitational field

(a) Consider any quantity $f(g_{ik}, \partial g_{ik}/\partial x^l)$ which depends only on the metric tensor and its first derivative. Around any event in spacetime we can construct a locally inertial coordinate system in which $g_{ik} = \eta_{ik}$ and $(\partial g_{ik}/\partial x^l) = 0$. Hence, the function f will become a constant at any event in the locally inertial frame. If the function is a scalar, then its value will remain the same in any other coordinate system. In other words, any covariant scalar function built out of the metric tensor and its derivatives will be a trivial constant.

Consider, next, a function built out of the metric, its first derivative and second derivative. If the function is linear in the second derivative then it is possible to express the integral for the action as one arising from a Lagrangian involving only the metric and its first derivative and another term which contributes only at the boundaries. By considering suitable variations of the metric tensor we can ignore the contribution from the boundary term and thus obtain equations of motion which only involve up to second derivatives in the metric tensor.

(b) We start with the expression for $\sqrt{-g}R$ written in the form

$$\sqrt{-g}R = \sqrt{-g}g^{ik}R_{ik} = \sqrt{-g}\left\{g^{ik}\frac{\partial\Gamma^l_{ik}}{\partial x^l} - g^{ik}\frac{\partial\Gamma^l_{il}}{\partial x^k} + g^{ik}\Gamma^l_{ik}\Gamma^m_{lm} - g^{ik}\Gamma^m_{il}\Gamma^l_{km}\right\}. \tag{5.89}$$

We rewrite the first two terms on the right-hand side separating out a total derivative by using

$$\sqrt{-g}g^{ik}\frac{\partial\Gamma^l_{ik}}{\partial x^l} = \frac{\partial}{\partial x^l}\left(\sqrt{-g}g^{ik}\Gamma^l_{ik}\right) - \Gamma^l_{ik}\frac{\partial}{\partial x^l}\left(\sqrt{-g}g^{ik}\right), \tag{5.90}$$

$$\sqrt{-g}g^{ik}\frac{\partial\Gamma^l_{il}}{\partial x^k} = \frac{\partial}{\partial x^k}\left(\sqrt{-g}g^{ik}\Gamma^l_{il}\right) - \Gamma^l_{il}\frac{\partial}{\partial x^k}\left(\sqrt{-g}g^{ik}\right). \tag{5.91}$$

Then (5.89) becomes

$$R\sqrt{-g} = \sqrt{-g}\mathcal{G} + \frac{\partial\sqrt{-g}w^i}{\partial x^i}, \tag{5.92}$$

where

$$\sqrt{-g}\mathcal{G} = \Gamma^m_{im}\frac{\partial}{\partial x^k}\left(\sqrt{-g}g^{ik}\right) - \Gamma^l_{ik}\frac{\partial}{\partial x^l}\left(\sqrt{-g}g^{ik}\right) - \left(\Gamma^m_{il}\Gamma^l_{km} - \Gamma^l_{ik}\Gamma^m_{lm}\right)g^{ik}\sqrt{-g} \tag{5.93}$$

and

$$w^i = g^{ab}\Gamma^i_{ab} - g^{ik}\Gamma^l_{kl}. \tag{5.94}$$

The expression for $\sqrt{-g}\mathcal{G}$ can be simplified using the identity derived in problem 5.5 and the relation

$$\frac{\partial g^{ik}}{\partial x^l} = -\Gamma^i_{ml}g^{mk} - \Gamma^k_{ml}g^{im}. \tag{5.95}$$

The first two terms of $\sqrt{-g}\mathcal{G}$ reduce to

$$\sqrt{-g}\left(2\Gamma^l_{ik}\Gamma^i_{lm}g^{mk} - \Gamma^m_{im}\Gamma^i_{kl}g^{kl} - \Gamma^l_{ik}\Gamma^m_{lm}g^{ik}\right) = \sqrt{-g}g^{ik}\left(2\Gamma^l_{mk}\Gamma^m_{li} - \Gamma^m_{lm}\Gamma^l_{ik} - \Gamma^l_{ik}\Gamma^m_{lm}\right)$$

$$= 2\sqrt{-g}g^{ik}\left(\Gamma^m_{il}\Gamma^l_{km} - \Gamma^l_{ik}\Gamma^m_{lm}\right). \tag{5.96}$$

Using this expression in (5.93) we finally obtain

$$\sqrt{-g}\mathcal{G} = \sqrt{-g}g^{ik}\left(\Gamma^m_{il}\Gamma^l_{km} - \Gamma^l_{ik}\Gamma^m_{lm}\right). \tag{5.97}$$

The full Lagrangian $R\sqrt{-g}$ can be written as

$$R\sqrt{-g} = \mathcal{L} + \frac{\partial}{\partial x^c}\left[\sqrt{-g}\left(g^{ik}\Gamma^c_{ik} - g^{kc}\Gamma^m_{km}\right)\right], \tag{5.98}$$

where

$$\mathcal{L} \equiv \sqrt{-g}\mathcal{G} = \sqrt{-g}\,g^{ik}\left(\Gamma^m_{i\ell}\Gamma^\ell_{km} - \Gamma^l_{ik}\Gamma^m_{\ell m}\right). \tag{5.99}$$

(c) From $\mathcal{L}$ we can define the field momenta to be $\pi^{abc} = \left(\partial\mathcal{L}/\partial g_{ab,c}\right)$, where $g_{ab,c} = \partial g_{ab}/\partial x^c$. Following the prescription of problem 2.1, we subtract

$\partial(g_{ab}\pi^{abc})/\partial x^c$ from $\mathscr{L}$ to obtain the equivalent Lagrangian $\mathscr{L}'$ with the end point momenta held fixed. From (5.99) we find, after some algebra,

$$g_{ab}\pi^{abc} = \sqrt{-g}\left(g^{kc}\Gamma^m_{km} - g^{ik}\Gamma^c_{ik}\right). \tag{5.100}$$

Comparing this with (5.98) we see that

$$\mathscr{L}' = \mathscr{L} - \frac{\partial}{\partial x^c}\left(g_{ab}\pi^{abc}\right) = R\sqrt{-g}. \tag{5.101}$$

Thus the Lagrangian density $R\sqrt{-g}$ is the one which is obtained when we decide to fix the field momenta on the boundary rather than the potentials g_{ab}.

The Lagrangian density $\mathscr{L}'$ is special in that it is a genuine scalar density, whereas $\mathscr{L}$ is not. Had we chosen g^{ab} as our field variables rather than g_{ik}, our prescription would *not* have given the invariant $R\sqrt{-g}$. Indeed, one may show that this process only leads to that invariant $\mathscr{L}'$ when the field variables chosen are homogeneous of degree one in the g_{ik}. This restricts the choice to g_{ik} or $\sqrt{-g}\,g^{ab}$ and, in fact, we find that

$$\mathscr{L} - \frac{\partial}{\partial x^c}\left[\sqrt{-g}\,g^{ab}\frac{\partial\mathscr{L}}{\partial(\sqrt{-g}\,g^{ab}_{,c})}\right] = R\sqrt{-g}. \tag{5.102}$$

(d) To the lowest non-vanishing order, we can calculate the quantity $\sqrt{-g}R$ by using the expressions obtained in problem 5.8(c). We have

$$\sqrt{-g}R \cong g^{00}R_{00} \cong (1-2\phi)\nabla^2\phi = \nabla^2\phi - 2\phi\nabla^2\phi = \nabla^2\phi - 2\nabla\cdot(\phi\nabla\phi) + 2(\nabla\phi)^2. \tag{5.103}$$

The first two terms are total divergences which may be ignored, leaving us with

$$\sqrt{-g}R \cong 2(\nabla\phi)^2. \tag{5.104}$$

The action in Newtonian gravity is $(8\pi G)^{-1}(\nabla\phi)^2$; in order to reproduce the correct Newtonian limit, we need to take the Lagrangian in general relativity to be $(16\pi G)^{-1}\sqrt{-g}R$.

To obtain the dynamical equations for the gravitational field, we have to vary the action with respect to the metric tensor g^{ik}. The gravitational part of the action gives the contribution

$$\delta\int R\sqrt{-g}\,d^4x = \delta\int g^{ik}R_{ik}\sqrt{-g}\,d^4x$$

$$= \int\left\{R_{ik}\sqrt{-g}\,\delta g^{ik} + R_{ik}g^{ik}\delta\sqrt{-g} + g^{ik}\sqrt{-g}\,\delta R_{ik}\right\}d^4x. \tag{5.105}$$

Using the fact that

$$\delta\sqrt{-g} = -\frac{1}{2\sqrt{-g}}\delta g = -\frac{1}{2}\sqrt{-g}g_{ik}\,\delta g^{ik}, \tag{5.106}$$

we obtain

$$\delta\int R\sqrt{-g}\,d^4x = \int\left(R_{ik} - \frac{1}{2}g_{ik}R\right)\delta g^{ik}\sqrt{-g}\,d^4x + \int g^{ik}\delta R_{ik}\sqrt{-g}\,d^4x. \tag{5.107}$$

To proceed further, we need to evaluate the quantity $g^{ik}\delta R_{ik}$. This is most easily done in a locally inertial frame in which $\Gamma^i_{kl} = 0$. Using the definition of R_{ik}, we find that

$$g^{ik}\delta R_{ik} = g^{ik}\left\{\frac{\partial}{\partial x^l}\delta\Gamma^l_{ik} - \frac{\partial}{\partial x^k}\delta\Gamma^l_{il}\right\} = g^{ik}\frac{\partial}{\partial x^l}\delta\Gamma^l_{ik} - g^{il}\frac{\partial}{\partial x^l}\delta\Gamma^k_{ik} = \frac{\partial w^l}{\partial x^l}, \quad (5.108)$$

with

$$w^l = g^{ik}\delta\Gamma^l_{ik} - g^{il}\delta\Gamma^k_{ik}. \quad (5.109)$$

We now use the fact that even though Γ^i_{kl} is not a tensor $\delta\Gamma^i_{kl}$ is a tensor. To see this, we only have to note that $\Gamma^i_{kl}A^k dx^l$ is the change in a vector under parallel displacement between two infinitesimally separated points; hence $\delta\Gamma^i_{kl}A^k dx^l$ is the difference between two vectors obtained by two parallel displacements (one with Γ^i_{kl} and the other with $(\Gamma^i_{kl} + \delta\Gamma^i_{kl})$) between two infinitesimal points. The difference between two vectors at the same point is a vector and hence $\delta\Gamma^i_{kl}$ is a tensor. Since it is a tensor we may write the above expression in any coordinate system by replacing $(\partial w^l/\partial x^l)$ by $w^l_{;l}$. This gives

$$g^{ik}\delta R_{ik} = \frac{1}{\sqrt{-g}}\frac{\partial}{\partial x^l}\left(\sqrt{-g}\,w^l\right) \quad (5.110)$$

and, consequently,

$$\int g^{ik}\delta R_{ik}\sqrt{-g}\,d^4x = \int \frac{\partial\left(\sqrt{-g}\,w^l\right)}{\partial x^l}\,d^4x. \quad (5.111)$$

This integral can be transformed into a surface integral. We will now assume that we shall only consider the variations δg^{ik} for which this surface term vanishes. In that case, this term does not contribute and the variation of the gravitational action gives

$$\delta\int R\sqrt{-g}\,d^4x = \int\left(R_{ik} - \frac{1}{2}g_{ik}R\right)\delta g^{ik}\sqrt{-g}\,d^4x. \quad (5.112)$$

The variation of the matter part of the action can be expressed in the form

$$\delta A_{\mathrm{m}} = \frac{1}{2}\int T_{ik}\,\delta g^{ik}\sqrt{-g}\,d^4x. \quad (5.113)$$

(This equation defines the second rank symmetric tensor T_{ik}.) Setting the total variation to zero, we obtain the equations for the gravitational field:

$$R_{ik} - \frac{1}{2}g_{ik}R = 8\pi G T_{ik}. \quad (5.114)$$

(e) The above equation shows that T^{ik} is the source of the gravitational field in Einstein's theory. It also follows from the definition that it is a symmetric tensor. In the Newtonian limit the component T^{00} should reduce to the mass density. This suggests that we interpret T^{ik} as the energy–momentum tensor of the matter field. In fact, equation (5.113) may be considered to be the definition of the stress-tensor in a fully relativistic situation.

To prove that $T^i_{k;i} = 0$, we shall proceed as follows. Consider an infinitesimal coordinate transformation from x^i to $x'^i = x^i + \xi^i(x)$, where the $\xi^i(x)$ are considered to be infinitesimal quantities. Under this transformation the metric tensor changes to

$$g'^{ik}(x'^l) = g^{lm}(x^l) \frac{\partial x'^i}{\partial x^l} \frac{\partial x'^k}{\partial x^m} = g^{lm}\left(\delta^i_l + \frac{\partial \xi^i}{\partial x^l}\right)\left(\delta^k_m + \frac{\partial \xi^k}{\partial x^m}\right)$$

$$\cong g^{ik}(x^l) + g^{im}\frac{\partial \xi^k}{\partial x^m} + g^{kl}\frac{\partial \xi^i}{\partial x^l}. \tag{5.115}$$

In this expression g'^{ik} is a function of x'^l, while g^{ik} is a function of x^l. We can express g'^{ik} as a function of the original coordinates x^l by expanding $g'^{ik}(x^l + \xi^l)$ in a Taylor series in powers of ξ^l and retaining up to linear order. This will give

$$g'^{ik}(x^l) = g^{ik}(x^l) - \xi^l \frac{\partial g^{ik}}{\partial x^l} + g^{il}\frac{\partial \xi^k}{\partial x^l} + g^{kl}\frac{\partial \xi^i}{\partial x^l}. \tag{5.116}$$

This expression provides the net change in the component g^{ik} treated as a function of the coordinates x^l. It is easy to verify that the last three terms can be combined to give $\xi^{i;k} + \xi^{k;i}$. Therefore, we see that the change in the metric tensor under an infinitesimal coordinate transformation can be expressed in the form

$$g'^{ik} = g^{ik} + \delta g^{ik}, \quad \delta g^{ik} = \xi^{i;k} + \xi^{k;i}. \tag{5.117}$$

Similarly,

$$g'_{ik} = g_{ik} + \delta g_{ik}, \quad \delta g_{ik} = -\xi_{i;k} - \xi_{k;i}. \tag{5.118}$$

Consider now the variation in the matter action A_m when the coordinates are changed by an infinitesimal amount. The change of coordinates will induce certain variation δq in the matter variables and a variation $\delta g^{ik} = \xi^{i;k} + \xi^{k;i}$ in the metric. The net change in the action is

$$\delta A_m = \left(\frac{\delta A_m}{\delta q}\right)_g \delta q + \left(\frac{\delta A_m}{\delta g^{ik}}\right)_q \delta g^{ik}. \tag{5.119}$$

However, when the equations of motion for the matter variables are satisfied, we know that $(\delta A/\delta q)_g = 0$. Further, since action is a scalar, we know that the total variation δA_m should vanish. Hence, we must have

$$0 = \delta A_m = \frac{1}{2}\int T_{ik}\, \delta g^{ik} \sqrt{-g}\, d^4x \tag{5.120}$$

when $\delta g^{ik} = \xi^{i;k} + \xi^{k;i}$. Using the symmetry of T_{ik}, we can write

$$0 = \frac{1}{2}\int T_{ik}\left(\xi^{i;k} + \xi^{k;i}\right)\sqrt{-g}\, d^4x = \int T_{ik}\xi^{i;k}\sqrt{-g}\, d^4x$$

$$= \int \left(T^k_i \xi^i\right)_{;k}\sqrt{-g}\, d^4x - \int T^k_{i;k}\xi^i \sqrt{-g}\, d^4x. \tag{5.121}$$

The first term can be expressed in the form

$$\int \frac{\partial}{\partial x^k} \left(\sqrt{-g} T_i^k \xi^i \right) d^4 x,$$ (5.122)

which could be converted into an integral over a hypersurface. Since ξ^i may be taken to vanish on this hypersurface, this integral drops out, allowing us to conclude that

$$T_{i;k}^k = 0.$$ (5.123)

(i) To work out the stress-tensor for the scalar field we shall consider the variation of the scalar field Lagrangian:

$$A_\phi = \frac{1}{2} \int g^{ik} \frac{\partial \phi}{\partial x^i} \frac{\partial \phi}{\partial x^k} \sqrt{-g} \, d^4 x$$ (5.124)

under the variation $g^{ik} \rightarrow g^{ik} + \delta g^{ik}$. We have

$$\delta A_\phi = \frac{1}{2} \int d^4 x \left\{ \delta g^{ik} \frac{\partial \phi}{\partial x^i} \frac{\partial \phi}{\partial x^k} \sqrt{-g} + g^{ab} \frac{\partial \phi}{\partial x^a} \frac{\partial \phi}{\partial x^b} \left(-\frac{1}{2} \sqrt{-g} g_{ik} \, \delta g^{ik} \right) \right\}$$
$$= \frac{1}{2} \int d^4 x \sqrt{-g} \, \delta g^{ik} \left[\frac{\partial \phi}{\partial x^i} \frac{\partial \phi}{\partial x^k} - \frac{1}{2} g_{ik} \left(g^{ab} \frac{\partial \phi}{\partial x^a} \frac{\partial \phi}{\partial x^b} \right) \right].$$ (5.125)

Comparing with our definition in (5.113) we find that

$$T_{ik} = \left[\frac{\partial \phi}{\partial x^i} \frac{\partial \phi}{\partial x^k} - \frac{1}{2} g_{ik} \left(g^{ab} \frac{\partial \phi}{\partial x^a} \frac{\partial \phi}{\partial x^b} \right) \right].$$ (5.126)

This condition $T^{ik}_{\ \ ;k} = 0$ is equivalent to the field equation $\phi^{;i}_{\ \ ;i} = 0$. Similar analysis is applicable even if the scalar field has a potential and is described by

$$L = \frac{1}{2} \left(\frac{\partial \phi}{\partial x^i} \frac{\partial \phi}{\partial x_i} \right) - V(\phi).$$ (5.127)

Then

$$T_{ik} = \left(\frac{\partial \phi}{\partial x^i} \right) \left(\frac{\partial \phi}{\partial x^k} \right) - g_{ik} L.$$ (5.128)

(ii) In the case of the electromagnetic field, the action is

$$A = -\frac{1}{16\pi} \int F_{ab} F^{ab} \sqrt{-g} \, d^4 x = -\frac{1}{16\pi} \int F_{ab} F_{dc} g^{ad} g^{cb} \sqrt{-g} \, d^4 x.$$ (5.129)

On varying g_{ab}, we obtain

$$\delta A = -\frac{1}{16\pi} \int d^4 x \left[2 F_{ab} F_{dc} g^{ad} \delta g^{bc} \sqrt{-g} + F_{mn} F^{mn} \delta(\sqrt{-g}) \right]$$
$$= -\frac{1}{16\pi} \int d^4 x \left[2 F_{ab} F^a_{\ c} \delta g^{bc} \sqrt{-g} + F_{mn} F^{mn} \frac{1}{2} \frac{(-1)}{\sqrt{-g}} (-g) g_{bc} \, \delta g^{bc} \right]$$
$$= \frac{1}{16\pi} \int d^4 x \sqrt{-g} \, \delta g^{bc} \left[-F_{ab} F^a_{\ c} + \frac{1}{4} F_{mn} F^{mn} g_{bc} \right],$$ (5.130)

so that

$$4\pi T_{bc} = -F_{ab}F^a{}_c + \frac{1}{4}F_{mn}F^{mn}g_{bc} \qquad (5.131)$$

which is the standard expression for the energy–momentum tensor of the electromagnetic field; T_0^0 denotes the energy density and T_0^m denotes the energy flux along the m-direction, etc. The conditions $T^a_{b;a} = 0$ can easily be shown to give the equations $F^a{}_{b;a} = 0$, which are Maxwell's equations in curved spacetime derived before. In this sense, Einstein's equations imply the equations of motion for the source term.

(f) The infinitesimal version of the transformation $g^{ik} \to \Omega^2 g^{ik}$ can be expressed as $g^{ik} \to [1 + \epsilon(x)]^2 g^{ik} \cong g^{ik} + 2\epsilon(x)g^{ik}$ so that $\delta g^{ik} = 2\epsilon(x)g^{ik}$. We are told that $\delta A_m = 0$ under such a variation. This gives

$$0 = \delta A_m = \int T_{ik}g^{ik}\epsilon(x)\sqrt{-g}\, d^4x = \int T^i_i \epsilon(x)\sqrt{-g}\, d^4x. \qquad (5.132)$$

Hence it follows that $T^i_i = 0$ if matter action is invariant under the conformal transformations.

(g) We have shown that for any valid matter action $T^i_{k;i} = 0$. Hence, the right-hand side of Einstein's equation has zero covariant divergence. We have seen earlier (see equation (5.79)) that the left-hand side also has vanishing divergence. Thus the ten equations in (5.114) are constrained by the four identities (5.79) leaving six independent equations. Also note that, among the ten variables g^{ik}, four can be assigned specific values by suitable choice of coordinates. Thus there are only six independent functions we need to solve for, which is the same as the number of independent equations available.

Structurally, Einstein's equations are ten second-order partial differential equations with the independent variable x^i. Thus, *a priori*, one would have thought that the values of the metric tensor g^{ik} and its first time derivatives ($\partial g^{ik}/\partial t$) need to be specified as initial conditions. This, however, is not true because of certain peculiar features in Einstein's equations.

It is clear from the definition of R_{iklm} that the second derivatives with respect to time are contained only in the components $R_{0\alpha0\beta}$, which they enter through the term $(-1/2)\ddot{g}_{\alpha\beta}$. This shows that the second derivatives of the metric components $g_{0\alpha}$ and g_{00} do not appear in R_{iklm} and hence in Einstein's equations. Further, even the second derivatives of $g_{\alpha\beta}$ appear only in the space–space part of Einstein's equations. The time–time part and the space–space part contain time derivatives *only up to first order*. To prove this claim, note that (5.79) can be written in the form

$$\left(R_i^0 - \frac{1}{2}\delta_i^0 R\right)_{;0} = -\left(R_i^\alpha - \frac{1}{2}\delta_i^\alpha R\right)_{;\alpha}, \qquad (5.133)$$

where $i = 0, 1, 2, 3$. The highest time derivatives appearing on the right-hand side of these equations are second derivatives; hence it follows that the quantities within the parentheses on the left-hand side can only contain first-order time derivatives. Hence, the time–time and time–space components G_0^0 and G_α^0 contain only the first time derivatives of the metric tensor.

Furthermore, the space–time and time–time equations do not contain the first derivatives $\dot{g}_{0\alpha}$ and $\dot{g}_{00}$ but only $\dot{g}_{\alpha\beta}$. This is because among all Christoffel symbols, only $\Gamma_{\alpha,00}$ and $\Gamma_{0,00}$ contain these quantities; but these appear only in the components $R_{0\alpha0\beta}$ which drop out from the time–time and time–space part of Einstein's equations.

The above considerations show that: (i) the space–time and the time–time parts of Einstein's equations only involve $\dot{g}_{\alpha\beta}$ as the highest-order time derivatives; (ii) the space–space part contains $\ddot{g}_{\alpha\beta}$; (iii) the time derivatives of g_{00} and $g_{0\alpha}$ do not appear in Einstein's equations.

It is therefore possible to assign as initial conditions the functions $g_{\alpha\beta}$ and $\dot{g}_{\alpha\beta}$ at some time $t = t_0$. The space–time and time–time parts of Einstein's equations will then determine the initial values of $g_{0\alpha}$ and g_{00}. (They are not freely specifiable and are determined by the constraint equations.) The initial values of $\dot{g}_{0\alpha}$ remain arbitrary. This is the valid initial data for integrating Einstein's equations.

The above considerations use an arbitrary coordinate system. If one uses liberty in the choice of coordinate system, it is possible to show that the gravitational field has only two genuine degrees of freedom per event. This can be seen as follows. We first use the four coordinate transformations $x^i \to x'^i$ to arrange that $g_{00} = 1$ and $g_{0\alpha} = 0$ in the neighbourhood of any event. This reduces the metric to the form

$$ds^2 = dt^2 + g_{\alpha\beta}\, dx^\alpha dx^\beta, \tag{5.134}$$

leaving six components of $g_{\alpha\beta}$ non-zero. Consider, now, the infinitesimal coordinate transformation $t \to t' = t + f$, $x^\alpha \to x'^\alpha = x^\alpha + \lambda^\alpha$. It is easy to verify that this will change the metric coefficients to

$$\begin{aligned}
g'_{00} &= 1 + 2\dot{f}, \\
g'_{0\alpha} &= f_{,\alpha} + g_{\alpha\beta}\dot{\lambda}^\beta, \\
g'_{\alpha\beta} &= g_{\alpha\beta} + \lambda^\gamma_{,\alpha}g_{\gamma\beta} + \lambda^\gamma_{,\beta}g_{\alpha\gamma}.
\end{aligned} \tag{5.135}$$

Demanding that the coordinate transformation should not change the conditions $g_{00} = 1$, $g_{0\alpha} = 0$ will lead to the constraints

$$\dot{f} = 0, \qquad f_{,\alpha} = -g_{\alpha\beta}\dot{\lambda}^\beta. \tag{5.136}$$

The first equation implies that f is a function of space alone; $f = f(\mathbf{x})$. Using this, the second equation can be integrated to give

$$\lambda^\alpha = p^\alpha(\mathbf{x}) - f_{,\beta} \int^t dt\, g^{\alpha\beta}, \tag{5.137}$$

where $p^\alpha(\mathbf{x})$ are arbitrary functions. Thus on *any given spatial hypersurface*, we have the freedom to choose the four functions $f(\mathbf{x})$ and $p^\alpha(\mathbf{x})$ in order to bring four of the six components of $g^{\alpha\beta}(\mathbf{x}, t)$ to preassigned values. Thus, only two components of $g_{\alpha\beta}$ remain arbitrary to be propagated forward in time.

5.10 Geodesic deviation

(a) From the definition of the covariant derivative and the relation $(\partial u^i / \partial v) = \partial v^i / \partial s$ (where u^i is the tangent vector to the geodesic), it follows that

$$u^i{}_{;k} v^k = v^i{}_{;k} u^k . \tag{5.138}$$

Consider now the variation of the separation vector v^i between two neighbouring geodesics. We have

$$\frac{D^2 v^i}{Ds^2} \equiv (v^i{}_{;k} u^k)_{;l} u^l = (u^i{}_{;k} v^k)_{;l} u^l = u^i{}_{;k;l} v^k u^l + u^i{}_{;k} v^k{}_{;l} u^l . \tag{5.139}$$

Using (5.138) in the second term and changing the order of the covariant derivatives in the first term we obtain

$$\frac{D^2 v^i}{Ds^2} = \left(u^i{}_{;l} u^l\right)_{;k} v^k + u^m R^i_{mkl} u^k v^l . \tag{5.140}$$

The first term vanishes since $u^i{}_{;l} u^l = 0$ along the geodesics and we are left with

$$\frac{D^2 v^i}{Ds^2} = R^i_{klm} u^k u^l v^m . \tag{5.141}$$

(b) Let us choose a coordinate system which is locally inertial along the trajectory of the first observer. In such a frame, the observer's four-velocity will be $u^i = \delta^i_0$ and the spatial component of the geodesic deviation equation becomes

$$g^\alpha \equiv \frac{D^2 v^\alpha}{Ds^2} = R^\alpha_{abi} u^a u^b v^i = R^\alpha_{00\beta} v^\beta . \tag{5.142}$$

(Note that $R^a_{b00} = 0$ due to antisymmetry in the last two indices and hence only the spatial part of v^β contributes on the right-hand side.) Taking the divergence of the relative acceleration, we obtain

$$\nabla_\mathbf{v} \cdot \mathbf{g} \equiv \frac{\partial g^\alpha}{\partial v^\alpha} = R^\alpha_{00\alpha} = -R_{00} = -8\pi G \left(T_{00} - \frac{1}{2} T \right) . \tag{5.143}$$

In the locally inertial coordinates we are using $(T_{00} - (1/2)T) = (\rho + T^\alpha_\alpha)$, where α is summed over the spatial indices. In the case of an ideal fluid, this equation becomes

$$\nabla \cdot \mathbf{g} = -4\pi G(\rho + 3p) . \tag{5.144}$$

This equation does not require any approximation regarding the gravitational field and is always valid in the frame of a freely falling observer. It shows that the effective gravitational mass in the frame of a freely falling observer is $(\rho + 3p)$.

5.11 Schwarzschild metric

(a) The most general spherically symmetric metric can be expressed in the form

$$ds^2 = A(r,\bar{t})\,d\bar{t}^2 + B(r,\bar{t})\,dr^2 + C(r,\bar{t})\,dr\,d\bar{t} - r^2\left(d\theta^2 + \sin^2\theta\,d\phi^2\right). \tag{5.145}$$

This line element is constructed so that surfaces of $t = $ constant, $r = $ constant have an area $4\pi r^2$. In fact, this criterion uniquely fixes the coordinate r. We can now make a transformation from $\bar{t}$ to t such that the quadratic form in $dr\,d\bar{t}$ is diagonalized. This will lead to a metric which can be written in the form

$$ds^2 = e^\nu dt^2 - e^\lambda dr^2 - r^2\left(d\theta^2 + \sin^2\theta\,d\phi^2\right), \tag{5.146}$$

where ν and λ are functions of r and t. The expression for G^i_k corresponding to this line element was worked out in problem 5.8. Using the results of that problem we can write Einstein's equation in empty space outside the matter to be

$$e^{-\lambda}\left(\frac{\nu'}{r} + \frac{1}{r^2}\right) - \frac{1}{r^2} = 0, \quad e^{-\lambda}\left(\frac{\lambda'}{r} - \frac{1}{r^2}\right) + \frac{1}{r^2}, \quad \dot\lambda = 0. \tag{5.147}$$

The last equation shows that λ is independent of time; further, adding the first two equations we see that $\lambda' + \nu' = 0$, implying $\lambda + \nu = f(t)$, where $f(t)$ is a function of time alone. In the metric (5.146), one still has the freedom of changing the time coordinate from t to any other arbitrary function of t. By using this freedom we can set $f(t) = 0$. Hence, we find that $\nu = -\lambda$, with both ν and λ depending only on r.

This result shows that all spherically symmetric solutions to source-free Einstein equations are time-independent. For example, consider a star which is collapsing *radially* in a *time-dependent* manner. Outside the star, the metric is independent of time. This result is known as Birkoff's theorem.

Integrating the first two equations in (5.147) we easily find that

$$e^\nu = e^{-\lambda} = 1 + \frac{\text{constant}}{r}. \tag{5.148}$$

At large distances from the spherical body, one expects the gravitational field to be weak and describable by Newtonian theory. In this limit, the metric should take the form $g_{00} \cong 1 + 2\phi = 1 - 2GM/r$. It follows that the constant in equation (5.148) should have the value $2GM$, where M is the mass of the central object. Hence, the metric can be expressed in the form

$$ds^2 = \left(1 - \frac{2GM}{r}\right)dt^2 - \left(1 - \frac{2GM}{r}\right)^{-1} dr^2 - r^2\left(d\theta^2 + \sin^2\theta\,d\phi^2\right). \tag{5.149}$$

Consider now the metric due to a thin spherical shell. Since spacetime inside and outside the shell are spherically symmetric and empty, the metric must have the form of (5.149) with (possibly) different constants M_{in} and M_{out} both inside and outside the shell. From the conditions at $r \to \infty$, it is clear that $M_{\text{out}} = M_{\text{shell}}$. The regularity at $r = 0$ will require $M_{\text{in}} = 0$. Hence, the spacetime inside a spherical shell is flat.

(b) The motion of a particle of mass m in a metric g^{ik} can be most easily tackled by using the relativistic Hamilton–Jacobi equation. We know that the momentum and energy of a particle can be expressed in terms of the derivatives of the action $A(x^i)$ of a particle treated as the function of the end points. In a relativistic situation, this relation has the form $p_i = (\partial A/\partial x^i)$. Since $p_i p^i = m^2$ the relativistic Hamilton–Jacobi equation becomes

$$g^{ik} \frac{\partial A}{\partial x^i} \frac{\partial A}{\partial x^k} - m^2 = 0. \tag{5.150}$$

In the case of the Schwarszschild metric, this equation is

$$\left(1 - \frac{2GM}{r}\right)^{-1} \left(\frac{\partial A}{\partial t}\right)^2 - \left(1 - \frac{2GM}{r}\right) \left(\frac{\partial A}{\partial r}\right)^2 - \frac{1}{r^2} \left(\frac{\partial A}{\partial \phi}\right)^2 = m^2. \tag{5.151}$$

We have assumed that the motion takes place in the plane $\theta = \pi/2$ and set $(\partial A/\partial \theta) = 0$. This Hamilton–Jacobi equation can be solved by the ansatz

$$A = -\mathscr{E}t + L\phi + A_r(r), \tag{5.152}$$

where L and $\mathscr{E}$ denote the angular momentum and energy of the particle. Substituting into the Hamilton–Jacobi equation and solving we find

$$A_r = \int \sqrt{\mathscr{E}^2 \left(1 - \frac{2GM}{r}\right)^{-2} - \left(m^2 + \frac{L^2}{r^2}\right) \left(1 - \frac{2GM}{r}\right)^{-1}} \, dr. \tag{5.153}$$

The trajectory of the particle is determined by the equations $(\partial A/\partial \mathscr{E}) = \text{constant}$ and $(\partial A/\partial L) = \text{constant}$. This gives $r(t)$ and $\phi(r)$ to be

$$t = \frac{\mathscr{E}}{m} \int dr \left(1 - \frac{2GM}{r}\right)^{-1} \left[\left(\frac{\mathscr{E}}{m}\right)^2 - \left(1 + \frac{L^2}{m^2 r^2}\right) \left(1 - \frac{2GM}{r}\right)\right]^{-1/2} \tag{5.154}$$

and

$$\phi = \int dr \left(\frac{L}{r^2}\right) \left[\mathscr{E}^2 - \left(m^2 + \frac{L^2}{r^2}\right) \left(1 - \frac{2GM}{r}\right)\right]^{-1/2}. \tag{5.155}$$

From equation (5.154) we obtain

$$\left(1 - \frac{2GM}{r}\right)^{-1} \frac{dr}{dt} = \frac{1}{\mathscr{E}} \left[\mathscr{E}^2 - V_{\text{eff}}^2(r)\right]^{1/2}, \tag{5.156}$$

with

$$V_{\text{eff}}^2(r) = m^2 \left(1 - \frac{2GM}{r}\right) \left(1 + \frac{L^2}{m^2 r^2}\right). \tag{5.157}$$

Similarly, from (5.155) we can determine $(d\phi/dr)$, which, when combined with the expression for (dr/dt), gives

$$r^2 \dot{\phi} = \left(\frac{L}{\mathscr{E}}\right) \left(1 - \frac{2GM}{r}\right). \tag{5.158}$$

Figure 5.1 gives a plot of (V_{eff}/m) against $(rc^2/GM) = r/M$ (in units with $G = c = 1$) for different values of L. Several important aspects of the motion can be deduced from this diagram.

(i) For a given value of L and $\mathscr{E}$ the nature of the orbit will, in general, be governed by the turning points in r, determined by the equation $V_{\text{eff}}^2(r) = \mathscr{E}^2$. For a given $L > 4GMm$ the function $V_{\text{eff}}(r)$ has one maxima and one minima. If the energy $\mathscr{E}$ of the particle is lower than m, then there will be two turning points. The particle will orbit the central body with a perihelion and an apehelion. This is similar to elliptic orbits in Newtonian gravity.

(ii) If $m < \mathscr{E} < V_{\text{max}}(L)$ there will only be one turning point. The particle will approach the central mass from infinity, reach a radius of closest approach and will travel back to infinity. This is similar to the hyperbolic orbits in Newtonian gravity.

(iii) If $\mathscr{E} > V_{\text{max}}(L)$, the particle 'falls to the centre'. This behaviour is in sharp contrast to Newtonian gravity, in which a particle with non-zero angular momentum can never reach $r = 0$.

(iv) As the angular momentum is lowered, V_{max} decreases and for $L = 4GMm$ the maximum value of the potential is at $V_{\text{max}} = m$. In this case all particles from infinity will fall to the origin. Particles with $\mathscr{E} < m$ will still have two turning points and will form bound orbits.

(v) When L is reduced still further, the maxima and minima of $V(r)$ approach each other. For $L \leq 2\sqrt{3}GMm$, there are no turning points in the $V(r)$ curve. Particles with lower angular momentum will fall to the origin irrespective of the energy.

(c) In order to determine the orbit of the particle it is convenient to begin from equation (5.155). We introduce the variables $u = GM/r$, $\epsilon = \mathscr{E}/m$ and $l = L/GMm$, in terms of which (5.155) becomes

$$\left(\frac{du}{d\phi}\right)^2 = \frac{1}{l^2}\left[\epsilon^2 - (1 - 2u)\left(1 + l^2 u^2\right)\right]. \tag{5.159}$$

Differentiating this equation with respect to ϕ we obtain

$$u'' + u = \frac{1}{l^2} + 3u^2, \tag{5.160}$$

where primes denote differentiation with respect to ϕ. For a large impact parameter (leading to small deflections) the right-hand side of (5.160) will be small. We can solve this equation perturbatively by assuming that the solution has the form

$$u = a\cos\phi + v(\phi), \tag{5.161}$$

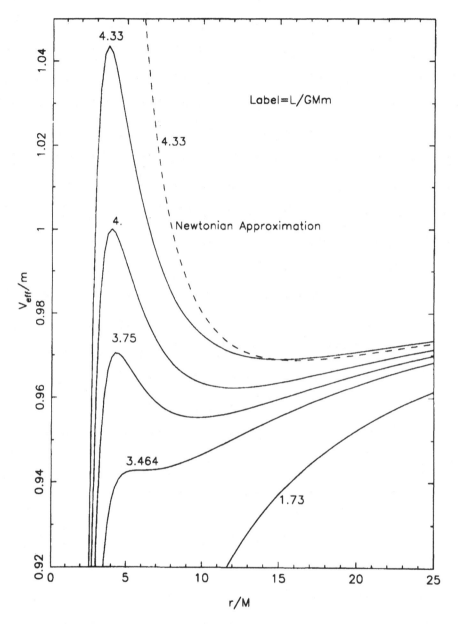

Fig. 5.1. Effective potential for motion in Schwarzschild metric.

where a is a constant and $v \ll a$. Substituting this ansatz in (5.160) we find that v satisfies the equation

$$v'' + v \cong \frac{1}{l^2} + 3a^2 \cos^2 \phi \qquad (5.162)$$

to the lowest order. The solution to this equation is

$$v = \frac{1}{l^2} + 2a^2 - a^2 \cos^2 \phi \tag{5.163}$$

and hence, to this order of accuracy,

$$u = a \cos \phi - a^2 \cos^2 \phi + \frac{1}{l^2} + 2a^2. \tag{5.164}$$

The first term on the right-hand side represents the unperturbed straight line motion and the other three terms give the lowest order corrections. When the particle is far away from the centre, $u \rightarrow 0$; so the asymptotes of the orbit are determined by the condition

$$0 = u = a^2 \cos^2 \phi - a \cos \phi - b, \quad b = \frac{1}{l^2} + 2a^2, \tag{5.165}$$

with the solution

$$\cos \phi = \frac{a - (a^2 + 4a^2 b)^{1/2}}{2a^2} \cong -\frac{b}{a} \tag{5.166}$$

when $b/a \ll 1$. To the same order of accuracy we find that $\phi \cong [(\pi/2) - (b/a)]$. Since there is equal deflection at both the asymptotes, the total angle of deflection is given by $\delta\phi = 2b/a$.

We can relate the deflection to the properties of the particle as follows. If R is the impact parameter then our ansatz shows that $a = GM/R$. Further, $l^2 = L^2/G^2 M^2 m^2$ with $L^2 = R^2 p_\infty^2$, where p_∞ is the initial momentum of the particle at infinity. Using $p_\infty^2 = \mathscr{E}^2 - m^2$ and $\mathscr{E} = m/\sqrt{1 - \beta^2}$ we find that

$$l^2 = \left(\frac{R}{GM}\right)^2 \frac{\beta^2}{1 - \beta^2}, \quad b = \left(\frac{GM}{R}\right)^2 \frac{1 + \beta^2}{\beta^2}. \tag{5.167}$$

Hence, the deflection is

$$\delta\phi = \frac{2GM}{R} \left(\frac{1 + \beta^2}{\beta^2}\right). \tag{5.168}$$

Clearly, $\delta\phi \rightarrow 4GM/R$ when $\beta \rightarrow 1$; this is twice the Newtonian value.

We shall next determine the deflection of a charged particle in the case of electromagnetism. The relevent equations in this case are

$$\frac{d}{dt} (m\gamma\dot{r}) = \frac{L^2}{m\gamma r^3} - \frac{Ze^2}{r^2}, \tag{5.169}$$

$$r^2\dot{\phi} = \frac{L}{m\gamma}, \quad m\gamma - \frac{Ze^2}{r} = E, \tag{5.170}$$

where E and L are the conserved energy and angular momentum of the particle and $\gamma = (1 - \beta^2)^{-1/2}$. Combining these equations, it is easy to obtain a differential equation for $u(\phi)$, where $u = 1/r$. We obtain

$$\frac{du^2}{d\phi^2} + \left(1 - \frac{Z^2 e^4}{L^2}\right) u = \frac{Ze^2 E}{L^2}. \tag{5.171}$$

This equation can be solved exactly to give

$$u = \frac{1}{R} \cos\left[\left(1 - \frac{Z^2 e^4}{L^2}\right)^{1/2} \phi\right] + \frac{E}{L^2}\left(\frac{Ze^2}{1 - Z^2 e^4/L^2}\right). \qquad (5.172)$$

In the case of small deflections, we may assume that $L^2 \gg Z^2 e^4$. Then the solution becomes

$$u = \frac{1}{R} \cos\phi + \frac{Ze^2 E}{L^2}. \qquad (5.173)$$

Determining the asymptotes as before, we find that the total deflection is given by $\delta\phi_{em} \cong (2RZe^2 E/L)$. Expressing E and L in terms of β, we obtain

$$\delta\phi_{em} = \frac{2Ze^2}{mR}\frac{(1 - \beta^2)}{\beta^2}. \qquad (5.174)$$

We note that the deflection vanishes as $\beta \to 1$. Since gravitational and electromagnetic fields transform differently under Lorentz transformation, the deflections have different β-dependences.

5.12 Gravitational lensing

(a) The geometry of gravitational lensing is shown in figure 5.2. We assume that the rays propagate in straight lines and are deflected instantaneously in the lens plane. From the geometry of the diagram it follows that $\alpha D_{LS} + \theta_s D_{OS} = \theta_i D_{OS}$. We shall project the lengths $\theta_s D_{OS}$, αD_{LS} and $\theta_i D_{OS}$ to the lens plane and obtain **s**, **d** and **i**. In that case,

$$D_{OS}\,\mathbf{s} + D_{LS}\,\mathbf{d} = D_{OS}\,\mathbf{i} \qquad (5.175)$$

or

$$\mathbf{s} = \mathbf{i} - \frac{D_{LS}}{D_{OS}}\mathbf{d}(\mathbf{i}). \qquad (5.176)$$

(b) Consider a bounded density distribution $\rho(\mathbf{x})$ producing a gravitational potential $\phi(\mathbf{x})$. If a light ray is moving along the z-axis, it will experience a transverse deflection by the amount

$$\frac{\mathbf{d}}{D_{OL}} = 2 \int_{-\infty}^{\infty} dz \left[\frac{^{(2)}\nabla\phi}{c^2}\right], \qquad (5.177)$$

where $^{(2)}\nabla \equiv (\partial/\partial x, \partial/\partial y)$ is the gradient in the x–y plane. The factor 2 is the correction to Newtonian deflection arising from general relativity obtained in the last problem. Consider now the 2-dimensional divergence $^{(2)}\nabla \cdot \mathbf{d}$. Using

$$^{(2)}\nabla^2\phi = \nabla^2\phi - \frac{\partial^2\phi}{\partial z^2} = 4\pi G\rho(\mathbf{x}) - \frac{\partial^2\phi}{\partial z^2}, \qquad (5.178)$$

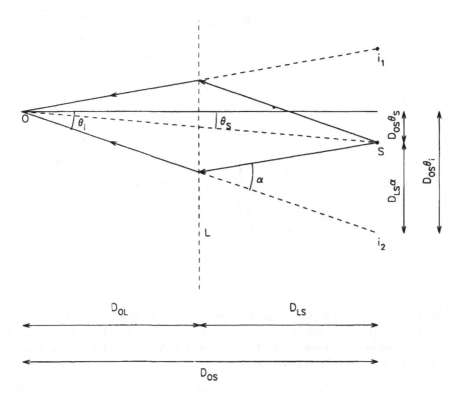

Fig. 5.2. The geometry of gravitational lensing. The source S is lensed by matter located at the plane L and produces two images i_1 and i_2 with respect to the observer O. We assume that all the deflection takes place at the plane of the lens.

we find

$$^{(2)}\nabla \cdot \mathbf{d} = \frac{2D_{\text{OL}}}{c^2} \int_{-\infty}^{\infty} dz \left[4\pi G\rho(x,y,z) - \frac{\partial^2 \phi}{\partial z^2} \right] = \frac{8\pi G D_{\text{OL}}}{c^2} \Sigma(x,y), \qquad (5.179)$$

where $\Sigma(x,y)$ is the surface mass density corresponding to $\rho(\mathbf{x})$. (The $\partial^2 \phi / \partial z^2$ term vanishes on integration.) This two-dimensional Poisson equation has the solution

$$\mathbf{d}(\mathbf{i}) = \frac{4G D_{\text{OL}}}{c^2} \int d^2x \, \Sigma(\mathbf{x}) \frac{(\mathbf{i} - \mathbf{x})}{|\mathbf{i} - \mathbf{x}|^2}, \qquad (5.180)$$

which gives the deflection $\mathbf{d}$ in terms of the surface density $\Sigma(\mathbf{x})$.

For a smooth, spherically symmetric distribution of density (like that due to a galaxy) we expect the deflection to fall as r^{-1} far away from the lens. Near the origin of the lens, $\rho \cong$ constant and $\phi \propto r^2$; so the deflection will be linear in r. Hence, we expect the quantity $(D_{\text{LS}}/D_{\text{OS}})|\mathbf{d}|$ to vary with $|\mathbf{i}|$, as shown in figure 5.3. The position of the images can be determined by finding the intersection of this curve with the line $y = (|\mathbf{i}| - |\mathbf{s}|)$.

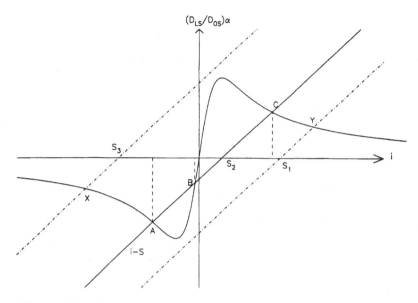

Fig. 5.3. Multiple images in the case of a spherical lens. When the source is near the centre of the lensing mass (at S_2, say), one obtains three images at the locations marked A, B and C. If the source is further away at S_1 or S_3, only one image (marked Y or X) is obtained.

It is clear that, when the source is far away from the centre of the lens, these curves intersect only at one point, giving rise to a single image. When the source is closer to the centre, these curves intersect at three different points, giving rise to three images.

(c) The magnification can most easily be calculated by considering the change in the image position δi for a small change in the source position δs. The amplification will be the determinant of the matrix of the transformation between the two:

$$A = \det \left| \frac{\partial i_a}{\partial s_b} \right|. \tag{5.181}$$

(d) (i) For $\Sigma(x, y) = \Sigma_0$, equation (5.179) has the solution $\mathbf{d}(\mathbf{i}) = A\mathbf{i}$ with $A = 4\pi G D_{\mathrm{OL}} \Sigma_0$. Using this in equation (5.176) we obtain

$$|\mathbf{s}| = |\mathbf{i}| \left(1 - \frac{\Sigma_0}{\Sigma_{\mathrm{crit}}} \right), \qquad \Sigma_{\mathrm{crit}} \equiv \frac{c^2 D_{\mathrm{OS}}}{4\pi G D_{\mathrm{OL}} D_{\mathrm{LS}}}. \tag{5.182}$$

The corresponding magnification is

$$A = \left(1 - \frac{\Sigma_0}{\Sigma_{\mathrm{crit}}} \right)^{-2}. \tag{5.183}$$

Note that when $\Sigma_0 = \Sigma_{\mathrm{crit}}$ the sheet focuses the beam on to the observer, leading to infinite amplification. Lower surface densities cannot focus the beam, while higher surface densities focus the beam before it reaches the observer (so that the beam is again diverging when it reaches the observer). In the cosmological context, we may take $D_{OL} \cong D_{OS} \cong D_{LS} \cong 3000$ Mpc; then the critical surface density is

$$\Sigma_{\text{crit}} \cong \frac{c^2}{4\pi G D_{OS}^2} \cong 1 \text{ g cm}^{-2}. \tag{5.184}$$

The surface density of galaxies is

$$\Sigma_{\text{gal}} \cong \frac{M_g}{R_g^2} \cong 0.3 \text{ g cm}^{-2} \left(\frac{M_g}{10^{11} M_\odot} \right) \left(\frac{R_g}{10 \text{ kpc}} \right)^{-2}, \tag{5.185}$$

which is on the borderline to produce multiple images. For clusters, the corresponding value is $\Sigma_{\text{clus}} \cong 0.03$ if $R \cong 3$ Mpc and $M \cong 10^{15} M_\odot$, which may not be very effective in producing multiple images.

(ii) For a point mass, $\Sigma(x, y) = M\delta(x)\delta(y)$; equation (5.180) gives

$$\mathbf{d(i)} = \frac{4GMD_{OL}}{c^2} \left(\frac{\mathbf{i}}{i^2} \right). \tag{5.186}$$

Substituting into (5.176), we find

$$\mathbf{s} = \mathbf{i} \left[1 - \frac{L^2}{i^2} \right], \qquad L^2 = \frac{4GMD_{LS}D_{OL}}{c^2 D_{OS}}. \tag{5.187}$$

The amplification is $A = (1 - L^4/i^4)^{-1}$, which diverges when $i = L$. In this case we obtain a circular ring of radius L and large intensity called the 'Einstein ring'. In the cosmological context, all the distances can be taken to be $D \cong 3000$ Mpc. Then the angular size of this ring will be

$$\theta_L \equiv \frac{L}{D} = \left(\frac{4GM}{c^2 D} \right)^{1/2} \cong 2 \text{ arc sec} \left(\frac{M}{10^{12} M_\odot} \right)^{1/2} \left(\frac{D}{3000 \text{ Mpc}} \right)^{-1/2}. \tag{5.188}$$

One can also think of θ_L as the typical angular separation between images produced by a lens of mass M located at cosmological distances. For a cluster $M \cong 10^{15} M_\odot$ and $\theta_L \cong 1$ arc min.

(iii) For an isothermal sphere, $M(r) = (2\sigma^2/G)r$ at large distances, where σ^2 is the velocity dispersion. The Newtonian potential for a mass distribution $M(r) = (2\sigma^2/G)r$ is

$$\phi(r) = \int^r \frac{GM(r)}{r^2} \, dr = 2\sigma^2 \ln r. \tag{5.189}$$

The deflection can be computed using (5.177):

$$\begin{aligned}
\frac{\mathbf{d}}{D_{OL}} &= \int_{-\infty}^{\infty} dz \left(\frac{4\sigma^2}{c^2} \right) {}^{(2)}\nabla \left(\ln \sqrt{x_\perp^2 + z^2} \right) \\
&= \frac{4\sigma^2}{c^2} \int_{-\infty}^{\infty} dz \frac{\mathbf{x}_\perp}{(x_\perp^2 + z^2)} = 4\pi \left(\frac{\sigma}{c} \right)^2 \cdot \left(\frac{\mathbf{x}_\perp}{x_\perp} \right),
\end{aligned} \tag{5.190}$$

where $\mathbf{x}_\perp = (x, y)$. Note that the angular deflection is constant in magitude and is given by

$$\alpha = 4\pi \left(\frac{\sigma}{c}\right)^2 \cong 2.6 \text{ arc sec} \left(\frac{\sigma}{300 \text{ km s}^{-1}}\right)^2. \tag{5.191}$$

The lensing equation (5.176) becomes

$$\mathbf{s} = \mathbf{i} - \frac{D_{LS}D_{OL}}{D_{OS}} \frac{\mathbf{i}}{|\mathbf{i}|}\alpha. \tag{5.192}$$

Taking the magnitudes, we find that

$$|\mathbf{i}| - |\mathbf{s}| = \frac{D_{LS}D_{OL}}{D_{OS}}\alpha. \tag{5.193}$$

Thus we obtain two images on the two sides at $|\mathbf{s}| \pm (D_{LS}D_{OL}/D_{OS})\alpha$. The amplification in this case is given by

$$A = \left(1 - \frac{D_{LS}D_{OL}}{D_{OS}} \frac{\alpha}{|\mathbf{i}|}\right)^{-1}. \tag{5.194}$$

(e) The gradient of $\psi(\mathbf{i})$ with respect to $\mathbf{i}$ is

$$\begin{aligned}
\nabla\psi &= \frac{4GD_{OL}D_{LS}}{c^2 D_{OS}} \int d^2x\, \Sigma(\mathbf{x}) \nabla \ln(|\mathbf{i} - \mathbf{x}|) \\
&= \frac{4GD_{OL}D_{LS}}{c^2 D_{OS}} \int d^2x\, \Sigma(\mathbf{x}) \frac{(\mathbf{i} - \mathbf{x})}{|\mathbf{i} - \mathbf{x}|^2}.
\end{aligned} \tag{5.195}$$

It follows that the equation $\nabla P = 0$ reduces to

$$\mathbf{i} = \frac{4GD_{OL}D_{LS}}{c^2 D_{OS}} \int d^2x\, \Sigma(\mathbf{x}) \frac{(\mathbf{i} - \mathbf{x})}{|\mathbf{i} - \mathbf{x}|^2}, \tag{5.196}$$

which is exactly (5.176) combined with (5.180), provided $\mathbf{s} = 0$. Hence, the extrema of the function $P(\mathbf{i})$ gives the image positions.

6

Friedmann model and thermal history

6.1 Maximally symmetric universe

(a) The assumption of homogeneity and isotropy of the three-space singles out a preferred class of observers, namely, those observers for whom the universe appears homogeneous and isotropic. Another observer, who is moving with a uniform velocity with respect to this fundamental class of observers, will perceive the universe to be anisotropic. The description of physics will be simplest if we use the coordinate system (t, x^α) appropriate to this fundamental class of observers. The general spacetime interval (ds^2) can be separated out as

$$ds^2 = g_{ab}\, dx^a dx^b = g_{00}\, dt^2 + 2g_{0\mu}\, dt\, dx^\mu - \sigma_{\mu\nu}\, dx^\mu dx^\nu, \qquad (6.1)$$

where $\sigma_{\mu\nu}$ is a positive definite spatial metric. Isotropy of space implies that the $g_{0\mu}$s must vanish; otherwise, they identify a particular direction in space related to the three-vector v_μ with components $g_{0\mu}$. Further, in the coordinate system determined by these fundamental observers, we may use the proper time of clocks carried by the observers to label the spacelike surfaces. This choice for the time coordinate t implies that $g_{00} = 1$, bringing the spacetime interval to the form

$$ds^2 = dt^2 - \sigma_{\mu\nu}\, dx^\mu dx^\nu \equiv dt^2 - dl^2. \qquad (6.2)$$

The problem now reduces to determining the three-metric $\sigma_{\mu\nu}$ of a three-space which, at any instant of time, is homogeneous and isotropic.

The assumption of isotropy – which implies spherical symmetry – allows us to write the line interval as

$$dl^2 = a^2 \left[\lambda^2(r)\, dr^2 + r^2(d\theta^2 + \sin^2\theta\, d\phi^2) \right] \qquad (6.3)$$

by separating out the $r^2 d\Omega^2$ part. Computing the scalar curvature ^3R for this three-dimensional space, using the formulas of chapter 5, we find that

$$^3R = \frac{3}{2a^2 r^3} \frac{d}{dr}\left[r^2 \left(1 - \frac{1}{\lambda^2}\right) \right]. \qquad (6.4)$$

Homogeneity implies that all geometrical properties are independent of r; hence ^3R must be a constant. Equating it to a constant and integrating the resulting equation we obtain

$$r^2\left(1 - \frac{1}{\lambda^2}\right) = c_1 r^4 + c_2, \quad c_1, c_2 = \text{constants}. \tag{6.5}$$

To avoid any singularity at $r = 0$, we need $c_2 = 0$. Thus we obtain $\lambda^2 = (1 - c_1 r^2)^{-1}$. When $c_1 \neq 0$, we can rescale r and make $c_1 = 1$ or -1. This leads to the full spacetime metric:

$$ds^2 = dt^2 - a^2(t)\left[\frac{dr^2}{1 - kr^2} + r^2(d\theta^2 + \sin^2\theta\, d\phi^2)\right]. \tag{6.6}$$

The prefactor a determines the overall scale of the spatial metric and, in general, can be a function of time: $a = a(t)$. This metric, called the Friedmann metric, describes a universe which is spatially homogeneous and isotropic at each instant of time.

(b) The spatial hypersurfaces of the Friedmann universe have positive, zero and negative curvatures for $k = +1, 0$ and -1 respectively; the magnitude of the curvature ^3R is $(6/a^2)$ when k is non-zero. To study the geometrical properties of these spaces it is convenient to introduce a coordinate χ, defined as

$$\chi = \int \frac{dr}{\sqrt{1 - kr^2}} = \begin{cases} \sin^{-1} r & (\text{for } k = 1) \\ r & (\text{for } k = 0) \\ \sinh^{-1} r & (\text{for } k = -1) \end{cases} \tag{6.7}$$

and a new time coordinate η with $d\eta = dt/a(t)$. In terms of $(\eta, \chi, \theta, \phi)$ the metric becomes

$$ds^2 = a^2(\eta)\left[d\eta^2 - d\chi^2 - f^2(\chi)(d\theta^2 + \sin^2\theta\, d\phi^2)\right], \tag{6.8}$$

where

$$f(\chi) = \begin{cases} \sin \chi & (\text{for } k = +1) \\ \chi & (\text{for } k = 0) \\ \sinh \chi & (\text{for } k = -1). \end{cases} \tag{6.9}$$

To obtain the second coordinate system given in the question, one notices that R must be set to $ra(t)$. In general, a transformation from (r, t) to $(R = ra(t), t)$ will introduce a non-trivial cross-term of the form $dR\,dt$. To eliminate this cross-term, one has to change to a new time coordinate $T = T(r, t)$ as well.

To obtain the explicit transformations, it is more convenient to proceed from (T, R) to (t, r). In the metric

$$ds^2 = e^\nu\, dT^2 - e^\lambda\, dR^2 - R^2\, d\Omega^2 \tag{6.10}$$

we substitute the ansatz

$$R = a(t)r, \quad dT = A(r)\,dr + B(t)\,dt, \tag{6.11}$$

obtaining

$$ds^2 = \left(e^\nu B^2 - e^\lambda r^2 \dot{a}^2\right) dt^2 + \left(e^\nu A^2 - e^\lambda a^2\right) dr^2 + 2\left(e^\nu AB - e^\lambda r a \dot{a}\right) dr\, dt - r^2 a^2 d\Omega^2.$$
(6.12)

We choose $AB = e^{(\lambda-\nu)} r a \dot{a}$ to eliminate the cross-term. The conditions $g_{tt} = 1$ and $g_{rr} = -a^2 \left(1 - kr^2\right)^{-1}$ lead to the equations

$$e^{-\nu} = \frac{\dot{a}B}{a}\left[\frac{aB}{\dot{a}} - Ar\right], \quad e^{-\lambda} = \frac{\dot{a}(1 - kr^2)}{aB}\left[\frac{aB}{\dot{a}} - Ar\right].$$
(6.13)

These equations can be satisfied if

$$A(r) = \frac{r}{1 - kr^2}, \quad B(t) = \frac{1}{a\dot{a}}.$$
(6.14)

Then

$$e^\lambda = \frac{1}{1 - kr^2 - \dot{a}^2 r^2}, \quad e^\nu = e^\lambda (a\dot{a})^2 \left(1 - kr^2\right).$$
(6.15)

In fact, a far more general class of transformations is possible. Since (6.11) involved only the differential dT, one can redefine T to be an arbitrary function of the original T and still obtain the required form of the metric. Thus the most general transformation is of the form $R = ra(t)$, $T = F(q)$, where

$$q \equiv \int^r \frac{x\, dx}{1 - kx^2} + \int^t \frac{dy}{a(y)\dot{a}(y)}$$
(6.16)

and F is an *arbitrary* function.

(c) For $k = 0$, the spatial part of the metric describes the familiar, flat, Euclidian three-space; the homogeneity and isotropy of this space are obvious.

For $k = 1$, the metric (6.8) represents a three-sphere of radius a embedded in an abstract, flat four-dimensional Euclidian space. Such a three-sphere is defined by the relation

$$x_1^2 + x_2^2 + x_3^2 + x_4^2 = a^2,$$
(6.17)

where (x_1, x_2, x_3, x_4) are the Cartesian coordinates of some abstract four-dimensional space. We can introduce angular coordinates (χ, θ, ϕ) on the three-sphere by the relations

$$x_1 = a \cos\chi \sin\theta \sin\phi, \quad x_2 = a \cos\chi \sin\theta \cos\phi,$$
$$x_3 = a \cos\chi \cos\theta, \quad x_4 = a \sin\chi.$$
(6.18)

The metric on the three-sphere can be determined by expressing the dx_is in terms of $d\chi, d\theta$ and $d\phi$ and substituting in the line element

$$dL^2 = dx_1^2 + dx_2^2 + dx_3^2 + dx_4^2.$$
(6.19)

This leads to the metric

$$dL^2_{\text{(three-sphere)}} = a^2[d\chi^2 + \sin^2\chi(d\theta^2 + \sin^2\theta\, d\phi^2)], \tag{6.20}$$

which is the same as (6.8) for $k = 1$.

The entire three-space of the $k = 1$ model is covered by the range of angles $[0 \leq \chi \leq \pi; 0 \leq \theta \lesssim \pi; 0 \leq \phi < 2\pi]$ and has a finite volume:

$$V = \int_0^{2\pi} d\phi \int_0^\pi d\theta \int_0^\pi d\chi \sqrt{g} = a^3 \int_0^{2\pi} d\phi \int_0^\pi \sin\theta\, d\theta \int_0^\pi \sin^2\chi\, d\chi = 2\pi^2 a^3. \tag{6.21}$$

The surface area of a two-sphere, defined by $\chi = $ constant, is $S = 4\pi a^2 \sin^2\chi$. As χ increases, S increases at first, reaches a maximum value of $4\pi a^2$ at $\chi = \pi/2$ and *decreases* thereafter. These are the properties of a three-space which is closed but has no boundaries.

In the case of $k = -1$, (6.8) represents the geometry of a hyperboloid embedded in an *abstract* four-dimensional Lorentzian space. Such a space is described by the line element

$$dL^2 = dx_1^2 + dx_2^2 + dx_3^2 - dx_4^2. \tag{6.22}$$

A three-dimensional hyperboloid, embedded in this space, is defined by the relation

$$x_4^2 - x_1^2 - x_2^2 - x_3^2 = a^2. \tag{6.23}$$

This three-space can be parametrized by the coordinates (χ, θ, ϕ), with

$$x_1 = a\sinh\chi\sin\theta\sin\phi, \quad x_2 = a\sinh\chi\sin\theta\cos\phi,$$
$$x_3 = a\sinh\chi\cos\theta, \quad x_4 = a\cosh\chi. \tag{6.24}$$

Expressing the dx_is in terms of $d\chi, d\theta$ and $d\phi$ and substituting into (6.22), the metric on the hyperboloid can be found to be

$$dL^2_{\text{(hyperboloid)}} = a^2\left[d\chi^2 + \sinh^2\chi(d\theta^2 + \sin^2\theta\, d\phi^2)\right], \tag{6.25}$$

which is the same as (6.8) for $k = -1$.

To cover this three-space, we need the range of coordinates to be $[0 \leq \chi \leq \infty; 0 \leq \theta \leq \pi; 0 \leq \phi < 2\pi]$. This space has infinite volume, just like the ordinary flat three-space. The surface area of a two-sphere, defined by $\chi = $ constant, is $S = 4\pi a^2 \sinh^2\chi$. This expression increases monotonically with χ.

(d) To obtain the effective Newtonian potential of the Friedmann universe, consider the transformation

$$R = ra(t), \quad T = t - t_0 + \frac{1}{2}a\dot{a}r^2 + \mathcal{O}(r^4), \tag{6.26}$$

in which only terms up to quadratic order in r are retained. Direct calculation, correct up to this order, transforms the Friedmann line element to the form

$$ds^2 \cong \left(1 - \frac{\ddot{a}}{a}R^2\right)dT^2 - \left(1 + \frac{k}{a^2}R^2 + \frac{\dot{a}^2}{a^2}R^2\right)dR^2 - R^2(d\theta^2 + \sin^2\theta\, d\phi^2). \tag{6.27}$$

Near $R = 0$, the space is flat and the deviations from flatness are of order $\mathcal{O}\left(\ddot{a}R^2/a\right)$. We know that, in the weak gravity limit, $g_{00} = (1 + 2\phi_N)$. We see, therefore, that the Friedmann universe has an effective gravitational potential

$$\phi_N\left(\mathbf{R}, t\right) = -\frac{1}{2}\frac{\ddot{a}}{a}R^2 \tag{6.28}$$

in the weak gravity limit, at scales $R^2 \ll \left(\ddot{a}/a\right)^{-1}$. This potential corresponds to a spatially uniform, time-dependent source.

6.2 Kinematics in Friedmann universe

(a) Since $a(t)$ multiplies the spatial coordinates, any proper distance $l(t)$ between spatial locations will change with time in proportion to $a(t)$:

$$l(t) = l_0 a(t) \propto a(t). \tag{6.29}$$

In particular, the proper separation between two observers, located at constant comoving coordinates, will change with time. Let the comoving separation between two such observers be δx, so that the proper separation is $\delta l = a(t)\delta x$. Each of the two observers will attribute to the other a velocity

$$\delta v = \frac{d}{dt}\delta l = \dot{a}\,\delta x = \left(\frac{\dot{a}}{a}\right)\delta l. \tag{6.30}$$

Consider a narrow pencil of electromagnetic radiation which crosses these two comoving observers. The time for transit will be $\delta t = \delta l$. Let the frequency of the radiation measured by the first observer be ω. Since the first observer sees the second one to be *receding* with velocity δv, he (she) will expect the second observer to measure a Doppler shifted frequency $(\omega + \delta\omega)$, where

$$\frac{\delta\omega}{\omega} = -\frac{\delta v}{c} = -\delta v = -\frac{\dot{a}}{a}\delta l = -\frac{\dot{a}}{a}\delta t = -\frac{\delta a}{a}. \tag{6.31}$$

(Since the observers are separated by an infinitesimal distance of first order, δl, we can introduce a locally inertial frame encompassing both the observers. The laws of special relativity can be applied in this frame.) This equation can be integrated to give

$$\omega(t)a(t) = \text{constant}. \tag{6.32}$$

In other words, the frequency of electromagnetic radiation changes due to expansion of the universe according to the law $\omega \propto a^{-1}$.

To obtain the same result from Maxwell's equations we may proceed as follows. The dynamics of the electromagnetic field in curved spacetime is described by the action

$$\mathscr{A}_{em} = \frac{1}{16\pi}\int F_{ik}F^{ik}\sqrt{-g}\,d^4x, \quad F_{ik} = \frac{\partial A_k}{\partial x^i} - \frac{\partial A_i}{\partial x^k}. \tag{6.33}$$

(See problem 5.6.) Consider the conformal transformation

$$A_i \to A_i, \quad x^i \to x^i, \quad g_{ik} \to \Omega^2 g_{ik}, \quad g^{ik} \to \Omega^{-2}g^{ik}. \tag{6.34}$$

Note that $(A^i)_{new} = g^{ik}_{new} (A_k)_{new} = \Omega^{-2} A^i_{old}$. Quite clearly,

$$F_{ik} F_{mn} g^{mi} g^{nk} \sqrt{-g} \to F_{ik} F_{mn} \left(\Omega^{-2} g^{mi}\right) \left(\Omega^{-2} g^{nk}\right) \left(\Omega^4 \sqrt{-g}\right) = F_{ik} F_{mn} g^{mi} g^{nk} \sqrt{-g}, \tag{6.35}$$

showing that $\mathscr{A}_{em}$ is invariant under conformal transformations. The $k = 0$ Friedmann universe in the $(\eta, \chi, \theta, \phi)$ coordinate system is conformally flat, with $g_{ik} = a^2(\eta) \eta_{ik}$. Since the electromagnetic field is conformally invariant, solution to the wave equation for A_i in the metric g_{ik} is the same as the solution in the flat spacetime metric η_{ik} and has the time dependence

$$A_i \propto \exp\left(-ik\eta\right) = \exp\left[-ik \int \frac{dt}{a(t)}\right]. \tag{6.36}$$

Since the time derivative of the phase of the wave defines the (instantaneous) frequency, we conclude that

$$\omega(t)a(t) = \text{constant}. \tag{6.37}$$

(This result is also valid in $k \neq 0$ models. Even in this case, the *time* dependence of A_i is the same as in (6.36).)

(b) The analysis proceeds exactly as before. The proper distance between two observers, with a comoving separation $n^\alpha \delta x$, along the path of the photon is

$$\delta l \cong a(t) \left[1 - \frac{1}{2} h_{\alpha\beta} n^\alpha n^\beta\right] (\delta x) + \mathcal{O}(h^2), \tag{6.38}$$

where we have only retained terms which are linear in $h_{\alpha\beta}$. Using an argument similar to the one which led to equation (6.31), we find that

$$\frac{\delta\omega}{\omega} = -\frac{\delta l}{c} \frac{d}{dt} \ln[\delta l] = -\frac{\delta l}{c} \left[\frac{\dot{a}}{a} - \frac{1}{2} \dot{h}_{\alpha\beta} n^\alpha n^\beta\right] = -\frac{\delta a}{a} + \frac{1}{2} \dot{h}_{\alpha\beta} n^\alpha n^\beta \delta t \tag{6.39}$$

or

$$\delta \ln(\omega a) = \frac{1}{2} \dot{h}_{\alpha\beta} n^\alpha n^\beta \delta t. \tag{6.40}$$

Integrating both sides along the photon path from the moment of emission t_1 to the moment of absorption t_2, we find

$$\ln\left(\frac{\omega_2 a_2}{\omega_1 a_1}\right) = \frac{1}{2} \int_{t_1}^{t_2} \dot{h}_{\alpha\beta} n^\alpha n^\beta \, dt. \tag{6.41}$$

(c) Consider two comoving observers separated by proper distance δl. Let a material particle pass the first observer with velocity v. When it has crossed the proper distance δl (in a time interval δt), it passes the second observer, whose velocity (relative to the first one) is

$$\delta u = \frac{\dot{a}}{a} \delta l = \frac{\dot{a}}{a} v \, dt = v \frac{\delta a}{a}. \tag{6.42}$$

The second observer will attribute to the particle the velocity

$$v' = \frac{v - \delta u}{1 - v \, \delta u} = v - (1 - v^2)\delta u + \mathcal{O}[(\delta u)^2] = v - (1 - v^2)v \frac{\delta a}{a}. \tag{6.43}$$

This follows from the special relativistic formula for addition of velocities which is valid in an infinitesimal region around the first observer. Rewriting this equation as

$$\delta v = -v(1 - v^2)\frac{\delta a}{a} \tag{6.44}$$

and integrating, we obtain

$$p = \frac{v}{\sqrt{1 - v^2}} = \frac{\text{constant}}{a}. \tag{6.45}$$

In other words, the magnitude of the three-momentum decreases as a^{-1} due to the expansion. If the particle is non-relativistic, then $v \propto p$ and velocity itself decays as a^{-1}.

(d) Consider a stream of particles propagating freely in spacetime. At some time t, a comoving observer finds dN particles in a proper volume $d\mathscr{V}$, all having momentum in the range $(\mathbf{p}, \mathbf{p}+d^3\mathbf{p})$. The phase space distribution function $f(\mathbf{x}, \mathbf{p}, t)$ for the particles is defined by the relation $dN = f \, d\mathscr{V} \, d^3\mathbf{p}$. At a later instant $(t+\delta t)$ the proper volume occupied by these particles would have increased by a factor $[a(t + \delta t)/a(t)]^3$, while the volume in the momentum space would be redshifted by $[a(t)/a(t+\delta t)]^3$, showing that the phase volume occupied by the particles does not change during the free propagation. Since the number of particles dN is also conserved, it follows that f is conserved along the streamline.

If $f(\mathbf{x}, \mathbf{p}, t)$ is the phase space density of photons, then the energy density du of photons in the frequency band $(\omega, \omega + d\omega)$ flowing into a particular solid angle $d\Omega$ is proportional to $f(\hbar\omega)(p^2 dp \, d\Omega)$. The intensity, $I(\omega)$, of the radiation (which is the energy density per unit bandwidth) is defined by the relation

$$du \equiv I(\omega) \, d\Omega \, d\omega = \hbar\omega \cdot f \cdot p^2 \, dp \, d\Omega. \tag{6.46}$$

Since $p^2 dp \propto \omega^2 d\omega$ for photons, we find that $I(\omega) \propto f\omega^3$, and, further, since f is conserved during the propagation, it follows that the quantity $[I(\omega)/\omega^3]$ is conserved during the propagation of radiation. If the intensity has the form $I(\omega) = \omega^3 G(\omega/T)$, then G is invariant during the propagation. From the relation $\omega \propto a^{-1}$ it follows that parameter T will effectively scale as $T \propto a^{-1}$.

Planck's spectrum has the form $I(\omega) = \omega^3 G(\omega/T)$, with T being the temperature. The above result shows that the Planck spectrum retains its shape as the universe expands with the temperature scaling as $T \propto a^{-1}$.

(e) (i) If the source has an intrinsic luminosity L, then it will emit an energy $L \, dt_1$ in a time interval dt_1. This energy, which will be received by the observer in a time interval $dt_0 = dt_1 [a(t_0)/a(t_1)]$, will have undergone a redshift by a factor $[a(t_1)/a(t_0)]$ and will be distributed over a sphere of radius $4\pi a^2(t_0)r_1^2$. The observed flux, therefore, will be

$$l = L\left(\frac{dt_1}{dt_0}\right) \cdot \frac{a(t_1)}{a(t_0)} \cdot \frac{1}{4\pi a^2(t_0)r_1^2} = \frac{L}{4\pi a_0^2 r_1^2} \cdot \left[\frac{a(t_1)}{a_0}\right]^2 = \frac{L}{4\pi a_0^2 r_1^2 (1 + z)^2}. \tag{6.47}$$

Defining a 'luminosity distance' $d_L(z)$ to the source (at redshift z) through the relation $l \equiv (L/4\pi d_L^2)$, we obtain

$$d_L(z) = a_0 r_1(t_1)(1+z). \tag{6.48}$$

To evaluate these expressions we have to relate r_1 and t_1. Suppose that the radiation was emitted at t_1 by a source at (r_1, θ, ϕ) and was detected at the origin at $t = t_0$. Along the radial geodesic taken by the radiation we must have $ds = 0$ and $d\theta = 0, d\phi = 0$. This gives

$$\int_0^{r_1} \frac{dr}{\sqrt{1-kr^2}} = -\int_{t_0}^{t_1} \frac{dt}{a(t)}. \tag{6.49}$$

Given the functional form of $a(t)$ we can compute this quantity.

(ii) If D is the physical size of an object which subtends an angle δ to the observer, then for small δ we have $D = r_1 a(t_1)\delta$. The 'angular diameter distance' $d_A(z)$ for the source is defined via the relation $\delta = (D/d_A)$; so

$$d_A(z) = r_1 a(t_1) = a_0 r_1(t_1)(1+z)^{-1}. \tag{6.50}$$

Clearly, $d_L = (1+z)^2 d_A$.

6.3 Dynamics of the Friedmann model

(a) The Friedmann metric contains a constant k and a function $a(t)$, both of which can be determined via Einstein's equations:

$$G_k^i = R_k^i - \frac{1}{2}\delta_k^i R = 8\pi G T_k^i \tag{6.51}$$

if the stress-tensor for the source is specified. The assumption of homogeneity and isotropy implies that the T_0^μ's must be zero and that the spatial components T_β^α must have a diagonal form with $T_1^1 = T_2^2 = T_3^3$. It is conventional to write such a stress-tensor as

$$T_k^i = \text{dia}\,[\rho(t), -p(t), -p(t), -p(t)]. \tag{6.52}$$

The tensor G_k^i was computed for the Friedmann metric in problem 5.8. The non-trivial components are

$$G_0^0 = \frac{3}{a^2}(\dot{a}^2 + k), \quad G_\nu^\mu = \frac{1}{a^2}(2a\ddot{a} + \dot{a}^2 + k)\delta_\nu^\mu. \tag{6.53}$$

Thus (6.51) gives two independent equations:

$$\frac{\dot{a}^2 + k}{a^2} = \frac{8\pi G}{3}\rho, \tag{6.54}$$

$$\frac{2\ddot{a}}{a} + \frac{\dot{a}^2 + k}{a^2} = -8\pi G p. \tag{6.55}$$

These two equations, combined with the equation of state $p = p(\rho)$, completely determine the three functions $a(t)$, $\rho(t)$ and $p(t)$.

(b) From equation (6.54) it follows that

$$\frac{k}{a^2} = \frac{8\pi G}{3}\rho - \frac{\dot{a}^2}{a^2} = \frac{\dot{a}^2}{a^2}\left[\frac{\rho}{(3H^2/8\pi G)} - 1\right]. \tag{6.56}$$

Evaluating this equation at $t = t_0$ and using the definition of Ω we obtain

$$\frac{k}{a_0^2} = H_0^2(\Omega - 1). \tag{6.57}$$

It is clear that $k = -1, 0$ or 1 depending on whether $\Omega < 1$, $\Omega = 1$ or $\Omega > 1$ respectively. Thus Ω determines the spatial geometry of the universe. Given H_0, Ω and that $k \neq 0$, we can find a_0 from (6.57):

$$a_0 = H_0^{-1}\left(|\Omega - 1|\right)^{-1/2}. \tag{6.58}$$

The value of $\dot{a}_0$ can be fixed by

$$\dot{a}_0 = H_0 a_0 = \left(|\Omega - 1|\right)^{-1/2}. \tag{6.59}$$

Hence (H_0, Ω) are valid initial conditions for determining $a(t)$.

If $k = 0$ Einstein's equations allow the scaling $a \to \mu a$. Hence, the normalization of a is arbitrary. It is conventional to take $a_0 = 1$ if $k = 0$; then the value of $(\dot{a}_0/a_0) = H_0 = \dot{a}_0$ determines $\dot{a}_0$.

(c) Substituting for $(\dot{a}^2 + k)/a^2$ in (6.55) from (6.54) and rearranging we obtain

$$\frac{\ddot{a}}{a} = -\frac{4\pi G}{3}(\rho + 3p). \tag{6.60}$$

For normal matter, $(\rho + 3p) > 0$ implying that $\ddot{a} < 0$. The $a(t)$ curve (which has positive $\dot{a}$ at the present epoch t_0) must be convex; in other words, a will have been smaller in the past and will have become zero at some time (in the past), say, at $t = t_{\text{sing}}$. It is also clear that $(t_0 - t_{\text{sing}})$ must be less than the value of the intercept $(\dot{a}/a)_0^{-1} = H_0^{-1}$. For convenience, we will choose the time coordinate such that $t_{\text{sing}} = 0$, that is, we take $a = 0$ at $t = 0$. In that case, the present 'age' of the universe t_0 satisfies the inequality $t_0 < t_{\text{univ}}$, where

$$t_{\text{univ}} \equiv H_0^{-1} = 3.1 \times 10^{17} h^{-1} \, \text{s} = 9.8 \times 10^9 h^{-1} \, \text{yr}. \tag{6.61}$$

As a becomes smaller the components of the curvature tensor R^i_{klm} become larger and when $a = 0$ these components diverge. Such a divergence (called a 'singularity') is an artifact of our theory. When the radius of curvature of the spacetime becomes comparable to the fundamental length $(G\hbar/c^3)^{1/2} \cong 10^{-33}$ cm constructed out of $G, \hbar$ and c, the quantum effects of gravity will become important, rendering the classical Einstein equations invalid. So, in reality, t_0 is the time which has elapsed from the moment at which the classical equations became valid.

The quantities ρ and p are *defined* in (6.52) as the T_0^0 and T_1^1 (say) components of the stress-tensor. The interpretation of p as 'pressure' depends on treating the source as an ideal fluid. The source for a Friedmann model should *always* have the form given in (6.52); but, if the source is not an ideal fluid, then it is not possible to interpret the spatial components of T_k^i as pressure. It is, therefore, quite possible that the equations of state for matter at high energies do not obey the condition $(\rho + 3p) > 0$. The violation of this condition may occur much before (i.e. at larger values of a) the quantum gravitational effects become important. If this happens, then the 'age of the universe' refers to the time interval since the breakdown of the condition $(\rho + 3p) > 0$.

(d) From equation (6.54), we see that $\rho a^3 = (3/8\pi G)a(\dot{a}^2 + k)$; differentiating this expression and using equation (6.55) we obtain

$$\frac{d}{dt}(\rho a^3) = -3a^2\dot{a}p = -p\frac{da^3}{dt} \tag{6.62}$$

or

$$\frac{d}{da}(\rho a^3) = -3a^2 p. \tag{6.63}$$

Given the equation of state $p = p(\rho)$, we can integrate (6.63) to obtain $\rho = \rho(a)$. Substituting this relation into (6.54) we can determine $a(t)$.

For an equation of state of the form $p = w\rho$, (6.63) gives $\rho \propto a^{-3(1+w)}$; in particular, for non-relativistic matter ($w = 0$) and radiation ($w = 1/3$) we find $\rho_{nr} \propto a^{-3}$ and $\rho_r \propto a^{-4}$. If $w = -1$, then we find that $\rho = $ constant as the universe expands. In this case the pressure $p = -\rho$ is negative (since we must have $\rho > 0$ to maintain $(\dot{a}^2/a^2 > 0)$) and the negative pressure allows for the energy inside a volume to increase even when the volume expands. If $w = 1$, then $p = \rho$ and $\rho \propto a^{-6}$. This is called a 'stiff' equation of state. In such a medium, the 'speed of sound' $(\partial p/\partial \rho) = 1$ is the same as the speed of light.

For $\rho \propto a^{-3(1+w)}$, the Friedmann equation becomes $(\dot{a}^2/a^2) \propto a^{-3(1+w)}$ or

$$\dot{a} \propto a^{-\frac{1}{2}(1+3w)}. \tag{6.64}$$

Integrating, we find:

$$\begin{aligned} a(t) &\propto t^{\frac{2}{3(1+w)}} \quad &\text{(for } w \neq -1\text{),} \\ &\propto \exp(\lambda t) \quad &\text{(for } w = -1\text{),} \end{aligned} \tag{6.65}$$

where λ is a constant. For $w = 0$, $a \propto t^{2/3}$; for $w = 1/3$, $a \propto t^{1/2}$; and for $w = 1$, $a \propto t^{1/3}$.

The last two equations of state $p = \pm\rho$ can arise for scalar fields with Lagrangians of the form

$$L = \frac{1}{2}\left(\frac{\partial\phi}{\partial x^i}\right)\left(\frac{\partial\phi}{\partial x_i}\right) - V(\phi). \tag{6.66}$$

The stress-tensor for this scalar field is (see problem 5.9(e))

$$T_{ik} = \left(\frac{\partial\phi}{\partial x^i}\right)\left(\frac{\partial\phi}{\partial x^k}\right) - g_{ik}L. \tag{6.67}$$

In a homogeneous universe, $\phi(t, \mathbf{x}) = \phi(t)$ and only the diagonal components of

T_k^i remain non-zero:

$$T_0^0 = \frac{1}{2}\dot{\phi}^2 + V(\phi), \quad T_1^1 = T_2^2 = T_3^3 = -\frac{1}{2}\dot{\phi}^2 + V(\phi), \qquad (6.68)$$

giving

$$\rho = \frac{1}{2}\dot{\phi}^2 + V(\phi), \quad p = \frac{1}{2}\dot{\phi}^2 - V(\phi). \qquad (6.69)$$

When the kinetic energy $(\dot{\phi}^2/2)$ of the field dominates over the potential energy $V(\phi)$, we obtain the equation of state $p = \rho$. If the potential $V(\phi)$ dominates over the kinetic energy $(\dot{\phi}^2/2)$, then $p = -\rho$. Thus the scalar field can exhibit both these equations of state in appropriate ranges.

(e) The energy density of the radiation at temperature T is given by $\rho_r = (\pi^2/15)\,T^4$. Dividing this by $\rho_{crit} \cong 1.88 \times 10^{-29} h^2 \mathrm{g\,cm}^{-3}$ we can find Ω_r. Taking $T = 2.73$ K we obtain

$$\Omega_r h^2 = 2.56 \times 10^{-5}. \qquad (6.70)$$

We have seen in part (d) that $\rho_{nr} \propto a^{-3}, \rho_r \propto a^{-4}$ and $\rho_V =$ constant, as the universe evolves. Hence the total energy density in the universe can be expressed as

$$
\begin{aligned}
\rho_{tot}(a) &= \rho_r(a) + \rho_{nr}(a) + \rho_V(a) \\
&= \rho_{crit}\left[\Omega_r\left(\frac{a_0}{a}\right)^4 + (\Omega_B + \Omega_{DM})\left(\frac{a_0}{a}\right)^3 + \Omega_V\right].
\end{aligned}
\qquad (6.71)
$$

Substituting this into Einstein's equation we obtain

$$\frac{\dot{a}^2}{a^2} + \frac{k}{a^2} = H_0^2\left[\Omega_r\left(\frac{a_0}{a}\right)^4 + \Omega_{nr}\left(\frac{a_0}{a}\right)^3 + \Omega_V\right], \qquad (6.72)$$

with $\Omega_{nr} = \Omega_B + \Omega_{DM}$. This equation can be cast in a more suggestive form. We write (k/a^2) as $(\Omega - 1)H_0^2(a_0/a)^2$ and move it to the right-hand side. Introducing a dimensionless time coordinate $\tau = H_0 t$ and writing $a = a_0 q(\tau)$ our equation becomes

$$\frac{1}{2}\left(\frac{dq}{d\tau}\right)^2 + V(q) = E, \qquad (6.73)$$

where

$$V(q) = -\frac{1}{2}\left[\frac{\Omega_r}{q^2} + \frac{\Omega_{nr}}{q} + \Omega_V q^2\right], \quad E = \frac{1}{2}(1 - \Omega). \qquad (6.74)$$

This equation has the structure of the first integral for motion of a particle with energy E in a potential $V(q)$.

(i) It is clear that in the early evolution of the universe, the radiation term (Ω_r/q^2) would dominate the dynamics. In this limit, the solution is given by

$$\left(\frac{a}{a_0}\right) \cong \sqrt{2}\,\Omega_r^{1/4}\,(H_0 t)^{1/2}. \qquad (6.75)$$

Thus for small a, we have $a \propto t^{1/2}$.

(ii) As the universe evolves the matter density will catch up with the radiation density and both a^{-4} and a^{-3} terms will be important. The equality of matter and radiation energies occurred at some time $t = t_{eq}$ in the past corresponding to a value $a = a_{eq}$ and redshift $z = z_{eq}$, with

$$(1 + z_{eq}) = \frac{a_0}{a_{eq}} = \frac{\Omega_{nr}}{\Omega_r} \cong 3.9 \times 10^4 (\Omega_{nr} h^2). \tag{6.76}$$

Since the temperature of the radiation grows as a^{-1}, the temperature of the universe at this epoch will be

$$T_{eq} = T_{now}(1 + z_{eq}) = 9.24(\Omega h^2) \, \text{eV} . \tag{6.77}$$

For $t \ll t_{eq}$ the energy density in the universe is dominated by radiation (with $p = (1/3)\rho$), while for $t \gg t_{eq}$, the energy density is dominated by matter (with $p \cong 0$). When both radiation and matter terms are taken into consideration, and other terms are ignored, equation (6.72) has the solution

$$H_{eq}t = \frac{2\sqrt{2}}{3} \left[\left(\frac{a}{a_{eq}} - 2 \right) \left(\frac{a}{a_{eq}} + 1 \right)^{1/2} + 2 \right] . \tag{6.78}$$

This equation gives $a(t)$ in terms of the two (known) parameters:

$$H_{eq}^2 \equiv \frac{16\pi G}{3} \rho_{eq} \equiv 2H_0^2 \Omega_r (1 + z_{eq})^4 = 2H_0^2 \Omega_{nr}(1 + z_{eq})^3 , \tag{6.79}$$

$$a_{eq} \equiv a_0(1 + z_{eq})^{-1} = H_0^{-1} |(\Omega - 1)|^{-1/2} (\Omega_r/\Omega_{nr}). \tag{6.80}$$

From (6.78) we can find the value of t_{eq}; setting $a = a_{eq}$ gives $H_{eq}t_{eq} \cong 0.552$, or

$$t_{eq} = \frac{2\sqrt{2}}{3} H_{eq}^{-1}(2 - \sqrt{2}) \cong 0.39 H_0^{-1} \Omega^{-\frac{1}{2}}(1 + z_{eq})^{-\frac{3}{2}} = 1.57 \times 10^{10}(\Omega h^2)^{-2} \, \text{s}. \tag{6.81}$$

From (6.78), we can also find two limiting forms for $a(t)$ valid for $t \gg t_{eq}$ and $t \ll t_{eq}$:

$$\left(\frac{a}{a_{eq}} \right) = \begin{cases} (3/2\sqrt{2})^{2/3}(H_{eq}t)^{2/3} \\ (3/\sqrt{2})^{1/2}(H_{eq}t)^{1/2} \end{cases} . \tag{6.82}$$

Thus $a \propto t^{2/3}$ in matter dominated phase (when the curvature is negligible) and $a \propto t^{1/2}$ in the radiation dominated phase.

(iii) At $z \ll z_{eq}$ we can ignore the radiation completely. The evolution now depends on the value of Ω_V and k. Let us take the case of $\Omega_V = 0$ first. The energy density of matter varies as $\rho = \rho_c \Omega(1 + z)^3$ while the curvature term grows as $a^{-2} = H_0^2 |(\Omega - 1)| (1 + z)^2$. If $\Omega < 1$, then the curvature term $(1 - \Omega)/2$ and the non-relativistic matter term, $\Omega_{nr}/2q$ will be equal at some $z = z_{curv}$, where

$$z_{curv} = \frac{1}{\Omega} - 2 . \tag{6.83}$$

(We have taken $\Omega \cong \Omega_{nr}$.) It is possible for the curvature term to dominate over matter density at sufficiently small z provided $\Omega < 0.5$. During this phase, the effect of ρ is ignorable and the expansion factor $a(t)$ grows as t. For $z \gg z_{curv}$ we can ignore the curvature term.

Equation (6.74) can be integrated exactly when both the curvature and matter terms are present but $\Omega_V = 0$ and Ω_r is ignorable. The solution is

$$H_0 t = \frac{\Omega}{2(\Omega - 1)^{3/2}} \left[\cos^{-1} \left(\frac{\Omega z - \Omega + 2}{\Omega z + \Omega} \right) - \frac{2(\Omega - 1)^{1/2}(\Omega z + 1)^{1/2}}{\Omega(1 + z)} \right] \qquad (6.84)$$

for $\Omega > 1$ and

$$H_0 t = \frac{\Omega}{2(1 - \Omega)^{3/2}} \left[\frac{2(1 - \Omega)^{1/2}(\Omega z + 1)^{1/2}}{\Omega(1 + z)} - \cosh^{-1} \left(\frac{\Omega z - \Omega + 2}{\Omega z + \Omega} \right) \right] \qquad (6.85)$$

for $\Omega < 1$. These expressions, together with the relation $a = a_0(1 + z)^{-1} = H_0^{-1} |(1 - \Omega)|^{-1/2} (1 + z)^{-1}$, completely determine $a(t)$.

(iv) Finally, consider the effect of Ω_V. If $\Omega_V > \Omega_{nr}$, then the Ω_V term will dominate over other terms at $z \ll z_V$ where $(1 + z_V) = (\Omega_V / \Omega_{nr})^{1/3}$. Keeping only the Ω_{nr} and Ω_V terms we can integrate the equation to determine $a(t)$. We obtain

$$\left(\frac{a}{a_0} \right)^3 = \left(\frac{\Omega_{nr}}{\Omega_V} \right) \sinh^2 \left[\frac{3}{2} \sqrt{\Omega_V} H_0 t \right]. \qquad (6.86)$$

When $\sqrt{\Omega_V} H_0 t \ll 1$, this reduces to the matter dominated evolution with $a^3 \propto t^2$; when $\sqrt{\Omega_V} H_0 t \gtrsim 1$, the growth is exponential with $a \propto \exp(\sqrt{\Omega_V} H_0 t)$.

(f) From (6.72) it follows that

$$d_H(z) = H_0^{-1} \left[\Omega_r (1 + z)^4 + \Omega_{nr}(1 + z)^3 + (1 - \Omega)(1 + z)^2 + \Omega_V \right]^{-1/2}. \qquad (6.87)$$

This has the limiting forms

$$d_H(z) \cong \begin{cases} H_0^{-1} \Omega_r^{-1/2}(1 + z)^{-2} & (z \gg z_{eq}) \\ H_0^{-1} \Omega_{nr}^{-1/2}(1 + z)^{-3/2} & (z_{eq} \gg z \gg z_{curv}; \Omega_V = 0) \\ H_0^{-1}(1 + z)^{-1}(1 + \Omega z)^{-1/2} & (z_{eq} \gg z; \Omega_{nr} \cong \Omega; \Omega_V = 0) \\ H_0^{-1} \Omega_{nr}^{-1/2} \left[(1 + z)^3 + \Omega_{nr}^{-1} - 1 \right]^{-1/2} & (z_{eq} \gg z; \Omega = \Omega_{nr} + \Omega_V = 1) \end{cases} \qquad (6.88)$$

during various epochs.

The physical length scale characterizing a region of size λ_0 today will evolve as $\lambda(z) = \lambda_0(1 + z)^{-1}$ with redshift. Since d_H increases faster with redshift, $\lambda(z) > d_H(z)$ at sufficiently large redshifts. For a given λ_0 we can assign a particular redshift z_{enter} such that $\lambda(z_{enter}) = d_H(z_{enter})$. For $z > z_{enter}$, the proper wavelength is bigger than the Hubble radius; while for $z < z_{enter}$ we have $\lambda < d_H$. It is conventional to say that the scale λ_0 'enters the Hubble radius' at the epoch z_{enter}. The exact relation between λ_0 and z_{enter} differs in the case of radiation dominated and matter dominated phases since $d_H(z)$ has different scalings in these two cases. Using equation (6.88) it is easy to verify that: (i) a scale $\lambda_{eq} \cong 14(\Omega_{nr}h^2)^{-1}$ Mpc enters the Hubble radius at $z = z_{eq}$; (ii) scales with $\lambda > \lambda_{eq}$ enter the Hubble radius in the matter dominated epoch with

$$z_{enter} \cong 900 \left(\Omega_{nr}h^2\right)^{-1} \left(\frac{\lambda_0}{100 \text{ Mpc}}\right)^{-2} ; \qquad (6.89)$$

(iii) scales with $\lambda < \lambda_{eq}$ enter the Hubble radius in the radiation dominated epoch with

$$z_{enter} \cong 4.55 \times 10^5 \left(\frac{\lambda}{1 \text{ Mpc}}\right)^{-1} . \qquad (6.90)$$

As the universe expands, the wavelength λ grows as $\lambda(t) = \lambda_0[a(t)/a_0]$ and the density of non-relativisitic matter decreases as $\rho(t) = \rho_0[a_0/a(t)]^3$. Hence, the mass $M(\lambda_0)$ contained inside a sphere of radius $(\lambda/2)$ remains constant as the universe expands:

$$M = \frac{4\pi}{3}\rho(t)\left[\frac{\lambda(t)}{2}\right]^3 = \frac{4\pi}{3}\rho_0\left(\frac{\lambda_0}{2}\right)^3 = 1.45 \times 10^{11} \text{ M}_\odot(\Omega_{nr}h^2)\left(\frac{\lambda_0}{1 \text{ Mpc}}\right)^3 . \quad (6.91)$$

This relation shows that a comoving scale $\lambda_0 \cong 1$ Mpc contains a typical galaxy mass; and $\lambda_0 \cong 10$ Mpc contains a typical cluster mass. From (6.90), we see that all these scales enter the Hubble radius in a radiation dominated epoch.

6.4 Optical depth due to galaxies

(a) The probability for a given line-of-sight to intersect a galaxy with redshift in the range of $(z, z + dz)$ is

$$dP = \pi r_G^2 n\, dl = \pi r_G^2 n(z)\frac{dl}{dz}\, dz \qquad (6.92)$$

where dl is the line-of-sight distance travelled by the light ray during the redshift interval dz. The quantity dl can be written as

$$dl = dt = \left|\frac{da}{\dot{a}}\right| = \frac{dz}{(1+z)}\left(\frac{a}{\dot{a}}\right) = \frac{dz}{(1+z)}d_H(z). \qquad (6.93)$$

Given a cosmological model, one can determine the function $d_H(z)$ and thus the quantity $(dl/dz) = d_H/(1+z)$. It follows that

$$\frac{dP}{dz} = \frac{\pi r_G^2 n(z)d_H(z)}{(1+z)} . \qquad (6.94)$$

The optical depth for intersection of objects up to some redshift z is given by the integrated probability:

$$\tau(z) = \int_0^z dP = \int_0^z dz\, \frac{\pi r_G^2 n(z)\, d_H(z)}{(1+z)} . \qquad (6.95)$$

For a $\Omega = 1$, matter dominated model, $d_H(z) = H_0^{-1}(1+z)^{-3/2}$; if galaxies are conserved, $n(z) = n_0(1+z)^3$. In this case, we obtain

$$\tau(z) = \frac{2}{3}\pi r_G^2 n_0 c H_0^{-1}(1+z)^{3/2}. \tag{6.96}$$

If $n_0 \cong 0.02\, h^3\, \text{Mpc}^{-3}$, $r_G \cong 10\, h^{-1}\, \text{kpc}$, we obtain $\tau \cong 10^{-2}(1+z)^{3/2}$. At $z = 1$, the fraction of the sky covered by galaxies is about $\tau(z = 1) \cong 0.04$. The optical depth reaches unity at a redshift of $z \cong 20$.

(b) The area corresponding to a solid angle $d\Omega$ is $dA = a^2 r^2\, d\Omega = a_0^2 r^2 (1 + z)^{-2}\, d\Omega$. The distance along a line-of-sight corresponding to a redshift interval dz is given by (6.93). Combining these, we find the proper volume element to be

$$dV = \left[\frac{d_H(z)}{(1+z)}dz \right]\left[\frac{(a_0 r)^2\, d\Omega}{(1+z)^2} \right] = \frac{a_0^2 r^2(z)\, d_H(z)}{(1+z)^3}\, dz\, d\Omega. \tag{6.97}$$

(Alternatively, one can express the proper volume as

$$dV = \frac{a\, dr}{\sqrt{1 - kr^2}} \cdot a^2 r^2\, d\Omega = \frac{a^3 r^2\, dr\, d\Omega}{\sqrt{1 - kr^2}}. \tag{6.98}$$

Using the relations (6.49) and (6.93), written in the form

$$\frac{dr}{\sqrt{1 - kr^2}} = \frac{dt}{a}, \quad dt = dz\frac{d_H(z)}{(1+z)}, \tag{6.99}$$

we can write dV as

$$dV = (a^3 r^2 d\Omega) \cdot \frac{dz}{a}\frac{d_H(z)}{(1+z)} = \frac{a_0^3 r^2(z)\, d_H(z)}{(1+z)^3}\, d\Omega\, dz, \tag{6.100}$$

which is the same as (6.97).) Assuming that the number density of galaxies evolves as $n = n_0(1+z)^3$, we find that the number count of galaxies per unit solid angle per redshift interval should vary as

$$\frac{dN}{d\Omega\, dz} = n(z)\frac{dV}{d\Omega\, dz} = \frac{n(z)a_0^2 r^2(z)\, d_H(z)}{(1+z)^3}. \tag{6.101}$$

To compute $(dN/d\Omega\, dz)$ explicitly, we need the function $r(z)$, which is determined by

$$\int_0^r \frac{dq}{\sqrt{1 - kq^2}} = \frac{1}{a_0}\int_0^z dz'd_H(z') \tag{6.102}$$

(see equation (6.99)). For a matter dominated universe with $\Omega_V = 0$, this integral gives

$$r(z) = \frac{2\Omega z + 2(\Omega - 2)\left(\sqrt{\Omega z + 1} - 1\right)}{H_0 a_0 \Omega^2 (1+z)}. \tag{6.103}$$

Using this and equation (6.88), we find

$$\frac{dN}{d\Omega\, dz} = \frac{4n(z)}{H_0^3(1+z)^6}\frac{\left[\Omega z + (\Omega - 2)\left(\sqrt{\Omega z + 1} - 1\right)\right]^2}{(\Omega z + 1)^{1/2}\Omega^4}. \tag{6.104}$$

The same procedure can be used for other cosmological models.

(c) Let the ray of light pass a lens, located at a redshift z_L within an impact parameter d. If the lens is an isothermal sphere, it induces a bending by an angle $\alpha = 4\pi(\sigma/c)^2$ which is independent of the impact parameter. From figure 5.2, we have $(d/D_{OL}) = (\alpha D_{LS}/D_{OS})$ or

$$d = \alpha \frac{D_{OL}D_{LS}}{D_{OS}}. \tag{6.105}$$

The probability for multiple imaging is the same as the probability for the line-of-sight to pass within a comoving distance d from a lens at redshift z_L. This can be computed using (6.94) by taking the effective cross-section for the process to be $A = \pi a^2 d^2$. We find that

$$\frac{dP}{dz_L} = 16\pi^3 a_0^2 \left(\frac{\sigma}{c}\right)^4 \left(\frac{D_{OL}D_{LS}}{D_{OS}}\right)^2 \frac{n(z_L)\,d_H\,(z_L)}{(1+z_L)^3}, \tag{6.106}$$

where we have used $a^2 = a_0^2(1+z)^{-2}$. The optical depth for gravitational lensing up to a redshift z_s can be obtained by integrating this expression in an interval $(0, z_s)$. Taking $n(z_L) = n_0(1+z_L)^3$, we obtain

$$\tau = 16\pi^3 \left(\frac{\sigma}{c}\right)^4 n_0 \int_0^{z_s} \left(\frac{D_{OL}D_{LS}}{D_{OS}}\right)^2 d_H(z_L)\,dz_L. \tag{6.107}$$

Given a cosmological model, expressions (6.106) and (6.107) can be evaluated explicitly. If $\Omega = 1$, then $D_{LS} = D_{OS} - D_{OL}$ and the expression $(D_{OL}D_{LS}/D_{OS})$ is maximum when $D_{OL} = D_{OS}/2$, that is, when the lens is midway between the source. Further, in this case, $d_H(z_L)dz_L = dr_{OL} = dD_{OL}$ (see equation (6.102) with $k = 0$, $a_0 = 1$); so we obtain

$$\tau = 16\pi^3 \left(\frac{\sigma}{c}\right)^4 \frac{n_0}{D_{OS}^2} \int_0^{D_{OS}} D_{OL}^2\,(D_{OS} - D_{OL})^2\,dD_{OL} = \frac{8\pi^3}{15} \left(\frac{\sigma}{c}\right)^4 n_0 D_{OS}^3. \tag{6.108}$$

For the $\Omega = 1$ case, $d_H(z) = H_0^{-1}(1+z)^{-3/2}$ and

$$D_{OS} = \int_0^{z_s} dz\,d_H(z) = 2H_0^{-1}\left[1 - \frac{1}{\sqrt{1+z_s}}\right]. \tag{6.109}$$

This gives

$$\tau = \frac{64\pi^3}{15} \left(\frac{\sigma}{c}\right)^4 n_0 H_0^{-3} \left[1 - \frac{1}{\sqrt{1+z_s}}\right]^3. \tag{6.110}$$

Given the distribution of n_0 and σ for the lenses, one can calculate τ. Taking $n_0 \cong 0.001\,h^3\,\mathrm{Mpc}^{-3}$, and $\sigma \cong 250\,\mathrm{km\,s^{-1}}$ as the values corresponding to galaxies, we find

$$\tau \cong 10^{-3} \left(\frac{n_0}{0.001\,h^3\,\mathrm{Mpc}^{-3}}\right) \left(\frac{\sigma}{250\,\mathrm{km\,s^{-1}}}\right)^4 \left[1 - \frac{1}{\sqrt{1+z}}\right]^3. \tag{6.111}$$

6.5 Radiative processes in an expanding universe

(a) Consider some emission process which takes place at a redshift of z for which the emissivity (which has units $\text{erg cm}^{-3}\,\text{s}^{-1}\,\text{Hz}^{-1}$) is $J_\omega(z)$. In an interval $(t, t+dt)$ this process will produce a spectral density of radiation $dI_\omega(z) = J_\omega(z)\,dt$. At the present epoch we will observe this radiation density as

$$dI_{\omega(1+z)^{-1}}(z=0) = \frac{J_\omega(z)\,dt}{(1+z)^3}, \qquad (6.112)$$

where we have used the results $\omega \propto (1+z)$ and $I \propto (1+z)^3$, derived earlier in problem 6.2. If the sources are distributed in a redshift interval (z_1, z_2), then the total spectral intensity of radiation observed at the present epoch will be

$$I[\omega_0; z=0] = \int_{z_1}^{z_2} \frac{J[\omega_0(1+z); z]}{(1+z)^3} \left|\frac{dt}{dz}\right| dz = \int_{z_1}^{z_2} \frac{J[\omega_0(1+z); z]}{(1+z)^4} d_{\text{H}}(z)\, dz.$$

$$(6.113)$$

The Ω-dependence of the result arises only through the d_{H} factor. From (6.88) we have

$$d_{\text{H}}(z) = \frac{1}{H_0(1+z)(1+\Omega z)^{1/2}} \qquad (6.114)$$

leading to

$$I(\omega_0; z=0) = \int_{z_1}^{z_2} \frac{J[\omega_0(1+z); z]}{H_0(1+z)^5(1+\Omega z)^{1/2}} dz. \qquad (6.115)$$

The flux of radiation $F(\omega) = (c/4\pi)I(\omega)$ is

$$F(\omega_0; z=0) = \frac{c}{4\pi H_0} \int_{z_1}^{z_2} \frac{J[\omega_0(1+z); z]\,dz}{(1+z)^5(1+\Omega z)^{1/2}}. \qquad (6.116)$$

(b) Suppose the observed frequency of a photon is v_0; then the frequency at redshift z will be $v = v_0(1+z)$. Further, if $x(z)$ is the fraction of hydrogen gas which is in ionized form at the redshift of z, then

$$n_{\text{H}} = n_0(1+z)^3[1-x(z)], \qquad (6.117)$$

where n_0 is the present number density of hydrogen atoms *and* ions. Substituting in (6.116) we find, for $v_0 < v_{\text{H}}$,

$$\begin{aligned}
F(v_0; z=0) &= \frac{3c(hv_0)}{16\pi H_0}(An_0) \int_0^\infty [1-x(z)] \frac{\delta[v_0(1+z) - v_{\text{H}}]}{(1+z)(1+\Omega z)^{1/2}} dz \\
&= \frac{3ch}{16\pi H_0}(An_0) \frac{[1-x(z_c)]}{(1+z_c)(1+\Omega z_c)^{1/2}}
\end{aligned} \qquad (6.118)$$

with $(1+z_c) \equiv (v_{\text{H}}/v_0)$. We will, therefore, observe the flux at all frequencies $v_0 < v_{\text{H}}$ and, of course, no flux at $v_0 > v_{\text{H}}$. The discontinuity at v_{H} will be by the amount

$$\Delta F = \frac{3ch}{16\pi H_0}(An_0)[1-x(z=0)] = \frac{3ch}{16\pi H_0}An_{\text{H}}(0), \qquad (6.119)$$

where $n_H(0)$ is the density of neutral atoms today. The observed value of ΔF can be used to estimate $n_H(0)$; when unobserved, bounds on ΔF can be converted into bounds on $n_H(0)$.

(c) Substituting the expression for Bremsstrahlung emissivity into (6.116), we obtain

$$F(v_0; z = 0) = \frac{cA}{4\pi H_0} n_e^2(0) \int_0^{z_{max}} dz \frac{(1+z)}{(1+\Omega z)^{1/2}} T_e^{-1/2}(z) \left[\exp\left(-\frac{h v_0(1+z)}{k T_e(z)} \right) \right].$$
(6.120)

This integral can be worked out if the temperature history of the gas $T_e(z)$ is known.

(d) To incorporate the effects of expansion on the optical depth property, we will proceed as follows. A photon received with frequency v_0 will have the frequency $v = v_0(1 + z)$ at the redshift z. During the interval $(t, t + dt)$ corresponding to the redshift range $(z, z + dz)$ the optical depth increases by $d\tau_v = \sigma(v)n(z)c\, dt = c\sigma(v)n(z)|(dt/dz)|\, dz$. Integrating this expression, we find the total optical depth due to matter in the redshift range $(0, z)$ to be

$$\tau[v_0; z] = \int_0^z \sigma(v)n(z)c \left| \frac{dt}{dz} \right| dz = \frac{c}{H_0} \int_0^z \frac{\sigma[v_0(1+z)]n(z)\, dz}{(1+z)^2(1+\Omega z)^{1/2}}.$$
(6.121)

As an example in the use of this formula, consider the absorption of radiation by neutral hydrogen at the wavelength of 21 cm. Substituting the expression for $\sigma(v)$ in (6.121) we obtain

$$\tau[v_0; z] = \frac{3A}{32\pi} \left(\frac{h v_H}{k T_{sp}} \right) \left(\frac{c}{v_H} \right)^3 \frac{n_H(0)}{H_0} \frac{(1+z_c)^2}{(1+\Omega z_c)^{1/2}} [1 - x(z_c)] \quad \text{(for } z > z_c\text{)},$$
(6.122)

where $(1 + z_c) = (v_H/v_0)$. Suppose we are receiving the radiation emitted by a class of sources at a typical redshift of $z \cong z_{source}$. Absorption can take place anywhere in the redshift range of $(0, z_{source})$; hence the intensity of the light received by us will be diminished at all frequencies in the band $[v_H(1 + z_{source})^{-1}, v_H]$. The discontinuity in τ at v_H is

$$\Delta\tau = \frac{3A}{32\pi} \left(\frac{h v_H}{k T_{sp}} \right) \left(\frac{c}{v_H} \right)^3 \frac{n_H(0)}{H_0} [1 - x(0)],$$
(6.123)

which allows us to determine (or put bounds on) the quantity $n_H(0)[1 - x(0)]$.

6.6 Statistical mechanics in an expanding universe

(a) To determine the form of $f_A(\mathbf{p}, t)$ we may reason as follows. The different species of particles will be interacting constantly through various forces, scattering off each other and exchanging energy and momentum. If the rate of these reactions, $\Gamma(t)$, is much higher than the rate of expansion of the universe, $H(t) = (\dot{a}/a)^{-1}$, then these interactions can produce (and maintain) thermodynamic equilibrium among the interacting particles with some temperature $T(t)$. All these interactions which occur between the particles have a short range. (The Coulomb force between charged particles has a long range; but, in a plasma, the process of Debye shielding reduces this range, making it effectively a short-range force.) Therefore, we may assume that the role of these interactions is limited to providing a mechanism for thermalization, and ignore their effects in deciding the *form* of the distribution function. In that case, the particles may be treated as an *ideal* Bose or Fermi gas with the distribution function

$$f_A(\mathbf{p}, t)\, d^3\mathbf{p} = \frac{g_A}{(2\pi)^3} \left\{ \exp[(E_\mathbf{p} - \mu_A)/T_A(t)] \pm 1 \right\}^{-1} d^3\mathbf{p}, \qquad (6.124)$$

where g_A is the spin degeneracy factor of the species, $\mu_A(T)$ is the chemical potential, $E(\mathbf{p}) = (\mathbf{p}^2 + m^2)^{1/2}$ and $T_A(t)$ is the temperature characterizing this species at time t. The upper sign $(+1)$ corresponds to fermions and the lower sign (-1) is for bosons.

At any instant in time, the universe will also contain a blackbody distribution of photons with some characteristic temperature $T_\gamma(t)$. If a particular species couples to the photon directly or indirectly, and if the rate of these A–γ interactions is high enough (i.e. $\Gamma_{A\gamma} \gg H$), then these particles will have the same temperature as photons: $T_A = T_\gamma$. Since this is usually the case, one often refers to the photon temperature by the term 'temperature of the universe'. Of course, any set of particle species $A, B, C, \ldots$ which are interacting among themselves at a high enough rate will also have the same temperature $T_A = T_B = T_C \ldots$. But, in general, this may not be the case and the universe could be populated by different species of particles, each with its own temperature.

As the universe evolves, the temperature $T(t)$ changes due to expansion on a timescale of the order of $H^{-1}(t) \equiv (\dot{a}/a)^{-1}$; the *rate* at which temperature is changing is given by $H(t)$. The rate of interaction (per particle) can be expressed as $\Gamma \equiv n \langle \sigma v \rangle$ where n is the number density of target particles, v is the relative velocity and σ is the interaction cross-section. Since σ is usually a function of energy, $\langle \sigma v \rangle$ denotes an average value for this combination. As long as $\Gamma \gg H$, the interactions can maintain equilibrium. In that case, f will evolve adiabatically, maintaining the form of the equilibrium distribution given by (6.124), with the temperature corresponding to the instantaneous value.

(b) From the distribution function (6.124), we can calculate the number density n, energy density ρ and pressure p. Suppressing the time dependence and the subscript A for simplicity, we have

$$n = \int f(\mathbf{k})\, d^3\mathbf{k} = \frac{g}{2\pi^2} \int_m^\infty \frac{(E^2 - m^2)^{1/2} E\, dE}{\exp[(E - \mu)/T] \pm 1}, \qquad (6.125)$$

$$\rho = \int E f(\mathbf{k})\, d^3\mathbf{k} = \frac{g}{2\pi^2} \int_m^\infty \frac{(E^2 - m^2)^{1/2} E^2\, dE}{\exp[(E - \mu)/T] \pm 1}, \qquad (6.126)$$

$$p = \int \frac{1}{3}\frac{|\mathbf{k}|^2}{E} f(\mathbf{k})\, d^3\mathbf{k} = \frac{g}{6\pi^2} \int_m^\infty \frac{(E^2 - m^2)^{3/2}\, dE}{\exp[(E - \mu)/T] \pm 1}. \qquad (6.127)$$

We will use the symbol k to denote the momentum when the pressure is denoted by the letter p. For a collection of relativistic particles, the pressure p corresponding to velocity v is $p = m(v^2/3\sqrt{1-v^2})$, which is the same as $p = (k^2/3E)$.

Differentiating (6.127) with respect to T, and treating μ as some specified function of T, we obtain

$$\frac{dp}{dT} = \frac{4\pi}{3} \int_0^\infty \frac{k^4 dk}{E} f^2 \left[\exp\frac{(E-\mu)}{T}\right] \left[\frac{E}{T^2} + \frac{d}{dT}\left(\frac{\mu}{T}\right)\right]. \tag{6.128}$$

Using the relation

$$\frac{df}{dk} = -\frac{k}{ET} f^2 \exp\frac{(E-\mu)}{T}, \tag{6.129}$$

we can rewrite (dp/dT) as

$$\frac{dp}{dT} = -\frac{4\pi}{3} \int_0^\infty dk(k^3 T) \left(\frac{df}{dk}\right) \left[\frac{E}{T^2} + \frac{d}{dT}\left(\frac{\mu}{T}\right)\right]. \tag{6.130}$$

Integrating by parts and using the definitions of ρ and p we find

$$\frac{dp}{dT} = \frac{1}{T}(\rho + p) + nT\frac{d}{dT}\left(\frac{\mu}{T}\right). \tag{6.131}$$

From the Friedmann equations we have the relation $d(\rho a^3) = -p\, d(a^3)$ (see equation (6.63)), which can be written as

$$\frac{d}{dT}[(\rho + p)a^3] = a^3 \frac{dp}{dT}. \tag{6.132}$$

Substituting for (dp/dT) from (6.131) and rearranging the terms, we finally obtain

$$d(sa^3) \equiv d\left\{\frac{a^3}{T}(\rho + p - n\mu)\right\} = \left(\frac{\mu}{T}\right) d(na^3). \tag{6.133}$$

In most cases of interest to us, either (na^3) will be approximately constant *or* we will have $\mu \ll T$. The above relation shows that, in either case, the quantity (sa^3) will be conserved.

When $\mu \ll T$, the expression for s reduces to $T^{-1}(\rho + p)$. Expanding the quantity $T d[a^3 T^{-1}(\rho + p)]$ and using the relation $(dp/dT) \cong T^{-1}(p + \rho)$ derived earlier, we obtain

$$T d(sa^3) = T d\left[\frac{(\rho + p)a^3}{T}\right] = d\left[(\rho + p)a^3\right] - (\rho + p)a^3 \frac{dT}{T}$$
$$\cong d[(\rho + p)a^3] - a^3 dp = d(\rho a^3) + p\, d(a^3). \tag{6.134}$$

Comparing with the familiar thermodynamic relation $T dS = dE + p\, dV$, we see that $s = T^{-1}(\rho + p)$ may be interpreted as the entropy density. Then (6.133) shows that entropy density $s \propto a^{-3}$ during expansion, provided $\mu \ll T$. Note that s is an additive quantity.

(c) When the particles are highly relativistic ($T \gg m$) and non-degenerate ($T \gg \mu$), we obtain

$$\rho \cong \frac{g}{2\pi^2} \int_0^\infty \frac{E^3 dE}{e^{E/T} \pm 1} = \begin{cases} g_B(\pi^2/30)T^4 & \text{(bosons)} \\ \dfrac{7}{8} g_F(\pi^2/30)T^4 & \text{(fermions)}. \end{cases} \tag{6.135}$$

(To relate the integrals for fermion and bosons, one can use the following trick. Let

$$I_n^{\pm} \equiv \int_0^\infty \frac{x^n dx}{e^x \pm 1}. \tag{6.136}$$

Then

$$I_n^- - I_n^+ = \int_0^\infty dx\, x^n \frac{2}{e^{2x} - 1} = \frac{1}{2^n} \int_0^\infty dy \frac{y^n}{e^y - 1} = 2^{-n} I_n^- \tag{6.137}$$

giving

$$I_n^+ = I_n^- \left(1 - \frac{1}{2^n}\right). \tag{6.138}$$

This accounts for the (7/8) factor when $n = 3$.) Thus the total energy density contributed by all the relativistic species together can be expressed as

$$\rho_{\text{tot}} = \sum_{i=B} g_i \left(\frac{\pi^2}{30}\right) T_i^4 + \sum_{i=F} \frac{7}{8} g_i \left(\frac{\pi^2}{30}\right) T_i^4 = g_{\text{tot}} \left(\frac{\pi^2}{30}\right) T^4, \tag{6.139}$$

where

$$g_{\text{tot}} \equiv \sum_B g_B \left(\frac{T_B}{T}\right)^4 + \sum_F \frac{7}{8} g_F \left(\frac{T_F}{T}\right)^4. \tag{6.140}$$

In writing g_{tot}, we have explicitly taken into account the possibility that even though all the species may have a thermal distribution they may not have the same temperature. If all species have the same temperature, then $g = g_B + (7/8)g_F$.

The pressure due to relativistic species is $p \cong (\rho/3) = g(\pi^2/90)T^4$; so the entropy density of the relativistic species of particles will be

$$s \cong \frac{1}{T}(\rho + p) = \frac{2\pi^2}{45} q T^3, \tag{6.141}$$

with

$$q \equiv q_{\text{tot}} = \sum_B g_B \left(\frac{T_B}{T}\right)^3 + \frac{7}{8} \sum_F g_F \left(\frac{T_F}{T}\right)^3. \tag{6.142}$$

Clearly, $q_{\text{tot}} = g_{\text{tot}}$ if all the particles have the same temperature. Our previous analysis shows that the quantity $S = qT^3a^3$ is conserved during the expansion.

The number density of relativistic particles can be computed in the same way:

$$n \cong \frac{g}{2\pi^2} \int_0^\infty \frac{E^2 dE}{e^{E/T} \pm 1} = \begin{cases} (\zeta(3)/\pi^2)g_B T^3 & \text{(boson)} \\ \dfrac{3}{4}(\zeta(3)/\pi^2)g_F T^3 & \text{(fermion)} \end{cases} \tag{6.143}$$

where $\zeta(3) \cong 1.202$ is the Riemann zeta function of order 3. Combining this with (6.135), we find that the mean energy of the particles $\langle E \rangle \equiv (\rho/n)$ is about $2.7T$ for bosons and $3.15T$ for fermions. Note that s is proportional to the number density of relativistic particles, if all species have the same temperature; in fact, $s \cong 1.8qn_\gamma$ where n_γ is the photon number density.

In the opposite limit of $T \ll m$, the exponential in (6.124) is large compared to unity. Then we obtain, for *both* bosons and fermions, the expression

$$n \cong \frac{g}{2\pi^2} \int_0^\infty p^2 dp \exp\left[-\frac{(m-\mu)}{T}\right] \exp\left(-\frac{p^2}{2mT}\right)$$
$$= g\left(\frac{mT}{2\pi}\right)^{3/2} \exp\left[-\frac{1}{T}(m-\mu)\right]. \qquad (6.144)$$

In this limit $\rho \cong nm$ and $p = nT \ll \rho$. A comparison of (6.143) and (6.144) shows that the number (and energy) density of non-relativistic particles are exponentially damped by the factor $\exp -(m/T)$ with respect to that of the relativistic particles.

For $t < t_{eq}$, in the radiation dominated phase, we may ignore the contribution of non-relativistic particles to ρ. We have seen earlier that, during the radiation dominated phase, $a(t) \propto t^{1/2}$; therefore

$$\left(\frac{\dot{a}}{a}\right)^2 = H^2(t) = \frac{1}{4t^2} = \frac{8\pi G}{3}\rho = \frac{8\pi G}{3}g\left(\frac{\pi^2}{30}\right)T^4. \qquad (6.145)$$

It is convenient to express these results in terms of the Planck energy $m_{Pl} = G^{-\frac{1}{2}}$ $= 1.22 \times 10^{19}\,\text{GeV}$:

$$H(T) \cong 1.66g^{1/2}\left(\frac{T^2}{m_{Pl}}\right), \qquad (6.146)$$

$$t \cong 0.3g^{-\frac{1}{2}}\left(\frac{m_{Pl}}{T^2}\right) \cong 1\text{s}\left(\frac{T}{1\,\text{MeV}}\right)^{-2}g^{-\frac{1}{2}}. \qquad (6.147)$$

(d) Once the species, A, is completely decoupled, each of the A-particles will be travelling along a geodesic in the spacetime. We have seen earlier that f is conserved during such free propagation. This allows one to obtain the function f_{dec} after the species has decoupled, from the known form of f_{eq} before decoupling. For simplicity, let us assume that the decoupling occurred instantaneously at some time $t = t_D$ when the temperature was T_D and the expansion factor was a_D. For $t < t_D$, the distribution function is given by (6.124). At some later time, $t > t_D$, let the distribution function be $f_{dec}(p, t)$. Because of the redshift in momentum, all particles with momentum p at time t must have had momentum $p[a(t)/a(t_D)]$ at $t = t_D$. Therefore,

$$f_{dec}(p, t) = f_{eq}\left(p\frac{a(t)}{a(t_D)}, t_D\right) \qquad \text{(for } t > t_D\text{)}, \qquad (6.148)$$

where f_{eq} is the equilibrium distribution function of (6.124). Thus, as long as the species A was in equilibrium at *some* time, we can determine its distribution function at all later times.

The expression in (6.148) simplifies considerably if the decoupling occurs either when the species is ultrarelativistic $(T_D \gg m)$ or when it is non-relativistic $(T_D \ll m)$. In the first case,

$$f_{dec}(p) = f_{eq}\left(p\frac{a(t)}{a(t_D)}, T_D\right) \cong \frac{g}{(2\pi)^3}\left[\exp\frac{1}{T_D}\left(p\frac{a(t)}{a(t_D)}\right) \pm 1\right]^{-1}. \quad (6.149)$$

This has the same form as the f_{eq} for a relativistic species with temperature

$$T(t) = T_D[a(t_D)/a(t)], \quad (6.150)$$

even though this species is not in thermodynamic equilibrium any longer. The 'temperature' in this distribution function falls *strictly* as a^{-1}; the entropy of these particles, $S_A = (s_A a^3)$ is conserved separately. Note that for the species which are still in thermal equilibrium, $T \propto q^{-\frac{1}{3}}a^{-1}$ falls more slowly.

The number density of these decoupled particles is given by (6.143):

$$n = g_{eff}\left(\frac{\zeta(3)}{\pi^2}\right)T_D^3\left(\frac{a_D}{a}\right)^3, \quad (6.151)$$

where $g_{eff} = (3g/4)$ for fermions and $g_{eff} = g$ for bosons. (Here g refers to the spin degeneracy factor of the particular species which has decoupled.) This number density will be comparable to the number density of photons at any given time. In particular, any such decoupled species will continue to exist in our universe today as a relic background, with number densities comparable to the number density of photons.

In estimating the density contributed by the decoupled species, the following point should be noted. Suppose that a species with mass m decouples at the temperature T_D with $T_D \gg m$. At the time of decoupling, most of these particles will be ultrarelativistic and their (mean) momentum $p(t_D)$ and energy $E(t_D) = (p^2(t_D) + m^2)^{1/2} \cong p(t_D)$ will be of order T_D. Their distribution function at $t = t_D$ is well approximated by the f_{eq} of zero-mass particles. Decoupling 'freezes' the distribution function in this form. At a later time $(t > t_D)$, the mean momentum of the particles will be redshifted to a value $p(t) = p(t_D)$ $(a_D/a) \cong T_D(a_D/a)$. For $t \gg t_D$, most of the particles will have momentum $p(t)$, which is much smaller than m. Thus the individual particles would have become non-relativistic when the universe expanded sufficiently which happens when the temperature of the universe drops below $T_{nr} \cong m$: that is, when $(a/a_D) \gtrsim (T_D/m)$. The energy of each of these particles will now be $E(t) = (p^2(t) + m^2)^{1/2} \cong m$. *But the distribution function (and the number density) of the particles will still be given by the (frozen-in) form which corresponds to relativistic particles.* Thus, for $t \gg t_D$, the number density of these particles will be similar to those of *relativistic* species but the energy density will be that of *non-relativistic* particles: $\rho_{dec} \cong nm$.

Consider, next, the other extreme case, that of a species which decouples when most of the particles are already non-relativistic: $(T_D \ll m)$. In this case,

$$f_{dec}(p) = f_{eq}\left(p\frac{a}{a_D}, T_D\right) \cong \frac{g}{(2\pi)^3} \exp\left[-\frac{(m-\mu)}{T_D}\right] \exp\left[-\frac{p^2}{2m}\frac{1}{T_D}\left(\frac{a}{a_D}\right)^2\right]$$

$$\cong \frac{g}{(2\pi)^3} e^{-m/T_D} \exp\left[-\frac{p^2}{2mT_D}\left(\frac{a}{a_D}\right)^2\right], \qquad (6.152)$$

where we have further assumed that $\mu \ll T_D$. This distribution function has the same form as that of a non-relativistic Maxwell–Boltzmann gas with a 'temperature' $T(t) \equiv T_D(a_D/a)^2$, which decreases as the *square* of the expansion factor. The corresponding number density is given by (6.144):

$$n = g\left(\frac{mT_D}{2\pi}\right)^{3/2}\left(\frac{a_D}{a}\right)^3 \exp -\frac{1}{T_D}(m-\mu)$$

$$\cong g\left(\frac{mT_D}{2\pi}\right)^{3/2}\left(\frac{a_D}{a}\right)^3 \exp -\left(\frac{m}{T_D}\right) \quad \text{(for } \mu \ll T_D\text{).} \qquad (6.153)$$

As is to be expected, $n \propto a^{-3}$. The energy density of these particles will be $\rho \cong nm$.

To any species of particle which is not being created or destroyed ($n \propto a^{-3}$), we can assign a conserved number $N \propto na^3$; since $a^3 \propto s^{-1}$, we can conveniently define this number to be $N \equiv (n/s)$. From our expressions (6.141), (6.143) and (6.144) (valid for $\mu \ll T$) it follows that

$$N = \begin{cases} [45\zeta(3)/2\pi^4]\,[g_{eff}/q] & \cong 0.28\,(g_{eff}/q) \\ [45/2\pi^4]\,(\pi/8)^{1/2}(g/q)\left(\frac{m}{T}\right)^{3/2} e^{-m/T} & \cong 0.15\,(g/q)\left(\frac{m}{T}\right)^{3/2} e^{-m/T} \end{cases}, \qquad (6.154)$$

where $g_{eff} = g$ for bosons and $g_{eff} = (3g/4)$ for fermions.

6.7 Relics of relativistic particles

(a) Since the rest mass of the electron $m_e \cong 0.5$ MeV $= 6 \times 10^9$ K, we expect a significant number density (i.e. number density comparable to that of photons) of ultrarelativistic electrons (e) and positrons ($\bar{e}$) at $T > 6 \times 10^9$ K. The only other particle species which could be relativistic at temperature 10^{12} K is the neutrino (v). It is known that there exist three kinds ('flavours') of neutrino, namely, electron-neutrino (v_e), muon-neutrino (v_μ) and the tau-neutrino (v_τ). The masses of these particles are uncertain; the experimental *upper bounds* on the masses are 13 eV, 0.25 MeV and 35 MeV for v_e, v_μ and v_τ respectively. We shall assume, for the time being, that $m_v = 0$ for all three species. The number densities of e, $\bar{e}$, v, $\bar{v}$ and γ must be comparable at these temperatures.

The neutrons and protons contained in the present-day universe must have existed at $T \cong 10^{12}$ K as well, since these particles could not have been produced at $T < 10^{12}$ K. The ratio between the number density of baryons (n_B) and the number density of photons (n_γ) remained approximately constant from $T \cong 10^{12}$ K till today. This number in the present-day universe is $(n_B/n_\gamma)_0 \cong (\rho_c \Omega_B /m_B n_\gamma)_0 \cong 10^{-8}$ to 10^{-10}. The smallness of this (conserved) number shows that we may ignore the effect of n_B on the overall dynamics of the radiation dominated universe.

Since photons are not conserved, the chemical potential for photons is identically zero. The reaction $e\bar{e} \leftrightarrow \gamma\gamma$ maintains the equilibrium between e, $\bar{e}$ and γ at this temperature. The conservation of chemical potential in this reaction implies that $(\mu_e + \bar{\mu}_e) = 0$, that is, $\bar{\mu}_e = -\mu_e$. (We will denote a particle–antiparticle pair by $A, \bar{A}$ and the corresponding chemical potentials by $\mu_A, \bar{\mu}_A$.) The excess of electrons over positrons will then be

$$n - \bar{n} = \frac{g}{2\pi^2} \int_m^\infty E(E^2 - m^2)^{1/2} \left[\frac{1}{\exp \frac{1}{T}(E - \mu) + 1} - \frac{1}{\exp \frac{1}{T}(E + \mu) + 1} \right] dE$$

$$\cong \frac{gT^3}{6\pi^2} \left[\pi^2 \left(\frac{\mu}{T} \right) + \left(\frac{\mu}{T} \right)^3 \right] \quad \text{(for } T \gg m\text{)}. \tag{6.155}$$

As the universe cools to temperature $T \ll m_e$, electrons and positrons will annihilate in pairs and only this small excess will survive.

The only other *charged* particle which will be present in the universe is the proton. Since our universe appears to be electrically neutral (the bound on the net number density of free charges being $(n_Q/s) \lesssim 10^{-27}$), the electron excess $(n - \bar{n})$ should be equal to the number density of protons n_p. Since $(n_p/n_\gamma) \cong 10^{-8}$ it follows that $[(n - \bar{n})/n_\gamma] \cong 10^{-8}$. Using (6.143) and (6.155), we can write

$$\frac{n - \bar{n}}{n_\gamma} \cong \left(\frac{g_e}{g_\gamma} \right) \frac{\pi^2}{6\zeta(3)} \left[\frac{\mu}{T} + \frac{1}{\pi^2} \left(\frac{\mu}{T} \right)^3 \right] \cong 1.33 \left(\frac{\mu}{T} \right) \cong 10^{-8}. \tag{6.156}$$

Clearly, $(\mu/T) \ll 1$ and we can set $\mu \cong 0$ for both electrons and positrons.

Similarly, from the reaction $\nu\bar{\nu} \leftrightarrow e\bar{e}$, it follows that $\mu_\nu + \bar{\mu}_\nu = \mu_e + \bar{\mu}_e = 0$. The excess of neutrinos over antineutrinos will again be given by an expression similar to (6.155). Unfortunately, the value of $[(n_\nu - \bar{n}_\nu)/n_\gamma]$ for our universe is not known. If this number is large, then our universe will have a large lepton number (L), which is far in excess of the baryon number B. Since our universe does not seem to have large values for any quantum number, a large value for L will require the special choice of initial conditions. This suggests that L should be small. In that case, $\mu_\nu = -\bar{\mu}_\nu \cong 0$. We will make this assumption in what follows.

The above arguments show that, at $T \cong 10^{12}$ K, the energy density of the universe is essentially contributed by e, $\bar{e}$, ν, $\bar{\nu}$ and photons. Since the interactions among them maintain the equilibrium, they all have the same temperature. Taking $g_B = g_\gamma = 2$, $g_e = \bar{g}_e = 2$, $g_\nu = \bar{g}_\nu = 1$ and including three flavours of neutrino, we find

$$g_{tot} = g_B + \frac{7}{8} g_F = 2 + \frac{7}{8}[2 + 2 + 2 \times 3] = \frac{43}{4} = 10.75 \,. \tag{6.157}$$

(The g-values for electrons and positrons represent the two-possible spin states for the massive, spin-1/2 fermions. Though photons have spin-1, they have only two accessible states (corresponding to two states of polarization) giving $g_\gamma = 2$. Massless spin-1/2 fermions, like neutrinos, exist only in left-handed or right-handed states, making $g_\nu = 1$.) From (6.146) and (6.147), we can find the precise time–temperature relationship for this phase of evolution:

$$H(T) \cong 5.44 \left(\frac{T^2}{m_{\text{Pl}}} \right), \quad t \cong 0.09 \left(\frac{m_{\text{Pl}}}{T^2} \right). \tag{6.158}$$

(b) Defining the 'Fermi coupling constant' $G_F = (\alpha/m_x^2) \cong 1.17 \times 10^{-5} (\text{GeV})^{-2} = (293\,\text{GeV})^{-2}$ and using the fact that $E \cong T$, we can write $\sigma \cong G_F^2 E^2 \cong G_F^2 T^2$. Since the number density of interacting particles is $n \cong (\zeta(3)g/\pi^2) T^3 \cong 1.3 T^3$ and $\langle v \rangle \cong c = 1$, the rate of interactions is given by

$$\Gamma = n\sigma|v| \cong 1.3 G_F^2 T^5. \tag{6.159}$$

The rate of expansion, from (6.158), is $H \cong 5.4\,(T^2/m_{\text{Pl}})$. So

$$\frac{\Gamma}{H} \cong 0.24 T^3 \left(\frac{m_{\text{Pl}}}{G_F^{-2}} \right) \cong \left(\frac{T}{1.4\,\text{MeV}} \right)^3 = \left(\frac{T}{1.6 \times 10^{10}\,\text{K}} \right)^3. \tag{6.160}$$

The interaction rate of neutrinos becomes lower than the expansion rate when the temperature drops below $T_D \cong 1\,\text{MeV}$. At lower temperatures, the neutrinos are completely decoupled from the rest of the matter.

(c) At $T_D > T \gtrsim m_e$, the νs have decoupled and their entropy is separately conserved; but the photons ($g = 2$) are in equilibrium with the electrons ($g = 2$) and positrons ($g = 2$). This gives $g(\gamma, e, \bar{e}) = 2 + (7/8) \times 4 = (11/2)$. For $T \ll m_e$, when the $e\bar{e}$ annihilation is complete, the only relativistic species left in this set is the photon ($g = 2$). The conservation of $S = q(Ta)^3$, applied to particles which are in equilibrium with radiation, shows that the quantity $q(T_\gamma a)^3 = g(aT_\gamma)^3$ remains constant during expansion. (Since γ, e and $\bar{e}$ all have the same temperature, $q = g$.) Because g decreases during the $e\bar{e}$ annihilation, the value of $(aT_\gamma)^3$ after the $e\bar{e}$ annihilation will be higher than its value before:

$$\left[\frac{(aT_\gamma)^3_{\text{after}}}{(aT_\gamma)^3_{\text{before}}} \right] = \left[\frac{g_{\text{before}}}{g_{\text{after}}} \right] = \frac{11}{4}. \tag{6.161}$$

The neutrinos, since they decoupled, do not participate in this process. They are characterized by a temperature $T_\nu(t)$ which falls *strictly* as a^{-1} and their entropy $(s_\nu a^3)$ is separately conserved. Let $T_\nu = Ka^{-1}$; originally, before $e\bar{e}$ annihilations began, the photons and neutrinos had the same temperature: $(aT_\nu)_{\text{before}} = (aT_\gamma)_{\text{before}} = K$. It follows that

$$(aT_\gamma)_{\text{after}} = \left(\frac{11}{4} \right)^{1/3} (aT_\gamma)_{\text{before}} = \left(\frac{11}{4} \right)^{1/3} (aT_\nu)_{\text{before}}$$

$$= \left(\frac{11}{4} \right)^{1/3} (aT_\nu)_{\text{after}} \cong 1.4 (aT_\nu)_{\text{after}}. \tag{6.162}$$

The first equality follows from (6.161), the second from the fact that $T_\gamma = T_\nu$ at $T \gtrsim m_e$ and the third from the strict constancy of (aT_ν). We see that the $e\bar{e}$ annihilations increase the temperature of photons compared to that of neutrinos by a factor $(11/4)^{1/3} \cong 1.4$.

Thus the species of particles which remain relativistic today will be photons $(g_\gamma = 2)$ with a temperature $T_\gamma \cong 2.7\,\text{K}$ and three flavours of massless neutrino and antineutrino $(g_F = 3 + 3 = 6)$ with a temperature $T_\nu = (4/11)^{1/3}T_\gamma$. From (6.140) and (6.142) we find

$$g(\text{now}) = 2 + \frac{7}{8} \times 6 \times \left(\frac{4}{11}\right)^{4/3} \cong 3.36$$

$$q(\text{now}) = 2 + \frac{7}{8} \times 6 \times \left(\frac{4}{11}\right) \cong 3.91.$$

(6.163)

The energy and entropy densities of these relativistic particles in the present-day universe are

$$\rho_r = \frac{\pi^2}{30}gT^4 = 8.09 \times 10^{-34}\,\text{g}\,\text{cm}^{-3}, \quad s = \frac{2\pi^2}{45}qT^3 \cong 2.97 \times 10^3\,\text{cm}^{-3}. \quad (6.164)$$

This ρ_r corresponds to the $\Omega_r = 4.3 \times 10^{-5}h^{-2}$. Note that $\rho_r = (g_{\text{tot}}/g_\gamma)\,\rho_\gamma \cong 1.68\rho_\gamma$; similarly, $\Omega_r = 1.68\,\Omega_\gamma$.

The matter density today is $\rho_{\text{nr}} = 1.88 \times 10^{-29}\,\Omega h^2\,\text{g}\,\text{cm}^{-3}$. The redshift z_{eq} at which matter and radiation have equal energy densities is determined by the relation $(1 + z_{\text{eq}}) = (\rho_{\text{nr}}/\rho_r)$. This quantity z_{eq} was calculated in problem 6.3 assuming that Ω_r is contributed by photons alone. The correct value, if there are three-massless neutrino species, is

$$(1 + z_{\text{eq}}) = \left(\frac{\Omega_{\text{nr}}}{\Omega_r}\right) = 2.3 \times 10^4(\Omega h^2). \quad (6.165)$$

This corresponds to the temperature

$$T_{\text{eq}} = T_0(1 + z_{\text{eq}}) = 5.5(\Omega h^2)\,\text{eV} \quad (6.166)$$

and time

$$t_{\text{eq}} \cong \frac{2}{3}H_0^{-1}\Omega^{-1/2}(1 + z_{\text{eq}})^{-3/2} = 5.84 \times 10^{10}(\Omega h^2)^{-2}\,\text{s}. \quad (6.167)$$

6.8 Relics of massive particles

Consider the case of a neutrino which decouples while it is still relativistic, that is, $T_D \gg m$, where m is the mass of the particle and T_D is the decoupling temperature. Such a particle is characterized by the conserved quantity

$$N = 0.28\left(\frac{g_{\text{eff}}}{q}\right)_{T=T_D} = 0.21\left(\frac{g}{q(T_D)}\right), \quad (6.168)$$

where we have set $g_{\text{eff}} = (3g/4)$ in (6.154). They will have a number density

$$n_0 = Ns_0 = 2.97 \times 10^3 N \, \text{cm}^{-3} \cong 619g[q(T_D)]^{-1} \, \text{cm}^{-3} \qquad (6.169)$$

in the present universe.

These particles would have become non-relativistic at some temperature $T_{\text{nr}} \cong m$ in the past. (We assume that $m > T_0$, where T_0 is the present radiation temperature, that is, $m \gtrsim 1.7 \times 10^{-4}$ eV. If this is not the case, the particles will be relativistic even today and will behave just like the massless neutrinos discussed in the last problem.) The energy contributed by each particle today is $E \cong m$; so the energy density of these particles today will be

$$\rho = n_0 m = 6.19 \times 10^3 g[q(T_D)]^{-1} \left(\frac{m}{10\,\text{eV}} \right) \text{eV cm}^{-3}, \qquad (6.170)$$

which corresponds to

$$(\Omega h^2) = 0.59 \left(\frac{m}{10\,\text{eV}} \right) \left[\frac{g}{q(T_D)} \right]. \qquad (6.171)$$

The precise value of the right-hand side depends on the value of q at the time of decoupling. Neutrinos with masses less than about 1 MeV decouple at $T_D \cong (1 - 3)$ MeV; at this temperature $q = 10.75$. So, for a single massive species with $g = 2$, we obtain

$$\Omega_\nu h^2 = \left(\frac{m_\nu}{91.5\,\text{eV}} \right). \qquad (6.172)$$

From observations, we know that Ωh^2 is at most of order unity. Using the bound $\Omega h^2 \lesssim 1$, we obtain

$$m_\nu \lesssim 91.5\,\text{eV}. \qquad (6.173)$$

6.9 Decoupling of matter and radiation

(a) The temperature, T_{rec}, at which hydrogen atoms are formed can be computed if we make two simplifying assumptions: (i) the system is in thermodynamic equilibrium; and (ii) the recombination proceeds through the electron and proton combining to form a hydrogen atom in the ground state. (Neither of these assumptions is quite correct; we will discuss the modifications which are required in a more realistic scenario at the end.)

Let n_e, n_p, n_H denote the number densities of electrons, protons and hydrogen atoms in thermal equilibrium. Introducing the 'fractional ionization', x_i, for each of the particle species and using the facts that $n_p = n_e$ and $n_p + n_H = n_B$, it follows that $x_p = x_e$ and $x_H = (n_H/n_B) = 1 - x_e$. The x_e can be determined using the results of problem 1.21(a); it is convenient to write the result in the form

$$\frac{1 - x_e}{x_e^2} = \frac{4\sqrt{2}\zeta(3)}{\sqrt{\pi}} \eta \left(\frac{T}{m_e} \right)^{3/2} \exp(B/T) \cong 3.84\eta \, (T/m_e)^{3/2} \exp(B/T), \quad (6.174)$$

where $\eta = 2.68 \times 10^{-8}(\Omega_B h^2)$ is the 'baryon-to-photon ratio' defined as (n_γ/n_B). We may define T_{rec} as the temperature at which 90% of the electrons have combined with protons, that is, when $x_e = 0.1$. This leads to the condition

$$(\Omega_B h^2)^{-1}\tau^{-\frac{3}{2}} \exp\left[-13.6\tau^{-1}\right] = 3.13 \times 10^{-18}, \qquad (6.175)$$

where $\tau = (T/1\,\mathrm{eV})$. For a given value of $(\Omega_B h^2)$, this equation can easily be solved by iteration. Taking logarithms and iterating once we find

$$\tau^{-1} \cong 3.084 - 0.0735 \ln(\Omega_B h^2), \qquad (6.176)$$

with the corresponding redshift $(1 + z_{rec}) = (T/T_0)$ given by

$$(1 + z_{rec}) = 1367[1 - 0.024 \ln(\Omega_B h^2)]^{-1}. \qquad (6.177)$$

For $\Omega_B h^2 = 1, 0.1, 0.01$ we obtain $T_{rec} \cong 0.324\,\mathrm{eV},\ 0.307\,\mathrm{eV},\ 0.292\,\mathrm{eV}$ respectively. These values correspond to redshifts of $1367, 1296$ and 1232.

(b) Since the above analysis was based on equilibrium densities, it is important to check that the rate of the reactions $p + e \leftrightarrow H + \gamma$ is fast enough to maintain equilibrium. The thermally averaged cross-section for the process of recombination is given by (see problem 4.13)

$$\frac{\langle\sigma v\rangle}{c} \cong 4.7 \times 10^{-24} \left(\frac{T}{1\,\mathrm{eV}}\right)^{-\frac{1}{2}} \mathrm{cm}^2. \qquad (6.178)$$

The reaction rate, therefore, will be

$$\begin{aligned}
\Gamma = n_p\langle\sigma v\rangle &= (x_e\eta n_\gamma)\langle\sigma v\rangle \\
&= 2.374 \times 10^{-10}\,\mathrm{cm}^{-1}\tau^{7/4}e^{-(6.8/\tau)}(\Omega_B h^2)^{1/2}. \qquad (6.179)
\end{aligned}$$

In arriving at this expression, we have approximated equation (6.174) by

$$x_e \cong \left[\frac{\pi}{4\sqrt{2}\zeta(3)}\right]^{1/2} \eta^{-\frac{1}{2}} \left(\frac{T}{m_e}\right)^{-\frac{3}{4}} \exp\left(-\frac{6.8}{\tau}\right), \qquad (6.180)$$

which is valid for $x_e \ll 1$. This Γ has to be compared with the expansion rate H of the universe. Taking the universe to be matter dominated at this temperature, we have

$$H = 3 \times 10^{-23}\mathrm{cm}^{-1}(\Omega h^2)^{1/2}\tau^{3/2}. \qquad (6.181)$$

Equating the two, we obtain

$$\tau^{-\frac{1}{4}}\exp(6.8/\tau) = 8.06 \times 10^{12}(\Omega_B/\Omega)^{1/2}. \qquad (6.182)$$

This equation can also be solved by taking logarithms and iterating the solution. We find that

$$\tau^{-1} \cong 4.316 - 0.074 \ln\left(\frac{\Omega}{\Omega_B}\right), \quad (1 + z) = 977\left[1 - 0.017 \ln\left(\frac{\Omega}{\Omega_B}\right)\right]^{-1}. \qquad (6.183)$$

For $\Omega \cong 10\Omega_B$, this gives $T_D \cong 0.24\,\text{eV}$. The fact that $T_D < T_{rec}$ justifies the assumption of thermal equilibrium used in the earlier calculation. (There are, however, some other difficulties with this assumption which will be discussed later.)

When the reaction rate falls below the expansion rate, the formation of neutral atoms ceases. The remaining electrons and protons have negligible probability of combining with each other. The residual fraction can be estimated as the fraction present at $T = T_D$, that is, $x_e(T_D)$. Combining (6.180) and (6.182), we find

$$x_e(T_D) \cong 7.4 \times 10^{-6} \left(\frac{T_D}{1\,\text{eV}} \right)^{-1} \left(\frac{\Omega^{1/2}}{\Omega_B h} \right). \tag{6.184}$$

For $T_D \cong 0.24\,\text{eV}$, this gives

$$x_e(T_D) \cong 3 \times 10^{-5} \left(\frac{\Omega^{1/2}}{\Omega_B h} \right). \tag{6.185}$$

Thus a small fraction ($\cong 10^{-5}$) of electrons and protons will remain free in the universe.

(c) The formation of atoms affects the photons, which were originally in thermal equilibrium with the rest of the matter through the various scattering processes described above. It is easy to verify that the timescales for Compton scattering and free–free absorption become much larger than the expansion timescale when x_e drops to its residual value. The only scattering which is still operational is the Thomson scattering; this process merely changes the direction of the photon without any energy exchange. Its only effect is to make any given photon perform a random walk. When the number density of charged particles decreases, even this interaction rate Γ of the photons drops and, eventually, at some $T = T_{dec}$, becomes lower than the expansion rate. For $T < T_{dec}$, the photons are decoupled from the rest of the matter. The rate of Thomson scattering is given by

$$\begin{aligned} \Gamma = \sigma n_e &= \sigma x_e n_B = \sigma x_e \eta n_\gamma \\ &= 3.36 \times 10^{-11} (\Omega_B h^2)^{1/2} \tau^{9/4} \exp{(-6.8/\tau)}\,\text{cm}^{-1}. \end{aligned} \tag{6.186}$$

Comparing this with the expansion rate in (6.181), we obtain the condition

$$\tau^{-\frac{3}{4}} e^{6.8/\tau} = 1.14 \times 10^{12} (\Omega_B/\Omega)^{1/2}. \tag{6.187}$$

where $\tau = (T_{dec}/1\,\text{eV})$. Solving this with one iteration we obtain

$$\tau^{-1} \cong 3.927 + 0.0735 \ln{(\Omega_B/\Omega)}, \tag{6.188}$$

which corresponds to the parameters

$$T_{dec} \cong 0.26\,\text{eV}, \quad (1 + z_{dec}) \cong 1100. \tag{6.189}$$

For $T \lesssim 0.2\,\text{eV}$, the neutral matter and photons evolve as uncoupled systems. The parameter T characterizing the Planck spectrum continues to fall as a^{-1} because of the redshift of photons. The neutral matter behaves as a gaseous mixture of hydrogen and helium.

It should be stressed that three distinct events take place in the universe around $T \cong (0.3 - 0.2)\,\text{eV}$. (i) Most of the protons and electrons combine to form H-atoms. (ii) The process of recombination stops, leaving a small fraction of free electrons and protons, when the interaction rate for $pe \leftrightarrow H\gamma$ drops below the expansion rate. (iii) The photon mean-free-path becomes larger than H^{-1}, decoupling radiation from matter. These events occur at almost the same epoch because $\eta \cong 10^{-8}$ and $\Omega h^2 \cong 1$, $\Omega_B \lesssim 1$. For a different set of values for these parameters, these events could occur at different epochs.

(d) After decoupling, the temperature of the neutral atoms falls faster than that of radiation. The decrease of matter temperature is governed by the equation

$$\frac{dT_m}{dt} + 2\frac{\dot{a}}{a}T_m = \left(\frac{x_e}{1+x_e}\right)\left(\frac{8\sigma_T a T^4}{3m_e c}\right)(T - T_m), \tag{6.190}$$

where T is the radiation temperature. The term $2(\dot{a}/a)\,T_m$ describes the cooling due to expansion, while the term on the right-hand side accounts for the energy transfer from radiation to matter (see problem 4.10(h)). At high temperatures $x_e \cong 1$ and $T_m \cong T$ to a high degree of accuracy. As x_e becomes smaller, the energy transfer from the radiation to matter becomes less and less effective. The adiabatic cooling term makes the matter temperature fall faster than the radiation temperature. Using the expression $(\dot{a}/a) = H(z) \cong H_0 \Omega^{1/2}(1+z)^{3/2}$, valid for the matter dominated phase, the ratio between cooling rate and expansion rate becomes

$$\frac{(\dot{T}_m/T_m)}{(\dot{a}/a)} = -2 + \left(\frac{8.4 \times 10^{-3}}{\Omega^{1/2}h}\right)\left(\frac{x_e}{1+x_e}\right)\left(\frac{T}{T_m} - 1\right)(1+z)^{5/2}$$

$$\equiv -2 + \mathscr{R}\left(\frac{T}{T_m} - 1\right). \tag{6.191}$$

If $x_e = 1$, then $\mathscr{R} \cong 1$ at $(1+z_c) \cong 9(\Omega h^2)^{1/5}$. For x_e given by (6.185), $\mathscr{R}$ reaches unity at $z_c \cong 600(\Omega_B h^2)^{2/5}$; for $\Omega_B h^2 \cong 0.015$ we obtain $z_c \cong 10^2$. At lower redshifts the matter temperature falls as $T_m \propto a^{-2}$. Notice that $T_m \cong T$ up to about $z_c \cong 10^2$, even though $z_{dec} \cong 10^3$.

(e) We shall now discuss the various approximations which have been made in the above calculation. To begin with, notice that we assumed a recombination process which directly produces a H-atom in the *ground* state. This will release a photon with energy of 13.6 eV in each recombination. If $n_\gamma\,(B)$ is the number density of photons in the background radiation with energy $B = 13.6\,\text{eV}$, then,

$$\frac{n_\gamma(B)}{n} \cong \frac{16\pi}{n}T^3 \exp\left(-\frac{B}{T}\right) \cong \frac{3 \times 10^7}{(\Omega_B h^2)}\exp\left(-\frac{13.6}{\tau}\right). \tag{6.192}$$

This ratio is unity at about 0.8 eV (i.e. at a redshift of $z \cong 3300$) and decreases rapidly at lower temperatures. Thus, at lower temperatures, the addition of

13.6 eV photons due to recombination significantly enhances the availability of ionizing photons. These energetic photons have a high probability of ionizing the neutral atoms formed a little earlier. (That is, the 'backward' reaction $H+\gamma \rightarrow p+e$ is enhanced.) Hence, this process is not very effective in producing a *net* number of neutral atoms.

The dominant process which actually operates is the one in which recombination proceeds through an excited state: $(e + p \rightarrow H^* + \gamma; H^* \rightarrow H + \gamma_2)$. This will produce two photons, each of which has *less* energy than the ionization potential of the hydrogen atom. The $2P$ and $2S$ levels provide the most rapid route for recombination; the decay from the $2P$ state produces a single photon, while the decay from the $2S$ state is through two-photons. Since the reverse process does not occur at the same rate, this is non-equilibrium recombination.

Because of the above complication, the recombination proceeds at a slower rate compared to that predicted by Saha's equation. The actual fractional ionization is higher than the value predicted by Saha's equation at temperatures below about 1300. For example, at $z = 1300$, these values differ by a factor of 3; at $z \cong 900$, they differ by a factor of 200. The values of T_{atom}, T_{dec}, etc., however, do not change significantly.

As an example, consider the value of T_{dec}. In the redshift range of $800 < z < 1200$, the fractional ionization is given (approximately) by the formula

$$x_e = 2.4 \times 10^{-3} \frac{(\Omega h^2)^{1/2}}{(\Omega_B h^2)} \left(\frac{z}{1000}\right)^{12.75}. \tag{6.193}$$

(This is obtained by fitting a curve to the numerical solution.) Using this expression, we can compute the optical depth for photons to be

$$\tau = \int_0^t n(t)x_e(t)\sigma_T \, dt = \int_0^z n(z)x_e(z)\sigma_T \left(\frac{dt}{dz}\right) dz \cong 0.37 \left(\frac{z}{1000}\right)^{14.25}, \tag{6.194}$$

where we have used the relation $H_0 \, dt \cong -\Omega^{-1/2} \, dz$, which is valid for $z \gg 1$. This optical depth is unity at $z_{dec} = 1072$. Our approximate calculation earlier gave a value of 1100, which is quite close to the exact value.

From the optical depth, we can also compute the probability that the photon was last scattered in the interval $(z, z + dz)$. This is given by $(\exp -\tau) \, (d\tau/dz)$, which can be expressed as

$$P(z) = e^{-\tau} \frac{d\tau}{dz} = 5.26 \times 10^{-3} \left(\frac{z}{1000}\right)^{13.25} \exp\left[-0.37 \left(\frac{z}{1000}\right)^{14.25}\right]. \tag{6.195}$$

This $P(z)$ can be well approximated by a Gaussian centred at $z \cong 1067$ with a width of about $\Delta z \cong 80$. It is therefore reasonable to assume that decoupling occurred at $z \cong 1070$ in an interval of about $\Delta z \cong 80$.

6.10 Perturbed Friedmann universe

(a) For the line element given in the question, we have

$$g^{\alpha\beta} = -a^{-2}\left(\delta_{\alpha\beta} + h_{\alpha\beta}\right), \quad \sqrt{-g} = a^3\left(1 - \frac{1}{2}h\right), \quad h = h_\alpha^\alpha. \tag{6.196}$$

(The equations in which tensor indices are mismatched in location are to be interpreted component-by-component.) Using these relations, the definition of δ and the expression for u^i, we find that

$$T^{00} = \rho_b(1 + \delta),$$
$$T^{\alpha\beta} = \frac{w\rho_b}{a^2}(1 + \delta)\left(\delta_{\alpha\beta} + h_{\alpha\beta}\right), \tag{6.197}$$
$$T_{\alpha\beta} = w\rho_b a^2(1 + \delta)\left(\delta_{\alpha\beta} - h_{\alpha\beta}\right).$$

To obtain the equation connecting δ and $h_{\alpha\beta}$, it is easiest to use the identity (see problem 5.5(d))

$$T^j_{i;j} = \frac{1}{\sqrt{-g}}\frac{\partial}{\partial x^j}\left(\sqrt{-g}\,T^j_i\right) - \frac{1}{2}\frac{\partial g_{jk}}{\partial x^i}T^{ik} = 0. \tag{6.198}$$

Using the form of the metric tensor and the stress-tensor, we can write this equation explicitly. The $i = 0$ component becomes, after some algebra,

$$\left(3\frac{\dot{a}}{a} - \frac{\dot{h}}{2}\right)\rho_b(1 + \delta) + \dot{\rho}_b(1 + \delta) + \rho_b\dot{\delta} = -3\frac{\dot{a}}{a}w\rho_b(1 + \delta) + \frac{1}{2}w\rho_b\dot{h}. \tag{6.199}$$

The unperturbed part of this equation gives

$$\dot{\rho}_b = -3\frac{\dot{a}}{a}(1 + w)\rho_b, \tag{6.200}$$

which is equivalent to the Friedmann equation. The first-order term gives, on using (6.200),

$$\dot{\delta} = \frac{1}{2}(1 + w)\dot{h}. \tag{6.201}$$

This equation relates the density contrast to the perturbation in the metric tensor.

(b) For the perturbed Friedmann metric, the non-vanishing Christoffel symbols are

$$\Gamma^0_{\alpha\beta} = a\dot{a}\left(\delta_{\alpha\beta} - h_{\alpha\beta}\right) - \frac{a^2}{2}\dot{h}_{\alpha\beta},$$
$$\Gamma^\alpha_{0\beta} = \frac{\dot{a}}{a}\delta_{\alpha\beta} - \frac{1}{2}\dot{h}_{\alpha\beta}, \tag{6.202}$$
$$\Gamma^\gamma_{\alpha\beta} = -\frac{1}{2}\left(h_{\alpha\gamma,\beta} + h_{\beta\gamma,\alpha} - h_{\alpha\beta,\gamma}\right).$$

In this case, R_{00} is

$$R_{00} = -3\frac{\ddot{a}}{a} + \frac{1}{2}\ddot{h} + \frac{\dot{a}}{a}\dot{h}. \tag{6.203}$$

Equating this to $8\pi G(T_{00} - g_{00}T/2)$ we obtain the equation

$$-3\frac{\ddot{a}}{a} + \frac{1}{2}\ddot{h} + \frac{\dot{a}}{a}\dot{h} = 4\pi G\rho_b(1 + 3w)(1 + \delta). \tag{6.204}$$

The unperturbed part of this equation gives

$$\frac{\ddot{a}}{a} = -\frac{4\pi G}{3}\rho_b(1 + 3w), \tag{6.205}$$

which is the same as one of the Friedmann equations. The first-order term gives

$$\ddot{h} + 2\frac{\dot{a}}{a}\dot{h} = 8\pi G\rho_b(1 + 3w)\delta. \tag{6.206}$$

Using (6.201) we can express $\dot{h}$ in terms of δ. Substituting into (6.206) we see that the density contrast satisfies the equation

$$\ddot{\delta} + 2\frac{\dot{a}}{a}\dot{\delta} = 4\pi G\rho_b(1 + w)(1 + 3w)\delta. \tag{6.207}$$

(c) To solve this equation, we need the background solution. When the source is an ideal fluid, with an equation of state $p = w\rho$, the background density evolves as $\rho_b \propto a^{-3(1+w)}$. In that case, the Friedmann equations lead to

$$a(t) \propto t^{[2/3(1+w)]}, \quad \rho_b = \frac{1}{6\pi G(1 + w)^2 t^2}, \tag{6.208}$$

provided $w \neq -1$. When $w = -1$, $a(t) \propto \exp(\lambda t)$ with a constant λ. (See problem 6.3.) We will consider the $w \neq -1$ case first. Substituting the solution for $a(t)$ and $\rho_b(t)$ into (6.207) we obtain

$$\ddot{\delta} + \frac{4}{3(1 + w)}\frac{\dot{\delta}}{t} = \frac{2}{3}\frac{(1 + 3w)}{(1 + w)}\frac{\delta}{t^2}. \tag{6.209}$$

This equation is homogeneous in t and hence admits power law solutions. Using an ansatz $\delta \propto t^n$, and solving the quadratic equation for n, we find the two linearly independent solutions to be

$$\delta_g \propto t^n, \quad \delta_d \propto \frac{1}{t}, \quad n = \frac{2}{3}\frac{(1 + 3w)}{(1 + w)}. \tag{6.210}$$

In the case of $w = -1$, $a(t) \propto e^{\lambda t}$ and the equation for δ reduces to

$$\ddot{\delta} + 2\lambda\dot{\delta} = 0. \tag{6.211}$$

This has the solution $\delta_g \propto \exp(-2\lambda t) \propto a^{-2}$.

All the above solutions can be expressed in a very interesting and unified manner. By direct substitution it can be verified that δ_g in all the above cases can be expressed as

$$\delta_g \propto \frac{1}{\rho_b a^2}. \tag{6.212}$$

We shall see later, in problem 7.6, that this result has an extremely simple interpretation.

7

Dynamics in the expanding universe

7.1 Trajectories in perturbed Friedmann universe

(a) The perturbed metric can be written in an equivalent form as

$$ds^2 = \left(1 + \frac{2\phi}{c^2}\right) c^2\, dt^2 - a^2(t) \left(1 - \frac{2\phi}{c^2}\right) \left(dx^2 + dy^2 + dz^2\right), \qquad (7.1)$$

where we have set $a(\eta)\, d\eta = dt$ and reintroduced the c factor. We saw earlier, in problem 6.1, that the unperturbed Friedmann metric can be transformed to the form

$$ds^2 \cong \left(1 - \frac{\ddot{a}}{a} \frac{R^2}{c^2}\right) c^2 dT^2 - d\mathbf{R}^2 \qquad (7.2)$$

if we restrict to quadratic order in (R/d_{H}). In the perturbed case, the gravitational potential at any point is due to two sources: (i) the effective gravitational potential of the background Friedmann model; and (ii) the gravitational potential $\phi(\mathbf{x}, t)$ of perturbed matter. In the limit of weak gravity ($\phi \ll c^2$), the superposition of these two potentials is allowed and we can write the effective metric as

$$ds_{\mathrm{pert}}^2 \cong \left(1 - \frac{\ddot{a}}{a} \frac{R^2}{c^2} + \frac{2\phi}{c^2}\right) c^2 dT^2 - dR^2 \cong \left(1 - \frac{\ddot{a}}{a} \frac{R^2}{c^2}\right) \left(1 + \frac{2\phi}{c^2}\right) c^2 dT^2 - dR^2.$$
$$(7.3)$$

Transforming back to (t, r) coordinates we will obtain (7.1) to lowest order. It is clear, therefore, that (7.1) represents the metric of a perturbed Friedmann universe in the Newtonian limit. The Newtonian superposition of potentials is valid for scales $\lambda \ll d_{\mathrm{H}}(t)$ and the effective Newtonian potential for the perturbed Friedmann universe is

$$\phi_{\mathrm{N}} = -\frac{1}{2} \left(\frac{\ddot{a}}{a}\right) R^2 + \phi. \qquad (7.4)$$

377

(b) The action for a particle of unit mass in the perturbed metric is

$$A = -\int ds = \int a \left[(1 + 2\phi) - (1 - 2\phi)|\dot{\mathbf{x}}|^2 \right]^{1/2} d\eta$$

$$\cong -\int a \, d\eta \left[\sqrt{1 - \dot{\mathbf{x}}^2} + \frac{1}{\sqrt{1 - \dot{\mathbf{x}}^2}} (\phi + \phi \dot{\mathbf{x}}^2) \right], \quad \dot{\mathbf{x}} = \frac{d\mathbf{x}}{d\eta} \quad (7.5)$$

to linear order in ϕ. In arriving at this expression we have also assumed that $\phi(1 - v^2)^{-1/2} \ll 1$, even though v^2 may be comparable to unity. The second term is a small perturbation to the Lagrangian in the unperturbed universe. Hence, the perturbed Hamiltonian will differ from the unperturbed one by the same amount (with opposite sign) to first order in ϕ and ψ, with the canonical momenta still determined by the unperturbed Lagrangian: $\mathbf{p} = (a\dot{\mathbf{x}})/(1 - \dot{\mathbf{x}}^2)^{1/2}$. Therefore, the perturbed Hamiltonian is

$$H(\mathbf{p}, \mathbf{x}) = a\sqrt{1 + p^2/a^2} + \left[\frac{a}{\sqrt{1 - \dot{\mathbf{x}}^2}} (\phi + \phi \dot{\mathbf{x}}^2) \right]_{\dot{\mathbf{x}} = \dot{\mathbf{x}}(\mathbf{p})}$$

$$= a\sqrt{1 + p^2/a^2} \left(1 + \phi + \phi \frac{p^2/a^2}{1 + p^2/a^2} \right). \quad (7.6)$$

This Hamiltonian leads to the equations of motion

$$\dot{\mathbf{x}} = \frac{\mathbf{p}}{\epsilon} \left[1 + \phi + \phi \left(\frac{2 + p^2/a^2}{1 + p^2/a^2} \right) \right], \quad (7.7)$$

$$\dot{\mathbf{p}} = -\epsilon \nabla \phi - \epsilon \left(\frac{p^2/a^2}{1 + p^2/a^2} \right) \nabla \phi, \quad \epsilon = a\sqrt{1 + \frac{p^2}{a^2}}. \quad (7.8)$$

In terms of velocity $\mathbf{v} = (\mathbf{p}/\epsilon)$, these equations become

$$\dot{\mathbf{x}} = \mathbf{v} \left[1 + 2\phi + \frac{\phi}{(1 + \epsilon^2 v^2/a^2)} \right], \quad (7.9)$$

$$\dot{\mathbf{p}} = -\epsilon (1 + v^2) \nabla \phi. \quad (7.10)$$

The energy of the particle is

$$E(\mathbf{x}, \mathbf{v}) = \epsilon \left[1 + \phi + \phi \frac{v^2 \epsilon^2/a^2}{1 + v^2 \epsilon^2/a^2} \right]. \quad (7.11)$$

The $(1 + v^2)$ term in (7.10) shows that relativistic particles (with $v \cong 1$) experience a gravitational force which is twice as much as in Newtonian theory. We have encountered this feature earlier, in problem 5.11. In the extreme relativistic (ER) limit we obtain

$$\dot{\mathbf{x}} \cong (1 + 2\phi) \frac{\mathbf{p}}{|\mathbf{p}|}, \quad \dot{\mathbf{p}} \cong -2|\mathbf{p}| \nabla \phi, \quad (7.12)$$

while in the non-relativistic (NR) limit we have

$$\dot{\mathbf{x}} = \frac{\mathbf{p}}{a}, \quad \dot{\mathbf{p}} = -a\nabla \phi. \quad (7.13)$$

The physical interpretation of these equations can be best presented in (t, r) coordinates. Using $dt = a(\eta)\, d\eta$, the ER equations can be written as

$$a|d\mathbf{x}| \cong (1 + 2\phi)\, dt,\tag{7.14}$$

$$|d\mathbf{p}| = -|\mathbf{p}|\nabla(2\phi)\, dt.\tag{7.15}$$

Equation (7.14) shows that the particles move on approximately null world lines with an effective gravitational potential which is (2ϕ). Similarly, (7.15) shows that the deflection of relativistic particles $\Delta\theta = (dp/p)$ is that due to the gradient of twice the Newtonian potential. (2ϕ).

The NR equation becomes, in the (t, r) coordinates,

$$a^2\left(\frac{d\mathbf{x}}{dt}\right) = \mathbf{p}, \qquad \frac{d\mathbf{p}}{dt} = -\nabla\phi.\tag{7.16}$$

The second equation is the familiar Newton law, while the first equation defines the momentum. The a^2 factor in the first equation arises because our definitions use comoving coordinates.

(c) Combining the two parts of equation (7.16) and using the (cosmic) time coordinate t with $dt = a(\eta)d\eta$ we obtain

$$\frac{d}{dt}\left(a^2\frac{d\mathbf{x}}{dt}\right) = a^2\frac{d^2\mathbf{x}}{dt^2} + 2a\dot{a}\frac{d\mathbf{x}}{dt} = \frac{d\mathbf{p}}{dt} = -\nabla\phi.\tag{7.17}$$

Or, on dividing by a^2,

$$\ddot{\mathbf{x}} + 2\frac{\dot{a}}{a}\dot{\mathbf{x}} = -\frac{1}{a^2}\nabla\phi,\tag{7.18}$$

where the dot denotes differentiation with respect to t. If the system is made up of N particles with positions $\mathbf{x}_i(t)$, then the full Newtonian evolution is described by the coupled equations

$$\ddot{\mathbf{x}}_i + \frac{2\dot{a}}{a}\dot{\mathbf{x}}_i = -\frac{1}{a^2}\nabla_\mathbf{x}\phi, \qquad \nabla_\mathbf{x}^2\phi = 4\pi Ga^2\rho_{bm}\delta,\tag{7.19}$$

where ρ_{bm} is the smooth background density of matter. Note that, in the non-relativistic limit, the perturbed potential ϕ satisfies the usual Poisson equation.

A simpler way of obtaining this equation is as follows. In the $(T, \mathbf{R})$ coordinate system of (7.3), one has the effective Newtonian potential ϕ_N of (7.4). The equation of motion, in the Newtonian limit, is

$$\frac{d^2\mathbf{R}}{dT^2} = \frac{d^2}{dT^2}(a\mathbf{x}) = -\nabla_\mathbf{R}\phi_N = \ddot{a}\mathbf{x} - \frac{1}{a}\nabla_\mathbf{x}\phi.\tag{7.20}$$

To the same order of accuracy we can replace (d^2/dT^2) by (d^2/dt^2) and obtain

$$a\ddot{\mathbf{x}} + 2\dot{a}\dot{\mathbf{x}} = -\frac{1}{a}\nabla_\mathbf{x}\phi.\tag{7.21}$$

Dividing by a, we obtain (7.18).

(d) We introduce a new 'time' coordinate $b = b(t)$ in place t, and the corresponding 'velocity' $\mathbf{w} \equiv (d\mathbf{x}/db) = (\dot{\mathbf{x}}/\dot{b}) = (\mathbf{v}/a\dot{b})$, where $\mathbf{v}$ is the original peculiar velocity $\mathbf{v} = a\dot{\mathbf{x}}$. The equation of motion can be written, in terms of b, as

$$\dot{b}\frac{d}{db}(\dot{b}\mathbf{w}) + 2\frac{\dot{a}}{a}\dot{b}\mathbf{w} = -\frac{1}{a^2}\nabla\phi. \tag{7.22}$$

Or

$$\dot{b}^2\frac{d\mathbf{w}}{db} + \left(\ddot{b} + 2\frac{\dot{a}}{a}\dot{b}\right)\mathbf{w} = -\frac{1}{a^2}\nabla\phi. \tag{7.23}$$

We now choose $b(t)$ to be the growing solution to the equation

$$\ddot{b} + \frac{2\dot{a}}{a}\dot{b} = 4\pi G\rho_{\rm bm}(t)b = \left(4\pi G\rho_{\rm bm}a^3\right)\frac{b}{a^3} = \frac{3}{2}H_0^2\left(\frac{a_0}{a}\right)^3 b. \tag{7.24}$$

This equation governs the growth of perturbations in linear theory and we will take $b(t)$ to be the fastest growing mode. (If $\Omega = 1$ and the universe is matter dominated, then the solution to (7.24) is $b(t) = a(t) = (t/t_0)^{2/3}$; see problem 6.10.) Using (7.24) in (7.23), we obtain

$$\frac{d\mathbf{w}}{db} + \frac{3}{2}\frac{H_0^2 a_0^3}{a^3}\left(\frac{b}{\dot{b}^2}\right)\mathbf{w} = -\frac{1}{a^2\dot{b}^2}\nabla\phi = -\frac{3}{2}\frac{H_0^2 a_0^3 b}{a^3\dot{b}^2}\nabla\left[\frac{2}{3}\left(\frac{a}{b}\right)\frac{\phi}{H_0^2 a_0^3}\right]. \tag{7.25}$$

We define a new potential ψ by

$$\psi = \frac{2}{3H_0^2 a_0^3}\left(\frac{a}{b}\right)\phi = \frac{1}{4\pi G\rho_{\rm bm}a^2}\cdot\frac{\phi}{b}. \tag{7.26}$$

Then

$$\nabla^2\psi = \frac{\nabla^2\phi}{4\pi G\rho_{\rm bm}a^2}\cdot\frac{1}{b} = \left(\frac{\delta}{b}\right) \tag{7.27}$$

and

$$\frac{d\mathbf{w}}{db} = -\frac{3}{2b}\left(\frac{H_0^2 a_0^3}{a^3}\frac{b^2}{\dot{b}^2}\right)[\mathbf{w} + \nabla\psi] = -\frac{3A}{2b}(\mathbf{w} + \nabla\psi), \tag{7.28}$$

where

$$A = \frac{H_0^2 a_0^3}{a^3}\left(\frac{b^2}{\dot{b}^2}\right) = \left(\frac{8\pi G}{3}\rho_{\rm bm}\right)\left(\frac{a}{\dot{a}}\right)^2\left[\frac{\dot{a}b}{a\dot{b}}\right]^2 = \left[\frac{\rho_{\rm bm}(t)}{\rho_{\rm c}(t)}\right]\left[\frac{\dot{a}b}{a\dot{b}}\right]^2. \tag{7.29}$$

Thus we arrive at the final set of equations:

$$\frac{d\mathbf{w}}{db} = -\frac{3A}{2b}(\mathbf{w} + \nabla\psi), \quad \nabla^2\psi = \left(\frac{\delta}{b}\right), \quad A = \left(\frac{\rho_{\rm bm}}{\rho_{\rm c}}\right)\left(\frac{\dot{a}b}{a\dot{b}}\right)^2. \tag{7.30}$$

This is the most convenient form to study several approximations which we will discuss later. For $\Omega = 1, a = b$ and $A = 1$.

7.2 Density contrast from the trajectories

The density $\rho(\mathbf{x}, t)$ due to a set of point particles, each of mass m, is given by

$$\rho(\mathbf{x}, t) = \frac{m}{a^3(t)} \sum_i \delta_{\text{Dirac}}(\mathbf{x} - \mathbf{x}_i(t)). \tag{7.31}$$

(To verify the a^{-3} normalization, we can calculate the average of $\rho(\mathbf{x}, t)$ over a large volume V. We obtain

$$\rho_{\text{b}}(t) \equiv \int \frac{d^3x}{V} \rho(\mathbf{x}, t) = \frac{m}{a^3(t)} \left(\frac{N}{V} \right) = \frac{M}{a^3 V} = \frac{\rho_0}{a^3}, \tag{7.32}$$

where N is the total number of particles inside the volume V and $M = Nm$ is the mass contributed by them. Clearly, $\rho_{\text{b}} \propto a^{-3}$, as it should.) The density contrast $\delta(\mathbf{x}, t)$ is related to $\rho(\mathbf{x}, t)$ by

$$1 + \delta(\mathbf{x}, t) \equiv \frac{\rho(\mathbf{x}, t)}{\rho_{\text{b}}} = \frac{m}{\rho_{\text{b}} a^3} \sum_i \delta_{\text{Dirac}}(\mathbf{x} - \mathbf{x}_i) = \frac{mV}{M} \sum_i \delta_{\text{Dirac}}(\mathbf{x} - \mathbf{x}_i). \tag{7.33}$$

Fourier transforming both sides and ignoring the $\mathbf{k} = 0$ mode, we obtain

$$\delta_k(t) \equiv \int \frac{d^3x}{V} \exp(i\mathbf{k} \cdot \mathbf{x}) \delta(\mathbf{x}, t) = \frac{1}{N} \sum_i \exp i\mathbf{k} \cdot \mathbf{x}_i(t). \tag{7.34}$$

Differentiating this expression, we find

$$\dot{\delta}_{\mathbf{k}} = \frac{1}{N} \sum_j (i\mathbf{k} \cdot \dot{\mathbf{x}}_j) \exp(i\mathbf{k} \cdot \mathbf{x}_j) \tag{7.35}$$

and

$$\ddot{\delta}_{\mathbf{k}} = \frac{1}{N} \sum_j \left[i\mathbf{k} \cdot \ddot{\mathbf{x}}_j - (\mathbf{k} \cdot \dot{\mathbf{x}}_j)^2 \right] \exp(i\mathbf{k} \cdot \mathbf{x}_j). \tag{7.36}$$

Using (7.18) to eliminate $\ddot{\mathbf{x}}_j$, one can obtain an equation for $\delta_k(t)$:

$$\ddot{\delta}_{\mathbf{k}} + 2\frac{\dot{a}}{a} \dot{\delta}_{\mathbf{k}} = -\frac{1}{N} \sum_j \left(\frac{i\mathbf{k} \cdot \nabla \phi_j}{a^2} \right) - B, \tag{7.37}$$

where

$$B = \frac{1}{N} \sum_j \left(\frac{m}{M} \right) (\mathbf{k} \cdot \dot{\mathbf{x}}_j)^2 \exp(i\mathbf{k} \cdot \mathbf{x}_j). \tag{7.38}$$

To proceed further, note that the potential ϕ_j acting on the jth particle is given by

$$\phi_j = -\frac{4\pi G}{aV} \sum_{l \neq j} \left(\frac{m}{p^2} \right) \exp(i\mathbf{p} \cdot (\mathbf{x}_i - \mathbf{x}_j)). \tag{7.39}$$

Hence

$$-\frac{1}{N}\sum_j \left(\frac{m}{M}\right)\left(\frac{i\mathbf{k}\cdot\nabla\phi_j}{a^2}\right) = \frac{4\pi G m^2}{MVa^3}\sum \left(\frac{\mathbf{k}\cdot\mathbf{p}}{p^2}\right)\exp\left(i\left[\mathbf{x}_i\cdot\mathbf{p}+\mathbf{x}_j\cdot(\mathbf{k}-\mathbf{p})\right]\right)$$

$$= 4\pi G\rho_b \sum_{\mathbf{p}}\left(\frac{\mathbf{k}\cdot\mathbf{p}}{p^2}\right)\delta_{\mathbf{k}}\delta_{\mathbf{k}-\mathbf{p}}. \tag{7.40}$$

In the sum, there is one term with $\mathbf{k}=\mathbf{p}$ which contributes at linear order, since $\delta_0 = 1$; this term gives $4\pi G\rho_b\delta_{\mathbf{k}}$. In the remaining set, one can change $\mathbf{p}$ to $(\mathbf{k}-\mathbf{p})$ and obtain the final result

$$\ddot{\delta}_{\mathbf{k}} + 2\frac{\dot{a}}{a}\dot{\delta}_{\mathbf{k}} = 4\pi G\rho_b\delta_{\mathbf{k}} + A - B, \tag{7.41}$$

with

$$A = 2\pi G\rho_b \sum_{\mathbf{k}'\neq 0,\mathbf{k}}\delta_{\mathbf{k}}\delta_{\mathbf{k}-\mathbf{k}'}\left\{\frac{\mathbf{k}\cdot\mathbf{k}'}{k'^2}+\frac{\mathbf{k}\cdot(\mathbf{k}-\mathbf{k}')}{|\mathbf{k}-\mathbf{k}'|^2}\right\}$$

$$B = \frac{m}{M}\sum_j (\mathbf{k}\cdot\dot{\mathbf{x}}_j)^2\exp\left[i\mathbf{k}\cdot\mathbf{x}_j(t)\right]. \tag{7.42}$$

This equation is exact but involves $\dot{\mathbf{x}}_i$ on the right-hand side. If the density contrasts are small and linear perturbation theory is valid, we can ignore the terms A and B.

7.3 Vlasov equation and the fluid limit

(a) Using the expressions for $\dot{\mathbf{x}}$ and $\dot{\mathbf{p}}$ from equation (7.16), we can write the collisionless Boltzmann equation (or 'Vlasov equation') in the NR limit:

$$\frac{df}{dt} = \frac{\partial f}{\partial t} + \mathbf{x}\cdot\frac{\partial f}{\partial\mathbf{x}} + \mathbf{p}\cdot\frac{\partial f}{\partial\mathbf{p}} = \frac{\partial f}{\partial t} + \frac{\mathbf{p}}{ma^2}\cdot\frac{\partial f}{\partial\mathbf{x}} - m\nabla_x\phi\cdot\frac{\partial f}{\partial\mathbf{p}} = 0. \tag{7.43}$$

The gravitational potential is related to f by

$$\nabla^2\phi = 4\pi Ga^2\rho_b\delta = 4\pi Ga^2\cdot\frac{m}{a^3}\int d^3\mathbf{p}\, f(t,\mathbf{x},\mathbf{p}) = \frac{4\pi Gm}{a}\int f\, d^3\mathbf{p} \tag{7.44}$$

if we interpret f as due to perturbed distribution. The explicit time dependence through the $a^2(t)$ term can be eliminated from the kinematic terms by introducing a time coordinate τ such that $d\tau = (dt/a^2)$. Then this equation becomes

$$\frac{\partial f}{\partial\tau} + \frac{\mathbf{p}}{m}\cdot\frac{\partial f}{\partial\mathbf{x}} = ma^2(\tau)\nabla\phi\cdot\frac{\partial f}{\partial\mathbf{p}}. \tag{7.45}$$

(b) The equations for a fluid can be obtained from the moments of the Vlasov equation. It is, however, more illuminating to derive these equations in the following manner: We start by noting that, in the proper coordinates, the equations describing the pressureless fluid are

$$\frac{\partial \rho_{\mathrm{m}}}{\partial t} + \nabla_r \cdot (\rho_{\mathrm{m}} \mathbf{U}) = 0, \tag{7.46}$$

$$\frac{\partial \mathbf{U}}{\partial t} + (\mathbf{U} \cdot \nabla) \mathbf{U} = -\nabla \phi_{\mathrm{tot}} = -\nabla \phi_{\mathrm{FRW}} - \nabla \phi, \tag{7.47}$$

where $\phi_{\mathrm{FRW}} = -(\ddot{a}/2a) r^2$ is the potential due to background expansion and ϕ is the potential due to the perturbations. (The coordinate $\mathbf{r} = a(t)\mathbf{x}$ is what was called R in problem 7.1(a). As noted in problem 7.1(c), we can set $T \cong t$ in the Newtonian limit.) The perturbed potential ϕ satisfies the equation

$$\nabla_r^2 \phi = 4\pi G (\rho_{\mathrm{m}} - \rho_{\mathrm{bm}}) = 4\pi G \rho_{\mathrm{bm}} \delta \tag{7.48}$$

or, in terms of comoving coordinates $\mathbf{x} = (\mathbf{r}/a)$,

$$\nabla_x^2 \phi = 4\pi G \rho_{\mathrm{bm}} a^2 \delta. \tag{7.49}$$

We shall now convert (7.46) and (7.47) to the $(t, \mathbf{x})$ coordinates using

$$\left(\frac{\partial}{\partial t}\right)_r = \left(\frac{\partial}{\partial t}\right)_x - H\mathbf{x} \cdot \frac{\partial}{\partial \mathbf{x}}, \quad \left(\frac{\partial}{\partial \mathbf{r}}\right)_t = \frac{1}{a}\left(\frac{\partial}{\partial \mathbf{x}}\right)_t, \quad H = \frac{\dot{a}}{a}. \tag{7.50}$$

We also define the peculiar velocity $\mathbf{v}$ by the relation $\mathbf{U} = H\mathbf{r} + \mathbf{v} = \dot{a}\mathbf{x} + \mathbf{v}$. Then,

$$\left(\frac{\partial \rho_{\mathrm{m}}}{\partial t}\right)_x + 3H\rho_{\mathrm{m}} + \frac{1}{a}\frac{\partial}{\partial x^i}\left(\rho_{\mathrm{m}} v^i\right) = 0, \tag{7.51}$$

$$\left(\frac{\partial \mathbf{v}}{\partial t}\right)_x + H\mathbf{v} + \frac{1}{a}\left[v^i \frac{\partial}{\partial x^i}\right]\mathbf{v} = -\frac{1}{a}\nabla_x \phi_{\mathrm{tot}} - \ddot{a}\mathbf{x}. \tag{7.52}$$

But

$$-\frac{1}{a}\nabla_x \phi_{\mathrm{tot}} = -\frac{1}{a}\nabla_x \phi - \frac{1}{a}\nabla_x \phi_{\mathrm{FRW}} = -\frac{1}{a}\nabla_x \phi + \frac{1}{a}\left(\frac{\ddot{a}a}{2}\right) \cdot 2\mathbf{x} = -\frac{1}{a}\nabla_x \phi + \ddot{a}\mathbf{x}. \tag{7.53}$$

Thus the $\ddot{a}\mathbf{x}$ terms cancel out on the right-hand side of (7.52) and we are left with the equations

$$\left(\frac{\partial \rho_{\mathrm{m}}}{\partial t}\right) + 3H\rho_{\mathrm{m}} + \frac{1}{a}\frac{\partial}{\partial x^i}\left(\rho_{\mathrm{m}} v^i\right) = 0 \tag{7.54}$$

$$\left(\frac{\partial v^i}{\partial t}\right) + \frac{1}{a}v^j \frac{\partial v^i}{\partial x^j} + Hv^i = -\frac{1}{a}\frac{\partial \phi}{\partial x^i}. \tag{7.55}$$

We now set $\rho_{\mathrm{m}} = \rho_{\mathrm{bm}} (1 + \delta)$ and transform from t to $b(t)$. The rest of the analysis proceeds exactly as in problem 7.1(d) and we define a velocity field $u^i = (v^i/a\dot{b})$. In terms of u^i we obtain

$$\frac{\partial \delta}{\partial b} + \partial_i \left[u^i(1 + \delta)\right] = 0, \quad \nabla^2 \psi = \left(\frac{\delta}{b}\right), \tag{7.56}$$

$$\frac{\partial u^i}{\partial b} + u^k \partial_k u^i = -\frac{3A}{2b}\left[\partial^i \psi + u^i\right], \quad A = \left(\frac{\rho_{\mathrm{bm}}}{\rho_{\mathrm{crit}}}\right)\left(\frac{\dot{a}b}{a\dot{b}}\right)^2. \tag{7.57}$$

The quantity $u^i = u^i(\mathbf{x}, b)$ is the velocity of the fluid at $(\mathbf{x}, b)$. Since we have ignored the velocity dispersion, u^i satisfies the same equation as the velocity w^i of individual particles (7.30).

(c) From (7.56) we find that

$$\frac{d\delta}{db} \equiv \frac{\partial\delta}{\partial b} + u^i\partial_i\delta = -(1 + \delta)\theta. \tag{7.58}$$

Taking the derivative of (7.57) with respect to x^i we obtain

$$\frac{d\theta}{db} + \frac{3A}{2b}\theta + \frac{1}{3}\theta^2 + \sigma^2 - 2\Omega^2 = -\frac{3A}{2b^2}\delta, \tag{7.59}$$

where $\sigma^2 \equiv \sigma^{ab}\sigma_{ab}$ and $\Omega^2 \equiv \Omega^i\Omega_i$. Combining (7.59) and (7.58) we find that δ satisfies the equation

$$\frac{d^2\delta}{db^2} + \frac{3A}{2b}\frac{d\delta}{db} - \frac{3A}{2b^2}\delta(1 + \delta) = \frac{4}{3}\frac{1}{(1 + \delta)}\left(\frac{d\delta}{db}\right)^2 + (1 + \delta)(\sigma^2 - 2\Omega^2). \tag{7.60}$$

From the last term on the right-hand side we see that shear contributes positively to $(d^2\delta/db^2)$, while rotation Ω^2 contributes negatively. Thus shear helps the growth of inhomogenities while rotation works against it.

To find the equation satisfied by Ω, we start with equation (7.57) and we use the identity

$$(\mathbf{u} \cdot \nabla)\mathbf{u} = \frac{1}{2}\nabla u^2 - (\mathbf{u} \times \Omega) \tag{7.61}$$

to write it as

$$\frac{\partial\mathbf{u}}{\partial b} = -\nabla\left[\frac{1}{2}u^2 + \frac{3A}{2b}\psi\right] - \frac{3A}{2b}\mathbf{u} + (\mathbf{u} \times \Omega). \tag{7.62}$$

Taking the curl of this equation we obtain

$$\frac{\partial\Omega}{\partial b} = -\frac{3A}{2b}\Omega + \nabla \times (\mathbf{u} \times \Omega). \tag{7.63}$$

This equation shows that if $\Omega = 0$ at $t = t_i$ then $\Omega = 0$ for $t > t_i$, keeping the flow irrotational.

(d) Using $u^i = \partial^i\Phi$ in equation (7.57) we obtain

$$\partial^i\left[\frac{\partial\Phi}{\partial b}\right] + \partial^k\Phi\partial_k\partial^i\Phi = -\frac{3A}{2b}\partial^i[\Phi + \Psi]. \tag{7.64}$$

Since $\partial^i(\partial^k\Phi)^2 = 2(\partial^k\Phi)[\partial_i\partial_k\Phi]$ this equation can be written as

$$\partial_i\left[\frac{\partial\Phi}{\partial b} + \frac{1}{2}(\partial_k\Phi)^2 + \frac{3A}{2b}(\Phi + \Psi)\right] = 0. \tag{7.65}$$

The expression in square brackets should clearly be a constant; this constant can be set to zero by a suitable boundary condition.

Now consider equation (7.56). Substituting for δ from the Poisson equation we obtain

$$\frac{\partial}{\partial b}\left[b\nabla^2\psi\right] + \partial_i\left[(1+b\nabla^2\psi)\,\partial^i\Phi\right] = 0 \tag{7.66}$$

or

$$\partial^i\left[\frac{\partial}{\partial b}(b\partial^i\psi) + (1+b\nabla^2\psi)\,\partial^i\Phi\right] = 0. \tag{7.67}$$

Thus the final result can be expressed in terms of the two scalar fields ψ, Φ:

$$\frac{\partial\Phi}{\partial b} + \frac{1}{2}(\partial_i\Phi)^2 = -\frac{3A}{2b}(\psi+\Phi), \tag{7.68}$$

$$\partial_i\left[\frac{\partial}{\partial b}(b\partial^i\psi) + (1+b\nabla^2\psi)\partial^i\Phi\right] = 0. \tag{7.69}$$

These equations contain the same information as (7.56) and (7.57) if the motion is irrotational.

7.4 Measures of inhomogeneity

(a) Expanding $f(\mathbf{x})$ in Fourier space, we have

$$\xi_f(\mathbf{x}) = \langle f(\mathbf{x}+\mathbf{y})f(\mathbf{y})\rangle = \int \frac{d^3k}{(2\pi)^3}\frac{d^3k'}{(2\pi)^3} f_\mathbf{k}f_{\mathbf{k}'}^* \exp(i\mathbf{k}\cdot\mathbf{x})\langle\exp(i(\mathbf{k}-\mathbf{k}')\cdot\mathbf{y})\rangle. \tag{7.70}$$

We can evaluate $\langle\cdots\rangle$ by averaging the expression over a cubical box of size $2L$. In any one dimension,

$$\langle\exp(i(k_x-k'_x)y)\rangle \equiv \int_{-L}^{L}\frac{dy}{2L}\exp\left(i(k_x-k'_x)y\right) = \frac{\sin\left(k_x-k'_x\right)L}{(k_x-k'_x)L}. \tag{7.71}$$

For sufficiently large L, we can use the limit

$$\lim_{L\to\infty}\left\{\frac{\sin kL}{kL}\right\} = \frac{2\pi}{(2L)}\delta_{\text{Dirac}}(k) \tag{7.72}$$

to obtain

$$\xi_f(\mathbf{x}) = \frac{(2\pi)^3}{V}\int \frac{d^3k}{(2\pi)^3}\frac{dk'}{(2\pi)^3} f_\mathbf{k}f_{\mathbf{k}'}^* e^{i\mathbf{k}\cdot\mathbf{x}}\delta_{\text{Dirac}}\left(\mathbf{k}-\mathbf{k}'\right)$$

$$= \int \frac{d^3k}{(2\pi)^3}\frac{|f_\mathbf{k}|^2}{V}e^{i\mathbf{k}\cdot\mathbf{x}} \equiv \int \frac{d^3k}{(2\pi)^3}P(\mathbf{k})e^{i\mathbf{k}\cdot\mathbf{x}}, \tag{7.73}$$

where $V = (2L)^3$ and $P(\mathbf{k}) = |f_\mathbf{k}|^2 V^{-1}$ is the 'power spectrum' of $f(\mathbf{x})$.

(b) In the Fourier space,

$$f_W(\mathbf{x}) = \int f(\mathbf{x}+\mathbf{y})W(\mathbf{y})\,d^3y \equiv \int \frac{d^3k}{(2\pi)^3} f_\mathbf{k}W_\mathbf{k}^* e^{i\mathbf{k}\cdot\mathbf{x}}. \tag{7.74}$$

The quantity $\langle f_W^2\rangle$ is the correlation function of f_W evaluated at $\mathbf{x} = 0$. So

$$\langle f_W^2(\mathbf{x})\rangle = \xi_{f_W}(\mathbf{x}=0) = \int \frac{d^3k}{(2\pi)^3}\frac{|f_\mathbf{k}|^2|W_\mathbf{k}|^2}{V} = \int \frac{d^3k}{(2\pi)^3}P(\mathbf{k})|W_\mathbf{k}|^2. \tag{7.75}$$

Using (7.73), this can be written as

$$\langle f_W^2 \rangle = \int \frac{d^3k}{(2\pi)^3} P(\mathbf{k}) |W_\mathbf{k}|^2 = \int \frac{d^3k}{(2\pi)^3} |W_\mathbf{k}|^2 \int d^3x \, \xi_f(\mathbf{x}) e^{-i\mathbf{k}\cdot\mathbf{x}}$$

$$= \int d^3x \, \xi_f(\mathbf{x}) \int \frac{d^3k}{(2\pi)^3} |W_\mathbf{k}|^2 e^{-i\mathbf{k}\cdot\mathbf{x}}. \tag{7.76}$$

We note that the **k**-integral defines the correlation function of the *window* function $W(\mathbf{x})$, that is,

$$\int \frac{d^3k}{(2\pi)^3} |W_\mathbf{k}|^2 e^{-i\mathbf{k}\cdot\mathbf{x}} = V \xi_W(-\mathbf{x}). \tag{7.77}$$

Therefore:

$$\langle f_W^2 \rangle = \int d^3x \, \xi_f(\mathbf{x}) \xi_W(-\mathbf{x}) V. \tag{7.78}$$

These calculations are often simplified by noting that, if $f(\mathbf{k}) = f(|\mathbf{k}|)$, then the Fourier transform becomes

$$f(x) = \int \frac{d^3k}{(2\pi)^3} f(\mathbf{k}) e^{i\mathbf{k}\cdot\mathbf{x}} = \int_0^\infty \frac{dk}{k} \frac{k^3 f(k)}{2\pi^2} \frac{\sin kx}{kx}. \tag{7.79}$$

(c) Spherical and Gaussian window functions are given by

$$W_{\text{sph}}(\mathbf{x}) = \left(\frac{4\pi R^3}{3} \right)^{-1} \theta \left(R - |\mathbf{x}| \right) \tag{7.80}$$

and

$$W_{\text{Gauss}}(\mathbf{x}) = \left[(2\pi)^{3/2} R^3 \right]^{-1} \exp \left(-\frac{x^2}{2R^2} \right). \tag{7.81}$$

The normalization constants are chosen so that the integral over all space of $W(\mathbf{x})$ is unity. The Fourier transforms of these functions are

$$W_{\text{sph}}(k) = \frac{3}{k^3 R^3} \left[\sin kR - kR \cos kR \right], \tag{7.82}$$

$$W_{\text{Gauss}}(k) = \exp \left(-\frac{1}{2} k^2 R^2 \right). \tag{7.83}$$

Using these in (7.75) we obtain

$$\sigma_{\text{sph}}^2(R) = \int \frac{d^3k}{(2\pi)^3} P(k) W_{\text{sph}}(k) = \int_0^\infty \frac{dk}{k} \left(\frac{k^3 P}{2\pi^2} \right) \left\{ \frac{3 (\sin kR - kR \cos kR)}{k^3 R^3} \right\}^2 \tag{7.84}$$

and

$$\sigma_{\text{Gauss}}^2(R) = \int_0^\infty \frac{dk}{k} \frac{k^3 P(k)}{2\pi^2} e^{-k^2 R^2}. \tag{7.85}$$

(d) Using

$$\xi(\mathbf{x}) \equiv \int \frac{d^3\mathbf{k}}{(2\pi)^3} P(\mathbf{k}) e^{i\mathbf{k}\cdot\mathbf{x}} = \int_0^\infty \frac{dk}{k} \left(\frac{k^3 P(k)}{2\pi^2} \right) \left(\frac{\sin kx}{kx} \right) \qquad (7.86)$$

and

$$\bar{\xi}(r) \equiv \frac{3}{r^3} \int_0^r dx\, x^2 \xi(x), \qquad (7.87)$$

we find that

$$\bar{\xi}(r) = \frac{3}{r^3} \int_0^\infty \frac{dk}{k^2} \left(\frac{k^3 P}{2\pi^2} \right) \int_0^r dx\,(x \sin kx) = \frac{3}{2\pi^2 r^3} \int_0^\infty \frac{dk}{k} P(k) \left[\sin kr - kr \cos kr \right]. \qquad (7.88)$$

To relate $\sigma_{\mathrm{sph}}^2(R)$ to $\xi(x)$ we can use (7.78). This requires the computation of $\xi_{\mathrm{W}}(x)$ for a spherical window function. We have, by definition,

$$\xi_{\mathrm{W}}(\mathbf{x}) = \left(\frac{3}{4\pi R^3} \right)^3 \langle \theta\,(R - |\mathbf{x} + \mathbf{y}|)\,\theta\,(R - |\mathbf{y}|) \rangle. \qquad (7.89)$$

The quantity in $\langle \cdots \rangle$ has a simple geometrical meaning. Consider two spheres, each of radius R, kept with their centres separated by vector $\mathbf{x}$. The spheres will encompass a common region when $|\mathbf{x}| < 2R$. If $\mathbf{y}$ is in this common region then $\theta(R - |\mathbf{x} + \mathbf{y}|)\,\theta(R - |\mathbf{y}|) = 1$ and if $\mathbf{y}$ is outside the common region one of the θ-functions will vanish. Hence $\langle \cdots \rangle$ merely gives the volume of the overlapping region between the spheres. From elementary geometry, this volume is

$$\mathscr{V} = \frac{4\pi}{3} R^3 \left(1 - \frac{x}{2R} \right)^2 \left(1 + \frac{x}{4R} \right) \qquad (7.90)$$

for $x < 2R$ and vanishes for $x > 2R$. Hence, $\xi_{\mathrm{W}} = (\mathscr{V}/V)$ and

$$\sigma_{\mathrm{sph}}^2(R) = \int d^3\mathbf{x}\, \xi(x)\, \xi_{\mathrm{W}}(-\mathbf{x}) V = \frac{3}{R^3} \int_0^{2R} x^2 dx\, \xi(x) \left(1 - \frac{x}{2R} \right)^2 \left(1 + \frac{x}{4R} \right). \qquad (7.91)$$

Finally, to obtain the relation between $\sigma_{\mathrm{sph}}^2(R)$ and $\bar{\xi}(x)$, we use

$$\xi(x) = \frac{1}{x^2} \frac{d}{dx} \left[\frac{1}{3} \bar{\xi}(x) x^3 \right] \qquad (7.92)$$

in (7.91) and integrate by parts to obtain

$$\sigma^2(R) = -\frac{3}{R^3} \int_0^{2R} dx \left[\frac{1}{3} \bar{\xi}(x) x^3 \right] \frac{d}{dx} \left[\left(1 + \frac{x}{4R} \right) \left(1 - \frac{x}{2R} \right)^2 \right]. \qquad (7.93)$$

The rest of the calculation is elementary and we find

$$\sigma^2(R) = \frac{3}{2} \int_0^{2R} \frac{dx}{(2R)} \bar{\xi}(x) \left(\frac{x}{R}\right)^3 \left[1 - \left(\frac{x}{2R}\right)^2\right]. \tag{7.94}$$

7.5 Gaussian random fields

(a) The correlation function is now defined as

$$\xi_\delta(\mathbf{x}) = \langle \delta(\mathbf{x}+\mathbf{y})\delta(\mathbf{y})\rangle = \int \frac{d^3k}{(2\pi)^3} \frac{d^3p}{(2\pi)^3} \langle \delta_{\mathbf{k}}\delta_{\mathbf{p}}^*\rangle \exp\left(i\mathbf{k}\cdot(\mathbf{x}+\mathbf{y})\right)\exp\left(-i\mathbf{p}\cdot\mathbf{y}\right), \tag{7.95}$$

where $\langle\cdots\rangle$ is the ensemble average. Using

$$\langle \delta_{\mathbf{k}}\delta_{\mathbf{p}}^*\rangle = (2\pi)^3 P(\mathbf{k})\delta_{\text{Dirac}}(\mathbf{k}-\mathbf{p}), \tag{7.96}$$

we obtain

$$\xi_\delta(\mathbf{x}) = \int \frac{d^3k}{(2\pi)^3} P(\mathbf{k})e^{i\mathbf{k}\cdot\mathbf{x}}, \tag{7.97}$$

which matches with (7.73) if we identify the two power spectra.

(b) Fourier transforming $\delta_{\text{W}}(\mathbf{x})$, we find that

$$\delta_{\text{W}}(\mathbf{x}) = \int \frac{d^3k}{(2\pi)^3} \delta_{\mathbf{k}} W_{\mathbf{k}}^* e^{i\mathbf{k}\cdot\mathbf{x}} \equiv \int \frac{d^3k}{(2\pi)^3} Q_{\mathbf{k}}. \tag{7.98}$$

If $\delta_{\mathbf{k}}$ is a Gaussian random variable, then $Q_{\mathbf{k}}$ is also a Gaussian random variable. Clearly, $\delta_{\text{W}}(\mathbf{x})$ – which is obtained by adding several Gaussian random variables $Q_{\mathbf{k}}$ – is also a Gaussian random variable. To find the probability distribution of $\delta_{\text{W}}(\mathbf{x})$ we only need to know the mean and variance of $\delta_{\text{W}}(\mathbf{x})$. These are

$$\langle\delta_{\text{W}}(\mathbf{x})\rangle = \int \frac{d^3k}{(2\pi)^3} \langle\delta_{\mathbf{k}}\rangle W_{\mathbf{k}}^* e^{i\mathbf{k}\cdot\mathbf{x}} = 0,$$
$$\langle\delta_{\text{W}}^2(\mathbf{x})\rangle = \int \frac{d^3k}{(2\pi)^3} P(\mathbf{k})|W_{\mathbf{k}}|^2 \equiv \Delta_{\text{W}}^2. \tag{7.99}$$

Hence, the probability of δ_{W} having a value of q at any location is given by

$$\mathscr{P}(q) = \frac{1}{(2\pi\Delta_{\text{W}}^2)^{1/2}} \exp\left(-\frac{q^2}{2\Delta_{\text{W}}^2}\right). \tag{7.100}$$

(c) The Gaussian nature of $\delta_{\mathbf{k}}$ cannot be maintained for $t > t_{\text{i}}$ if the evolution couples the modes for different values of $\mathbf{k}$. Equation (7.41), which describes the evolution of $\delta_{\mathbf{k}}(t)$, shows that the modes do mix with each other as time goes on. Thus, in general, the Gaussian nature of the $\delta_{\mathbf{k}}$s cannot be maintained.

But if, $\delta(t,\mathbf{x}) \ll 1$, then one can describe the evolution of $\delta(t,\mathbf{x})$ by linear perturbation theory. (In this limit we ignore the A and B terms in (7.41), which couples different modes.) Then each mode $\delta_{\mathbf{k}}(t)$ will evolve independently and we can write

$$\delta_{\mathbf{k}}(t) = T_{\mathbf{k}}(t, t_{\text{i}})\delta_{\mathbf{k}}(t_{\text{i}}), \tag{7.101}$$

where $T_{\mathbf{k}}(t, t_i)$ depends on the dyanamics. In that case,

$$\langle \delta_{\mathbf{k}}(t) \delta_{\mathbf{p}}^*(t) \rangle = T_{\mathbf{k}}(t, t_i) T_{\mathbf{p}}^*(t, t_i) \langle \delta_{\mathbf{k}}(t_i) \delta_{\mathbf{p}}^*(t_i) \rangle = (2\pi)^3 |T_{\mathbf{k}}(t, t_i)|^2 P(\mathbf{k}, t_i) \delta_{\text{Dirac}}(\mathbf{k} - \mathbf{p}) \tag{7.102}$$

and the Gaussian nature of $\delta_{\mathbf{k}}$ is preserved by evolution with the power spectrum evolving as

$$P(\mathbf{k}, t) = |T_{\mathbf{k}}(t, t_i)|^2 P(\mathbf{k}, t_i). \tag{7.103}$$

7.6 Linear evolution in the smooth fluid limit

(a) Consider a spherical region of radius $\lambda (\gg d_{\text{H}})$, containing energy density $\rho_1 = \rho_{\text{b}} + \delta\rho$, embedded in a $k = 0$ Friedmann universe of density ρ_{b}. It follows from spherical symmetry that the inner region is not affected by the matter outside; hence the inner region evolves as a $k \neq 1$ Friedmann universe. Therefore, we can write, for the two regions

$$H^2 = \frac{8\pi G}{3}\rho_{\text{b}}, \quad H^2 + \frac{k}{a^2} = \frac{8\pi G}{3}(\rho_{\text{b}} + \delta\rho). \tag{7.104}$$

The change of density from ρ_{b} to $\rho_{\text{b}} + \delta\rho$ is accommodated by adding a spatial curvature term (k/a^2). If this condition is to be maintained at all times, we must have

$$\frac{8\pi G}{3}\delta\rho = \frac{k}{a^2}, \tag{7.105}$$

or

$$\frac{\delta\rho}{\rho_{\text{b}}} = \frac{3}{8\pi G(\rho_{\text{b}}a^2)}. \tag{7.106}$$

If $(\delta\rho/\rho_{\text{b}})$ is small, $a(t)$ on the right-hand side will only differ slightly from the expansion factor of the unperturbed universe. This allows one to find how $(\delta\rho/\rho_{\text{b}})$ scales with a. Since $\rho_{\text{b}} \propto a^{-4}$ in the radiation dominated phase $t < t_{\text{eq}}$ and $\rho_{\text{b}} \propto a^{-3}$ in the matter dominated phase $t > t_{\text{eq}}$ we obtain

$$\left(\frac{\delta\rho}{\rho}\right) \propto \begin{cases} a^2 & (\text{for } t < t_{\text{eq}}) \\ a & (\text{for } t > t_{\text{eq}}). \end{cases} \tag{7.107}$$

Thus, the amplitude of the mode with $\lambda > d_{\text{H}}$ always grows: as a^2 in the radiation dominated phase and as a in the matter dominated phase. (The same result was obtained in problem 6.10 by perturbing the Friedmann model.)

(b) The exact equation satisfied by $\delta_{\mathbf{k}}$ is (7.41). In the linear limit we can ignore the δ^2 and $\dot{\mathbf{x}}^2$ terms; that is, we can drop A and B. Then the perturbation equation becomes

$$\ddot{\delta}_{\text{DM}} + \frac{2\dot{a}}{a}\dot{\delta}_{\text{DM}} \cong 4\pi G \rho_{\text{DM}} \delta_{\text{DM}}. \tag{7.108}$$

The expansion of the background universe is due to both radiation and dark matter:

$$\frac{\dot{a}^2}{a^2} = \frac{8\pi G}{3}(\rho_{rad} + \rho_{DM}).$$ (7.109)

Introducing the variable $x \equiv (a/a_{eq})$ and using (7.109) in (7.108), this equation becomes

$$2x(1+x)\frac{d^2\delta_{DM}}{dx^2} + (2+3x)\frac{d\delta_{DM}}{dx} = 3\delta_{DM}, \qquad x = \frac{a}{a_{eq}}.$$ (7.110)

One solution to this equation can be written down by inspection:

$$\delta_{DM} = 1 + \frac{3}{2}x.$$ (7.111)

In other words $\delta_{DM} \cong$ constant for $a \ll a_{eq}$ (no growth in the radiation dominated phase) and $\delta_{DM} \propto a$ for $a \gg a_{eq}$ (growth proportional to a in the matter dominated phase).

We now have to find the second solution. Given the growing solution, the decaying solution Δ can be found by the Wronskian condition $(Q'/Q) = -[(2 + 3x)/2x(1 + x)]$, where $Q = \delta_{DM}\Delta' - \delta'_{DM}\Delta$. Writing the decaying solution as $\Delta = f(x)\,\delta_{DM}(x)$ and substituting in this equation, we find

$$\frac{f''}{f'} = -\frac{2\delta'_{DM}}{\delta_{DM}} - \frac{2+3x}{2x(1+x)},$$ (7.112)

which can be integrated to give

$$f = -\int \frac{dx}{x(1+3x/2)^2(1+x)^{1/2}}.$$ (7.113)

The integral is straightforward and the decaying solution is

$$\Delta = f\delta_{DM} = \left(1 + \frac{3x}{2}\right)\ln\left[\frac{(1+x)^{1/2}+1}{(1+x)^{1/2}-1}\right] - 3(1+x)^{1/2}.$$ (7.114)

Thus the general solution to the perturbation equation, for a mode which is inside the Hubble radius, is the linear superposition $\delta = A\delta_{DM} + B\Delta$; the asymptotic forms of this solution are

$$\delta_{gen}(x) = A\delta_{DM}(x) + B\Delta(x) = \begin{cases} A + B\ln(4/x) & (x \ll 1) \\ (3/2)Ax + (4/5)Bx^{(-3/2)} & (x \gg 1). \end{cases}$$ (7.115)

This result shows that dark matter perturbations can grow only logarithmically during the epoch $a_{enter} < a < a_{eq}$. During this phase the universe is dominated by radiation which is unperturbed. Hence, the damping term due to expansion $(2\dot{a}/a)\dot{\delta}$ in equation (7.108) dominates over the gravitational potential term on the right-hand side and restricts the growth of perturbations. In the matter dominated phase with $a \gg a_{eq}$, the perturbations grow as a. This result, combined with that of part (a), shows that in the matter dominated phase all the modes (i.e. modes which are inside or outside the Hubble radius) grow in proportion to the expansion factor.

(c) The general solution after the mode has entered the Hubble radius is given by (7.115). The constants A and B have to be fixed by matching this solution to the growing solution, which was valid when the mode was bigger than the Hubble radius. Since the latter solution is given by $\delta(x) = x^2$ in the radiation dominated phase, the matching conditions are

$$x_{\text{enter}}^2 = [A\delta_{\text{DM}}(x) + B\Delta(x)]_{x=x_{\text{enter}}}$$
$$2x_{\text{enter}} = [A\delta'_{\text{DM}}(x) + B\Delta'(x)]_{x=x_{\text{enter}}}. \tag{7.116}$$

This determines the constants A and B in terms of $x_{\text{enter}} = (a_{\text{enter}}/a_{\text{eq}})$ which, in turn, depends on the wavelength of the mode through a_{enter}.

As an example, we consider a mode for which $x_{\text{enter}} \ll 1$. The 'decaying' solution has the asymptotic form $\Delta(x) \cong \ln(4/x)$ for $x \ll 1$. Using this, the matching conditions become

$$x_{\text{enter}}^2 = A\left(1 + \frac{3}{2}x_{\text{enter}}\right) + B\ln\frac{4}{x_{\text{enter}}},$$
$$2x_{\text{enter}} = \frac{3}{2}A - Bx_{\text{enter}}^{-1}. \tag{7.117}$$

Solving these to determine A and B (to the lowest order in x_{enter}), we find

$$A = x_{\text{enter}}^2\left[1 + 2\ln\left(\frac{4}{x_{\text{enter}}}\right)\right], \qquad B = -2x_{\text{enter}}^2, \tag{7.118}$$

so that the properly matched mode, inside the Hubble radius, is

$$\delta(x) = x_{\text{enter}}^2\left[1 + 2\ln\left(\frac{4}{x_{\text{enter}}}\right)\right]\left(1 + \frac{3x}{2}\right) - 2x_{\text{enter}}^2\ln\frac{4}{x}. \tag{7.119}$$

During the radiation dominated phase, that is, till $a \lesssim a_{\text{eq}}$, $(x \lesssim 1)$, this mode can grow by a factor

$$\frac{\delta(x \cong 1)}{\delta(x_{\text{enter}})} = \frac{1}{x_{\text{enter}}^2}\delta(x \cong 1) \cong 5\ln\left(\frac{1}{x_{\text{enter}}}\right) = 5\ln\left(\frac{a_{\text{eq}}}{a_{\text{enter}}}\right) = \frac{5}{2}\ln\left(\frac{t_{\text{eq}}}{t_{\text{enter}}}\right). \tag{7.120}$$

Since the time t_{enter} for a mode with wavelength λ is fixed by the condition $\lambda a_{\text{enter}} \cong \lambda t_{\text{enter}}^{1/2} \cong d_{\text{H}}(t_{\text{enter}}) \cong t_{\text{enter}}$, it follows that $\lambda \propto t_{\text{enter}}^{1/2}$. Hence,

$$\frac{\delta_{\text{final}}}{\delta_{\text{enter}}} = 5\ln\left(\frac{\lambda_{\text{eq}}}{\lambda}\right) = \frac{5}{3}\ln\left(\frac{M_{\text{eq}}}{M}\right) \tag{7.121}$$

for a mode with wavelength $\lambda \ll \lambda_{\text{eq}}$. (Here, M is the mass contained in a sphere of radius λ; see problem 6.3.) The growth in the radiation dominated phase, therefore, is logarithmic. Note that the matching procedure has brought in an amplification factor *which depends on the wavelength.*

(d) Let $\rho(t)$ be a solution to the background Friedmann model dominated by pressureless dust. Consider now the function $\rho_1(t) \equiv \rho(t + \tau)$, where τ is some constant. Since the Friedmann equations contain t only through the derivative, $\rho_1(t)$ is also a valid solution. If we now take τ to be small, then $[\rho_1(t) - \rho(t)]$ will be a small perturbation to the density. The corresponding density contrast is

$$\delta(t) = \frac{\rho_1(t) - \rho(t)}{\rho(t)} = \frac{\rho(t+\tau) - \rho(t)}{\rho(t)} \cong \tau \frac{d\ln\rho}{dt} = -3\tau H(t), \qquad (7.122)$$

where the last relation follows from the fact that $\rho \propto a^{-3}$ and $H = (\dot{a}/a)$. Since τ is a constant, it follows that $H(t)$ is a solution to the perturbation equation. This curious fact, of course, can be verified directly. From the equations describing the Friedmann model, it follows that $\dot{H} + H^2 = (-4\pi G\rho/3)$. Differentiating this relation and using $\dot{\rho} = -3H\rho$, we immediately obtain $\ddot{H} + 2H\dot{H} - 4\pi G\rho H = 0$. Thus H satisfies the same equation as δ.

Since $\dot{H} = -H^2 - (4\pi G\rho/3)$, we know that $\dot{H} < 0$, that is, H is a decreasing function of time, and the solution $\delta = H \equiv \delta_d$ is a decaying mode. The growing solution ($\delta \equiv \delta_i$) can easily be found by using the fact that, for any two linearly independent solutions of equation (7.108), the Wronskian $(\dot{\delta}_i\delta_d - \dot{\delta}_d\delta_i)$ has a value a^{-2}. This implies

$$\delta_i = \delta_d \int \frac{dt}{a^2\delta_d^2} = H(t) \int \frac{dt}{a^2H^2(t)}. \qquad (7.123)$$

Thus we see that the $H(t)$ of the background spacetime allows one to completely determine the evolution of density contrast.

Since these equations hold for all $\mathbf{k}$, the $\delta(\mathbf{x}, t)$ in real space also grows as $b(t)$, preserving the form. Hence $\delta(\mathbf{x}, t) = b(t)q(\mathbf{x})$. Since $\nabla^2\psi = (\delta/b)$ it follows that $\psi(\mathbf{x}, t) = \psi(\mathbf{x})$ is a constant in time. Finally, in the linear order, equation (7.57) becomes

$$\frac{\partial u^i}{\partial b} \cong -\frac{3A}{2b}\left[\partial^i\psi + u^i\right]. \qquad (7.124)$$

Since $\partial^i\psi$ is independent of t (or b), this equation has the solution

$$\mathbf{u}(\mathbf{x}) = -\nabla\psi(\mathbf{x}), \qquad (7.125)$$

for which both sides vanish identically. Thus, in the linear approximation, $\psi(t, \mathbf{x}) = \psi(\mathbf{x})$ completely specifies the solution:

$$\delta(t, \mathbf{x}) = b(t)\nabla^2\psi(\mathbf{x}), \quad \mathbf{u}(t, \mathbf{x}) = \mathbf{u}(\mathbf{x}) = -\nabla\psi(\mathbf{x}). \qquad (7.126)$$

(e) The true gravitational potential is $\phi = (4\pi G\rho_b a^2 b)\psi$ and the true peculiar velocity is $\mathbf{v} = a\dot{b}\mathbf{u}$. In terms of ϕ and $\mathbf{v}$ equation (7.125) becomes

$$\frac{\mathbf{v}}{a\dot{b}} = -\frac{1}{(4\pi G\rho_b a^2 b)}\nabla\phi \qquad (7.127)$$

or

$$\mathbf{v} = -\left(\frac{\dot{b}}{4\pi G\rho_b ba}\right)\nabla\phi = \left(\frac{\dot{b}}{4\pi G\rho_b b}\right)\mathbf{g} = \frac{2}{3H\Omega}\mathbf{g}\left[\frac{\dot{b}a}{b\dot{a}}\right], \qquad (7.128)$$

where $\mathbf{g} = -a^{-1}\nabla\phi, H = (\dot{a}/a)$ and $\Omega = [\rho_b(t)/\rho_c(t)]$. Writing $(\dot{b}a/b\dot{a}) \equiv f$ we obtain

$$\mathbf{v} = \frac{2f}{3H\Omega}\mathbf{g} = -\frac{2f}{3H\Omega}\frac{1}{a}\nabla\phi, \qquad (7.129)$$

where ϕ is the Newtonian potential generated by the excess density $\rho_b\delta$. Thus, the peculiar velocity $\mathbf{v}$ is proportional to the peculiar acceleration $\mathbf{g}$ with the coefficient of proportionality $(2f/3H\Omega)$, where

$$f(a) = \frac{a}{\delta}\frac{d\delta}{da} = -\frac{(1+z)}{\delta}\frac{d\delta}{dz}. \qquad (7.130)$$

The peculiar velocities and accelerations observed in our universe at the present epoch are related by the value of f at $z = 0$, which, in turn, will depend only on Ω. Though $f(\Omega, z = 0)$ can be calculated exactly from (7.123), its functional form is not convenient for further manipulations. It turns out, however, that $f(\Omega, z = 0)$ is very well approximated by the power law $f(\Omega) \cong \Omega^{0.6}$, which is often used, instead of the exact form, for estimates. With this approximation,

$$\mathbf{v} = \frac{2}{3H\Omega}\Omega^{0.6}\mathbf{g}, \qquad (7.131)$$

where all the quantities are evaluated at present and $\mathbf{g}$ is the peculiar acceleration generated by the density contrast.

(f) For $\Omega \neq 1$, we have the relations

$$a(z) = a_0(1+z)^{-1}, \qquad H(z) = H_0(1+z)(1+\Omega z)^{1/2} \qquad (7.132)$$

and

$$H_0 dt = -(1+z)^{-2}(1+\Omega z)^{-\frac{1}{2}}dz. \qquad (7.133)$$

Taking $\delta_d = H(z)$, we obtain

$$\delta_g = \delta_d(z)\int a^{-2}\delta_d^{-2}(z)\left(\frac{dt}{dz}\right)dz$$

$$= (a_0 H_0)^{-2}(1+z)(1+\Omega z)^{1/2}\int_z^\infty dx(1+x)^{-2}(1+\Omega x)^{-\frac{3}{2}}. \qquad (7.134)$$

This integral can be expressed in terms of elementary functions:

$$\delta_g = \frac{1+2\Omega+3\Omega z}{(1-\Omega)^2} - \frac{3}{2}\frac{\Omega(1+z)(1+\Omega z)^{1/2}}{(1-\Omega)^{5/2}}\ln\left[\frac{(1+\Omega z)^{1/2}+(1-\Omega)^{1/2}}{(1+\Omega z)^{1/2}-(1-\Omega)^{1/2}}\right]. \qquad (7.135)$$

Thus $\delta_g(z)$ for an arbitrary Ω can be given in closed form. The solution in (7.135) is not normalized in any manner; normalization can be achieved by multiplying δ_g by some constant, depending on the context.

For large z (i.e. early times), $\delta_g \propto z^{-1}$. This is to be expected because for large z, the curvature term can be ignored and the Friedmann universe can be approximated as a $\Omega = 1$ model. (The large z expansion of the logarithm in

(7.135) has to be taken up to $O(z^{-5/2})$ to obtain the correct result; it is easier to obtain the asymptotic form directly from the integral.) For $\Omega \ll 1$, one can see that $\delta_g \cong$ constant for $z \ll \Omega^{-1}$. This is the curvature dominated phase, in which the growth of perturbations is halted by rapid expansion.

7.7 Linear evolution of distribution function

(a) In the collisionless Boltzman equation

$$\frac{\partial f}{\partial \tau} + \left(\frac{\mathbf{p}}{m}\right) \cdot \frac{\partial f}{\partial \mathbf{x}} = ma^2(\tau) \nabla \phi \cdot \frac{\partial f}{\partial \mathbf{p}}, \qquad (7.136)$$

we will substitute $f(\mathbf{x}, \mathbf{p}, \tau) = f_0(\mathbf{p}, \tau) + f_1(\mathbf{x}, \mathbf{p}, \tau)$ and linearize the equation in f_1. Since ϕ is due to perturbed density contrast, it is already a first-order quantity, and we can replace $\nabla \phi \cdot (\partial f/\partial \mathbf{p})$ by $\nabla \phi \cdot (\partial f_0/\partial \mathbf{p})$ to the lowest order. Then the equation becomes

$$\frac{\partial f_1}{\partial \tau} + \left(\frac{\mathbf{p}}{m}\right) \cdot \frac{\partial f_1}{\partial \mathbf{x}} = ma^2(\tau) \nabla \phi \cdot \frac{\partial f_0}{\partial \mathbf{p}}. \qquad (7.137)$$

Since ϕ is linear in f_1, this is a linear equation in $f_1(\mathbf{x}, \mathbf{p}, \tau)$. The $\mathbf{x}$-dependence can be tackled by Fourier transforming f_1 and ϕ in $\mathbf{x}$. If we set

$$f_1(\mathbf{x}, \mathbf{p}, \tau) = \int \frac{d^3k}{(2\pi)^3} f_\mathbf{k}(\mathbf{p}, \tau) e^{i\mathbf{k}\cdot\mathbf{x}}, \quad \phi(\tau, \mathbf{x}) = \int \frac{d^3k}{(2\pi)^3} \phi_\mathbf{k}(\tau) e^{i\mathbf{k}\cdot\mathbf{x}}, \qquad (7.138)$$

the equation satisfied by $f_\mathbf{k}(\mathbf{p}, \tau)$ will be

$$\frac{\partial f_\mathbf{k}}{\partial \tau} + \left(\frac{i\mathbf{k}\cdot\mathbf{p}}{m}\right) f_\mathbf{k} = ma^2(\tau) \left(i\mathbf{k}\cdot\frac{\partial f_0}{\partial \mathbf{p}}\right) \phi_\mathbf{k}(\tau). \qquad (7.139)$$

Changing this equation into the form

$$\frac{\partial}{\partial \tau}\left[f_\mathbf{k}(\mathbf{p}, \tau) \exp\left(\frac{i\mathbf{k}\cdot\mathbf{p}}{m}\tau\right)\right] = ma^2(\tau) \left(i\mathbf{k}\cdot\frac{\partial f_0}{\partial \mathbf{p}}\right) \phi_\mathbf{k}(\tau) \exp\left(\frac{i\mathbf{k}\cdot\mathbf{p}}{m}\tau\right), \qquad (7.140)$$

we can integrate both sides from some τ_i to τ. This gives

$$f_\mathbf{k}(\mathbf{p}, \tau) \exp\left(\frac{i\mathbf{k}\cdot\mathbf{p}}{m}\tau\right) - f_\mathbf{k}(\mathbf{p}, \tau_i) \exp\left(\frac{i\mathbf{k}\cdot\mathbf{p}}{m}\tau_i\right)$$

$$= mi\mathbf{k} \cdot \int_{\tau_i}^{\tau} ds \left(\frac{\partial f_0}{\partial \mathbf{p}}\right) \phi_\mathbf{k}(s) a^2(s) \exp\left(\frac{i\mathbf{k}\cdot\mathbf{p}}{m}s\right). \qquad (7.141)$$

Or

$$f_\mathbf{k}(\mathbf{p}, \tau) = f_\mathbf{k}(\mathbf{p}, \tau_i) \exp\left[-\left(\frac{i\mathbf{k}\cdot\mathbf{p}}{m}\right)(\tau - \tau_i)\right]$$

$$+ mi\mathbf{k} \cdot \left(\frac{\partial f_0}{\partial \mathbf{p}}\right) \int_{\tau_i}^{\tau} ds\, \phi_\mathbf{k}(s) a^2(s) \exp\left[-\left(\frac{i\mathbf{k}\cdot\mathbf{p}}{m}\right)(\tau - s)\right]. \qquad (7.142)$$

This is an integral equation for $f_\mathbf{k}$ since $\phi_\mathbf{k}$ depends on $f_\mathbf{k}$.

(b) From the relation

$$\rho(\tau, \mathbf{x}) = \frac{m}{a^3} \int d^3\mathbf{p}\, f(\tau, \mathbf{x}, \mathbf{p}) = \rho_b(\tau)[1 + \delta(\tau, \mathbf{x})], \qquad (7.143)$$

it follows that

$$\delta(\tau, \mathbf{x}) = \frac{m}{\rho_b\, a^3} \int d^3\mathbf{p}\, f_1(\tau, \mathbf{x}, \mathbf{p}) = \frac{1}{n_0} \int d^3\mathbf{p}\, f_1(\tau, \mathbf{x}, \mathbf{p}), \qquad (7.144)$$

where $n_0 = (\rho_b a^3/m) = (\Omega \rho_c/m)$ is the (constant) comoving number density of particles. So, the Fourier transform $\delta_{\mathbf{k}}(\tau)$ of the density contrast can be obtained by integrating $[f_{\mathbf{k}}(\mathbf{p}, \tau)/n_0]$ over $\mathbf{p}$. Performing this integral in (7.142) and ignoring the first term, which only adds an initial condition, we obtain

$$\delta_{\mathbf{k}}(\tau) = \frac{mik}{n_0} \cdot \int_{\tau_i}^{\tau} ds\, a^2(s)\, \phi_{\mathbf{k}}(s) \int d^3\mathbf{p} \left(\frac{\partial f_0}{\partial \mathbf{p}}\right) \exp\left[-\left(\frac{i\mathbf{p}\cdot\mathbf{k}}{m}\right)(\tau - s)\right]. \qquad (7.145)$$

In the $d^3\mathbf{p}$ integral, we integrate by parts and assume that $f_0(\mathbf{p})$ vanishes for $|\mathbf{p}| \to \infty$. Then

$$\delta_{\mathbf{k}}(\tau) = -\frac{k^2}{n_0} \int_{\tau_i}^{\tau} ds\,(\tau - s)\, a^2(s)\, \phi_{\mathbf{k}}(s) \int d^3\mathbf{p}\, f_0(\mathbf{p}) \exp\left[-\left(\frac{i\mathbf{p}\cdot\mathbf{k}}{m}\right)(\tau - s)\right]$$

$$= -k^2 \int_{\tau_i}^{\tau} ds\,(\tau - s)\, a^2(s)\, \phi_{\mathbf{k}}(s)\, \mathscr{G}\left[\mathbf{k}(\tau - s)/m\right], \qquad (7.146)$$

where $\mathscr{G}$ is the Fourier transform of (f_0/n_0):

$$\mathscr{G}(\mathbf{q}) \equiv \frac{1}{n_0} \int d^3\mathbf{p}\, f_0(\mathbf{p})\, e^{-i\mathbf{p}\cdot\mathbf{q}}. \qquad (7.147)$$

(c) If $f_0(\mathbf{p}) = f_0(|\mathbf{p}|)$ this integral can be written as

$$\mathscr{G}(\mathbf{q}) = \mathscr{G}(q) = \frac{4\pi}{n_0 q} \int_0^{\infty} dp\, p f_0(p) \sin pq, \qquad (7.148)$$

so that (7.146) becomes

$$\delta_{\mathbf{k}}(\tau) = -k^2 \int_{\tau_i}^{\tau} ds\,(\tau - s)\, a^2(s)\, \phi_{\mathbf{k}}(s) \cdot \frac{4\pi}{n_0} \cdot \frac{m}{k(\tau - s)} \int_0^{\infty} dp\, p f_0(p) \sin \frac{pk}{m}(\tau - s)$$

$$= -\frac{4\pi k m}{n_0} \int_{\tau_i}^{\tau} ds\, a^2(s)\, \phi_{\mathbf{k}}(s)\, I_k(\tau - s), \qquad (7.149)$$

with

$$I_k(\tau - s) \equiv \int_0^{\infty} dp\, p f_0(p) \sin \frac{pk}{m}(\tau - s). \qquad (7.150)$$

In this equation ϕ_k is the *total* gravitational potential. If external sources are present, then ϕ_k is due to both the external sources and the dark matter. In that case, the Poisson equation in Fourier space is

$$-k^2\phi_k = 4\pi Ga^2\rho_b\left[\delta_k + \delta_k^{ext}\right] \qquad (7.151)$$

and equation (7.149) becomes

$$\delta_k(\tau) = -\frac{4\pi mk}{n_0}\int_{\tau_i}^{\tau} ds\, a^2(s)\left[-\frac{4\pi Ga^2(s)\rho_b}{k^2}\right]\left[\delta_k + \delta_k^{ext}\right] I_k(\tau - s)$$

$$= \frac{16\pi^2 Gm\rho_{b0}}{n_0 k}\int_{\tau_i}^{\tau} ds\, a(s)\left[\delta_k + \delta_k^{ext}\right] I_k(\tau - s)$$

$$= \frac{16\pi^2 Gm^2}{k}\int_{\tau_i}^{\tau} ds\, a(s)\left[\delta_k + \delta_k^{ext}\right] I_k(\tau - s). \qquad (7.152)$$

In arriving at the last equation we have used the fact that $\rho_{b0} = (n_0 m)$.

(d) An equivalent form of this equation, when $\delta_k^{ext} = 0$, is

$$\delta_k(\tau) = \frac{3}{2}\frac{H_0^2}{k}\int_{\tau_i}^{\tau} ds\, a(s)\delta_k(s)J_k(\tau - s), \qquad (7.153)$$

with

$$J_k(\tau - s) \equiv \frac{4\pi m}{n_0}\int_0^{\infty} dp\, pf_0(p)\sin\left[\frac{pk(\tau - s)}{m}\right]. \qquad (7.154)$$

For long wavelengths $(k \to 0)$, this expression becomes

$$J_k(\tau - s) \cong \frac{k(\tau - s)}{n_0}\int_0^{\infty} 4\pi p^2 f_0(p)\, dp = k(\tau - s)\frac{\rho_b a^3}{mn_0} = k(\tau - s). \qquad (7.155)$$

Hence, for a matter dominated universe with $a(\tau) = (4/H_0^2\tau^2)$, the equation for δ_k becomes

$$\delta_k(\tau) = \frac{3}{2}H_0^2\int_{-\infty}^{\tau} ds\, a(s)\delta_k(s)(\tau - s) = 6\int_{-\infty}^{\tau} ds\left(\frac{\tau - s}{s^2}\right)\delta_k(s). \qquad (7.156)$$

(We have taken t_i to be sufficiently small so that $a(t_i) \cong 0$; this corresponds to $\tau_i = -\infty$.) It can be directly verified that (7.156) has the solution $\delta(\tau) \propto \tau^{-2} = A\tau^{-2}$:

$$6\int_{-\infty}^{\tau} ds\left(\frac{\tau - s}{s^2}\right)\delta_k(s) = 6A\int_{-\infty}^{\tau} ds\left[\frac{\tau}{s^4} - \frac{1}{s^3}\right] = \frac{A}{\tau^2} = \delta_k(\tau). \qquad (7.157)$$

So, in the $k \to 0$ limit, $\delta_k \propto \tau^{-2} \propto a(\tau)$; this agrees with the result obtained earlier in problem 7.6. In this limit, the velocity dispersion is not significant and both the Vlasov equation and the pressureless fluid equation give the same result.

7.8 Free streaming

(a) This effect is most easily seen using (7.142). The first term in (7.142) is kinematic in the sense that, if the initial distribution is homogeneous $[f_{\mathbf{k}}(\mathbf{p}, \tau_i) \propto \delta_{\text{Dirac}}(\mathbf{k}) f_0(\mathbf{p})]$, then this term is zero for $\mathbf{k} \neq 0$. Consider the second term with $s = q - \tau_i$. The relevant part of the integral is

$$I = \int_0^{\tau - \tau_i} dq \, a^2(q - \tau_i) \phi_{\mathbf{k}}(q - \tau_i) \exp\left[-\left(\frac{i\mathbf{k} \cdot \mathbf{p}q}{m}\right)\right]. \tag{7.158}$$

For particles with low velocities we will have $[(\mathbf{k} \cdot \mathbf{p})q/m] \ll 1$ for the entire range of integration (or, equivalently, $(\tau - \tau_i)v \ll |k|^{-1}$); these particles do not move too far away from the crests or troughs during the time available. If a perturbation is set up initially, its shape will be maintained to reasonable accuracy.

On the other hand, if $[(\mathbf{k} \cdot \mathbf{p})q/m] \gg 1$ (or, equivalently, if $(\tau - \tau_i)v \gg k^{-1}$), then the phase of the integral oscillates rapidly and the integral makes very little contribution. Physically, the particles which were originally in the crests move to troughs and vice versa (within the time available), thereby wiping out the perturbation. We can conclude that the amplitude of the perturbation will be drastically reduced for small wavelengths with $k^{-1} \ll v(\tau - \tau_i)$.

(b) To convert (7.146) into a second-order differential equation for $\delta_{\mathbf{k}}$ we have to obtain the $\ddot{\delta}_{\mathbf{k}}, \dot{\delta}_{\mathbf{k}}$ terms, where the dot denotes derivatives with respect to τ. The first derivative $\dot{\delta}_{\mathbf{k}}$ is given by

$$\dot{\delta}_{\mathbf{k}} = -k^2 \int_{\tau_i}^{\tau} ds \, a^2(s) \phi_{\mathbf{k}}(s) \frac{d}{dz} [z\mathcal{G}(z)], \tag{7.159}$$

with $z \equiv |\mathbf{k}|(\tau - s)/m$. (The derivative arising from the upper limit of the integral vanishes since $z\mathcal{G}(z)$ vanishes at $z = 0$.) Differentiating again and setting $[d(z\mathcal{G})/dz]_{z=0} = \text{constant} = B$, we obtain

$$\ddot{\delta}_{\mathbf{k}} = -k^2 a^2(\tau) \phi_{\mathbf{k}}(\tau) B - \frac{k^3}{m} \int_{\tau_i}^{\tau} ds \, a^2(s) \phi_{\mathbf{k}}(s) \frac{d^2}{dz^2} [z\mathcal{G}(z)]. \tag{7.160}$$

If the second derivative of $z\mathcal{G}(z)$ can be expressed as a linear combination of the first derivative and the function $\mathcal{G}(z)$, then we can obtain a closed differential equation for $\delta_{\mathbf{k}}$. This, of course, is possible only for a special class of $\mathcal{G}(z)$. As an example, consider the function $\mathcal{G}(z) = \exp(-\alpha z)$. In the original $\mathbf{p}$-space, this corresponds to a Lorentzian distribution function $f_0(\mathbf{p}) \propto (\alpha^2 + p^2)^{-2}$. For this $\mathcal{G}(z)$, we have $B = 1$ and

$$\frac{d^2}{dz^2}(z\mathcal{G}) = -2\alpha \frac{d}{dz}(z\mathcal{G}) - \alpha^2 z\mathcal{G}. \tag{7.161}$$

Using this in (7.160) and converting integrals involving $d(z\mathscr{G})/dz$ and $z\mathscr{G}$ with the help of (7.159) and (7.146), we find a closed equation for δ_k:

$$\ddot{\delta}_k + \frac{2\alpha k}{m}\dot{\delta}_k + \frac{\alpha^2 k^2}{m^2}\delta_k = -k^2 a^2 \phi_k. \tag{7.162}$$

Converting back to t with $(d/d\tau) = a^2(d/dt)$ we obtain

$$\frac{d^2\delta_k}{dt^2} + 2\left(\frac{\dot{a}}{a} + \frac{\alpha k}{a^2}\right)\frac{d\delta_k}{dt} + \frac{\alpha^2 k^2}{a^4}\delta_k = -\frac{k^2 \phi_k}{a^2} = 4\pi G\rho_b\left(\delta_k + \delta_k^{\text{ext}}\right). \tag{7.163}$$

The last equality uses the Poisson equation to relate ϕ_k to $(\delta_k + \delta_k^{\text{ext}})$. If $\delta_k^{\text{ext}} = 0$, then

$$\frac{d^2\delta_k}{dt^2} + 2\left(\frac{\dot{a}}{a} + \frac{\alpha k}{a^2}\right)\frac{d\delta_k}{dt} + \left(\frac{\alpha^2 k^2}{a^4} - 4\pi G\rho_b\right)\delta_k = 0. \tag{7.164}$$

The quantity $(\alpha/a) \equiv c_s$ can be thought of as the effective velocity dispersion of the background particles (which scales as a^{-1} since the mean momentum $\langle|\mathbf{p}|\rangle \propto a^{-1}$). Then we obtain

$$\ddot{\delta}_k + 2\left(\frac{\dot{a}}{a} + c_s\frac{k}{a}\right)\dot{\delta}_k + \left(\frac{c_s^2 k^2}{a^2} - 4\pi G\rho_b\right)\delta_k = 0, \tag{7.165}$$

with dots denoting (d/dt). This is quite similar to equation (7.108) except for two additional terms. The damping term $(2c_s k/a)\dot{\delta}_k$ can be significant within one expansion time-scale if $(c_s k/a)\,H^{-1} \gg 1$, that is, for physical wavelengths with $k^{-1}a \ll c_s H_0^{-1}$. (This is exactly the conclusion we reached earlier from equation (7.158).) The term $(c_s^2 k^2/a^2)$ acts like an effective pressure gradient term. For the collisionless system we are considering, this term arises from the velocity dispersion of the particles.

(c) For $(c_s k/a) \gg H(t)$, we can ignore the $(\dot{a}/a)$ and $4\pi G\rho_b$ terms in (7.165). To the same order of approximation, we can ignore the time dependence of the $(c_s k/a)$ term. In this limit the equation becomes

$$\ddot{\delta}_k + \frac{2c_s k}{a}\dot{\delta}_k + \frac{c_s^2 k^2}{a^2}\delta_k \cong 0, \tag{7.166}$$

which has the two solutions

$$\delta_1 \propto t\,\exp\left(-\frac{c_s k}{a}t\right), \quad \delta_2 \propto \exp\left(-\frac{c_s kt}{a}\right). \tag{7.167}$$

Both these solutions die down for large values of $c_s kt/a$. At any given time t, small scale perturbations (for which $k > (a/c_s t)$) will have very little power in a collisionless system.

(d) The proper distance travelled by a particle in time t can be written as

$$l_{\text{FS}}(t) = a(t)\int_0^t \frac{v(t')}{a(t')}\,dt'. \tag{7.168}$$

(This arises from the fact that $a\,dL = v\,dt$ defines the proper velocity $v(t)$.) During $0 < t < t_{nr}$, the dark matter particles are relativistic and $v \cong 1$; since $a(t) \propto t^{1/2}$, this gives

$$l_{FS}(t) = a \int_0^t \frac{dt'}{a_{nr}} \cdot \left(\frac{t_{nr}}{t'}\right)^{1/2} = a(t)\left[\frac{2t_{nr}^{1/2}t^{1/2}}{a_{nr}}\right] = 2t \propto a^2 \quad \text{(for } t < t_{nr}). \quad (7.169)$$

For $t_{nr} < t < t_{eq}$, $v \propto a^{-1}$ and we obtain

$$l_{FS}(t) = \left[\frac{l_{FS}(t_{nr})}{a_{nr}} + \int_{t_{nr}}^t \frac{dt'}{a(t')} \cdot \frac{a_{nr}}{a(t')}\right] a(t)$$

$$= \left[\frac{2t_{nr}}{a_{nr}} + \frac{2t_{nr}}{a_{nr}} \ln \frac{a}{a_{nr}}\right] a = \frac{2t_{nr}a}{a_{nr}}\left[1 + \ln \frac{a}{a_{nr}}\right] \quad (t_{nr} < t < t_{eq}). \quad (7.170)$$

Finally, for $t > t_{eq}$, $a(t) \propto t^{2/3}$. So

$$l_{FS}(t) = \left[\frac{l_{FS}(t_{eq})}{a_{eq}} + \int_{t_{eq}}^t \frac{a_{nr}}{a_{eq}^2}\left(\frac{t_{eq}}{t'}\right)^{4/3} dt'\right] a(t)$$

$$= \left[\frac{2t_{nr}}{a_{nr}}\left(1 + \ln \frac{a_{eq}}{a_{nr}}\right) + \frac{3t_{nr}}{a_{nr}}\left(1 - \frac{a_{eq}^{1/2}}{a^{1/2}}\right)\right] a(t). \quad (7.171)$$

Thus we find that

$$\frac{l_{FS}(t)}{a(t)} = \begin{cases} (2t_{nr}/a_{nr}^2)a = (2t/a) & (t < t_{nr}) \\ (2t_{nr}/a_{nr})\left[1 + \ln(a/a_{nr})\right] & (t_{nr} < t < t_{eq}), \\ (2t_{nr}/a_{nr})\left[(5/2) + \ln(a_{eq}/a_{nr})\right] & (t_{eq} \ll t) \end{cases} \quad (7.172)$$

(In arriving at the last equation, we have used the fact that, for $t \gg t_{eq}$, the second term inside the square bracket in (7.171) becomes a constant.) We now have to determine the range of wavelengths for which the condition $\lambda(t) \leq l_{FS}(t)$ is satisfied, or, equivalently, the range for which $(\lambda/a) \leq (l_{FS}(t)/a)$. Since (λ/a) is a constant independent of time we only have to consider the evolution of (l_{FS}/a). For $t < t_{nr}$, $(l_{FS}/a) \propto (t/a) \propto a$; during $(t_{nr} < t < t_{eq})$, (l_{FS}/a) grows only logarithmically; for $t > t_{eq}$, (l_{FS}/a) grows still more slowly and l_{FS} saturates at the value

$$\lambda_{FS} \equiv l_{FS}(t_0) = \left(\frac{a_0}{a_{nr}}\right)(2t_{nr})\left(\frac{5}{2} + \ln \frac{a_{eq}}{a_{nr}}\right). \quad (7.173)$$

Since this is the largest value of l_{FS}, all proper wavelengths $\lambda > \lambda_{FS}$ will survive the process of free streaming. To be rigorous, we should evaluate λ_{FS} at some $t = t_{nl}$ at which non-linear effects become important. In practice, this makes very little difference. Numerically, this corresponds to

$$\lambda_{FS} \cong 28 \, \text{Mpc}\left(\frac{m_\nu}{30 \, \text{eV}}\right)^{-1} \quad (7.174)$$

if the dark matter is made of massive neutrinos of mass m_v. This lengthscale contains a mass of

$$M_{FS} \cong 4 \times 10^{15} \, M_\odot \left(\frac{m_v}{30 \, \text{eV}}\right)^{-2}, \qquad (7.175)$$

which corresponds to large clusters. Thus, if dark matter is made of massive neutrinos, it will have very little power at scales with $\lambda < \lambda_{FS}$.

7.9 Power spectra for dark matter

(a) Let $\delta_\lambda(t_i)$ denote the amplitude of the dark matter perturbation corresponding to some wavelength λ at the initial instant t_i. To each λ, we can associate a wavenumber $k \propto \lambda^{-1}$ and a mass $M \propto \lambda^3$; accordingly, we may label the perturbation as $\delta_M(t)$ or $\delta_k(t)$, as well, with the scalings $M \cong \lambda^3$, $k \cong \lambda^{-1}$. We are interested in the value of $\delta_\lambda(t)$ at some $t \gtrsim t_{dec}$.

To begin with, note that the process of free streaming will wipe out the perturbations at all scales smaller than λ_{FS} corresponding to a mass M_{FS}. So we have the first result:

$$\delta_M(t) \cong 0 \qquad (\text{for } M < M_{FS}; \lambda < \lambda_{FS}). \qquad (7.176)$$

Consider next the range of wavelengths $\lambda_{FS} < \lambda < \lambda_{eq}$. These modes enter the Hubble radius in the radiation dominated phase; however, their growth is suppressed in the radiation dominated phase by the rapid expansion of the universe; therefore, they do not grow significantly until $t = t_{eq}$, giving $\delta_\lambda(t_{eq}) \cong \delta_\lambda(t_{enter})$. After matter begins to dominate, the amplitude of these modes grows in proportion to the scale factor a. Thus,

$$\delta_M(t) = \delta_M(t_{enter}) \left(\frac{a}{a_{eq}}\right) \qquad (\text{for } M_{FS} < M < M_{eq}). \qquad (7.177)$$

Consider next the modes with $\lambda_{eq} < \lambda < \lambda_H$ where $\lambda_H \equiv H^{-1}(t)$ is the Hubble radius at the time t when we are studying the spectrum. These modes enter the Hubble radius in the matter dominated phase and grow proportional to a afterwards. So,

$$\delta_M(t) = \delta_M(t_{enter}) \cdot \left(\frac{a}{a_{enter}}\right) \qquad (\text{for } M_{eq} < M < M_H), \qquad (7.178)$$

which may be rewritten as

$$\delta_M(t) = \delta_M(t_{enter}) \left(\frac{a_{eq}}{a_{enter}}\right) \left(\frac{a}{a_{eq}}\right). \qquad (7.179)$$

But notice that, since t_{enter} is fixed by the condition $\lambda a_{enter} \propto t_{enter} \propto \lambda t_{enter}^{2/3}$, we have $t_{enter} \propto \lambda^3$. Further, $(a_{eq}/a_{enter}) = (t_{eq}/t_{enter})^{2/3}$, giving

$$\left(\frac{a_{eq}}{a_{enter}}\right) = \left(\frac{\lambda_{eq}}{\lambda}\right)^2 = \left(\frac{M_{eq}}{M}\right)^{2/3}. \qquad (7.180)$$

Substituting (7.180) in (7.179), we obtain

$$\delta_M(t) = \delta_M(t_{\text{enter}}) \left(\frac{\lambda_{\text{eq}}}{\lambda}\right)^2 \left(\frac{a}{a_{\text{eq}}}\right) = \delta_M(t_{\text{enter}}) \left(\frac{M_{\text{eq}}}{M}\right)^{2/3} \left(\frac{a}{a_{\text{eq}}}\right). \qquad (7.181)$$

Comparing (7.177) and (7.181) we see that the mode which enters the Hubble radius after t_{eq} has its amplitude decreased by a factor $M^{-2/3}$, compared to its original value.

Finally, consider the modes with $\lambda > \lambda_H$ which are still outside the Hubble radius at t and will enter the Hubble radius at some *future* time $t_{\text{enter}} > t$. During the time interval (t, t_{enter}), they will grow by a factor (a_{enter}/a). Thus,

$$\delta_\lambda(t_{\text{enter}}) = \delta_\lambda(t) \left(\frac{a_{\text{enter}}}{a}\right) \qquad (7.182)$$

or

$$\delta_\lambda(t) = \delta_\lambda(t_{\text{enter}}) \left(\frac{a}{a_{\text{enter}}}\right) = \delta_M(t_{\text{enter}}) \left(\frac{M_{\text{eq}}}{M}\right)^{2/3} \left(\frac{a}{a_{\text{eq}}}\right) \quad (\lambda > \lambda_H). \quad (7.183)$$

(The last equality follows from the preceding analysis.) Thus the behaviour of the modes is the same for the cases $\lambda_{\text{eq}} < \lambda < \lambda_H$ and $\lambda_H < \lambda$, that is, for all $\lambda > \lambda_{\text{eq}}$. Combining all these pieces of information, we can state the final result as follows:

$$\delta_\lambda(t) = \begin{cases} 0 & (\lambda < \lambda_{\text{FS}}) \\ \delta_\lambda(t_{\text{enter}})(a/a_{\text{eq}}) & (\lambda_{\text{FS}} < \lambda < \lambda_{\text{eq}}) \\ \delta_\lambda(t_{\text{enter}})(a/a_{\text{eq}})(\lambda_{\text{eq}}/\lambda)^2 & (\lambda_{\text{eq}} < \lambda) \end{cases} \qquad (7.184)$$

or, equivalently,

$$\delta_M(t) = \begin{cases} 0 & (M < M_{\text{FS}}) \\ \delta_M(t_{\text{enter}})(a/a_{\text{eq}}) & (M_{\text{FS}} < M < M_{\text{eq}}) \\ \delta_M(t_{\text{enter}})(a/a_{\text{eq}})(M_{\text{eq}}/M)^{2/3} & (M_{\text{eq}} < M). \end{cases} \qquad (7.185)$$

Thus the amplitude at late times is completely fixed by the amplitude of the modes when they enter the Hubble radius. A more relevent quantity characterizing the density inhomogeneity is $\Delta_k^2 \equiv (k^3 P(k)/2\pi^2)$. This quantity behaves as

$$\Delta_k^2 = \begin{cases} 0 & (\text{for } k_{\text{FS}} < k) \\ \Delta_k^2(t_{\text{enter}})(a/a_{\text{eq}})^2 & (\text{for } k_{\text{eq}} < k < k_{\text{FS}}) \\ \Delta_k^2(t_{\text{enter}})(a/a_{\text{eq}})^2(k/k_{\text{eq}})^4 & (\text{for } k < k_{\text{eq}}). \end{cases} \qquad (7.186)$$

Let us next determine $\Delta_k^2(t_{\text{enter}})$. The initial power spectrum, when the mode was much larger than the Hubble radius, was a power law with $\Delta_k^2 \propto k^3 P(k) \propto k^{n+3}$. This mode was growing as a^2 while it was bigger than the Hubble radius (in the radiation dominated phase). Hence $\Delta_k^2(t_{\text{enter}}) \propto a_{\text{enter}}^4 k^{n+3}$. In the radiation dominated phase, we can relate a_{enter} to λ by noting that $\lambda a_{enter} \propto t_{enter} \propto a_{enter}^2$; so $\lambda \propto a_{enter} \propto k^{-1}$. Therefore,

$$\Delta_k^2(t_{enter}) \propto a_{enter}^4 k^{n+3} \propto k^{n-1}. \tag{7.187}$$

Using this in (7.186) we find that

$$\Delta_k^2 = \begin{cases} 0 & \text{(for } k_{FS} < k) \\ k^{n-1}(a/a_{eq})^2 & \text{(for } k_{eq} < k < k_{FS}) \\ k^{n+3}(a/a_{eq})^2 & \text{(for } k < k_{eq}). \end{cases} \tag{7.188}$$

This is the shape of the power spectrum for $a > a_{eq}$. It retains its initial primordial shape at very large scales ($k < k_{eq}$ or $\lambda > \lambda > \lambda_{eq}$). At smaller scales, its amplitude is reduced by four powers (from k^{n+3} to k^{n-1}). This arises because the small wavelength modes enter the Hubble radius earlier on and their growth is suppressed more severely during the phase $a_{enter} < a < a_{eq}$.

Note that the index $n = 1$ is rather special. In this case, $\Delta_k^2(t_{enter})$ is independent of k and all the scales enter the Hubble radius with the same amplitude. The above analysis suggests that if $n = 1$, then all scales in the range $k_{eq} < k < k_{FS}$ will have the same power. In reality, this is not true since the modes do grow by a small logarithmic factor during the radiation dominated phase. Small scales will have slightly more power than the large scales if $k_{FS} \ll k_{eq}$.

The power is exponentially suppressed at scales smaller than the free-streaming scale. It follows that the actual shape of the spectrum depends crucially on the ratio

$$\frac{M_{FS}}{M_{eq}} = 0.05(\Omega h^2)^4 \left(\frac{m}{100\,\text{eV}}\right)^{-4}. \tag{7.189}$$

If neutrinos with $m \cong 30\,\text{eV}$ constitute the dark matter (called 'hot dark matter') then $(M_{FS}/M_{eq}) \cong 4(\Omega h^2)^4$, which is around the range of unity. Thus the spectrum will have a relatively sharp peak around M_{FS}. If, on the other hand, the dark matter particle is heavier (say 1 MeV or so; called 'cold dark matter'), then $M_{FS} \ll M_{eq}$, and the spectrum will be relatively flat between M_{FS} and M_{eq}.

(b) Figures 7.1(a) and (b) show various measures of inhomogeneity for the two spectra in (i) and (ii). The dot–dash line is $\Delta_k = (k^3 P/2\pi^2)^{1/2}$ and is a direct measure of power in the logarithmic interval. The dashed line is $\sigma(R) = <(\delta M/M)^2>^{1/2}$, which gives the root mean square fluctuation in the mass contained in a sphere of radius R. The thick line is the square root of the mean correlation function $\bar{\xi}(R)^{1/2}$.

When $k^3 P(k)$ varies smoothly, all three measures are approximately equal (see figure 7.1(b) for CDM). But when $P(k)$ vanishes exponentially for small scales, there will be a considerable difference between Δ_k and the other two measures of dispersion. From equations (7.84) and (7.88) it is easy to see that $\sigma^2(R)$ and $\bar{\xi}(R)$

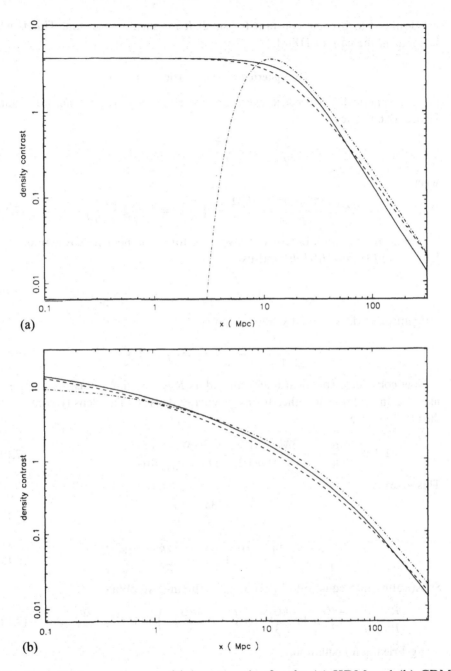

Fig. 7.1. Various measures of inhomogeneity for the (a) HDM and (b) CDM spectra.

are designed to become constant for small R for such a spectrum. This is what happens in the case of HDM.

7.10 Spherically symmetric evolution

(a) For spherically symmetric evolution the shear σ_{ab} and rotation Ω^a vanish. Hence, (7.60) gives

$$\delta'' + \frac{3A}{2b}\delta' = \frac{4}{3}\frac{(\delta')^2}{(1+\delta)} + \frac{3A}{2b^2}\delta(1+\delta),\tag{7.190}$$

where

$$A = \frac{8\pi G\rho_b b^2}{3\dot{b}^2} = \frac{H_0^2\Omega_m a_0^3}{a^3}\left(\frac{b}{\dot{b}}\right)^2 = \left(\frac{\rho_b}{\rho_c}\right)\left(\frac{\dot{a}b}{a\dot{b}}\right)^2.\tag{7.191}$$

To obtain the equation in terms of t, we transform the independent variable from b to t with $(d/db) = (d/\dot{b}\,dt)$ and use

$$\ddot{b} + \frac{2\dot{a}}{a}\dot{b} = 4\pi G\rho_b b, \quad \frac{\ddot{b}}{\dot{b}^2} + \frac{2a'}{a} = 4\pi G\rho_b \frac{b}{\dot{b}^2}.\tag{7.192}$$

Straightforward, but lengthy, algebra gives

$$\ddot{\delta} - \frac{4}{3}\frac{\dot{\delta}^2}{(1+\delta)} + \frac{2\dot{a}}{a}\dot{\delta} = 4\pi G\rho_b\delta\,(1+\delta).\tag{7.193}$$

Now consider a spherical region of radius $R(t)$ containing dust-like matter of mass M in addition to other forms of energy density. The density contrast for dust will be given by

$$1 + \delta = \frac{\rho}{\rho_b} = \frac{3M}{4\pi R^3(t)}\frac{1}{\rho_b(t)} = \frac{2GM}{\Omega_m H_0^2 a_0^3}\left[\frac{a(t)}{R(t)}\right]^3 \equiv \lambda\frac{a^3}{R^3}.\tag{7.194}$$

This gives

$$\frac{\dot{\delta}}{1+\delta} = \frac{3\dot{a}}{a} - \frac{3\dot{R}}{R},\tag{7.195}$$

$$\frac{\ddot{\delta}}{1+\delta} = \frac{3\ddot{a}}{a} - \frac{3\ddot{R}}{R} + \frac{6\dot{a}^2}{a^2} + \frac{12\dot{R}^2}{R^2} - \frac{18\dot{a}}{a}\frac{\dot{R}}{R}.\tag{7.196}$$

Substituting into equation (7.193) and simplifying, we obtain

$$\frac{\ddot{R}}{R} = \frac{\ddot{a}}{a} - \frac{4\pi G}{3}\rho + \frac{4\pi G\rho_b}{3} = \left(\frac{\ddot{a}}{a} + \frac{4\pi G}{3}\rho_b\right) - 4\pi G\rho_b\cdot\frac{\lambda a^3}{R^3}.\tag{7.197}$$

Using Friedmann equations,

$$\frac{\ddot{a}}{a} = -\frac{4\pi G}{3}(\rho_{tot} + 3p_{tot}) = -\frac{4\pi G}{3}\rho_b - \frac{4\pi G}{3}(\rho + 3p)_{rest}.\tag{7.198}$$

this can be written as

$$\ddot{R} = -\frac{4\pi G}{3}(\rho + 3p)_{rest}R - \frac{1}{R^2}\cdot\frac{2GM}{\Omega_m H_0^2 a_0^3}\cdot 4\pi G\rho_0 a_0^3.\tag{7.199}$$

Or, finally,

$$\ddot{R} = -\frac{GM}{R^2} - \frac{4\pi G}{3}(\rho + 3p)_{\text{rest}} R. \tag{7.200}$$

This equation could have been written down 'by inspection' using the relations $\ddot{R} = -\nabla\phi_{\text{tot}}$ and $\phi_{\text{tot}} = \phi_{\text{FRW}} + \delta\phi = -(\ddot{a}/2a)R^2 - G\delta M/R$. Note that this equation is valid for perturbed 'dust-like' matter in *any* background spacetime with density ρ_{rest} and pressure p_{rest} contributed by the rest of the matter.

For reference, we also give the equation using the time coordinate

$$\tau = k\int \frac{dt}{a^2}, \qquad k^2 = 4\pi G\rho_b a^3 = \frac{3}{2}a_0^3\Omega_m H_0^2, \tag{7.201}$$

for which we obtain

$$\delta_{\tau\tau} - \frac{4}{3}\delta_\tau^2(1+\delta)^{-1} = a\delta(1+\delta). \tag{7.202}$$

This equation has a very simple structure since the linear term in δ_τ is eliminated.

(b) For a $\Omega = 1$, matter dominated universe, we have to solve

$$\frac{d^2 R}{dt^2} = -\frac{GM}{R^2}. \tag{7.203}$$

The first integral of this equation is

$$\frac{1}{2}\left(\frac{dR}{dt}\right)^2 - \frac{GM}{R} = E, \tag{7.204}$$

where E is a constant of integration. At the initial instant t_i, the kinetic and potential energies at $t = t_i$ are

$$K_i \equiv \left(\frac{\dot{R}^2}{2}\right)_{t=t_i} = \frac{H_i^2 r_i^2}{2}. \tag{7.205}$$

$$|U| = \left(\frac{GM}{R}\right)_{t=t_i} = G\frac{4\pi}{3}\rho_b(t_i)r_i^2(1+\delta_i) = \frac{1}{2}H_i^2 r_i^2(1+\delta_i) = K_i(1+\delta_i). \tag{7.206}$$

The total energy of the shell is, therefore,

$$E = K_i - K_i(1+\delta_i) = -K_i\delta_i. \tag{7.207}$$

Since $E < 0$ for $\delta_i > 0$, the overdense region will expand to a maximum radius r_m and then collapse. The maximum radius r_m which such a shell attains can easily be derived. At the instant of maximum expansion, we have $\dot{R} = 0$, giving

$$E = -GM/r_m = -(r_i/r_m)K_i(1+\delta_i). \tag{7.208}$$

Equating this expression for E with the one in (7.207), we obtain

$$\frac{r_m}{r_i} = \frac{(1+\delta_i)}{\delta_i}. \tag{7.209}$$

Clearly, $r_m \gg r_i$ if δ_i is small; shells which are only slightly overdense will expand much further and can take a long time to collapse.

(i) The time evolution of the shell can be found by integrating the equations of motion. The solution to equation (7.204), for $E < 0$, is given in a parametric form by

$$R = A(1 - \cos \theta), \quad t + T = B(\theta - \sin \theta), \quad A^3 = GMB^2, \tag{7.210}$$

where A and B are constants related to each other as shown. The parameter θ increases with increasing t, while R increases to a maximum value before decreasing to zero. The constant T allows us to set the initial condition that at $t = t_i$, $R = r_i$. A shell enclosing mass M and initially expanding with the background universe will slow down progressively, reach a maximum radius at $\theta = \pi$, 'turn-around' and collapse. The epoch of maximum radius is also referred to as the epoch of 'turn-around'. At the 'turn-around', $dR/dt = 0$ and $R = r_m$.

The constants A and B can be determined by using (7.209). At $\theta = \pi$, $R(\pi) = r_m = 2A$; comparing with (7.209), we obtain

$$A = \frac{r_i}{2} \left(\frac{1 + \delta_i}{\delta_i} \right). \tag{7.211}$$

Using $A^3 = GMB^2$, $M = (4\pi/3) r_i^3 \rho_b(t_i)(1 + \delta_i)$ we find B to be

$$B = \frac{1}{2H_i} \frac{(1 + \delta_i)}{\delta_i^{3/2}}. \tag{7.212}$$

The value of T can be fixed by setting $R = r_i$ at $t = t_i$. At $t = t_i$ we have to satisfy the conditions

$$r_i = \frac{r_i}{2} \left(\frac{1 + \delta_i}{\delta_i} \right) (1 - \cos \theta_i), \tag{7.213}$$

$$t_i + T = \frac{1}{2H_i} \left(\frac{1 + \delta_i}{\delta_i^{3/2}} \right) (\theta_i - \sin \theta_i). \tag{7.214}$$

From (7.213), we obtain $\cos \theta_i = (1 - \delta_i)(1 + \delta_i)^{-1}$. Since δ_i is expected to be quite small, we can approximate this relation as $\cos \theta_i \cong 1 - 2\delta_i$, obtaining $\theta_i^2 = 4\delta_i$. Substituting in (7.214), we obtain

$$H_i(t_i + T) = \frac{2}{3}(1 + \delta_i). \tag{7.215}$$

Or, since $H_i t_i = (2/3)$ for the $\Omega = 1$ universe, $H_i T = (2/3)\delta_i$. This shows that $(T/t_i) = \delta_i \ll 1$. Hence, we will ignore T in what follows.

Equation (7.210), with the constants A and B fixed by (7.211) and (7.212), gives the complete information about how each perturbed mass shell evolves. These equations can be used to work out all the characteristics of a spherical perturbation.

Consider, for example, the evolution of density within the shell. The density within the shell is

$$\rho(t) = \frac{3M}{4\pi R^3} = \frac{3M}{4\pi A^3 (1 - \cos \theta)^3}, \tag{7.216}$$

while the background density of the universe is

$$\rho_b(t) = \frac{1}{6\pi G t^2}. \tag{7.217}$$

Hence the density contrast is

$$\frac{\rho(t)}{\rho_b(t)} = 1 + \delta(t) = \frac{3M}{4\pi A^3} \frac{6\pi G B^2 (\theta - \sin\theta)^2}{(1 - \cos\theta)^2}, \tag{7.218}$$

where we have used the relation between t and θ given in equation (7.210) and set $T = 0$. Since $A^3 = GMB^2$ it follows that

$$\delta = \frac{9}{2} \frac{(\theta - \sin\theta)^2}{(1 - \cos\theta)^3} - 1. \tag{7.219}$$

The linear evolution for the average density contrast $\delta \propto t^{2/3}$ is recovered in the limit of small t. We find that, to the leading order,

$$\delta = \frac{3}{5}\delta_i \left(\frac{t}{t_i}\right)^{2/3} \propto a(t). \tag{7.220}$$

This is the correct growth law ($\delta \propto t^{2/3}$) for the purely growing mode in the linear regime if the initial peculiar velocity is zero. Assuming that δ_i is small compared to unity, and retaining only the leading terms of δ_i in A and B we can write

$$A \cong \frac{r_i}{2\delta_i}, \quad B \cong \frac{3t_i}{4\delta_i^{3/2}}. \tag{7.221}$$

For further discussion, it is convenient to use two other variables x and δ_0 in place of r_i and δ_i. The quantity x is the comoving radius: $x = r_i[a(t_0)/a(t_i)]$ corresponding to r_i; the parameter δ_0 is defined as: $\delta_0 = (a(t_0)/a(t_i))(3\delta_i/5) = (3/5)\delta_i(1 + z_i)$. This is the present value of the density contrast, as predicted by the linear theory, if the density contrast was δ_i at the redshift z_i. In terms of x and δ_0, we have

$$A = \frac{3x}{10\delta_0}, \quad B = \left(\frac{3}{5}\right)^{3/2} \frac{3t_0}{4\delta_0^{3/2}}. \tag{7.222}$$

Collecting all our results, the evolution of a spherical overdense region can be summarized by the following equations:

$$R(t) = \frac{r_i}{2\delta_i}(1 - \cos\theta) = \frac{3x}{10\delta_0}(1 - \cos\theta), \tag{7.223}$$

$$t = \frac{3t_i}{4\delta_i^{3/2}}(\theta - \sin\theta) = \left(\frac{3}{5}\right)^{3/2} \frac{3t_0}{4\delta_0^{3/2}}(\theta - \sin\theta), \tag{7.224}$$

$$\rho(t) = \rho_b(t)\frac{9(\theta - \sin\theta)^2}{2(1 - \cos\theta)^3}. \tag{7.225}$$

The density can be expressed in terms of the redshift by using the relation $(t/t_i)^{2/3} = (1 + z_i)(1 + z)^{-1}$. This gives

$$(1 + z) = \left(\frac{4}{3}\right)^{2/3} \frac{\delta_i(1 + z_i)}{(\theta - \sin \theta)^{2/3}} = \left(\frac{5}{3}\right)\left(\frac{4}{3}\right)^{2/3} \frac{\delta_0}{(\theta - \sin \theta)^{2/3}}, \quad \cdot \quad (7.226)$$

$$\delta = \frac{9}{2} \frac{(\theta - \sin \theta)^2}{(1 - \cos \theta)^3} - 1. \quad (7.227)$$

Given an initial density contrast δ_i at redshift z_i, these equations define (implicitly) the function $\delta(z)$ for $z > z_i$. Equation (7.226) defines θ in terms of z (implicitly); equation (7.227) gives the density contrast at that $\theta(z)$. For comparison note that linear evolution gives the density contrast δ_L, where

$$\delta_L = \frac{\bar{\rho}_L}{\rho_b} - 1 = \frac{3}{5} \frac{\delta_i(1 + z_i)}{1 + z} = \frac{3}{5} \left(\frac{3}{4}\right)^{2/3} (\theta - \sin \theta)^{2/3}. \quad (7.228)$$

(ii) We can estimate the accuracy of the linear theory by comparing $\delta(z)$ and $\delta_L(z)$. To begin with, for $z \gg 1$, we have $\theta \ll 1$ and we obtain $\delta(z) \cong \delta_L(z)$. When $\theta = (\pi/2)$, $\delta_L = (3/5)(3/4)^{2/3}(\pi/2 - 1)^{2/3} = 0.341$, while $\delta = (9/2)(\pi/2 - 1)^2 - 1 = 0.466$; thus the actual density contrast is about 40% higher. When $\theta = (2\pi/3)$, $\delta_L = 0.568$ and $\delta = 1.01 \cong 1$. If we interpret $\delta = 1$ as the transition point to non-linearity, then such a transition occurs at $\theta = (2\pi/3)$, $\delta_L \cong 0.57$. From (7.226), we see that this occurs at the redshift $(1 + z_{nl}) = 1.06\delta_i(1 + z_i) = (\delta_0/0.57)$.

(iii) The spherical region reaches the maximum radius of expansion at $\theta = \pi$. From our equations, we find that the redshift z_m, the proper radius of the shell r_m and the average density contrast δ_m at 'turn-around' are

$$(1 + z_m) = \frac{\delta_i(1 + z_i)}{\pi^{2/3}(3/4)^{2/3}} = 0.57(1 + z_i)\delta_i = \frac{5}{3} \frac{\delta_0}{(3\pi/4)^{2/3}} \cong \frac{\delta_0}{1.062},$$

$$r_m = \frac{3x}{5\delta_0}, \quad (7.229)$$

$$\left(\frac{\bar{\rho}}{\rho_b}\right)_m = 1 + \bar{\delta}_m = \frac{9\pi^2}{16} \cong 5.6.$$

The first equation gives the redshift at turn-around for a region, parametrized by the (hypothetical) linear density contrast δ_0 at the present epoch. If, for example, $\delta_i \cong 10^{-3}$ at $z_i \cong 10^4$, such a perturbation would have turned around at $(1 + z_m) \cong 5.7$ or when $z_m \cong 4.7$ The second equation gives the maximum radius reached by the perturbation. The third equation shows that the region under consideration is nearly six times denser than the background universe, at turn-around. This corresponds to a density contrast of $\delta_m \cong 4.6$, which is definitely in the non-linear regime. The linear evolution gives $\delta_L = 1.063$ at $\theta = \pi$.

(c) After the spherical overdense region turns around it will continue to contract. Equation (7.225) suggests that at $\theta = 2\pi$ all the mass will collapse to a point. However, long before this happens, the approximation that matter is distributed in spherical shells and that the random velocities of the particles are small will break down. The collisionless component of density, namely, the dark matter, will reach virial equilibrium by violent relaxation. (See problem 2.13.) The collisionless (dark matter) component will relax to a configuration with radius r_{vir}, velocity dispersion v and density ρ_{coll}. After virialization of the collapsed shell, the potential energy U and the kinetic energy K will be related by $|U| = 2K$ so that the total energy $\mathscr{E} = U + K = -K$. At $t = t_m$ all the energy was in the form of potential energy. For a spherically symmetric system with constant density, $\mathscr{E} \cong -3GM^2/5r_m$. The 'virial velocity' v and the 'virial radius' r_{vir} for the collapsing mass can be estimated by the equations

$$K \equiv \frac{Mv^2}{2} = -\mathscr{E} = \frac{3GM^2}{5r_m}, \quad |U| = \frac{3GM^2}{5r_{vir}} = 2K = Mv^2. \quad (7.230)$$

We obtain

$$v = (6GM/5r_m)^{1/2}, \quad r_{vir} = r_m/2. \quad (7.231)$$

(i) The time taken for the fluctuation to reach virial equilibrium, t_{coll}, is essentially the time corresponding to $\theta = 2\pi$. From equation (7.226), we find that the redshift at collapse, z_{coll}, is

$$(1 + z_{coll}) = \frac{\delta_i(1 + z_i)}{(2\pi)^{2/3}(3/4)^{2/3}} = 0.36\delta_i(1 + z_i) = 0.63(1 + z_m) = \frac{\delta_0}{1.686}. \quad (7.232)$$

The density of the collapsed object can also be determined fairly easily. Since $r_{vir} = (r_m/2)$, the mean density of the collapsed object is $\rho_{coll} = 8\rho_m$ where ρ_m is the density of the object at turn-around.

(ii) We have $\rho_m \cong 5.6\rho_b(t_m)$ and $\rho_b(t_m) = (1 + z_m)^3 (1 + z_{coll})^{-3}\rho_b(t_{coll})$. Combining these relations, we obtain

$$\rho_{coll} \cong 2^3\rho_m \cong 44.8\rho_b(t_m) \cong 170\rho_b(t_{coll}) \cong 170\rho_0(1 + z_{coll})^3, \quad (7.233)$$

where ρ_0 is the present cosmological density. This result determines ρ_{coll} in terms of the redshift of formation of a bound object. Once the system has virialized, its density and size does not change. Since $\rho_b \propto a^{-3}$, the density contrast δ increases as a^3 for $t > t_{coll}$.

7.11 Scaling laws for spherical evolution

(a) Given an initial density profile $\rho_i(r)$, we can calculate the mass $M(r_i)$ and energy $E(r_i)$ of each shell labelled by the initial radius r_i. In spherically symmetric evolution, M and E are conserved and each shell will be described by equations (7.203) and (7.204). Assuming that the average density contrast $\bar{\delta}_i(r_i)$ decreases with r_i, the shells will never cross during the evolution. Each shell will evolve in accordance with equations (7.223) and (7.224), with δ_i replaced by the mean initial density contrast $\bar{\delta}_i(r_i)$ characterising the shell of initial radius r_i. Equation

(7.225) gives the mean density inside each of the shells from which the density profile can be computed at any given instant.

(b) If the energy of a shell containing mass M is given by

$$E(M) = E_0 \left(\frac{M}{M_0}\right)^{2/3 - \epsilon} < 0, \qquad (7.234)$$

then the turn-around radius and turn-around time are given by

$$r_m(M) = -\frac{GM}{E(M)} = -\frac{GM_0}{E_0}\left(\frac{M}{M_0}\right)^{\frac{1}{3} + \epsilon}, \qquad (7.235)$$

$$t_m(M) = \frac{\pi}{2}\left(\frac{r_m^3}{2GM}\right)^{1/2} = \frac{\pi GM}{(-E_0/2)^{3/2}}\left(\frac{M}{M_0}\right)^{3\epsilon/2}. \qquad (7.236)$$

To avoid shell crossing, we must have $\epsilon > 0$ so that outer shells with more mass turn around at later times. In such a scenario, the inner shells expand, turn around, collapse and virialize first and the virialization proceeds progressively to the outer shells. We shall assume that each virialized shell settles down to a final radius which is a fixed fraction of the maximum radius. Then the density in the virialized part will scale as (M/r^3), where M is the mass contained inside a shell whose turn-around radius is r. Using (7.235) to relate the turn-around radius and mass, we find that

$$\rho(r) \propto \frac{M(r_m = r)}{r^3} \propto r^{3/(1 + 3\epsilon)} r^{-3} \propto r^{-9\epsilon/(1 + 3\epsilon)}. \qquad (7.237)$$

Two special cases of scaling relations are worth mentioning. If the energy of each shell is dominated by a central mass m located at the origin, then $E \propto Gm/r \propto M^{-1/3}$. In that case, the density profile of the virialized region falls as $r^{-9/4}$. The situation corresponds to an accretion on to a massive object.

If $\epsilon = 2/3$ then the binding energy E is the same for all shells. In that case we obtain $\rho \propto r^{-2}$, which corresponds to an isothermal sphere.

7.12 Self-similar evolution of Vlasov equation

(a) We will first introduce a set of new coordinates by the definitions $T = \ln(t/t_0)$, $\mathbf{q} = \mathbf{x}(t_0/t)^{\alpha}$, $\mathbf{K} = (t_0/t)^{\alpha + \frac{1}{3}}(a^2\dot{\mathbf{x}})$, where α and t_0 are constants. In terms of these coordinates the Vlasov equation (7.43) becomes

$$\frac{\partial f}{\partial T} + (\mathbf{K}t_0 - \alpha\mathbf{q}) \cdot \frac{\partial f}{\partial \mathbf{q}} - \frac{1}{t_0}\nabla V \cdot \frac{\partial f}{\partial \mathbf{K}} - \left(\alpha + \frac{1}{3}\right)\mathbf{K} \cdot \frac{\partial f}{\partial \mathbf{K}} = 0, \quad \nabla_q^2 V = \frac{2}{3}\delta. \quad (7.238)$$

If this equation admits a solution of the form $f(\mathbf{q}, \mathbf{K}, T) = e^{-\mu T}\hat{f}(\mathbf{q}, \mathbf{K})$, then we must have

$$-\mu\hat{f} + (\mathbf{K}t_0 - \alpha\mathbf{q})\frac{\partial\hat{f}}{\partial\mathbf{q}} - \left(\alpha + \frac{1}{3}\right)\mathbf{K} \cdot \frac{\partial\hat{f}}{\partial\mathbf{K}} = \frac{1}{t_0}\nabla V \cdot \frac{\partial\hat{f}}{\partial\mathbf{K}}. \qquad (7.239)$$

The left-hand side is independent of T and hence $V(\mathbf{q}, T)$ must be independent of T. This requires $\delta(\mathbf{q}, T) = \delta(\mathbf{q})$. But δ and f are related by

$$1 + \delta(\mathbf{q}, T) = \frac{1}{\rho_b a^3} \int f(\mathbf{x}, \mathbf{p}, t) d^3\mathbf{p} \propto t^{3\alpha+1} \int d^3\mathbf{K} e^{-\mu T} \hat{f}(\mathbf{q}, \mathbf{K})$$

$$\propto e^{(3\alpha+1-\mu)T} \int d^3\mathbf{K} \hat{f}(\mathbf{q}, \mathbf{K}), \qquad (7.240)$$

which is independent of T if $\mu = (3\alpha + 1)$. Hence, the Vlasov equation admits solutions of the form

$$f(t, \mathbf{x}, \mathbf{p}) = t^{-(3\alpha+1)} \hat{f}\left(\frac{\mathbf{x}}{t^\alpha}, \frac{\mathbf{p}}{t^{\alpha+1/3}}\right). \qquad (7.241)$$

(b) If the solution to $f(t, \mathbf{x}, \mathbf{p})$ is self-similar, then $\delta^2(\mathbf{x}, t)$ must be a function of $(\mathbf{x}/t^\alpha)$ alone. Hence, $\delta^2(\mathbf{x}, t) = \delta^2(\mathbf{x}/t^\alpha) \equiv \delta^2(s)$, where $s = (x/t^\alpha)$. In the linear theory, $k^3 P(k, t) \propto k^{n+3}$ at constant t; and $k^3 P(k, t) \propto a^2 \propto t^{4/3}$ at constant k. Hence, $\sigma_L^2(x, t) \propto k^3 P(k, t) \propto t^{4/3} x^{-[n+3]}$. This can indeed be written in the form $\sigma_L^2 \propto s^{-(n+3)}$ with $\alpha = [4/3(n+3)]$.

In the non-linear regime we expect $\sigma^2 \propto a^3 F[ax]$, where F is an undetermined function. This result can be obtained as follows. Consider any fixed proper radius $r = ax$. In the extreme non-linear limit, structures with fixed r do not participate in the cosmic expansion. The contribution of such structures to $(1 + \delta)^2 \cong \delta^2 = \langle(\rho^2/\rho_b^2)\rangle$ will grow as a^6 when ρ_b decreases as a^{-3}. However, when the average over all space is computed, only a fraction proportional to a^{-3} will contain such structures. Thus the δ^2 due to structures with a fixed $r = ax$ must grow as $(a^6/a^3) = a^3$. Hence, $\sigma_{NL}^2 \propto a^3 F(r) \propto a^3 F(ax)$ in the non-linear regime.

If the linear and non-linear ends should match, we must have $F(ax) \propto (ax)^{-m}$ since the linear theory has a power law behaviour in x. So $\sigma_{NL}^2 \propto t^2 (ax)^{-m} \propto t^{2-2m/3} x^{-m}$. This can be expressed as a function of (x/t^α) only if $\alpha = (2/m) - 2/3$. Equating the expressions for α found in the linear and non-linear ends, it follows that

$$\frac{4}{3(n+3)} = \alpha = \frac{2}{m} - \frac{2}{3}. \qquad (7.242)$$

Or

$$m = \frac{3(3+n)}{5+n}. \qquad (7.243)$$

This shows that, in the non-linear regime,

$$\delta_{\mathbf{k}}^2 \propto t^{2/3(3-m)} k^m \propto a^{\left(\frac{6}{5+n}\right)} k^{\frac{3(3+n)}{5+n}}, \qquad (7.244)$$

with m given by (7.242).

A geometrical description of this result is given in figure 7.2. The diagram shows $\ln \sigma^2(a, x)$ as a function of $\ln(ax)$ at two epochs $a = a_1$ and $a = a_1 e$. In

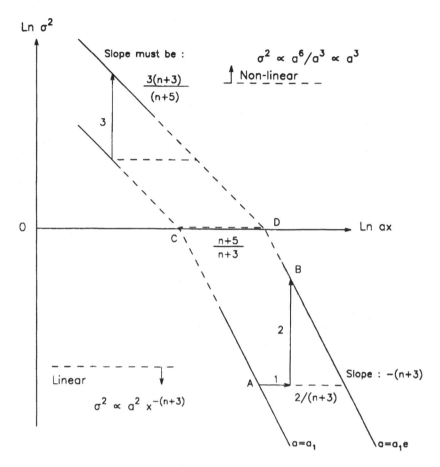

Fig. 7.2. Schematic diagram showing the relation between the slope of σ^2 in the extreme linear and non-linear ends. In the linear end (bottom part of the diagram), the two lines have the slope $[-(n+3)]$. During the linear evolution, the point A moves to B when $a(t)$ increases by a factor e (base of the natural logarithm). Because of the stretching of the scales, A moves to the right by one unit; the growth in the amplitude of σ^2, which is proportional to a^2 in the linear regime, causes the point to move up by two units. Knowing the slope at the linear end, we can determine the length of CD by elementary geometry. In the non-linear end, structures do not expand but the amplitude grows as a^3. This causes each point to move up by three units. Knowing CD and the vertical shift, we can find the slope at the non-linear end.

the linear end, point A moves to point B; the stretching of (ax) due to expansion moves it by one unit to the right and the linear growth $\sigma^2 \propto a^2$ at fixed x moves it up by two units. Since linear slope is $[-(n+3)]$, the line CD must have a length of $(n+5)/(n+3)$ units. In the non-linear end, σ^2 at fixed (ax) merely moves up by three units. This fixes the slope at the non-linear end at $3(n+3)/(n+5)$.

7.13 Non-linear evolution from a scaling ansatz

(a) In the extreme non-linear limit $(\bar{\xi} \gg 1)$, bound structures do not expand with Hubble flow. To maintain a stable structure, the relative pair velocity $v_p(a, x)$ of particles separated by x should balance the Hubble velocity $Hr = \dot{a}x$; hence, $v_p = -\dot{a}x$ or $h(a, x) \cong 1$.

The behaviour of $h(a, x)$ for $\bar{\xi} \ll 1$ is more complicated and can be derived as follows. Let the peculiar velocity field be $v(x)$ (we will suppress the a dependence since we will be working at constant a). The mean relative velocity at a separation $r = (x - y)$ is given by

$$v_{rel}(r) \equiv \langle [v(x) - v(y)] [1 + \delta(x)] [1 + \delta(y)] \rangle$$
$$= \langle [v(x) - v(y)] \delta(x) \rangle + \langle [v(x) - v(y)] \delta(y) \rangle \qquad (7.245)$$

to lowest order, since the δ^2-term is higher order and $\langle v(x) - v(y) \rangle = 0$. Denoting $(v(x) - v(y))$ by v_{xy} and writing $x = y + r$, the radial component of relative velocity is

$$v_{xy} \cdot r = \int v(k) \cdot r \left[e^{ik \cdot (r+y)} - e^{ik \cdot y} \right] \frac{d^3k}{(2\pi)^3}, \qquad (7.246)$$

where $v(k)$ is the Fourier transform of $v(x)$. This quantity is related to δ_k by

$$v(k) = iHa \left(\frac{\delta_k}{k^2} \right) k. \qquad (7.247)$$

Using this in (7.246) and writing $\delta(x), \delta(y)$ in Fourier space we find that

$$v_{xy} \cdot r [\delta(x) + \delta(y)]$$
$$= iHa \int \frac{d^3k}{(2\pi)^3} \int \frac{d^3p}{(2\pi)^3} \left(\frac{k \cdot r}{k^2} \right) \delta_k \delta_p^* e^{i(k-p) \cdot y} \left[e^{ik \cdot r} - 1 \right] \left[e^{-ip \cdot r} + 1 \right]. \qquad (7.248)$$

We average this expression using $\langle \delta_k \delta_p^* \rangle = (2\pi)^3 \delta_D(k - p) P(k)$, to obtain

$$v_{rel} \cdot r \equiv \langle v_{xy} \cdot r [\delta(x) + \delta(y)] \rangle = iHa \int \frac{d^3k}{(2\pi)^3} \frac{P(k)}{k^2} (k \cdot r) \left[e^{ik \cdot r} - e^{-ik \cdot r} \right]$$
$$= -2Ha \int \frac{d^3k}{(2\pi)^3} \frac{P(k)}{k^2} (k \cdot r) \sin(k \cdot r). \qquad (7.249)$$

From the symmetries in the problem, it is clear that $v_{rel}(r)$ is in the direction of r. So $v_{rel} \cdot r = v_{rel} r$. The angular integrations are straightforward and give

$$r v_{rel} = \langle v_{xy} \cdot r [\delta(x) + \delta(y)] \rangle = \frac{Ha}{r\pi^2} \int_0^{\infty} \frac{dk}{k} P(k) [kr \cos kr - \sin kr]. \qquad (7.250)$$

Using expression (7.88) for $\bar{\xi}(r)$ this can be written as

$$r v_{rel}(r) = -\frac{2}{3} (Har^2) \bar{\xi}. \qquad (7.251)$$

Dividing by r and $Hr_{\text{prop}} = Har$, we obtain

$$h = -\frac{v_{\text{rel}}(r)}{Hr_{\text{prop}}} = -\frac{v_{\text{rel}}(r)}{aHr} = \frac{2}{3}\bar{\xi}. \qquad (7.252)$$

(b) To do this we shall first obtain an equation connecting h and $\bar{\xi}$. By solving this equation, one can relate $\bar{\xi}$ and $\bar{\xi}_{\text{L}}$.

The mean number of neighbours within a distance x of any given particle is

$$N(x,t) = (na^3)\int_0^x 4\pi y^2 dy[1 + \xi(y,t)], \qquad (7.253)$$

when n is the comoving number density. Hence, the conservation law for pairs implies

$$\frac{\partial\xi}{\partial t} + \frac{1}{ax^2}\frac{\partial}{\partial x}[x^2(1+\xi)v] = 0, \qquad (7.254)$$

where $v(t,x)$ denotes the mean relative velocity of pairs at separation x and epoch t. Using

$$(1 + \xi) = \frac{1}{3x^2}\frac{\partial}{\partial x}[x^3(1+\bar{\xi})] \qquad (7.255)$$

in (7.254), we obtain

$$\frac{1}{3x^2}\frac{\partial}{\partial x}\left[x^3\frac{\partial}{\partial t}(1+\bar{\xi})\right] = -\frac{1}{ax^2}\frac{\partial}{\partial x}\left[\frac{v}{3}\frac{\partial}{\partial x}[x^2(1+\bar{\xi})]\right]. \qquad (7.256)$$

Integrating, we find

$$x^3\frac{\partial}{\partial t}(1+\bar{\xi}) = -\frac{v}{a}\frac{\partial}{\partial x}[x^3(1+\bar{\xi})]. \qquad (7.257)$$

The integration would allow the addition of an arbitrary function of t on the right-hand side. We have set this function to zero so as to reproduce the correct limiting behaviour (see below). It is now convenient to change the variables from t to a, thereby obtaining an equation for $\bar{\xi}$:

$$a\frac{\partial}{\partial a}[1 + \bar{\xi}(a,x)] = \left(\frac{v}{-\dot{a}x}\right)\frac{1}{x^2}\frac{\partial}{\partial x}[x^3(1+\bar{\xi}(a,x))] \qquad (7.258)$$

or, defining $h(a,x) = -(v/\dot{a}x)$,

$$\left(\frac{\partial}{\partial\ln a} - h\frac{\partial}{\partial\ln x}\right)(1+\bar{\xi}) = 3h(1+\bar{\xi}). \qquad (7.259)$$

This equation shows that the behaviour of $\bar{\xi}(a,x)$ is essentially decided by h, the dimensionless ratio between the mean relative velocity v and the Hubble velocity $\dot{a}x = (\dot{a}/a)x_{\text{prop}}$, both evaluated at scale x.

We shall now assume that

$$h(x,a) = H(\bar{\xi}(x,a)). \qquad (7.260)$$

This assumption, of course, is consistent with the extreme linear limit $h = (2/3)\bar{\xi}$ and the extreme non-linear limit $h = 1$. When $h(x,a) = H[\bar{\xi}(x,a)]$, it is possible to find a solution to (7.260) which reduces to the form $\bar{\xi} \propto a^2$ for $\bar{\xi} \ll 1$ as follows. Let $A = \ln a$, $X = \ln x$ and $D(X, A) = \ln(1 + \bar{\xi})$. We define curves ('characteristics') in the X, A, D space which satisfy

$$\left.\frac{dX}{dA}\right|_c = -H(D[X, A]), \tag{7.261}$$

that is, the tangent to the curve at any point (X, A, D) is constrained by the value of H at that point. Along this curve, the left-hand side of (7.259) is a total derivative allowing us to write

$$\left(\frac{\partial D}{\partial A} - H(D)\frac{\partial D}{\partial X}\right)_c = \left(\frac{\partial D}{\partial A} + \frac{\partial D}{\partial X}\frac{dX}{dA}\right)_c \equiv \left.\frac{dD}{dA}\right|_c = 3H. \tag{7.262}$$

This determines the variation of D along the curve. Integrating:

$$\exp\left(\frac{1}{3}\int\frac{dD}{DH(D)}\right) = \exp(A + c) \propto a. \tag{7.263}$$

Squaring and determining the constant from the initial conditions at a_0, in the linear regime,

$$\exp\left(\frac{2}{3}\int_{\bar{\xi}(a_0, l)}^{\bar{\xi}(x)}\frac{d\bar{\xi}}{H(\bar{\xi})(1 + \bar{\xi})}\right) = \frac{a^2}{a_0^2} = \frac{\bar{\xi}_L(a, l)}{\bar{\xi}_L(a_0, l)}. \tag{7.264}$$

We now need to relate the scales x and l. Equation (7.261) can be written, using equation (7.262), as

$$\frac{dX}{dA} = -H = \frac{1}{3D}\frac{dD}{dA}, \tag{7.265}$$

giving

$$3X + \ln D = \ln[x^3(1 + \bar{\xi})] = \text{constant}. \tag{7.266}$$

Using the initial condition in the linear regime,

$$x^3(1 + \bar{\xi}) = l^3. \tag{7.267}$$

This shows that $\bar{\xi}_L$ should be evaluated at $l = x(1 + \bar{\xi})^{1/3}$. It can be checked directly that (7.267) and (7.264) satisfy (7.259). The final result can therefore be summarized by the following equation (equivalent to (7.264) and (7.267)):

$$\bar{\xi}_L(a, l) = \exp\left(\frac{2}{3}\int^{\bar{\xi}(a,x)}\frac{d\mu}{H(\mu)(1 + \mu)}\right), \quad l = x(1 + \bar{\xi}(a, x))^{1/3}. \tag{7.268}$$

Given the function $H(\bar{\xi})$, this relates $\bar{\xi}_L$ and $\bar{\xi}$. The lower limit of the integral is chosen to give $\ln\bar{\xi}$ for small values of $\bar{\xi}$ in the linear regime. It may be mentioned that equation (7.265) and its integral (7.267) are independent of the ansatz $h(a, x) = H[\bar{\xi}(a, x)]$.

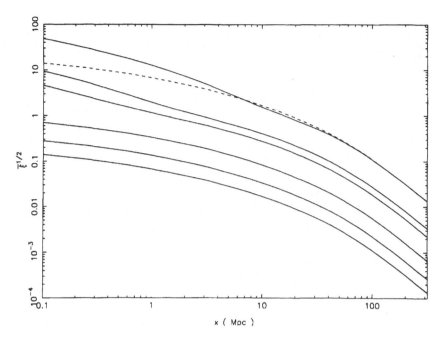

Fig. 7.3. The density contrast of a CDM model at different redshifts. The density contrast is defined as the square root of the mean correlation function. The dashed line is the corresponding quantity from the linear theory.

(c) Figure 7.3 shows that $\bar{\xi}^{1/2}(a, x)$ for CDM at the redshifts of $z = 0, 3, 5, 20, 50$ and 100 (from top to bottom). The broken line gives the $\bar{\xi}_L^{1/2}(a, x)$ at $z = 0$.

(d) The relation between σ_{NL}^2 and σ_L^2 can be approximated as the same relation between $\bar{\xi}_{NL}$ and $\bar{\xi}_L$. Then, in the quasilinear and non-linear regimes,

$$\sigma_{NL}^2(a, x) = \begin{cases} 0.7[\sigma_L^2(a, l)]^3 & \text{(for } 1.2 < \sigma_{NL}^2 < 195) \\ 11.7[\sigma_L^2(a, l)]^{3/2} & \text{(for } 195 < \sigma_{NL}^2) \end{cases}, \tag{7.269}$$

with $l = x(1 + \sigma_{NL}^2)^{1/3} \cong x\sigma_{NL}^{2/3}(a, x)$. The first relation gives, for a power law spectra,

$$\sigma_{NL}^2(a, x) = 0.7\left[a^2 l^{-(n+3)}\right]^3 = 0.7a^6 x^{-3(n+3)}\sigma_{NL}^{-2(n+3)} \tag{7.270}$$

or

$$\sigma_{NL}(a, x) = (0.7)^{\frac{1}{2(n+4)}} a^{\frac{3}{n+4}} x^{-\frac{3}{2}\left(\frac{n+3}{n+4}\right)}. \tag{7.271}$$

Similarly, in the non-linear regime we obtain

$$\sigma_{NL}(a, x) = (11.7)^{\frac{1}{n+5}} a^{\frac{3}{n+5}} x^{-\frac{3}{2}\left(\frac{n+3}{n+5}\right)}. \tag{7.272}$$

(The a- and x-dependence in the non-linear regime are the same as those obtained in problem 7.12; but the numerical coefficient allows us to make actual estimates.)

Equation (7.271) shows that $n = -1$ is special. In this case,

$$\sigma_{\text{NL}}^2(a, x) = (0.7)^{1/3} a^2 x^{-2} \cong \sigma_{\text{L}}^2(a, x). \qquad (7.273)$$

Thus linear evolution is valid all the way up to $\sigma_{\text{NL}}^2 \cong 200$.

The results for power law spectra can also be expressed in another way. Since there is no scale in the problem we can express $\sigma_{\text{NL}}^2(a, x)$ as a function of $[x/x_{\text{NL}}(a)]$, where $x_{\text{NL}}(a)$ is the scale at which $\sigma_{\text{L}}^2 = 1$. This gives $x_{\text{NL}} = a^{2/n+3}$ or $a = x_{\text{NL}}^{(n+3)/2}$. Equations (7.271) and (7.272) become

$$\sigma_{\text{NL}}(a, x) = \begin{cases} (0.7)^{\frac{1}{2(n+3)}} \left(x/x_{\text{NL}}(a) \right)^{-\frac{3}{2}\left(\frac{n+3}{n+4}\right)} & (\text{for } 1 \lesssim \sigma_{\text{NL}}^2 < 195) \\ (11.7)^{\frac{1}{n+3}} \left(x/x_{\text{NL}}(a) \right)^{-\frac{3}{2}\left(\frac{n+3}{n+5}\right)} & (\text{for } 195 < \sigma_{\text{NL}}^2) \end{cases}. \qquad (7.274)$$

7.14 Zeldovich approximation and pancakes

(a) In the linear limit, the velocities of the particles are given by $w^i = u^i(\mathbf{q}) = \partial^i \psi(\mathbf{q})$. The Zeldovich approximation consists of assuming that each particle moves with this constant initial velocity, so that its position at any later time is given by

$$\mathbf{x}(b) = \mathbf{q} + b\mathbf{u}(\mathbf{q}) = \mathbf{q} + b[\nabla \psi(\mathbf{q})]_i. \qquad (7.275)$$

This ansatz satisfies (7.68) by making both sides of it vanish identically. The right-hand side is zero by construction, leading to the equation

$$\frac{\partial \Phi}{\partial b} + \frac{1}{2}(\nabla \Phi)^2 = 0. \qquad (7.276)$$

This is the Hamilton–Jacobi equation for the 'action' Φ for a free particle. From the theory of partial differential equations, we know that the most general solution to this equation is of the form $\Phi = \mathbf{u} \cdot \mathbf{x} - (1/2)u^2 b + A(\mathbf{u})$, where A is an arbitrary function and $\mathbf{u}$ is a function of $\mathbf{x}$ and b determined by the condition $0 = \partial \Phi / \partial \mathbf{u} = \mathbf{x} - \mathbf{u}b + (\partial A/\partial \mathbf{u})$. Setting $(\partial A/\partial \mathbf{u}) = \mathbf{q}$ and treating $\mathbf{u}$ as an arbitrary function of $\mathbf{q}$, we find $\mathbf{x} = \mathbf{q} + \mathbf{u}(\mathbf{q})b$, which is the same as (7.275). The ansatz satisfies (7.69) only approximately.

(b) In the Zeldovich approximation, the proper Eulerian position $\mathbf{r}$ of a particle is related to its Lagrangian position $\mathbf{q}$ by

$$\mathbf{r}(t) \equiv a(t)\,\mathbf{x}(t) = a(t)[\mathbf{q} + b(t)\mathbf{p}(\mathbf{q})], \qquad (7.277)$$

where $\mathbf{x}(t)$ is the comoving Eulerian coordinate. This equation gives the comoving position $(\mathbf{x})$ and proper position $(\mathbf{r})$ of a particle at time t, given that at some time in the past it had the comoving position $\mathbf{q}$. If the initial, unperturbed density is $\bar{\rho}$ (which is independent of $\mathbf{q}$), then the conservation of mass implies that the perturbed density will be

$$\rho(\mathbf{r}, t)\, d^3\mathbf{r} = \bar{\rho}\, d^3\mathbf{q}. \qquad (7.278)$$

Therefore,

$$\rho(\mathbf{r}, t) = \bar{\rho}\det\left(\frac{\partial q_i}{\partial r_j}\right) = \frac{\bar{\rho}/a^3}{\det(\partial x_j/\partial q_i)} = \frac{\rho_b(t)}{\det(\delta_{ij} + b(t)(\partial p_j/\partial q_i))}, \tag{7.279}$$

where we have set $\rho_b(t) = [\bar{\rho}/a^3(t)]$. Since $\mathbf{p}(\mathbf{q})$ is a gradient of a scalar function, the Jacobian in (7.279) is a real symmetric matrix. This matrix can be diagonalized at every point $\mathbf{q}$, to yield a set of eigenvalues and principal axes as a function of $\mathbf{q}$. If the eigenvalues of $(\partial p_j/\partial q_i)$ are $[-\lambda_1(\mathbf{q}), -\lambda_2(\mathbf{q}), -\lambda_3(\mathbf{q})]$ then the perturbed density is given by

$$\rho(\mathbf{r}, t) = \frac{\rho_b(t)}{(1 - b(t)\lambda_1(\mathbf{q}))(1 - b(t)\lambda_2(\mathbf{q}))(1 - b(t)\lambda_3(\mathbf{q}))}, \tag{7.280}$$

where $\mathbf{q}$ can be expressed as a function of $\mathbf{r}$ by solving (7.277). This expression describes the effect of deformation of an infinitesimal, cubical volume (with the faces of the cube determined by the eigenvectors corresponding to λ_n) and the consequent change in the density.

For a growing perturbation, $b(t)$ increases with time; therefore, a positive λ denotes collapse and a negative λ signals expansion. In an overdense region the density will become infinite if one of the terms in parentheses in the denominator of (7.280) becomes zero. In the generic case, these eigenvalues will be different from each other; let $\lambda_1 \geq \lambda_2 \geq \lambda_3$. At any particular value of $\mathbf{q}$, one of them, say λ_1, will be maximum. Then the density will diverge for the first time when $(1 - b(t)\lambda_1) = 0$; at this instant the material contained in a cube in the $\mathbf{q}$-space becomes compressed to a sheet in the $\mathbf{r}$-space, along the principal axis corresponding to λ_1. Thus sheetlike structures, or 'pancakes', will be the first non-linear structures to form when gravitational instability amplifies density perturbations.

(c) If the density inhomogeneity is spherically symmetric, then the three eigenvalues $(\lambda_1, \lambda_2, \lambda_3)$ will be equal. Hence, we will have

$$\rho(x, t) = \frac{\rho_b(t)}{[1 - b(t)\lambda(q)]^3}. \tag{7.281}$$

In the linear limit, this gives

$$\delta_L(x, t) = \frac{\rho}{\rho_b} - 1 \cong 3b(t)\lambda(q). \tag{7.282}$$

But we know that, in the linear theory, $\delta_L(x, t) = b(t)f(x)$, where $f(x)$ denotes the initial density contrast at some epoch. Hence, $b(t)\lambda(q) = b(t)f(x)/3$; using this in (7.281), we obtain

$$\rho(x, t) = \rho_b(t)\left[1 - \frac{1}{3}b(t)f(x)\right]^{-3} = \rho_b(t)\left[1 - \frac{1}{3}\delta_L(x, t)\right]^{-3}. \tag{7.283}$$

This shows that $\rho \to \infty$ as $\delta_L \to 3$; we saw earlier that, in the exact theory $\rho \to \infty$ as $\delta_L \to 1.68$. Figure 7.4 compares Zeldovich and spherical top-hat approximations.

(d) In one dimension, equations (7.68) and (7.69) reduce to

$$\frac{\partial \Phi}{\partial b} + \frac{1}{2}\left(\frac{\partial \Phi}{\partial x}\right)^2 = -\frac{3A}{2b}(\Phi + \psi), \qquad (7.284)$$

$$\frac{\partial}{\partial x}\left[\frac{\partial}{\partial b}\left(b\frac{\partial \psi}{\partial x}\right) + \frac{\partial \Phi}{\partial x} + b\left(\frac{\partial^2 \psi}{\partial x^2}\right)\left(\frac{\partial \Phi}{\partial x}\right)\right] = 0. \qquad (7.285)$$

In Zeldovich approximation, we set $\psi = -\Phi$ and choose Φ to be the solution of

$$\frac{\partial \Phi}{\partial b} + \frac{1}{2}\left(\frac{\partial \Phi}{\partial x}\right)^2 = 0. \qquad (7.286)$$

Hence, (7.284) is satisfied trivially. Using $\psi = -\Phi$, (7.285) becomes

$$\frac{\partial}{\partial x}\left[-b\frac{\partial^2 \Phi}{\partial b \partial x} - b\frac{\partial^2 \Phi}{\partial x^2}\frac{\partial \Phi}{\partial x}\right] = 0. \qquad (7.287)$$

This can be rewritten as

$$-b\frac{\partial^2}{\partial x^2}\left[\frac{\partial \Phi}{\partial b} + \frac{1}{2}\left(\frac{\partial \Phi}{\partial x}\right)^2\right] = 0, \qquad (7.288)$$

which is identically satisfied due to (7.286). Therefore, Zeldovich 'approximation' is exact in one dimension.

7.15 Accuracy of Zeldovich approximation

The accuracy of Zeldovich approximation can be estimated as follows. From the relation $x_\alpha = a(t)\left[q_\alpha - b(t)p_\alpha(\mathbf{q})\right]$, where $p_\alpha(\mathbf{q}) = (\partial \Phi/\partial q_\alpha)$, it follows that $\dot{x}_\alpha = (\dot{a}/a)x_\alpha - ab\dot{p}_\alpha$. Differentiating this relation once again we can rewrite it in the form

$$\ddot{\mathbf{x}} = \frac{\ddot{a}}{a}\mathbf{x} - (2\dot{a}\dot{b} + a\ddot{b})\mathbf{p}. \qquad (7.289)$$

We next need to calculate the divergence of this expression with respect to $\mathbf{x}$. The first term gives $3(\ddot{a}/a)$; to evaluate $\nabla_x \cdot \mathbf{p}$ proceed as follows. We note that

$$\nabla_x \cdot \mathbf{p} = \frac{\partial p_\alpha}{\partial x^\beta} = \frac{\partial p_\alpha}{\partial q^\beta}\cdot\frac{\partial q^\beta}{\partial x^\alpha} = \left(\frac{\partial^2 \Phi}{\partial q^\alpha \partial q^\beta}\right)\left[\frac{\partial x^\alpha}{\partial q^\beta}\right]^{-1} = d_{\alpha\beta}\left[a(\delta_{\alpha\beta} - b\,d_{\alpha\beta})\right]^{-1}, \quad (7.290)$$

where we have defined a matrix $d_{\alpha\beta} \equiv (\partial^2 \Phi/\partial q^\alpha \partial q^\beta)$. In the locally diagonal system $d_{ik} = \text{dia}(\lambda_1, \lambda_2, \lambda_3)$ and $[a(\delta_{ik} - bd_{ik})]^{-1} = a^{-1}\text{dia}\left[(1-b\lambda_1)^{-1}, (1-b\lambda_2)^{-1},\right.$

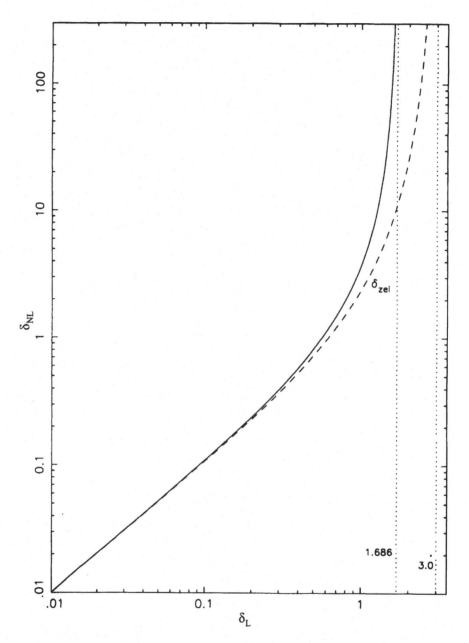

Fig. 7.4. The non-linear density contrast is plotted against the linear density contrast in the spherical top-hat model and spherical Zeldovich model.

$(1 - b\lambda_3)^{-1}]$. In this frame we have

$$
\begin{aligned}
\nabla_x \cdot \mathbf{p} &= \frac{1}{a} \left[\frac{\lambda_1}{1 - b\lambda_1} + \frac{\lambda_2}{1 - b\lambda_2} + \frac{\lambda_3}{1 - b\lambda_3} \right] \\
&= \frac{1}{a} \frac{(I_1 - 2bI_2 + 3b^2 I_3)}{(1 - bI_1 + b^2 I_2 - b^3 I_3)},
\end{aligned} \tag{7.291}
$$

where I_1, I_2 and I_3 are the invariants of $d_{\alpha\beta}$. Using $4\pi G\rho = 4\pi G\rho_0 a^{-3} (1 - bI_1 + b^2 I_2 - b^3 I_3)^{-1} \equiv 4\pi G\rho_0 a^{-3} h^{-1}$, we can compute $Q = \nabla \cdot \ddot{\mathbf{x}} + 4\pi G\rho$. We obtain

$$
hQ = 3\frac{\ddot{a}}{a}(1 - bI_1 + b^2 I_2 - b^3 I_3) - (2\frac{\dot{a}}{a}\dot{b} + \ddot{b})(I_1 - 2bI_2 + 3b^2 I_3) + \frac{4\pi G\rho_0}{a^3}, \tag{7.292}
$$

which can be rewritten as

$$
\begin{aligned}
hQ = 3 \left(\frac{\ddot{a}}{a} + \frac{4\pi G\rho_0}{3a^3} \right) - \left(\ddot{b} + 2\frac{\dot{a}}{a}\dot{b} + 3\frac{\ddot{a}}{a}b \right) (I_1 - 2bI_2 + 3b^2 I_3) \\
+ 3\frac{\ddot{a}}{a} \left[-b^2 I_2 + 2b^3 I_3 \right],
\end{aligned} \tag{7.293}
$$

where $h = (1 - b\lambda_1)(1 - b\lambda_2)(1 - b\lambda_3)$. The first term vanishes identically due to the Friedmann equations. The second term vanishes when $b(t)$ satisfies the linear perturbation equation. Hence, the fractional accuracy of Zeldovich approximation is of the order of

$$
\left(\frac{g}{h} \right) \cong \frac{b^2 I_2 - 2b^3 I_3}{(1 - b\lambda_1)(1 - b\lambda_2)(1 - b\lambda_3)}. \tag{7.294}
$$

8

Structure formation

8.1 Perturbations in baryons

(a) The analysis for baryonic perturbations proceeds in a manner similar to that in problem 7.3. In the proper coordinates $(t, \mathbf{r})$ with $\mathbf{r} = a(t)\mathbf{x}$, the equations describing the baryonic fluid can be written as

$$\frac{\partial \rho}{\partial t} + \nabla_{\mathbf{r}} \cdot (\rho \mathbf{U}) = 0$$

$$\frac{\partial \mathbf{U}}{\partial t} + (\mathbf{U} \cdot \nabla)\mathbf{U} = -\nabla \Psi_{\text{tot}} - \frac{1}{\rho} \nabla p \qquad (8.1)$$

$$\nabla^2 \Psi_{\text{tot}} = 4\pi G \rho_{\text{tot}}.$$

Comparing these equations with the basic equations in problem 7.3(b), we see that the *only* difference is the additional term $(\nabla p / \rho)$ in Euler's equation which is due to the pressure support. We now introduce the perturbed potential ϕ, perturbed density contrast of baryons δ_B, perturbed pressure δp and the perturbed velocity $\mathbf{v}$, using the equations

$$\phi = \Psi - \phi_{\text{FRW}}, \qquad \mathbf{v} = \mathbf{U} - H\mathbf{r},$$

$$\delta_B = \frac{(\rho - \bar{\rho}_B)}{\bar{\rho}_B}, \qquad \delta p = c_s^2 (\delta \rho). \qquad (8.2)$$

The perturbed potential and velocity are defined by subtracting out the corresponding quantities for the homogeneous Friedmann model. The perturbed density contrast is defined using the background density of *baryons*; the perturbed pressure is related to the perturbed density through c_s^2. When photons and baryons are tightly coupled, we would expect $c_s^2 \cong (\partial p_{\text{rad}} / \partial \rho_{\text{rad}}) = (1/3)$. After the matter has decoupled from radiation, c_s^2 will be equal to the adiabatic sound speed in the gas.

We now linearize the equations in the perturbed variables and change from proper to comoving coordinates using the relations derived in problem 7.3. The analysis proceeds exactly as in that problem and we obtain

$$\ddot{\delta}_k^{(B)} + 2\frac{\dot{a}}{a}\dot{\delta}_k^{(B)} + \frac{k^2 c_s^2}{a^2}\delta_k^{(B)} = 4\pi G \bar{\rho}_B \delta_k^{(B)} + 4\pi G \bar{\rho}_{\text{DM}} \delta_k^{\text{DM}}. \qquad (8.3)$$

This differs from the equation for dark matter perturbations in two respects:
(i) the term $(k_s^2 c_s^2/a^2)$ arises due to the pressure gradient which was not present
for the dark matter; (ii) on the right-hand side we have now retained terms which
arise from both baryonic and dark matter perturbations. When dark matter
perturbations dominate over baryonic perturbations (which is often the case), the
baryonic driving term can be ignored on the right-hand side.

This equation can be rewritten in the form

$$\ddot{\delta}_k^{(B)} + 2\frac{\dot{a}}{a}\dot{\delta}_k^{(B)} + \left(\frac{k^2 c_s^2}{a^2} - 4\pi G\bar{\rho}_B\right)\delta_k^{(B)} = 4\pi G\bar{\rho}_{DM}\delta_k^{DM}, \qquad (8.4)$$

which is similar to that of a forced, damped, harmonic oscillator. The $(\dot{a}/a)$ term
represents the damping due to expansion and the term on the right-hand side
provides the driving force.

In the absence of dark matter, the nature of the solution is essentially determined by the sign of the third term on the left-hand side. If $k^2 > k_J^2$, where

$$k_J^2 = \frac{4\pi G\bar{\rho}_B}{c_s^2}, \qquad (8.5)$$

then the 'frequency' of the oscillator is real and we obtain oscillatory solutions.
The perturbation in the baryonic component oscillates as an acoustic vibration.
On the other hand, if $k^2 < k_J^2$, then the 'frequency' is imaginary and we obtain
growing and decaying amplitudes. In such a case, the perturbations in the
baryonic component can grow.

When both dark matter and baryons are present, then $\bar{\rho}_{DM}\delta_{DM}$ is usually
dominant over $\bar{\rho}_B\delta_B$. The growth in δ_{DM} can induce a corresponding growth in
δ_B at large scales. At small scales, however, the $(k^2 c_s^2/a^2)$ term will reduce this
effect significantly (see problem 8.2(b)).

(b) By definition, the Jeans mass scales as

$$M_J \propto \rho_B \lambda_J^3 \propto \rho_B\left(\frac{c_s^3}{\rho_B^{3/2}}\right) \propto \left(\frac{c_s^3}{\rho_B^{1/2}}\right). \qquad (8.6)$$

During $a_{eq} < a < a_{dec}$, the baryons and photons are tightly coupled and $c_s \cong$
constant. Hence, $M_J \propto \rho_B^{-1/2} \propto a^{3/2}$. For $a > a_{dec}$, $c_s \propto a^{-1}$ since the velocity
dispersion decreases as the inverse of the expansion factor for an adiabatically
expanding gas. Then $M_J \propto (a^{-3}/a^{-3/2}) \propto a^{-3/2}$. After decoupling, the pressure
support provided by the photons vanishes and the baryonic gas can resist gravity
only by normal gas pressure. At $T = T_{dec}$, the radiation pressure is given
by $p_{rad} \cong n_\gamma k T_{dec}$, while the gas pressure is given by $p_{gas} \cong n_B k T_{dec}$. Since
$n_B \cong 10^{-8} n_\gamma$, the pressure drops drastically at $t = t_{dec}$. In other words, the
sound speed c_s (as well as the Jeans length and Jeans mass) drops by orders of
magnitude at $t = t_{dec}$. Putting in numbers, one can easily see that the Jeans mass

before decoupling is

$$M_J^{(1)} = M_J(t < t_{dec}) \cong 10^{16}\, M_\odot \left(\frac{\Omega_B}{\Omega}\right) (\Omega h^2)^{-1/2} \qquad (8.7)$$

and the Jeans mass after decoupling is about

$$M_J^{(2)} = M_{JB}(t \gtrsim t_{dec}) \cong 10^4\, M_\odot \left(\frac{\Omega_B}{\Omega}\right) (\Omega h^2)^{-1/2}. \qquad (8.8)$$

Perturbations carrying masses with $M > M_J^{(1)}$ will always be growing since the pressure support cannot prevent their growth. Perturbations containing masses in the range $M_J^{(2)} < M < M_J^{(1)}$ cannot grow until $a = a_{dec}$ since for these scales the pressure support can withstand gravity. They can, however, grow after decoupling since the Jeans mass would now have dropped to a low value.

8.2 Acoustic oscillations of baryons

(a) When the pressure term dominates over gravity and $c_s^2 = (k_B T / m_p) = (k_B T_0 / m_p)(1/a)$, the baryonic perturbation equation (8.4) becomes

$$\ddot{\delta}_B + 2\frac{\dot{a}}{a}\dot{\delta}_B + \left(\frac{k_B T_0}{m_p}\right)\left(\frac{k^2}{a^3}\right)\delta_B \cong 0. \qquad (8.9)$$

Changing the independent variable from t to $x = (a/a_0)^{1/2}$, we obtain

$$\frac{d^2\delta}{dx^2} + \frac{2}{x}\frac{d\delta}{dx} + \frac{\omega^2\delta}{x^2} = 0, \quad \omega^2 \equiv \frac{4k^2}{H^2 a^3}\left(\frac{k_B T_0}{m_p}\right), \quad H(t) = \frac{\dot{a}}{a} \qquad (8.10)$$

or

$$\frac{1}{x^2}\frac{d}{dx}\left(x^2\frac{d\delta}{dx}\right) + \frac{\omega^2\delta}{x^2} = 0. \qquad (8.11)$$

Note that ω^2 is a constant because $H^2 \propto \rho \propto a^{-3}$. Further, $\omega \gg 1$ since we are considering modes for which $\lambda_{phy} \ll d_H$. This equation has the solution $\delta = x^n$, with $n \cong [(-1/2) \pm i\omega]$ for $\omega \gg 1$. Therefore, using $x \propto a^{1/2} \propto t^{1/3}$, the density contrast can be written as

$$\delta = t^{-1/6} \exp\left(\pm\frac{i\omega}{3}\ln t\right), \qquad (8.12)$$

which decays as $t^{-1/6}$.

This decay law has a simple interpretation. The oscillations in the baryon density can be thought of as acoustic vibrations in a medium. The frequency of these vibrations scales as $v = (v/\lambda) \propto a^{-1/2}a^{-1} \propto a^{-3/2} \propto t^{-1}$, while the energy of the vibrations in a volume V will vary as $E \propto \rho_b(v\delta)^2 V \propto v^2\delta^2 \propto \delta^2 a^{-1} \propto \delta^2 t^{-2/3}$. In the case of $(kv \gg Ha)$ there will be several acoustic oscillations within one expansion time-scale, and, hence, the expansion may be treated as (adiabatically) slow. In such a case, the quantity (E/v), which is an adiabatic invariant, will remain constant. Since $v \propto t^{-1}$ and $E \propto t^{-2/3}\delta^2$, it follows that $t^{1/3}\delta^2 = $ constant or $\delta \propto t^{-1/6}$.

(b) To study the evolution of baryonic perturbations at $a > a_{\text{dec}}$ we shall approximate equation (8.4) as follows. (i) The gravitational force is dominated by the dark matter and we shall ignore $4\pi G \bar{\rho}_B \delta_B$ in comparison with $4\pi G \bar{\rho}_{DM}\delta_{DM}$. (ii) We shall take $c_s^2 = (k_B T/m_p)$ with $T \propto a^{-1}$. In other words, the pressure support is provided by the gas, the temperature of which is maintained as equal to radiation temperature. (iii) The dark matter density contrast grows as $\delta_{DM} = Bt^{2/3} \propto a$.

Under these assumptions, equation (8.4) can be written as

$$\ddot{\delta} + \frac{4}{3t}\dot{\delta} + \left(\frac{k_B T}{m_p}\right)\frac{k^2}{a^3}\delta = \frac{2}{3}\frac{\delta_{DM}}{t^2}, \tag{8.13}$$

where δ stands for the baryonic density contrast corresponding to wave vector k; the subscripts are omitted to simplify the notation. Since the third term on the left-hand side varies as $a^{-3} = (t_0/t)^2$, the left-hand side is homogeneous in t and is driven by δ_{DM}. In that case the solution can be found by substituting the ansatz $\delta = Ct^{2/3}$. This leads to the relation

$$C\left[1 + \frac{3}{2}\left(\frac{k_B T_0}{m_p}\right)t_0^2\right] = B. \tag{8.14}$$

So we obtain

$$\delta_k^{(B)}(t) = \frac{\delta_k^{(DM)}(t)}{1 + Ak^2}, \tag{8.15}$$

where

$$A = \frac{3}{2}\left(\frac{k_B T_0}{m_p}\right)\left(\frac{t^2}{a^3}\right) = 1.5 \times 10^{-6}h^{-2}\text{ Mpc}^2. \tag{8.16}$$

It is clear that at large scales (i.e. for small Ak^2) the dark matter perturbations induce corresponding perturbations in the baryons. For $Ak^2 \gtrsim 1$, the baryonic perturbations are suppressed in amplitude with respect to the dark matter perturbations because of the pressure support.

8.3 Silk damping

The simplest way to estimate the effect of photon diffusion is as follows. At $t \ll t_{\text{dec}}$, photons and baryons are very tightly coupled due to Thomson scattering. The proper length corresponding to the photon mean-free-path at some time t is

$$l(t) = \frac{1}{x_e n_e \sigma} \cong 1.3 \times 10^{29}\text{ cm }x_e^{-1}(1+z)^{-3}(\Omega_B h^2)^{-1}, \tag{8.17}$$

where x_e is the electron ionization fraction. For wavelengths $\lambda \lesssim l$, the photon streaming will clearly damp any perturbation almost instantaneously. But, actually, the damping effect is felt at even larger scales. Photons can slowly diffuse out of overdense to underdense regions, dragging the tightly coupled charged particles. Though not much decay occurs during one oscillation period of the perturbation, significant loss can take place within an expansion time-scale. The characteristic distance that an average photon can diffuse till $t = t_{dec}$ can be computed as follows.

Consider a time interval Δt in which a photon suffers $N = (\Delta t / l(t))$ collisions. Between successive collisions it travels a proper distance $l(t)$, or, equivalently, a coordinate distance $[l(t)/a(t)]$. Because of this random walk, it acquires a mean square coordinate displacement:

$$(\Delta x)^2 = N \left(\frac{l}{a}\right)^2 = \frac{\Delta t}{l(t)} \frac{l^2}{a^2} = \frac{\Delta t}{a^2} l(t). \tag{8.18}$$

The total mean square coordinate distance travelled by a typical photon till the time of decoupling is

$$x^2 \equiv \int_0^{t_{dec}} \frac{dt}{a^2(t)} l(t) = \frac{3}{5} \frac{t_{dec} l(t_{dec})}{a^2(t_{dec})}, \tag{8.19}$$

where we have used the earlier expression for $l(t)$ and the fact that $a \propto t^{2/3}$. This corresponds to the proper distance

$$l_S = a(t_{dec})x = \left[\frac{3}{5} t_{dec} l(t_{dec})\right]^{1/2} \cong 3.5 \, \text{Mpc} \left(\frac{\Omega}{\Omega_B}\right)^{1/2} (\Omega h^2)^{-\frac{3}{4}}. \tag{8.20}$$

If we assume that baryons are tightly coupled to photons before t_{dec}, it follows that baryons will be dragged along with photons. Then all perturbations at wavelengths $\lambda < l_S$ will be wiped out. This length l_S corresponds to the mass

$$M_S \cong 6.2 \times 10^{12} \, M_\odot \left(\frac{\Omega}{\Omega_B}\right)^{3/2} (\Omega h^2)^{-5/4}. \tag{8.21}$$

Using the expression (8.19) for an arbitrary instant of time t, we can easily work out the scaling of various quantities with a. We find that $l_S \propto a^{9/4}$ and $M_S \propto a^{15/4}$ during the period $a_{eq} < a < a_{dec}$. For $a > a_{dec}$ this process, of course, ceases to exist. Thus the Silk mass rises very steeply to its final value at $a = a_{dec}$.

8.4 Overall evolution of baryonic perturbations

During very early epochs, when the mode was outside the Hubble radius, pressure gradients had negligible influence in its dynamics. In that case, all components evolve in the same manner and baryonic perturbations also grow as a^2.

At some time $t = t_{enter}$ in the radiation dominated epoch, a perturbation enters the Hubble radius. At this time, we may assume that baryons and dark matter perturbations have equal amplitudes. Once inside the Hubble radius, processes such as pressure support become operational and prevent the growth of baryonic perturbations. Hence, δ_B oscillates as an acoustic wave with a decaying amplitude during the entire period $a_{enter} < a < a_{eq}$. (Note that dark matter perturbations, unaffected by pressure gradients, grow logarithmically during $a_{enter} < a < a_{eq}$

and grow linearly during $a_{eq} < a < a_{dec}$.) Modes for which $M > M_J \cong 10^{16} M_\odot$ could have grown after they entered the Hubble radius but these modes are not of much cosmological interest.

For $a > a_{dec}$, the baryonic perturbations can grow. During this epoch, their growth is driven by the already existing perturbations in the dark matter component and $\delta_B \to \delta_{DM}$ for large a.

We have also seen that small scale power in the baryonic component will be wiped out during $a < a_{dec}$ because of Silk damping. However, this power can be regenerated for $a > a_{dec}$ due to the driving by the dark matter. Thus, for $a \gtrsim a_{dec}$, baryonic perturbations will closely follow those of dark matter.

8.5 Non-linear collapse of baryons

(a) When we ignore the effects of cooling, the dynamical evolution of baryons is similar to that of dark matter. The baryonic component expands more slowly compared to the background universe, turns around and collapses. During the collapse, the gaseous mixture of hydrogen and helium develops shocks and becomes reheated to a temperature at which pressure balance can prevent further collapse. At this stage the thermal energy will be comparable to the gravitational potential energy. The temperature of the gas, T_{vir}, is related to the velocity dispersion v^2 by $3\rho_{gas} T_{vir}/2\mu = \rho_{gas} v^2/2$, where ρ_{gas} is the gas density and μ is its mean molecular weight. This gives $T_{vir} = \mu v^2/3$. It is useful to express the above results with typical numbers for the various quantities shown explicitly. If the He fraction is Y by weight and the gas is fully ionized, then

$$\mu = \frac{(m_H n_H + m_{He} n_{He})}{(2n_H + 3n_{He})} = \frac{m_H}{2} \left(\frac{1+Y}{1+0.375Y} \right) \cong 0.57 m_H, \qquad (8.22)$$

for $Y = 0.25$. Apart from the cosmological parameters, two other parameters need to be specified to determine the evolution. These may be chosen to be the mass M of the overdense region and the redshift of formation z_{coll}. Using the results of problem 7.9 we can express r_{vir}, v and T_{vir} in terms of z_{coll} [or $\delta_0 = 1.68(1 + z_{coll})$] and the mass M of the structure:

$$r_{vir} = 258 (1 + z_{coll})^{-1} \left(\frac{M}{10^{12} M_\odot} \right)^{1/3} h_{0.5}^{-2/3} \text{ kpc} = 434 \delta_0^{-1} h_{0.5}^{-2/3} M_{12}^{1/3} \text{ kpc},$$

$$v = 100 (1 + z_{coll})^{1/2} \left(\frac{M}{10^{12} M_\odot} \right)^{1/3} h_{0.5}^{1/3} \text{ km s}^{-1} = 77 \delta_0^{1/2} M_{12}^{1/3} h_{0.5}^{1/3} \text{ km s}^{-1},$$

$$T_{vir} = 2.32 \times 10^5 (1 + z_{coll}) \left(\frac{M}{10^{12} M_\odot} \right)^{2/3} h_{0.5}^{2/3} \text{ K} = 1.36 \times 10^5 \delta_0 M_{12}^{2/3} h_{0.5}^{2/3} \text{ K}.$$

$$(8.23)$$

We have used

$$x = 0.92 (\Omega h^2)^{-1/3} (M/10^{12} M_\odot)^{1/3} \text{ Mpc} \qquad (8.24)$$

to relate M and the comoving scale, and have set $\Omega = 1$, $h = 0.5$. Also note that

$$t_{\text{coll}} = t_0 (1 + z_{\text{coll}})^{-3/2}, \quad (1 + z_m) = 1.59 \, (1 + z_{\text{coll}}). \qquad (8.25)$$

The above results can be used to estimate the typical parameters of collapsed objects once we are given M and the collapse redshift. For example, if objects with $M = 10^{12} \, M_\odot$ (which is typical of galaxies) collapse at a redshift of, say, 2, then one obtains $r_{\text{vir}} \cong 86 \, \text{kpc}$, $t_{\text{coll}} \cong 1.2 \times 10^9 \, \text{yr}$, $v \cong 173 \, \text{km s}^{-1}$, $T_{\text{vir}} \cong 7 \times 10^5 \, \text{K}$. The density contrast of the galaxy at present will be $(\rho_{\text{coll}}/\rho_0) \cong 170 \, (1 + z_{\text{coll}})^3 \cong 4.6 \times 10^3$.

(b) Consider a gas cloud of mass M and radius R. The evolution of such a cloud will depend crucially on the relative values of the cooling time-scale:

$$t_{\text{cool}} = \frac{E}{\dot{E}} \cong \frac{3\rho kT}{2\mu \Lambda(T)} = 8 \times 10^6 \, \text{yr} \left(\frac{n}{1 \, \text{cm}^{-3}}\right)^{-1} \left[\left(\frac{T}{10^6 \, \text{K}}\right)^{-\frac{1}{2}} + 1.5 \left(\frac{T}{10^6 \, \text{K}}\right)^{-\frac{3}{2}}\right]^{-1} \qquad (8.26)$$

and the dynamical time-scale:

$$t_{\text{dyn}} \cong \frac{\pi}{2} \left[\frac{2GM}{R^3}\right]^{-1/2} = 5 \times 10^7 \, \text{yr} \left(\frac{n}{1 \, \text{cm}^{-3}}\right)^{-1/2}. \qquad (8.27)$$

Here ρ is the average *baryonic* density and $\Lambda(T)$ gives the cooling rate of the gas at temperature T. Note that we have taken t_{dyn} to be the freefall time of a uniform density sphere of radius R.

There are three possibilities which should be distinguished as regards the evolution of such a cloud. Firstly, if t_{cool} is greater than the Hubble time, H_0^{-1}, then the cloud could not have evolved much since its formation. On the other hand, if $H_0^{-1} > t_{\text{cool}} > t_{\text{dyn}}$, the gas can cool; but as it cools the cloud can retain the pressure support by adjusting its pressure distribution. In this case the collapse of the cloud will be quasistatic on a time-scale of order t_{cool}. Finally, there is the possibility that $t_{\text{cool}} < t_{\text{dyn}}$. In this case the cloud will cool rapidly (compared to its dynamical time-scale) to a minimum temperature. This will lead to the loss of pressure support and the gas will undergo an almost freefall collapse. Fragmentation into smaller units can now occur because, as the collapse proceeds, smaller and smaller mass scales will become gravitationally unstable.

The criterion $t_{\text{cool}} < t_{\text{dyn}}$ can determine the masses of galaxies. Only when this condition is satisfied can a gravitating gas cloud collapse appreciably and fragment into stars. Further, in any heirarchical theory of galaxy formation, unless a gas cloud cools within a dynamical time-scale and becomes appreciably bound, collapse on a larger scale will disrupt it. In these theories, galaxies are the first structures which have resisted such disruption by being able to satisfy the above criterion.

Let us first consider the evolution of baryons, ignoring the dark matter component. We are interested in the ratio $\tau = (t_{\text{cool}}/t_{\text{dyn}})$. The condition $\tau = 1$ defines a curve on the $\rho - T$ space, which marks out the region of parameter space in

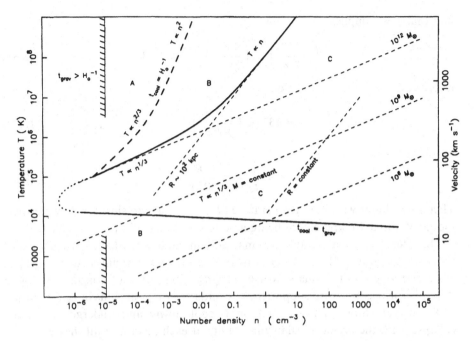

Fig. 8.1   Cooling diagram for baryons. See text for discussion.

which cooling occurs rapidly within a dynamical time, from the region of weak cooling (see figure 8.1; the solid line corresponds to $\tau = 1$).

For $T < T^*$, when line cooling is dominant, we have $t_{cool} \propto (T^{3/2}/\rho)$ and $t_{dyn} \propto \rho^{-1/2}$, giving $\tau \propto (T^{3/2}/\rho^{1/2}) \propto M$; hence the $\tau = 1$ curve will be parallel to the lines of constant mass in the ρ–T plane. Substituting the numbers and using the expression for the cooling time from (8.26) we find that $\tau = 1$ implies

$$\left(\frac{T}{10^6 \, \text{K}}\right)^{3/2} \left(\frac{n}{\text{cm}^{-3}}\right)^{-1/2} = 4.28 \,. \tag{8.28}$$

Expressing the mass of the cloud as

$$M = \frac{5RT}{G\mu} = 2.1 \times 10^{11} \, M_\odot \left(\frac{T}{10^6 \, \text{K}}\right)^{3/2} \left(\frac{n}{\text{cm}^{-3}}\right)^{-1/2}, \tag{8.29}$$

we can write

$$\tau = \frac{t_{cool}}{t_{dyn}} \cong \frac{M}{9 \times 10^{11} \, M_\odot}, \tag{8.30}$$

if $\mu = 0.57$. Thus the criterion for efficient cooling can be satisfied for masses below a critical mass of about $10^{12} \, M_\odot$, provided $T < 10^6 \, \text{K}$.

On the other hand, for $T > T^*$, when Bremsstrahlung dominates the cooling process, $t_{cool} \propto (T^{1/2}/\rho)$ and $t_{dyn} \propto \rho^{-1/2}$. So $\tau \propto (T^{1/2}/\rho^{1/2}) \propto R$, and the curve $\tau = 1$ will be parallel to the lines of constant radius in the ρ–T space. We now find that $\tau = 1$ implies

$$\left(\frac{T}{10^6 \, \text{K}}\right)^{1/2} \left(\frac{n}{\text{cm}^{-3}}\right)^{-1/2} = 6.43 . \qquad (8.31)$$

Expressing the radius of the cloud as

$$R = \frac{GM\mu}{5T} = 13 \, \text{kpc} \left(\frac{T}{10^6 \, \text{K}}\right)^{1/2} \left(\frac{n}{\text{cm}^{-3}}\right)^{-1/2} , \qquad (8.32)$$

we obtain

$$\tau = \frac{t_{\text{cool}}}{t_{\text{dyn}}} \simeq \frac{R}{80 \, \text{kpc}} . \qquad (8.33)$$

Therefore, clouds with high temperature ($T > T^*$) have to shrink below a critical radius of about 10^2 kpc before being able to cool efficiently to form galaxies.

These features are illustrated schematically in figure 8.1, which is usually called a 'cooling diagram'. The ρ–T space is divided into three regimes A, B and C. A gas cloud with constant mass evolves roughly along lines of constant M_J, with $T \propto \rho^{1/3}$, if it is pressure supported. Gas clouds in region A have $t_{\text{cool}} > t_{\text{Hubble}}$ and cool very little. Those in region B cool slowly and undergo quasistatic collapse, with the pressure balancing gravity at each instant, until they enter the region C, where $\tau < 1$. Gas clouds in C can cool efficiently to form galaxies because they have masses below $10^{12} \, M_\odot$ or radii below 10^2 kpc. These masses and radii compare well with the scales characteristic of galaxies.

Let us now consider the effects of including the dark matter component. The dynamical time-scale is now determined by the total density of dark matter and baryons, whereas the cooling time still depends only on the density of the baryonic gas. In this case, the gas will not be at the virial temperature initially. It is only during collapse that the gas becomes heated up by shocks produced when different bits of gas run into each other. If the cooling time-scale of the shocked gas is larger than the dynamical time-scale in which the cloud settles down to equilibrium, then the gas will eventually become heated up to the virial temperature. On the other hand, if the cooling time were shorter, the gas may *never* reach such a pressure supported equilibrium. Efficient cooling will result in the gas sinking to the centre of the dark matter potential well which is being formed, until halted by rotation or fragmentation into stars.

Clearly, it is again the ratio of the cooling time to the dynamical time of the object which governs the evolution. Further, note that smaller mass clumps are disrupted as larger masses turn around and collapse. However, if the gas component can cool efficiently enough, it may shrink sufficiently close to the centre of the dark matter potential and thus resist further disruption. This process will break the hierarchy. Galaxies could again be thought of as the first structures that have survived the disruption due to hierarchical clustering.

The spherical model can be used to estimate the relevant dynamical time-scale. We assume t_{dyn} to be comparable to ($t_{\text{coll}}/2$), the time taken for a spherical top-hat fluctuation to collapse after turning around. This expression is the same as the t_{dyn} given in (8.27) above, provided we identify R in (8.27) with the radius of turn-around r_m. Then

$$t_{dyn} \cong \frac{t_{coll}}{2} \cong 1.5 \times 10^9 \left(\frac{M}{10^{12} \, M_\odot} \right)^{-1/2} \left(\frac{r_m}{200 \, kpc} \right)^{3/2} yr. \qquad (8.34)$$

For estimating the cooling time-scale, we use (8.26) and assume that the gas makes up a fraction F of the total mass and is uniformly distributed within a radius $r_m/2$. The gas temperature is taken to be of the order of the virial temperature obtained in the spherical model, that is, $T_{vir} \cong (\mu v^2/3)$, where $v^2 \cong (6GM/5r_m)$. This corresponds to the temperature achieved by heating by shocks which have a velocity of the order of the virial velocity. In that case,

$$t_{cool} \cong 2.4 \times 10^9 \left(\frac{F}{0.1} \right)^{-1} \left(\frac{M}{10^{12} \, M_\odot} \right)^{1/2} \left(\frac{r_m}{200 \, kpc} \right)^{3/2} yr. \qquad (8.35)$$

We have assumed that the line cooling dominates at the temperature $T = T_{vir}$ relevant to the galaxies, and have adopted a typical value of $F \cong 0.1$. Note that the collapse, in general, is likely to be highly inhomogeneous and the above estimates are only supposed to give a rough idea of the numbers involved. From the last two equations we obtain

$$\tau = \left(\frac{t_{cool}}{t_{dyn}} \right) \cong 1.6 \left(\frac{F}{0.1} \right)^{-1} \left(\frac{M}{10^{12} \, M_\odot} \right), \qquad (8.36)$$

so that efficient cooling (with $\tau < 1$) requires

$$M < M_{crit} \cong 6.4 \times 10^{11} \, M_\odot \left(\frac{F}{0.1} \right). \qquad (8.37)$$

It is clear that masses of the order of galactic masses are again picked out preferentially, even when the dark matter is included.

The procedure outlined above can be used to analyse any particular theory of structure formation involving heirarchical clustering. The starting point will be the cooling diagram, in which the $\tau = 1$ curve is plotted. Given the power spectrum of density fluctuations, one can work out the density contrast at various scales $\delta_0 = v\sigma(M)$. Then the various properties, like ρ and T of the collapsed objects which are formed, can be estimated using the spherical model. We saw that these properties depend only on one parameter M, once the density contrast δ_0 is fixed. Thus, for each value of v one obtains a curve on the ρ–T plane, giving the properties of collapsed objects. These curves assume that the proto-condensations have virialized, but that the gas has not cooled and condensed. Cooling moves points on these curves to higher densities. In the same diagram one can also plot, for comparison, the observed positions of galaxies, groups and clusters of galaxies.

(c) The cooling rate of a gas with electron density n_e and temperature T embedded in a blackbody radiation field of density ρ_{rad} and temperature T_{rad} is given by

$$\Lambda_{\text{Comp}} = \frac{4\sigma_T n_e \rho_{\text{rad}}(T - T_{\text{rad}})}{m_e}. \tag{8.38}$$

(See problem 4.10.) The cooling time for matter, due to inverse Compton scattering off the cosmic background photons, will therefore be

$$t_{\text{Comp}} = \frac{3 m_p m_e (1+z)^{-4}}{8 \mu \sigma_T \Omega_{\text{rad}} \rho_c} \cong 2.1 \times 10^{12} (1+z)^{-4} \, \text{yr}. \tag{8.39}$$

Here we have assumed that $T \gg T_{\text{rad}}$ and have used $\rho_{\text{rad}}(z) = \Omega_{\text{rad}} \rho_c (1+z)^4$ to take into account the expansion of the universe. Comparing t_{Comp} with the dynamical time in (8.34) we obtain

$$\tau_{\text{Comp}} = \frac{t_{\text{Comp}}}{t_{\text{dyn}}} \cong 2 \times 10^2 (1 + z_{\text{coll}})^{-5/2}. \tag{8.40}$$

This ratio is less than unity for $z_{\text{coll}} > 7$, independently of the mass of the collapsing object. So Compton cooling can efficiently cool an object only if it collapses at a redshift higher than $z \cong 10$, whatever is its mass.

8.6 Angular momentum of galaxies

(a) Let us first ignore the presence of massive dark halos around galaxies, and assume that a protogalaxy is just a self-gravitating cloud of baryonic gas. The binding energy of the protogalaxy will be $|E| \cong GM^2/R$, where R is its characteristic radius. Since M is constant during collapse, $|E| \propto R^{-1}$ and so $\lambda \propto R^{-1/2}$. The gas cloud has to collapse by a factor of about $(\lambda_d/\lambda_i)^2 \cong (0.5/0.05)^2 \cong 100$, before it can spin up sufficiently to form a rotationally supported system, where λ_i is the initial value of λ produced by tidal torques. Therefore, to form a rotationally supported galactic disc of mass $10^{11} \, M_\odot$ and radius $10 \, \text{kpc}$, matter needs to collapse from an initial radius of $1 \, \text{Mpc}$. This process would take an inordinately long time, about $t_{\text{coll}} = (\pi/2)(R^3/2GM)^{1/2} \cong 5.3 \times 10^{10} \text{yr}$, much longer than the age of the universe. Note that even the material in the core, with a scale length of $r_c \cong 3 \, \text{kpc}$, would have to collapse from a distance of $300 \, \text{kpc}$ and would take about $10^{10} \, \text{yr}$.

(b) This difficulty is easily avoided if a massive halo exists. In the presence of a massive dark halo, the initial spin parameter of the system, before collapse of the gas, can be written as $\lambda_i = (L|E|^{1/2}/GM^{5/2})$, where the various quantities, L, E and M refer to the *combined* dark matter–gas system, although the contribution from the gas is negligible compared to that of the dark matter. After the collapse, the gas becomes self-gravitating and the spin parameter of the resulting disc galaxy will be $\lambda_d = (L_d|E_d|^{1/2}/GM_d^{5/2})$, where the parameters now refer to the disc. So we find that

$$\frac{\lambda_d}{\lambda_i} = \left(\frac{L_d}{L}\right) \left(\frac{|E_d|}{|E|}\right)^{1/2} \left(\frac{M_d}{M}\right)^{-5/2}. \tag{8.41}$$

The energy of the virialized dark matter–gas system, assuming that the gas has not yet collapsed, can be written as $|E| = k_1(GM^2/R_c)$, while that of the disc is given by $|E_d| = k_2(GM_d^2/r_c)$. Here R_c and r_c are the characteristic radii associated with the combined system and the disc respectively, while k_1, k_2 are constants of order unity which depend on the precise density profile and geometry of the two systems. The ratio of the binding energy of the collapsed disc to that of the combined system is then

$$\frac{|E_d|}{|E|} = \frac{k_2}{k_1}\left(\frac{M_d}{M}\right)^2\left(\frac{r_c}{R_c}\right)^{-1}. \tag{8.42}$$

Further, the total angular momentum (per unit mass) acquired by the gas, destined to form the disc, should be the same as that of the dark matter. This is because all the material in the system experiences the same external torques before the gas separates out due to cooling. Assuming that the gas conserves its angular momentum during the collapse, we have $(L_d/M_d) = (L/M)$. Hence,

$$\frac{\lambda_d}{\lambda_i} = \left(\frac{M_d}{M}\right)\left(\frac{k_2}{k_1}\right)^{1/2}\left(\frac{M_d}{M}\right)\left(\frac{R_c}{r_c}\right)^{1/2}\left(\frac{M}{M_d}\right)^{5/2} = \left(\frac{k_2}{k_1}\right)^{1/2}\left(\frac{R_c}{r_c}\right)^{1/2}\left(\frac{M}{M_d}\right)^{1/2}, \tag{8.43}$$

where we have used (8.42) to simplify (8.41). The gas originally occupied the same region as the halo before collapsing and so had a precollapse radius of R_c. Hence the collapse factor for the gas is

$$\frac{R_c}{r_c} = \left(\frac{k_1}{k_2}\right)\left(\frac{M_d}{M}\right)\left(\frac{\lambda_d}{\lambda_i}\right)^2. \tag{8.44}$$

We see that the required collapse factor for the gas to attain rotational support has been reduced by a factor (M_d/M), from what was required in the absence of a dominant dark halo. For a typical galaxy with a halo which is ten times as massive as the disc, one needs a collapse by only a factor of about 10 or so before the gas can spin up sufficiently to attain rotational support.

8.7 Angular pattern of MBR anisotropies

A physical process, characterized by a length scale L at $z = z_{dec}$, will subtend an angle $\theta(L) = [L/d_A(z_{dec})]$ in the sky today, where $d_A(z)$ is the angular diameter distance. We saw in problem 6.2(e) that $d_A = a_0 r_1(z)(1+z)^{-1}$. The expression for $r_1(z)$ was worked out in problem 6.4(b). For large z, this expression gives $r_1(z) \cong 2(H_0 a_0 \Omega)^{-1}$. So, for $z \gg 1$, the angular diameter distance becomes $d_A(z) \cong 2H_0^{-1}(\Omega z)^{-1}$, giving

$$\theta(L) \cong \left(\frac{\Omega}{2}\right)\left(\frac{Lz}{H_0^{-1}}\right) = 34.4''(\Omega h)\left(\frac{\lambda_0}{1\,\text{Mpc}}\right). \tag{8.45}$$

(As always, we quote numerical values with length scales by extrapolating them to the present epoch. Thus $\lambda_0 = L(1 + z_{dec}) \cong Lz_{dec}$ is the proper length *today* which would have been L at the redshift of z_{dec}. We will not bother to indicate this fact with a subscript '0' when no confusion is likely to arise.) In particular, consider the angle subtended by the region which has size equal to that of the Hubble radius at z_{dec}, that is, we take L to be $d_H(z_{dec}) = H^{-1}[z = z_{dec}] = H_0^{-1}(\Omega z_{dec})^{-1/2}z_{dec}^{-1}$ so that $\lambda = d_H(z_{dec})\,[1 + z_{dec}] \cong d_H(z_{dec})z_{dec} \cong H_0^{-1}(\Omega z_{dec})^{-1/2}$. Then

$$\theta_H \equiv \theta(d_H) \cong 0.87° \, \Omega^{1/2} \left(\frac{z_{dec}}{1100}\right)^{-\frac{1}{2}} \cong 1°. \tag{8.46}$$

Therefore, angular separation of more than one degree in the sky would correspond to regions which were bigger than the Hubble radius at the time of decoupling. (For comparison, note that the scale λ_{eq} subtends an angle of about $0.13°h^{-1}$.)

Let us now estimate the magnitude of the anisotropy $(\Delta T/T)$ due to processes (1), (2) and (3) given in the question. We begin with the intrinsic anisotropies in the radiation field. We shall first consider large angular scales, $\theta > \theta_H$.

If baryons and photons are tightly coupled, then we expect the anisotropy in the energy density of the radiation, δ_{rad}, to be comparable to δ_B. The exact relation between these two depends on the initial process which generated the fluctuations. Most popular models generate fluctuations which are 'adiabatic', in the sense that the entropy per baryon is unaffected by the fluctuation. Since the entropy is mostly contributed by the radiation, adiabatic fluctuations are characterized by

$$0 = \delta\left[\frac{s_{rad}}{n_B}\right] = \frac{s_{rad}}{n_B}\left[\frac{\delta s_{rad}}{s_{rad}} - \frac{\delta \rho_B}{\rho_B}\right]. \tag{8.47}$$

Using $s_{rad} \propto T^3$ and $\rho_{rad} \propto T^4$, it follows that $(\delta s_{rad}/s_{rad}) = (3/4)\delta_{rad}$. Therefore, for adiabatic fluctuations we will have $(3/4)\delta_{rad} = \delta_B$ or

$$\left(\frac{\Delta T}{T}\right)_{int} = \frac{1}{3}\delta_B(t_{dec}). \tag{8.48}$$

We can relate $\delta_B(t_{dec})$ to $\delta_{DM}(t_{dec})$ in the following manner: Since the baryons were tightly coupled to the photons, $\delta_B(t_{dec}) = \delta_B(t_{enter}) = \delta_{DM}(t_{enter})$. But the dark matter perturbations were growing during the phase $a_{eq} < a < a_{dec}$. So $\delta_{DM}(t_{dec}) = (a_{dec}/a_{eq})\delta_{DM}(t_{enter})$. It follows that $\delta_B(t_{dec}) \cong (a_{dec}/a_{eq})\,\delta_{DM}(t_{enter}) \cong (\delta_{DM}/20\Omega h^2)$. We can thus relate the intrinsic anisotropy in the radiation field to the density contrast of *dark matter* perturbation at decoupling:

$$\left(\frac{\Delta T}{T}\right)_{int} \cong \frac{1}{60\,\Omega h^2}\delta_{DM}(t_{dec}). \tag{8.49}$$

To determine the angular scale over which this effect is significant, we proceed as follows: Each Fourier mode of $\delta_{DM}(k)$, labelled by a wave vector k, will correspond to a wavelength $\lambda \propto k^{-1}$ and will contribute at an angle θ given by equation (8.45):

$$\theta(L) \cong \left(\frac{\Omega}{2}\right)\left(\frac{Lz}{H_0^{-1}}\right) = 34.4''(\Omega h)\left(\frac{\lambda_0}{1\,\text{Mpc}}\right). \tag{8.50}$$

The mean square fluctuation in temperature will therefore scale with the angle θ as

$$\left(\frac{\Delta T}{T}\right)^2_{\text{int}} \propto d^3k\, P(k) \propto \frac{dk}{k}P(k)k^3 \propto \lambda^{-(3+n)} \propto \theta^{-(3+n)} \tag{8.51}$$

for $\theta > \theta_\text{H}$. (If $n = 1$, then $(\Delta T/T)_{\text{in}} \propto \theta^{-2}$.)

Consider next the contribution to $(\Delta T/T)$ from the velocity of the scatterers in the last scattering surface (LSS). If a particular photon which we receive was last scattered by a charged particle moving with a velocity v, then the photon will suffer a Doppler shift of the order (v/c). This will contribute $(\Delta T/T)^2 \cong (v^2/c^2)$. At large scales (for $\lambda \gtrsim d_\text{H}$), the velocity of the baryons is essentially due to the gravitational force exerted by the dark matter potential wells. By an analysis similar to the above, we conclude that

$$\left(\frac{\Delta T}{T}\right)^2_{\text{Dopp}} \propto d^3k\,k^2\frac{P}{k^4} \propto \frac{dk}{k}kP(k) \propto \lambda^{-(1+n)} \propto \theta^{-(1+n)} \tag{8.52}$$

for $\theta > \theta_\text{H}$. (If $n = 1$, then $(\Delta T/T)_{\text{Dopp}} \propto \theta^{-1}$.)

Finally, let us consider the contribution to $(\Delta T/T)$ from the gravitational potential along the path of the photons. When the photons travel through a region of gravitational potential ϕ, they undergo a redshift of the order of (ϕ/c^2). In the $\Omega = 1$, matter dominated universe, the perturbed gravitational potential is independent of time, that is, $\phi(t, \mathbf{x}) = \phi(\mathbf{x})$. We may therefore estimate the temperature anisotropies due to gravitational potential as $(\Delta T/T) \cong (\phi/c^2)$, where ϕ is the gravitational potential *on* the LSS. This contribution to temperature anisotropy is usually called 'Sachs–Wolfe' effect. Its angular dependence will be

$$\left(\frac{\Delta T}{T}\right)^2_{\text{SW}} \propto d^3k\,\frac{P(k)}{k^4} \propto \frac{dk}{k} \propto \begin{cases} \lambda^{5-n} & (\text{for } \lambda < \lambda_{\text{eq}}) \\ \lambda^{1-n} & (\text{for } \lambda > d_\text{H}) \end{cases}$$

$$\propto \begin{cases} \theta^4 & (\text{for } \theta < \theta_{\text{eq}}) \\ \text{constant} & (\text{for } \theta > \theta_\text{H}). \end{cases} \tag{8.53}$$

(The second line assumes that $n = 1$.)

Let us now consider the anisotropies at small angular scales $\theta < \theta_\text{H}$. The contribution $(\Delta T/T)^2_{\text{SW}}$ varies as θ^4 and hence is not very significant at small θ. As regards $(\Delta T/T)^2_{\text{int}}$ and $(\Delta T/T)^2_{\text{Dopp}}$, we should note the following. At small scales, the evolution of δ_B is governed by the pressure support in the baryonic fluid and the density contrast δ_B and velocity v will oscillate as an acoustic wave with a wavelength $\lambda \cong \lambda_\text{J} \cong \lambda_\text{H}$. Hence $(\Delta T/T)$ at small angular scales will have crests and troughs separated at wavenumbers $\Delta k \cong (\pi/\lambda_\text{J}) \cong 0.03h\,\text{Mpc}^{-1}$. This corresponds to an angular separation $\Delta\theta \cong 0.3°$ in the sky.

The amplitudes of all the three contributions will be comparable around $\theta \cong \theta_{eq}$. At smaller θ, the potential contribution is ignorable and the other two contributions will be comparable in magnitude but (approximately) opposite in phase.

There is, however, another effect which needs to be taken into account at small angular scales. The above analysis assumes that decoupling took place at a sharp value of redshift z_{dec}. In reality, the LSS has a finite thickness $\Delta z \cong 80$. Any photon detected today has a probability $\mathscr{P}(z)dz$ of having last been scattered in the redshift interval $(z, z + dz)$. The observed $(\Delta T/T)$ has to be computed as

$$\left(\frac{\Delta T}{T}\right)_{obs} = \int dz \left\{ \begin{array}{l} (\Delta T/T) \text{ if the last} \\ \text{scattering was at } z \end{array} \right\} \times \mathscr{P}(z). \qquad (8.54)$$

We saw in problem 6.9(e) that $\mathscr{P}(z)$ is very well approximated by a Gaussian peaked at $z = z_{dec}$ with a width $\Delta z \cong 80$. This width corresponds to a line-of-sight (comoving) distance of

$$\Delta l = c \left(\frac{dt}{dz}\right) \Delta z \cdot (1 + z_{dec}) \cong H_0^{-1} \frac{\Delta z}{\Omega^{1/2} z_{dec}^{3/2}} \cong 8 \left(\Omega h^2\right)^{-1/2} \text{ Mpc}. \qquad (8.55)$$

The integral in (8.54) is equivalent to multiplying each Fourier mode of $(\Delta T/T)_k$ by the Gaussian $\exp\left[-k^2(\Delta l)^2/2\right] = \exp\left[-(k/k_T)^2\right]$ with $k_T \cong 0.2h\,\text{Mpc}^{-1}$. In other words, the anisotropies at lengthscales smaller than Δl will be exponentially suppressed due to the finite thickness of LSS. This lengthscale corresponds to an angular scale of about $\theta_{\Delta z} \cong 3.8'\Omega^{1/2}$.

Taking this effect into account, the final pattern of anisotropies will look like the one in figure 8.2. The contribution at large angular scales due to intrinsic anisotropy, the Doppler effect and the Sachs–Wolfe effect are shown by the dotted line, the dot–dash line and the dashed line respectively. (For the sake of simplicity, the oscillations at small scales are not shown in the individual components.) The sum of the three contributions will look something like that shown by the unbroken line. The dot–triple dash line indicates the Gaussian cut-off due to the finite thickness of the LSS. The thick line clearly shows the Doppler peaks and the small scale suppression.

The overall picture regarding the angular dependence of $(\Delta T/T)$ is as follows. At large scales, the anisotropies are dominated by the Sachs–Wolfe effect and are independent of the angular scale. As we move to smaller scales, other contributions add up making $(\Delta T/T)$ increase. At scales $\theta \lesssim \theta_{eq}$, the oscillations in δ_B and v make their presence felt and we see characteristic peaks called Doppler peaks. At still smaller scales, the effects due to the finite thickness of LSS drastically reduce the amplitude of the fluctuations.

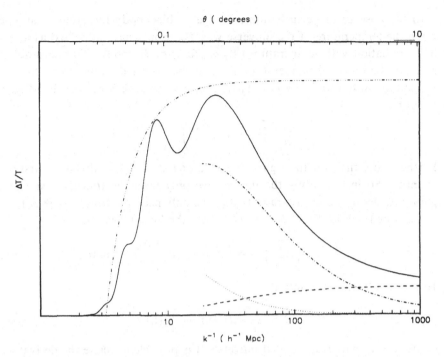

Fig. 8.2. Schematic diagram showing the expected anisotropies in the CMBR temperature at different angular scales.

8.8 Derivation of MBR anisotropies

Since the temperature anisotropy δT is proportional to the frequency shift $\Delta \omega$, the result of problem 6.2 can be written as

$$\ln \left(\frac{T_{\text{ob}} a_{\text{ob}}}{T_{\text{em}} a_{\text{em}}} \right) = \frac{1}{2} \int_{t_{\text{em}}}^{t_{\text{ob}}} \dot{h}_{\alpha\beta} \, [t, x^{\mu}(t)] \, n^{\alpha} n^{\beta} \, dt \,. \tag{8.56}$$

(To avoid possible misunderstanding, we stress the following point. The quantity $\dot{h}_{\alpha\beta} \, [t, x^{\mu}(t)]$ is obtained by taking the partial derivative of $h_{\alpha\beta}(t, \mathbf{x})$ with respect to t (at constant x^{i}), *followed* by the replacement of $\mathbf{x}$ by the function $\mathbf{x}(t) = \mathbf{n}\eta(t)$. We cannot, therefore, write the result of the integration as $(h_{\alpha\beta}^{(2)} - h_{\alpha\beta}^{(1)})$.) The path of the photon can be written as follows. Consider a photon which was emitted at $t = t_{\text{e}}$ and reaches an observer at $t = t_{0}$ travelling along a null geodesic. Let the *unit* vector pointing along this null ray, *from* the observer *to* the source, be $\mathbf{n}$. Assuming the observer is at the origin, the location of the photon at some time t will be at

$$\mathbf{x} = \mathbf{n}\eta(t), \quad \eta(t) = \int_{t}^{t_{0}} \frac{dt'}{a(t')} \,. \tag{8.57}$$

This is the path of propagation in the unperturbed metric; as we shall see, we do not need the perturbed trajectory.

In the absence of perturbations, $T \propto a^{-1}$; blackbody radiation at an epoch when the temperature of the universe was T_{em} would have appeared as a blackbody radiation with temperature $(T_{em}a_{em}/a_{ob}) \equiv T_0$ today. The deviation ΔT between the observed temperature T_{ob} and the expected temperature T_0 will be a small quantity caused by $h_{\alpha\beta}$. Therefore, we can replace the left-hand side of (8.56) by

$$\ln \left(\frac{T_0 + \delta T_{ob}}{T_0 + \delta T_{em}} \right) \cong \left(\frac{\delta T}{T} \right)_{obs} - \left(\frac{\delta T}{T} \right)_{em}. \tag{8.58}$$

Further, note that, on the right-hand side of (8.56), $h_{\alpha\beta}$ is already a first-order quantity. So in evaluating the integral we only need the trajectory $\mathbf{x}(t)$ of the photon to zeroth order accuracy, that is, we only need the trajectory (8.57) in the unperturbed metric. Therefore, we can write the final answer as

$$\left(\frac{\delta T}{T_0} \right) - \left(\frac{\delta T}{T_0} \right)_{int} = \frac{1}{2}n^{\alpha}n^{\beta} \int_{t_e}^{t_0} dt \left[\frac{\partial}{\partial t}h_{\alpha\beta}(t, \mathbf{n}\eta(t)) \right]. \tag{8.59}$$

To proceed further, we note that

$$\dot{\delta} = \frac{\partial}{\partial t}[a(t)f(\mathbf{x})] = \frac{\dot{a}}{a}\delta(t, \mathbf{x}) \tag{8.60}$$

in the $\Omega = 1$, matter dominated universe. It is possible to relate the derivative of the metric perturbation $\dot{h}_{\alpha\beta}$ to the perturbed gravitational potential by a simple trick. We know that the peculiar velocity field is related to the gravitational potential by (see problem 7.6(e))

$$v_{\alpha} = -\frac{\dot{a}}{4\pi G\rho_b a^2} \frac{\partial \phi}{\partial x^{\alpha}}. \tag{8.61}$$

Consider now two particles with comoving separation $\delta x^{\beta} = n^{\beta}\delta x$. The relative velocity of these two particles is given by

$$\delta v = n^{\alpha}\delta v_{\alpha} = -\frac{\dot{a}}{4\pi G\rho_b a^2} \frac{\partial^2 \phi}{\partial x^{\alpha}\partial x^{\beta}} n^{\alpha}n^{\beta}\delta x. \tag{8.62}$$

But we have seen in problem 6.2 that the same quantity can be expressed as

$$\delta v = -\frac{1}{2}\dot{h}_{\alpha\beta}n^{\alpha}n^{\beta}\delta l = -\frac{1}{2}\dot{h}_{\alpha\beta}n^{\alpha}n^{\beta}a\delta x. \tag{8.63}$$

Equating the two expressions for δv we find

$$\dot{h}_{\alpha\beta} = \frac{\dot{a}}{2\pi G\rho_b a^3} \frac{\partial^2 \phi}{\partial x^{\alpha}\partial x^{\beta}}. \tag{8.64}$$

It follows that

$$n^{\alpha}n^{\beta}\dot{h}_{\alpha\beta} = \frac{\dot{a}}{Q}n^{\alpha}n^{\beta} \frac{\partial^2 \phi}{\partial x^{\alpha}\partial x^{\beta}}, \tag{8.65}$$

where $Q \equiv 2\pi G\rho_b a^3$ is a constant.

The integrand in (8.59) involves this quantity evaluated along the photon path $\mathbf{x} = \mathbf{n}\eta(t)$. Consider any function $f(x^\alpha)$ evaluated along a path $x^\alpha = n^\alpha\eta(t)$. Clearly, $(df/d\eta) = (\partial f/\partial x^\alpha)n^\alpha$. Therefore, along the path, we can write

$$n^\alpha n^\beta \frac{\partial^2 \phi}{\partial x^\alpha \partial x^\beta} = \left(\frac{d^2\phi}{d\eta^2}\right), \tag{8.66}$$

where $\phi = \phi(x^\alpha = n^\alpha\eta) = \phi(\eta)$ is now treated as a function of η. From (8.57), it follows that $dt = -a\,d\eta$ with the minus sign arising from the lower limit of the integral. Converting the variable from t to η, (8.59) becomes

$$\left(\frac{\delta T}{T_0}\right)_{\text{path}} \equiv \frac{\delta T}{T_0} - \left(\frac{\delta T}{T_0}\right)_{\text{int}} = \frac{1}{2Q}\int dt\,\dot{a}\left(\frac{d^2\phi}{d\eta^2}\right) = \frac{1}{2Q}\int d\eta\left(\frac{da}{d\eta}\right)\left(\frac{d^2\phi}{d\eta^2}\right). \tag{8.67}$$

Integrating by parts, we obtain

$$\left(\frac{\delta T}{T_0}\right)_{\text{path}} = \left[\frac{1}{2Q}\left(\frac{da}{d\eta}\right)\left(\frac{d\phi}{d\eta}\right)\right]_{\eta_1}^{\eta_1} - \frac{1}{2Q}\int d\eta \frac{d^2 a}{d\eta^2}\left(\frac{d\phi}{d\eta}\right). \tag{8.68}$$

In the matter dominated case, $a \propto t^{2/3}$ so that $\eta \propto t^{1/3}$ and $a \propto \eta^2$. Therefore, $(d^2 a/d\eta^2)$ in the second term is a constant. Further note that, in the expression

$$\frac{1}{2Q}\left(\frac{d^2 a}{d\eta^2}\right)\left(\frac{d\phi}{d\eta}\right)d\eta = \frac{1}{(4\pi G\rho_b a^3)}\left[a\frac{d}{dt}(a\dot{a})\right]\phi\,dt, \tag{8.69}$$

a appears three times both in the numerator and the denominator. Thus the proportionality constant in the $a \propto t^{2/3}$ relation is irrelevant. Taking just $a = t^{2/3}$, we find the right-hand side of (8.69) to be

$$\frac{\phi\,dt}{(4\pi G\rho_b t^2)}\left[t^{2/3}\cdot\frac{d}{dt}\left(t^{2/3}\cdot\frac{2}{3}t^{-1/3}\right)\right] = \frac{2}{9}\frac{\phi\,dt}{(4\pi G\rho_b t^2)} = \frac{1}{3}\phi\,dt. \tag{8.70}$$

In arriving at the last step, we have used the relation $6\pi G\rho_b t^2 = 1$ for the matter dominated phase. Consider now the first term in equation (8.68), which can be written as

$$-\frac{1}{(4\pi G\rho_b a^3)}(a\dot{a})\left(\frac{d\phi}{d\eta}\right) = -\frac{(\dot{a}/a)}{(4\pi G\rho_b)}\cdot\frac{1}{a}\cdot n^\alpha\left(\frac{\partial\phi}{\partial x^\alpha}\right) = \mathbf{v}\cdot\mathbf{n} \tag{8.71}$$

if we use (8.61). Substituting this result and (8.70) in (8.68) we find that the temperature fluctuation becomes

$$\frac{\delta T}{T_0} = \left(\frac{\delta T}{T}\right)_{\text{int}} + \mathbf{n}\cdot(\mathbf{v}_{\text{ob}} - \mathbf{v}_{\text{em}}) - \frac{1}{3}\int\phi\,dt$$

$$= \left(\frac{\delta T}{T}\right)_{\text{int}} + \mathbf{n}\cdot(\mathbf{v}_{\text{ob}} - \mathbf{v}_{\text{em}}) - \frac{1}{3}\left[\phi(0) - \phi(x_{\text{em}})\right]. \tag{8.72}$$

Each of these terms has a simple interpretation: $(\mathbf{n}\cdot\mathbf{v}_{\text{ob}})$ is the Doppler shift due to the motion of the observer with respect to the comoving frame; $(-\mathbf{n}\cdot\mathbf{v}_{\text{em}})$ is the corresponding Doppler effect in the emitting surface; the $(1/3)[\phi(0) - \phi(x_{\text{em}})]$ is caused by local variations in the gravitational potential at the source and the observer. The term $(1/3)\,\phi(0)$ merely adds a constant to $(\delta T/T_0)$ and will not introduce any directional dependence in $(\delta T/T_0)$. The other three terms, of course, depend on $\mathbf{n}$ and hence vary from direction to direction, causing angular anisotropies in the sky.

8.9 Sachs–Wolfe effect and COBE

(a) Let $\mathbf{n}$ and $\mathbf{m}$ be two directions in the sky with an angle α between them. The two-point correlation function of the temperature fluctuations in the sky can be defined as

$$\mathscr{C}(\alpha) = \langle S(\mathbf{n})S(\mathbf{m})\rangle = \sum\sum \langle a_{lm}a_{l'm'}^*\rangle Y_{lm}(\mathbf{n})Y_{l'm'}^*(\mathbf{m}). \tag{8.73}$$

Since the sources of temperature fluctuations are related linearly to the density inhomogeneities, we may think of a_{lm} as Gaussian random fields with some power spectrum. In that case $< a_{lm}a_{l'm'}^* >$ will be non-zero only if $l = l'$ and $m = m'$. Writing

$$\langle a_{lm}a_{l'm'}^*\rangle = C_l\delta_{ll'}\delta_{mm'} \tag{8.74}$$

and using the addition theorem of spherical harmonics, we find that

$$\mathscr{C}(\alpha) = \sum_l \frac{(2l+1)}{4\pi}C_lP_l(\cos\alpha), \tag{8.75}$$

with $C_l = < |a_{lm}|^2 >$. To evaluate the C_ls we may proceed as follows. At large angular scales, where the Sachs–Wolfe effect dominates the temperature anisotropy, we have

$$\left(\frac{\Delta T}{T}\right) = -\frac{1}{3}\phi(\mathbf{x}_{\text{em}}) = -\frac{1}{3}G\rho_{\text{b}}(t_0)\int d^3x' \frac{\delta(\mathbf{x}',t_0)}{|\mathbf{x}_{\text{em}}-\mathbf{x}'|}$$

$$= -\frac{1}{2}H_0^2 \int \frac{d^3k}{(2\pi)^3}\frac{\delta_k}{k^2}\exp(-i\mathbf{k}\cdot\mathbf{x}_{\text{em}}). \tag{8.76}$$

In arriving at the last expression, we have used $G\rho_{\text{b}} = (3H^2/8\pi)$ and introduced the Fourier components. From the definition of a_{lm}, we have

$$\frac{\Delta T}{T} = \sum_{l,m}^{\infty} a_{lm}Y_{lm}(\theta,\psi). \tag{8.77}$$

Comparing (8.76) and (8.77) and using the fact that $\mathbf{x}_{\text{em}} = 2H_0^{-1}\mathbf{n}$, we can express $< |a_{lm}|^2 >$ in terms of $|\delta_k|^2$. We obtain, after a straightforward calculation,

$$\langle |a_{lm}|^2\rangle = C_l = \frac{H_0^4}{2\pi}\int_0^\infty dk\frac{|\delta_k|^2}{k^2}\left|j_l\left(2H_0^{-1}k\right)\right|^2, \tag{8.78}$$

where j_l is the spherical Bessel function of order l. This equation, along with (8.75), expresses the temperature correlation function in terms of the power spectrum of the theory. The mean square fluctuation in the temperature is

$$\left(\frac{\Delta T}{T}\right)^2_{\text{rms}} = \mathscr{C}(0) = \frac{1}{4\pi}\sum_{l=2}^{\infty}(2l+1)\,C_l,\tag{8.79}$$

where we have omitted the dipole contribution. The leading term in this sum arises from the quadrapole and has the magnitude

$$\left(\frac{\Delta T}{T}\right)^2_{Q} = \frac{5}{4\pi}C_2.\tag{8.80}$$

(b) The above formulas, especially (8.79), are correct provided the instrument can probe all values of l with equal sensitivity. But, in practise, any instrument will have a finite angular resolution, θ_c. This implies that the response of the instrument will decrease significantly for modes with $l > l_c$ where $l_c \cong \theta_c^{-1}$. This effect can be taken into account by introducing a Gaussian response profile for the detector. The modified formulas for the anisotropy will be

$$\left(\frac{\Delta T}{T}\right)^2_{\text{rms}} = \frac{1}{4\pi}\sum_{l=2}^{\infty}(2l+1)C_l\exp\left(-\frac{l^2\theta_c^2}{2}\right),\tag{8.81}$$

$$\left(\frac{\Delta T}{T}\right)^2_{Q} = \frac{5}{4\pi}C_2\exp(-2\theta_c^2).\tag{8.82}$$

For COBE experiment, $\theta_c \cong 10°$. Assuming that the power spectrum has the form $P(k) = Ak$, we obtain from (8.78)

$$C_2 = \frac{AH_0^4}{24\pi},\quad C_l = \frac{6C_2}{l(l+1)}.\tag{8.83}$$

Substituting these relations into (8.81) and (8.82) we obtain

$$\left(\frac{\Delta T}{T}\right)^2_{Q} \cong \left(\frac{5}{96\pi^2}\right)(AH_0^4) = (5.28\times10^{-3})(AH_0^4)\tag{8.84}$$

and

$$\left(\frac{\Delta T}{T}\right)^2 = \frac{1}{4\pi}\sum_{l=2}^{\infty}(2l+1)\left[\frac{6C_2}{l(l+1)}\right]\exp\left(-\frac{1}{2}l^2\theta_c^2\right) = \frac{C_2}{4\pi}\times28.45 = 0.03(AH_0^4).\tag{8.85}$$

The quantity (AH_0^4) is directly related to the fluctuations in the gravitational potential $\Phi^2 = (k^3|\phi_k|^2/2\pi^2)$ at large scales. Since $\phi_k = (4\pi G\rho_b)\,(\delta_k/k^2) = (3/2)H^2(\delta_k/k^2)$ we find that

$$\Phi^2(k) = \frac{k^3|\phi_k|^2}{2\pi^2} = \frac{9}{4}\left(\frac{H}{k}\right)^4\left(\frac{k^3|\delta_k|^2}{2\pi^2}\right) = \frac{9}{8\pi^2}(AH_0^4)\tag{8.86}$$

if $|\delta_k|^2 \cong Ak$. Therefore, $AH_0^4 = (8\pi^2/9)\Phi^2$ and we can re-express (8.84) and (8.85) as

$$\left(\frac{\Delta T}{T}\right)_Q \cong 0.22\Phi, \quad \left(\frac{\Delta T}{T}\right)_{\text{rms}} \cong 0.51\Phi. \tag{8.87}$$

We can now compare the theoretical results with the COBE observations. To begin with, the above equation predicts that $(\Delta T_{\text{rms}}/\Delta T_Q) \cong 2.3$ if the spectrum has $n = 1$. The COBE results allow this ratio to fall between 1.43 and 3.94 with a mean value of 2.29. This is consistent with the assumption of $n = 1$, though the uncertainty is rather large.

The parameter Φ and the amplitude A can now be determined by comparing (8.87) with the COBE result. Ideally, of course, both $(\Delta T/T)_{\text{rms}}$ and $(\Delta T/T)_Q$ should lead to the same value for Φ and A. However, because of instrumental and systematic errors, we obtain slightly different values. Comparing the root-mean-square values we find

$$\begin{aligned}
\Phi_{\text{rms}} &\cong 1.96(\Delta T/T)_{\text{rms}} \cong 2.2 \times 10^{-5}, \\
A_{\text{rms}} &\cong (8\pi^2/9)\Phi^2 R_{\text{H}}^4 \cong 4.1 \times 10^{-9} R_{\text{H}}^4 \cong (24\ h^{-1}\ \text{Mpc})^4.
\end{aligned} \tag{8.88}$$

Similarly, by comparing the quadrapole anisotropy, we obtain

$$\begin{aligned}
\Phi_Q &\cong 4.55(\Delta T/T)_Q \cong 2.2 \times 10^{-5}, \\
A_Q &\cong (8\pi^2/9)\Phi^2 R_{\text{H}}^4 \cong 4.2 \times 10^{-9} R_{\text{H}}^4 \cong (24\ h^{-1}\ \text{Mpc})^4.
\end{aligned} \tag{8.89}$$

Within the error bars, we may take $\Phi \cong 2.2 \times 10^{-5}$ and $A = (24h^{-1}\ \text{Mpc})^4$.

9
High redshift objects

9.1 Gunn–Peterson effect

The procedure for calculating the optical depth τ has been described in problem 6.5. The absorption cross-section for a Lyman-α photon is

$$\sigma(v) = \frac{\pi e^2}{m_e} f \delta(v - v_\alpha), \tag{9.1}$$

where $f = 0.416$ is the oscillator strength and v_α is the frequency corresponding to the Lyman-α photon (see problem 4.12). An analysis similar to that carried out in problem 6.5 will now give

$$\tau = \frac{n_H(\bar{z})}{(1 + \bar{z})(1 + \Omega\bar{z})^{1/2}} \frac{\pi e^2}{m_e v_\alpha} f \frac{1}{H_0} \cong 4.14 \times 10^{10} h^{-1} \frac{n_H(\bar{z})}{(1 + \bar{z})(1 + \Omega\bar{z})^{1/2}}, \tag{9.2}$$

where $(1 + \bar{z}) = (v_\alpha/v)$. The absorption will lead to a dip in the intensity of the quasar spectrum at wavelengths $\lambda < \lambda_\alpha$. The limit on the optical depth from the observations is $\tau \lesssim 0.05$ for quasars with a mean redshift of $\bar{z} \cong 2.6$. This leads to the upper limit

$$n_H(z = 2.64) \lesssim 8.4 \times 10^{-12} h \, \text{cm}^{-3}, \tag{9.3}$$

for a universe with $\Omega = 1$. This value should be compared with the baryon number density at this redshift:

$$n_B = 1.1 \times 10^{-5}(1 + z)^3 \Omega_B h^2 \, \text{cm}^{-3} \cong 2 \times 10^{-4}(\Omega_B h^2) \, \text{cm}^{-3}. \tag{9.4}$$

So we see that n_H is much smaller than n_B. (Note that the optical depth is $\tau \cong 10^4$ for this value of n_B; it is this large optical depth which allows us to put such a strong bound.) Assuming that the bound on the optical depth is typically like $\tau \lesssim 0.1$ in the redshift range $0 < z < 5$, we may write the above bound as

$$n_H \lesssim 2 \times 10^{-12} \Omega^{1/2} h(1 + z)^{3/2} \, \text{cm}^{-3}, \tag{9.5}$$

or, equivalently, as

$$\Omega(n_H) \lesssim 2 \times 10^{-7} \Omega^{1/2} h^{-1}(1 + z)^{-3/2}. \tag{9.6}$$

443

In arriving at the last two expressions, we have set

$$(1 + z)(1 + \Omega z)^{1/2} \cong \Omega^{1/2}(1 + z)^{3/2}. \tag{9.7}$$

This is, of course, exact if $\Omega = 1$ but remains a reasonable approximation for $z \gtrsim 1$ if $\Omega \gtrsim 0.3$.

9.2 Ionization of IGM

(a) In the case of collisional ionization, we need to maintain equilibrium between the reaction $H + e \rightarrow p + e + e$ and $p + e \rightarrow H + \gamma$. This gives

$$\langle \sigma_{\text{coll}} v \rangle n_e n_H = \alpha n_p n_e , \tag{9.8}$$

where n_e, n_p and n_H are the number densities of free electrons, protons and hydrogen atoms, $\sigma_{\text{coll}} \cong 2a_o^2$ is the collisional cross-section where a_0 is the Bohr radius, v is the typical velocity of the electron at $T \cong 10^6$ K and α is the recombination coefficient. We saw in problem 4.13 that

$$\alpha \cong 2 \times 10^{-13} \left(\frac{T}{10^4 \text{ K}} \right)^{-1/2} \text{cm}^3 \text{ s}^{-1}. \tag{9.9}$$

For $T \cong 10^6$ K, we have $\sigma_{\text{coll}} v \cong 3 \times 10^{-8}$ cm^3 s^{-1} and we obtain

$$\frac{n_H}{n_p} \cong 5 \times 10^{-7} \left(\frac{T}{10^6 \text{ K}} \right)^{-1/2}. \tag{9.10}$$

Using the Gunn–Peterson bound in equation (9.6), this becomes

$$\Omega_{\text{IGM}} = \frac{n_p}{n_{\text{crit}}} \lesssim 0.4(1 + z)^{-3/2} \Omega^{1/2} h^{-1} \left(\frac{T}{10^6 \text{ K}} \right)^{1/2}. \tag{9.11}$$

(The recombination coefficient in (9.9) was based on a single transition. A more sophisticated analysis gives $\alpha \cong 4.36 \times 10^{-13} T_4^{-0.7}$cm^3 s^{-1}. We shall continue to use (9.9) since we are only interested in order of magnitude estimates.)

(b) The number density of photons per logarithmic interval in frequency can be related to the flux J_ν as follows:

$$n_\gamma \equiv \frac{dN}{dV \, d(\ln \nu)} = \nu \frac{dN}{dA \, (c \, dt) \, d\nu} = \frac{1}{hc} \frac{dE}{dA \, dt \, d\nu} = \frac{4\pi}{hc} \left(\frac{dE}{dA \, dt \, d\nu \, d\Omega} \right) = \frac{4\pi}{hc} J_\nu . \tag{9.12}$$

Numerically,

$$n_\gamma = 6 \times 10^{-5} J_0 \left(\frac{\nu_0}{\nu} \right) \text{ cm}^{-3}. \tag{9.13}$$

Comparing with (9.4), we find that $(n_\gamma / n_B) \cong 0.1 J_0 (\Omega_B h^2)^{-1}$ at $z = 3$. For $\Omega_B h^2 \cong 1$, this is hardly sufficient to ionize the IGM. For lower values of $\Omega_B h^2$, this may be marginally sufficient, though detailed modelling shows that it is not quite adequate.

The photoionization cross-section for hydrogen was worked out in problem 4.13 and is given by

$$\sigma_I = \sigma_0 \left(\frac{\omega_0}{\omega}\right)^{7/2}, \tag{9.14}$$

where $\sigma_0 \cong 8 \times 10^{-18}\,\mathrm{cm}^2$ and $\omega_0 \cong 2 \times 10^{16}$ Hz. The rate of ionization per hydrogen atom due to the J_ν will be

$$\mathcal{R} = \int_{\nu_0}^{\infty} \frac{4\pi J_\nu}{h\nu}\sigma_I\, d\nu \cong 3 \times 10^{-12} J_0\,\mathrm{s}^{-1}. \tag{9.15}$$

In this case, the equilibrium between photoionization and recombination will lead to the condition

$$\mathcal{R}n_H = \alpha n_e n_p = \alpha n_p^2, \tag{9.16}$$

if $n_e = n_p$. In reality, we have to take into account the fact that the plasma may not fill space uniformly but could be clumped. If we assume that a fraction $(1/f)$ of space is uniformly filled by plasma, with the rest of the region being empty, we need to multiply the right-hand side of the above equation by a factor f. In that case, we obtain

$$\frac{n_H}{n_p} = \frac{\alpha f n_p}{\mathcal{R}} \cong 10^{-6}\frac{f\Omega_{IGM}h^2(1+z)^3}{J_0}\left(\frac{T}{10^4\,\mathrm{K}}\right)^{-1/2}. \tag{9.17}$$

Using the Gunn–Peterson bound we can translate this equation to

$$\Omega_{IGM} \lesssim 0.4\frac{\Omega^{1/4}J_0^{1/2}}{h^{3/2}f^{1/2}}\left(\frac{T}{10^4\,\mathrm{K}}\right)^{1/4}(1+z)^{-9/4}. \tag{9.18}$$

This is comparable to the bound obtained in part (a) above.

9.3 Re-ionization and CMBR

Let us suppose that the re-ionization occurred at a redshift of z_{ion} and that the intergalactic medium remained fully ionized for all $z < z_{ion}$. Then the optical depth due to Thomson scattering up to a redshift of z ($< z_{ion}$) will be

$$\tau = \int \sigma_T n_e dt = \left(\frac{\sigma_T \Omega_B \rho_c}{m_p}\right)\int_0^z H_0^{-1}\, dz\, (1+z)^3 \frac{1}{(1+z)^2(1+\Omega z)^{1/2}}$$
$$= \frac{\sigma_T H_0}{4\pi G m_p}\cdot\frac{\Omega_B}{\Omega}\cdot\left[2 - 3\Omega + (1+\Omega z)^{1/2}(\Omega z + 3\Omega - 2)\right]. \tag{9.19}$$

This expression has the limiting forms

$$\tau = \frac{\sigma_T H_0}{4\pi G m_p}\cdot\frac{3}{2}(\Omega_B z) = 0.026(\Omega_B z) \quad (\text{for } \Omega z \ll 1),$$

$$= \frac{\sigma_T H_0}{4\pi G m_p}\left(\frac{\Omega_B}{\Omega^{1/2}}\right)z^{3/2} = 0.017\left(\frac{\Omega_B}{\Omega^{1/2}}\right)z^{3/2}(\text{for } \Omega z \gg 1). \tag{9.20}$$

If $\tau \cong 1$ for some $z \leq z_{\text{ion}}$, then the photons will be significantly scattered by the re-ionized plasma. This will partially wipe out information about the original $z = z_{\text{dec}}$ surface. For universes with $\Omega \cong 1$, $\Omega_B \cong 0.1$, $\tau = 1$ occurs near $z \cong 70$, while for $\Omega \cong \Omega_B \cong 0.2$, $\tau = 1$ can occur at much lower redshifts: $z \cong 25$; thus in standard dark matter models re-ionization has to occur at fairly high redshifts (greater than 70) to produce any effect. If such re-ionization occurred at a redshift of z_{ion}, its primary effect would be to wipe out the original $(\delta T/T)$ at scales smaller than the Hubble radius $d_H(z_{\text{ion}})$ at z_{ion}. From the results of problem 8.8 we see that d_H subtends the angle

$$\theta_{\text{ion}}(z_{\text{ion}}) \cong 3°\Omega^{1/2}\left(\frac{100}{z_{\text{ion}}}\right)^{1/2} \tag{9.21}$$

in the sky. Since $\tau = 1$ corresponds to $z \cong 15.08\Omega^{1/3}(\Omega_B h)^{-2/3}$, the maximum value of θ_{ion} will be $\theta_m(z_m) \cong 7.4°(\Omega_B \Omega h)^{1/3}$. If such re-ionization occurs, the trace of primordial anisotropies will remain only at angles larger than about 8°.

9.4 Mass functions

(a) Let us consider a density field $\delta_R(\mathbf{x})$ smoothed by a window function W_R of radius R. We have seen earlier in problem 7.5 that the probability that this field will have a value δ at any chosen point is

$$P(\delta, t) = \left[\frac{1}{2\pi\sigma^2(R, t)}\right]^{1/2} \exp\left(-\frac{\delta^2}{2\sigma^2(R, t)}\right), \tag{9.22}$$

where

$$\sigma^2(R, t) = \int \frac{d^3k}{(2\pi)^3} |\delta_k(t)|^2 W_k^2(R). \tag{9.23}$$

Let $\delta_i(t, t_i)$ be the density contrast needed at time t_i so that $\delta = \delta_c = 1.68$ by time t. For the $\Omega = 1$, matter dominated universe, $\delta_i(t, t_i) = \delta_c(t_i/t)^{2/3}$. As a first approximation, we may assume that a region with $\delta > \delta_i(t, t_i)$ (when smoothed on the scale R at time t_i) will form a gravitationally bound object with mass $M \propto \bar{\rho}R^3$ by time t. Therefore, the fraction of bound objects with mass greater than M will be

$$F(M) = \int_{\delta_i(t,t_i)}^{\infty} P(\delta, R, t_i)\, d\delta = \frac{1}{\sqrt{2\pi}} \frac{1}{\sigma(R, t_i)} \int_{\delta_i(t,t_i)}^{\infty} \exp\left(-\frac{\delta^2}{2\sigma^2(R, t_i)}\right) d\delta$$

$$= \frac{1}{2}\text{erfc}\left(\frac{\delta_i(t, t_i)}{\sqrt{2}\sigma(R, t_i)}\right), \tag{9.24}$$

where $\text{erfc}(x)$ is the complementary error function. This expression can be written in a more convenient form using

$$\frac{\delta_i(t, t_i)}{\sigma(R, t_i)} = \frac{\delta_c(t_i/t)^{2/3}}{\sigma(R, t_i)} = \frac{\delta_c}{\sigma_L(R, t)}, \tag{9.25}$$

where $\sigma_L(R,t) \equiv \sigma(R,t_i)(t/t_i)^{2/3}$ is the *linear* density contrast at t. So

$$F(M) = \frac{1}{2}\text{erfc}\left[\frac{\delta_c}{\sqrt{2}\sigma_L(R,t)}\right]. \tag{9.26}$$

The mass function $f(M)$ is just $(\partial F/\partial M)$; the (comoving) number density $N(M,t)$ can be found by dividing this expression by (M/ρ_c). Carrying out these operations we obtain

$$N(M,t)\,dM = -\left(\frac{\rho_c}{M}\right)\left(\frac{1}{2\pi}\right)^{1/2}\left(\frac{\delta_c}{\sigma}\right)\left(\frac{1}{\sigma}\frac{d\sigma}{dM}\right)\exp\left(-\frac{\delta_c^2}{2\sigma^2}\right)dM. \tag{9.27}$$

Given the power spectrum $|\delta_k|^2$ and a window function W_R one can explicitly compute the right-hand side of this expression.

There is, however, one fundamental difficulty with equation (9.24). The integral of $f(M)$ over all M should give unity; but it is easy to see that, for the expression in (9.24),

$$\int_0^\infty f(M)\,dM = \int_0^\infty dF = \frac{1}{2}. \tag{9.28}$$

This arises because we have not taken into account the underdense regions correctly.

To see the origin of this difficulty more clearly, consider the interpretation of (9.24). If a point in space has $\delta > \delta_c$ when filtered at scale R, then that point should correspond to a system with mass greater than $M(R)$; this is taken care of correctly by equation (9.24). However, consider those points which have $\delta < \delta_c$ under this filtering. There is a *non-zero* probability that such a point will have $\delta > \delta_c$ when the density field is filtered with a radius $R_1 > R$. Therefore, to be consistent with the interpretation in (9.24), such points should *also* correspond to a region with mass greater than M. But (9.24) ignores these points completely and thus *underestimates* $F(M)$ (by a factor $(1/2)$).

To correct this mistake, we should replace (9.24) by the relation

$$F(M) = \int_{\delta_c}^\infty P(\delta,R)\,d\delta + \int_{-\infty}^{\delta_c} C(\delta_c,\delta)\,d\delta, \tag{9.29}$$

where the second term represents the probability P_u that a point which has $\delta < \delta_c$ at the filter scale R has the density $\delta > \delta_c$ at a larger filter scale $R_1 > R$. For a sequence of filter scales $R_1, R_2, \ldots, R_n$, we obtain a sequence of Gaussian random fields parametrized by the dispersions $\Delta_1, \Delta_2, \ldots, \Delta_n$. The probability that a point *remains underdense* (i.e. $\delta < \delta_c$) for all these filter scales is given by

$$P_{\text{survive}} \equiv P_s = \int_{-\infty}^{\delta_c} d\delta_1 \int_{-\infty}^{\delta_c} d\delta_2 \ldots \int_{-\infty}^{\delta_c} d\delta_n\, P_J(\delta_1,\delta_2,\ldots,\delta_n), \tag{9.30}$$

where $P_J[\delta_i]$ is the joint probability distribution that the Gaussian variables δ_i take the set of values simultaneously. Obviously, $(1-P_s)$ gives the probability that a point becomes overdense somewhere along the sequence of filterings $(R_1, \ldots, R_n)$.

The Gaussian variables obtained by different filtering scales, unfortunately, are not independent. We can see that

$$\langle \delta_a \delta_b^* \rangle = \int \frac{d^3\mathbf{k}}{(2\pi)^3} \frac{d^3\mathbf{p}}{(2\pi)^3} W_k(R_a) W_p^*(R_b) < \delta_k \delta_p^* > e^{i(\mathbf{k}-\mathbf{p})\cdot\mathbf{x}}$$

$$= \int \frac{d^3\mathbf{k}}{(2\pi)^3} W_k(R_a) W_k^*(R_b) \sigma_k^2, \qquad (9.31)$$

is, in general, non-zero. Hence, calculating (9.30) is a non-trivial task.

We can look upon this process in a different, but equivalent, manner. Consider any one fixed location in space. When the filtering scale is some large value R_1 (with a dispersion Δ_1), let us assume that this point has a density contrast δ_1. When we reduce the scale to R_2, we will have a *new* probability distribution for δ; let the value of density contrast at our chosen point now be δ_2. As we go through a sequence of filtering scales, $R_1, R_2, \ldots, R_n$ (in *decreasing* order), the density contrast performs a random walk through the points $(\delta_1, \delta_2, \ldots, \delta_n)$. Suppose the *first* instance when δ crosses the value δ_c occurs at the k-th step. Then we will attribute the chosen point to a mass $M_k \propto \bar{\rho} R_k^3$. Note that, since $\delta < \delta_c$ for all the *higher* filtering scales, that is, for all $(R_1, R_2, \ldots, R_{k-1})$, this point *does not* belong to any higher mass. (This takes care of the original difficulty in (9.24).) The random walk concept merely translates the content of (9.30) into a pictorial form. This random walk problem is equally difficult to solve because the steps are not independent. In fact, the answer will clearly depend on the correlation between the steps; and from (9.31) it follows that the answer will critically depend on the form of the window function.

(b) Consider a window function which is sharply truncated in k-space, that is, for $W_k(R) = \theta(R^{-1} - k)$, which acts as a low-pass-filter in k-space. From (9.31) it follows that, for this window function,

$$\langle \delta_a \delta_{a+1} \rangle = \sigma_a^2, \quad \langle \delta_a \delta_b \rangle = \sigma_a^2 \quad \text{(for } a \leq b\text{)}. \qquad (9.32)$$

The step lengths of the random walks are $l_1 \equiv (\delta_2 - \delta_1)$, $l_2 = (\delta_3 - \delta_2)$, $\ldots, l_a = (\delta_{a+1} - \delta_a)$, etc. Each of these is a Gaussian variable with the dispersion

$$\langle l_a^2 \rangle = \langle (\delta_{a+1} - \delta_a)^2 \rangle = \sigma_{a+1}^2 + \sigma_a^2 - 2\langle \delta_{a+1} \delta_a \rangle = \sigma_{a+1}^2 - \sigma_a^2, \qquad (9.33)$$

and *zero* cross-correlation

$$\langle l_a l_b \rangle = \langle (\delta_{a+1} - \delta_a)(\delta_{b+1} - \delta_b) \rangle$$
$$= \langle \delta_{a+1} \delta_{b+1} \rangle - \langle \delta_{a+1} \delta_b \rangle - \langle \delta_a \delta_{b+1} \rangle + \langle \delta_a \delta_b \rangle \qquad (9.34)$$
$$= \sigma_{a+1}^2 - \sigma_{a+1}^2 - \sigma_a^2 + \sigma_a^2 = 0$$

for $(a+1) < b$; the other cases can be considered in a similar manner and can be shown to vanish. In other words, a sharp filter in the k-space produces a random walk in which each step l_a is independent and is drawn from a Gaussian variable with dispersion $(\sigma_{a+1}^2 - \sigma_a^2)$. In the continuum limit, this random walk is described by a diffusion equation. The probability $P(\delta, \sigma^2)$ that the particle is at $(\delta, \delta + d\delta)$ when the dispersion is σ^2 obeys the diffusion equation

$$\frac{\partial P}{\partial \sigma^2} = \frac{1}{2} \frac{\partial^2 P}{\partial \delta^2}. \qquad (9.35)$$

We are interested in the probability that the trajectory reaches (δ, σ) without exceeding δ_c earlier, that is, at smaller σ. This is equivalent to solving (9.35) with the boundary condition that there exists an absorbing barrier at $\delta = \delta_c$. This is straightforward and the answer is

$$P(\delta, \sigma^2) = \frac{1}{\sigma\sqrt{2\pi}} \left[\exp\left(-\frac{\delta^2}{2\sigma^2}\right) - \exp\left(-\frac{(\delta - 2\delta_c)^2}{2\sigma^2}\right) \right]. \qquad (9.36)$$

Integrating this expression from δ_c to ∞ and differentiating with respect to M, we obtain

$$dF(M) = \sqrt{\frac{2}{\pi}} \cdot \frac{\delta_c}{\sigma^2} \cdot \left(-\frac{\partial \sigma}{\partial M}\right) \exp\left(-\frac{\delta_c^2}{2\sigma^2}\right) dM, \qquad (9.37)$$

or

$$N(M)\, dM = -\frac{\bar{\rho}}{M} \left(\frac{2}{\pi}\right)^{1/2} \frac{\delta_c}{\sigma^2} \left(\frac{\partial \sigma}{\partial M}\right) \exp\left(-\frac{\delta_c^2}{2\sigma^2}\right) dM, \qquad (9.38)$$

which is precisely *twice* the value obtained by using (9.24). Of course, the normalization problem is solved automatically. (The quantity σ here refers to the linearly extrapolated density σ_L; the subscript L is omitted to simplify the notation.) The corresponding result for (9.26) is larger by a factor of two:

$$F(M, z) = \text{erfc}\left[\frac{\delta_c}{\sqrt{2}\sigma_L(M, z)}\right] = \text{erfc}\left[\frac{\delta_c(1+z)}{\sqrt{2}\sigma_0(M)}\right], \qquad (9.39)$$

where $\sigma_0(M)$ is the linearly extrapolated density contrast today, and we have used the fact $\sigma_L(M, z) \propto (1+z)^{-1}$. Note that, by definition, $F(M, z)$ gives the Ω contributed by the collapsed objects with mass larger than M at redshift z; equation (9.39) shows that this can be calculated given only the linearly extrapolated $\sigma_0(M)$.

(c) Let us take the mass of Abell clusters to be $M = 5 \times 10^{14} \alpha \, M_\odot$, where α quantifies the uncertainty. Similarly, we take the abundance to be $\mathscr{A} = 4 \times 10^{-6} \beta h^3 \, \text{Mpc}^{-3}$. The contribution of the Abell clusters to the density of the universe is

$$F = \Omega_{\text{clus}} = \frac{M\mathscr{A}}{\rho_c} \cong 8\alpha\beta \times 10^{-3}. \qquad (9.40)$$

Assuming that $\alpha\beta$ varies between 0.1 to 3 we obtain

$$\Omega_{\text{clus}} \cong \left(8 \times 10^{-4} - 2.4 \times 10^{-2}\right). \qquad (9.41)$$

In figure 9.1(a) we have plotted F calculated from (9.39) as a function of σ_0. (The curves are for $z = 0, 1, 2, 3, 4, 5$ with the topmost curve being for $z = 0$. We shall concentrate on this curve for the purpose of this problem.) The fractional abundance of (9.41), at $z = 0$, requires a $\sigma \cong 0.5 - 0.78$ at the cluster scales. All we need to determine now is whether a particular model has this range of σ for $M \cong 10^{15}\,M_\odot$.

This is done in part (b) of figure 9.1. The plots in 9.1(b) give the relation between $\sigma_0(M)$ (at $z = 0$, calculated by a spherical window function) and M and are for $\Gamma = 0.2, 0.25, 0.3, 0.4, 0.5, 0.7$. (The lowest curve is for the lowest value of Γ.) For $M \cong (10^{14} - 10^{15})\,M_\odot$, we need $\sigma \cong 0.5$–0.78. This region is marked by a box in figure 9.1(b). We see that $\Gamma \cong 0.25 - 0.4$ is consistent with the cluster abundance. (The curves for these Γ-values go through the box.) A larger value of Γ has higher σ at cluster scales and will overproduce clusters. Any other model for structure formation can be analysed along similar lines.

9.5 Abundance of quasars

(a) We have seen in problem 1.4 that the luminosity of a system accreting at maximum possible value can be related to the mass of the central object by

$$L_E = \frac{4\pi G m_p c M_{BH}}{\sigma_T} \cong 1.3 \times 10^{47} \left(\frac{M_{BH}}{10^9\,M_\odot} \right) \text{erg s}^{-1}. \tag{9.42}$$

One can infer a characteristic black hole mass from (9.42) using the observed luminosity of the quasar. The quasars with $z > 4$ have typical luminosities of $10^{47}\,\text{erg s}^{-1}$, in a universe with $\Omega = 1$ and $h = 0.5$. This requires black holes of mass $M_{BH} \cong 10^9\,M_\odot$.

From the luminosity, one can also estimate the amount of fuel that must be present to power the quasar for a lifetime t_Q. If ϵ is the efficiency with which the rest mass energy of the fuel is converted into radiation, then the fuel mass is

$$M_f = \frac{L t_Q}{\epsilon} \cong 2 \times 10^9\,M_\odot \left(\frac{L}{10^{47}\,\text{erg s}^{-1}} \right) \left(\frac{t_Q}{10^8\,\text{yr}} \right) \left(\frac{\epsilon}{0.1} \right)^{-1}. \tag{9.43}$$

So if the lifetime is about 10^8 yr, and the efficiency is about 10%, the required fuel mass is comparable to the mass of the central black hole.

The mass estimated above corresponds to that involved in the central engine of the quasar. This mass will, in general, be a small fraction F, of the mass of the host galaxy. We can write F as a product of three factors: a fraction f_b of matter in the universe which is baryonic; some fraction f_{ret} of the baryons originally associated with the galaxy, which was retained when the galaxy was formed (the remaining mass could be expelled via a supernova-driven wind); a fraction f_{hole} of the baryons is retained, which participates in the collapse to form the compact central object. In standard cold dark matter models, $f_b \cong 0.1$ for an $\Omega = 1$ universe, while f_{ret} will depend crucially on the depth of the potential well of the galaxy or, equivalently, on the circular velocity v_c; $f_{ret} \cong 0.1$ for $v_c \lesssim 100\,\text{km s}^{-1}$. The quantity f_{hole} depends on the way central mass accumulates

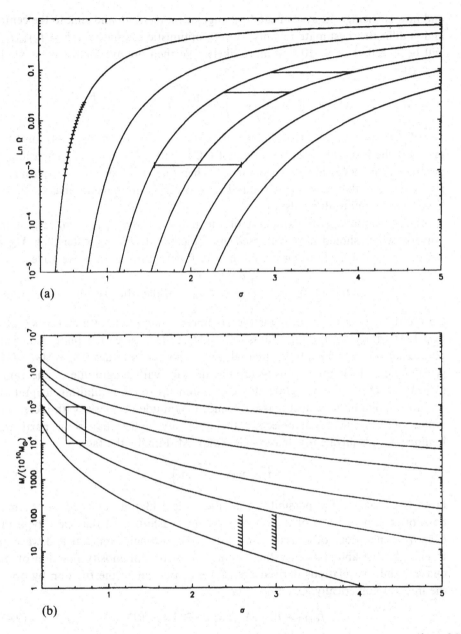

Fig. 9.1. (a) The Ω contributed by collapsed objects with mass greater than M plotted against $\sigma(M)$ at different values of z. The curves are for $z = 0, 1, 2, 3, 4$ and 5, from top to bottom. The constraint arising from cluster abundance at $z = 0$ (see problem 9.4(c)), quasar abundance at $z = 2-4$ (see problem 9.5) and the abundance of damped Lyman-α systems at $z = 2-3$ (see problem 9.7) are marked. (b) The $M-\sigma$ relation in a class of CDM-like models; see text for discussion.

and how efficiently the system can lose angular momentum and sink to the centre, and is difficult to estimate reliably. A very optimistic estimate for $F = f_b f_{ret} f_{hole}$ will be $F \cong 0.01$; $F \cong 10^{-3}$ is more likely. We then obtain, for the mass of the host galaxy,

$$M_G = 2 \times 10^{11} \, M_\odot \left(\frac{L}{10^{47} \, \mathrm{erg \, s^{-1}}} \right) \left(\frac{t_Q}{10^8 \, \mathrm{yr}} \right) \left(\frac{\epsilon}{0.1} \right)^{-1} \left(\frac{F}{0.01} \right)^{-1}. \qquad (9.44)$$

For $(F/0.01) \cong (0.1\text{--}1)$, with the earlier value being more likely, (9.44) implies a mass for the host galaxy in the range of $(10^{11} - 10^{12}) \, M_\odot$ if other dimensionless parameters in (9.43) are of order unity. Therefore, the existence of quasars at $z > 4$ suggests that a reasonable number of objects with galactic masses should have formed before this redshift.

(b) An estimate of the fraction of the mass density, $f_{coll}(> M_G, z)$, in the universe which should have collapsed into objects with mass greater than M_G at redshift z can be made using the luminosity function of quasars. We have

$$\rho(z) f_{coll}(> M, z) = \int_{-\infty}^{M_B(M)} \tau M_{host}(M_B) \phi(M_B, z) \, dM_B. \qquad (9.45)$$

On the right-hand side, $M_B(M)$ is the bolometric magnitude of a quasar situated in a host galaxy with a total mass M; $\phi(M_B, z) \, dM_B$ gives the number density of quasars in the magnitude interval dM_B. So the product $M_{host}(M_B) \phi \, dM_B$ gives the collapsed mass density around quasars with magnitude in the range $(M_B, M_B + dM_B)$. We integrate this expression up to some limiting magnitude M_B corresponding to a host mass M. Since magnitude decreases with increasing luminosity, this picks out objects with luminosity higher than some particular luminosity or, equivalently, mass higher than M. Finally, the factor

$$\tau = \max \left(\frac{t(z)}{t_Q}, 1 \right) \qquad (9.46)$$

takes into account the possibility that the typical lifetime of a quasar t_Q may be shorter than the age of the universe at the redshift z. In that case, typically $(t(z)/t_Q)$ generations of quasars can exist at the particular redshift and we need to enhance the abundance by this factor. Given the luminosity function of the quasars and the relations derived above, we can calculate the fraction by doing the integral numerically. One finds that

$$f_{coll} \left(> 10^{12} \, M_\odot, z \gtrsim 2 \right) \cong 1.4 \times 10^{-3}. \qquad (9.47)$$

From the results of problem 9.4 we can write the collapse fraction as

$$f_{coll} (> M, z) = \mathrm{erfc} \left[\frac{\delta_c(1+z)}{\sqrt{2} \sigma(M)} \right], \qquad (9.48)$$

where erfc (z) is the complimentary error function and $\sigma(M)$ is the linearly extrapolated density contrast today at the mass scale M. Figure 9.1(a) plots the fraction f as a function of σ for various redshifts. The quasar abundance is also marked for reference. This shows that at the scales corresponding to the host mass of quasars (which could be above $10^{12} \, M_\odot$), we must have $\sigma \cong 1.4$–2.6. Any theoretical model for structure formation should produce at least these values of σ to be viable. In figure 9.1(b), we have marked the region with mass $(10^{11}$–$10^{12}) \, M_\odot$ and $\sigma \cong 2.6$. Models with higher σ are viable as far as quasar abundance is concerned. Note that this requires about $\Gamma \gtrsim 0.25$.

9.6 Lyman-alpha forest

(i) The mean proper distance between the Lyman-α forest clouds along the line-of-sight is given by

$$L = \frac{c \, dt}{dN} = \frac{c \, dt}{dz} \frac{dz}{dN} \cong \frac{c}{H_0 \Omega^{1/2} B (1+z)^{\gamma+5/2}} = 10^3 \left(\Omega h^2 \right)^{-1/2} (1+z)^{-5.25} \, \text{Mpc}$$

$$\cong 0.6 \left(\Omega h^2 \right)^{-1/2} \, \text{Mpc} \qquad (\text{at } z = 3). \tag{9.49}$$

(We have used (9.7) to obtain the Ω-dependence; this will be used in the rest of the problem as well.)

(ii) The estimate for the density of hydrogen atoms depends on the size of the clouds and the column density. Taking $\Sigma = 10^{14} \sigma_{\text{H}} \, \text{cm}^{-2}$ and $l = 10 \, l_0 h^{-1} \, \text{kpc}$, where σ_{H} and l_0 are scaling parameters, we obtain

$$n_{\text{H}} \cong \frac{\Sigma}{l} \cong 3 \times 10^{-9} h l_0^{-1} \sigma_{\text{H}} \, \text{cm}^{-3}. \tag{9.50}$$

This corresponds to a mass of neutral hydrogen

$$M_{\text{H}} \cong m_{\text{p}} \Sigma l^2 \cong 100 l_0^2 h^{-2} \sigma_{\text{H}} \, M_\odot. \tag{9.51}$$

(iii) Let $\langle n_{\text{H}} \rangle$ be the mean number density of hydrogen atoms due to clouds with a mean surface density $\langle \Sigma \rangle$ along a line of sight. In that case, we have the relation $\langle n_{\text{H}} \rangle (c \, dt/dz) = \langle \Sigma \rangle (dN/dz)$. The density of hydrogen atoms within a cloud is $n_{\text{H}} = \langle \Sigma \rangle / l$; from these relations it follows that the fraction of space filled by the clouds is

$$f = \frac{\langle n_{\text{H}} \rangle}{n_{\text{H}}} = \frac{\langle \Sigma \rangle \left(dN/c \, dt \right)}{\langle \Sigma \rangle / l} = \frac{l}{L} = 10^{-5} (1+z)^{5.25} l_0 \Omega^{1/2}$$

$$= 0.02 \, l_0 \Omega^{1/2} \qquad (\text{at } z = 3). \tag{9.52}$$

(A more transparent way of deriving this result is as follows. If $n(z)$ is the number density of clouds, then

$$\frac{dN}{dz} = \left(\pi l^2 \right) n(z) \frac{c \, dt}{dz}. \tag{9.53}$$

The fraction of space filled by clouds in some large volume V is

$$f = \frac{(4\pi/3) l^3 n(z) V}{V} = \frac{4\pi}{3} n(z) l^3. \tag{9.54}$$

Substituting for $n(z)$ from (9.53) we obtain

$$f = \frac{4\pi}{3} l^3 \cdot \left(\frac{c\, dt}{dz} \right)^{-1} \left(\frac{dN}{dz} \right) \left(\frac{1}{\pi l^2} \right) \cong \left(\frac{dN}{dz} \right) \left(\frac{dz}{c\, dt} \right) l = \frac{l}{L}, \qquad (9.55)$$

where L is defined by (9.49).) The number density of clouds is

$$n_{\text{cloud}} \cong \frac{f}{l^3} = 2 \times 10^4 \frac{h^3 \Omega^{1/2}}{l_0^2} \; \text{Mpc}^{-3}. \qquad (9.56)$$

This corresponds to a mean intercloud separation of about $40\, h^{-1} l_0^{2/3}$ kpc at $z = 3$. Notice that if $l_0 \cong 1$, then the mean distance between the clouds is of the order of cloud size. The Lyman-α forest clouds are fairly densely packed in the intergalactic medium.

(iv) By an analysis similar to that in problem 9.2, we find that

$$\frac{n_{\text{H}}}{n_{\text{p}}} = \frac{\alpha n_{\text{p}}}{\mathcal{R}} \qquad (9.57)$$

or

$$n_{\text{p}} = \left(\frac{\mathcal{R}}{\alpha} n_{\text{H}} \right)^{1/2} = \left(\frac{\mathcal{R}}{\alpha} \frac{\Sigma}{l} \right)^{1/2}$$

$$\cong 2 \times 10^{-4} \left(\frac{T}{10^4 \, \text{K}} \right)^{1/4} \left(\frac{\Sigma}{10^{14} \, \text{cm}^{-2}} \right)^{1/2} \left(\frac{J_0 h}{l_0} \right)^{1/2} \text{cm}^{-3}. \qquad (9.58)$$

The fraction of neutral hydrogen is

$$F = \frac{\alpha}{\mathcal{R}} n_{\text{p}} = \left(\frac{\alpha \Sigma}{\mathcal{R} l} \right)^{1/2} \cong 2 \times 10^{-5} \left(\frac{h}{J_0 l_0} \right)^{1/2} \left(\frac{\Sigma}{10^{14} \, \text{cm}^{-2}} \right)^{1/2} \left(\frac{T}{10^4 \, \text{K}} \right)^{-1/4}. \qquad (9.59)$$

The neutral hydrogen fraction is fairly low in these clouds.

(v) The total *baryonic* mass of the plasma will be about

$$M_{\text{B}} \cong n_{\text{p}} l^3 \cong 10^7 \left(\frac{T}{10^4 \, \text{K}} \right)^{1/4} \left(\frac{l_0}{h} \right)^{5/2} \left(\frac{\Sigma}{10^{14} \, \text{cm}^{-2}} \right)^{1/2} J_0^{1/2} \, M_{\odot}. \qquad (9.60)$$

(vi) Consider the clouds which have a typical size l, neutral hydrogen mass M_{H}, column density in the range $\Sigma, \Sigma + d\Sigma$ and number density $n(z)$. Then we have

$$\frac{dN}{d\Sigma\, dz}\, d\Sigma \equiv g(\Sigma, z)\, d\Sigma = \pi l^2 n(z) \frac{c\, dt}{dz} \qquad (9.61)$$

and

$$M_{\text{H}} \cong m_{\text{p}} \pi l^2 \Sigma. \qquad (9.62)$$

The density contributed by neutral hydrogen in these clouds is

$$\rho_{\text{H}} = n(z) M_{\text{H}} = \left[g\, d\Sigma \frac{1}{\pi l^2} \frac{dz}{c\, dt} \right] \left[m_{\text{p}} \pi l^2 \Sigma \right] = m_{\text{p}} \Sigma g(\Sigma, z) \left(\frac{dz}{c\, dt} \right) d\Sigma. \qquad (9.63)$$

Note that this result is independent of the size l of the clouds. The total ρ_H contributed by clouds with column densities in the range $(\Sigma_{min}, \Sigma_{max})$ can be obtained by integrating this expression between the two limits. Dividing by $\rho_c(1+z)^3$ one can find the Ω contributed by the neutral hydrogen in these clouds:

$$\Omega_H = \frac{m_p}{\rho_c(1+z)^3}\left(\frac{dz}{c\,dt}\right)\int_{\Sigma_{min}}^{\Sigma_{max}}\Sigma g(\Sigma,z)\,d\Sigma$$

$$= \left(\frac{m_p}{\rho_c}\right)\left(\frac{H_0\Omega}{c}\right)^{1/2}\frac{1}{(1+z)^{1/2}}\int_{\Sigma_{min}}^{\Sigma_{max}}\Sigma g(\Sigma,z)\,d\Sigma. \tag{9.64}$$

Using this expression, one can compute the Ω contributed by clouds with any column density range. Taking $g = 91.2(\Sigma/\Sigma_0)^{-\beta}\Sigma_0^{-1}$ with $\Sigma_0 = 10^{14}\,\text{cm}^{-2}$, $\beta = 1.46$, $\Sigma_{max} \cong 10^{22}\,\text{cm}^{-2}$, $\Sigma_{min} \cong 10^{14}\,\text{cm}^{-2}$, we obtain, at $z = 3$,

$$\Omega_H \cong 2\times 10^{-3}\Omega^{1/2}h^{-1}\left(\frac{\Sigma_{max}}{10^{22}\,\text{cm}^{-2}}\right)^{2-\beta}. \tag{9.65}$$

For $\Sigma_{max} \cong 10^{15}\,\text{cm}^{-2}$, $\Sigma_{min} \cong 10^{13}\,\text{cm}^{-2}$, which is appropriate for Lyman-α forest clouds, this will give

$$\Omega_H(\text{Lyman forest}) \cong 3\times 10^{-7}\Omega^{1/2}h^{-1} \tag{9.66}$$

at $z = 3$.

Let us next consider the Ω contributed by the plasma. In a cloud with column density Σ, the plasma contributes a mass which is F^{-1} larger than that of neutral hydrogen. So the $d\Omega_B$ due to clouds with column density in the range $(\Sigma, \Sigma + d\Sigma)$ will be

$$d\Omega_B = \frac{d\Omega_H}{F} = \left(\frac{\mathscr{R}l}{\alpha\Sigma}\right)^{1/2}\left(\frac{d\Omega_H}{d\Sigma}\right)d\Sigma. \tag{9.67}$$

The integral is now

$$\Omega_B = \left(\frac{\mathscr{R}l}{\alpha}\right)\left(\frac{m_p}{\rho_c}\right)\left(\frac{H_0\Omega^{1/2}}{c}\right)\frac{1}{(1+z)^{1/2}}\int_{\Sigma_{min}}^{\Sigma_{max}}\Sigma^{1/2}g(\Sigma,z)\,d\Sigma. \tag{9.68}$$

Since the exponent β of $g \propto \Sigma^{-\beta}$ is close to 1.5, the integral is close to a logarithm. Taking $\beta = 1.5$ for simplicity, we obtain, at $z = 3$,

$$\Omega_B \cong 6.3\times 10^{-3}\left(\frac{J_0 l_0\Omega}{h^3}\right)^{1/2}\left(\frac{T}{10^4\,\text{K}}\right)^{1/4}\ln\left(\frac{\Sigma_{max}}{\Sigma_{min}}\right). \tag{9.69}$$

If we take $\Sigma_{max} = 10^{15}\,\text{cm}^{-2}$ and $\Sigma_{min} = 10^{13}\,\text{cm}^{-2}$ the logarithm is about 4.6 and we find that

$$\Omega_B \cong 3\times 10^{-2}\left(\frac{J_0 l_0\Omega}{h^3}\right)^{1/2}\left(\frac{T}{10^4\,\text{K}}\right)^{1/4}. \tag{9.70}$$

9.7 Abundance of damped Lyman-alpha clouds

(i) For $(dN/dz) = B \cong 0.3$, we obtain

$$L = \frac{c\,dt}{dz}\frac{dz}{dN} = \frac{c}{H_0\Omega^{1/2}B(1+z)^{5/2}} = 10^4(\Omega h^2)^{-1/2}(1+z)^{-5/2}$$

$$= 300(\Omega h^2)^{-1/2}\,\mathrm{Mpc} \qquad (\text{at } z = 3). \tag{9.71}$$

(ii) For $\Sigma \cong 10^{22}\,\mathrm{cm}^{-2}$, $l \cong 10\,l_0 h^{-1}\,\mathrm{kpc}$,

$$n_\mathrm{H} \cong \frac{\Sigma}{l} = 0.3hl_0^{-1}\,\mathrm{cm}^{-3}$$

$$M_\mathrm{H} \cong m_p\Sigma l^2 \cong 10^{10}l_0^2 h^{-2}\,M_\odot \tag{9.72}$$

(iii) The fraction of space filled by damped Lyman-α clouds is

$$f = \frac{l}{L} = 3.3 \times 10^{-5}l_0\Omega^{1/2} \qquad (\text{at } z = 3). \tag{9.73}$$

The number density of clouds will be

$$n_\mathrm{cloud} = \frac{f}{l^3} = 33h^3\Omega^{1/2}l_0^{-2}\,\mathrm{Mpc}^{-3}, \tag{9.74}$$

with a mean intercloud separation of $n_\mathrm{cloud}^{-1/3} \cong 0.3h^{-1}l_0^{2/3}\,\mathrm{Mpc}$. This is nearly 30 times larger than the size of damped Lyman-α systems if $l_0 \cong 1$.

(iv),(v) The high column density of these clouds implies that they will be mostly neutral. The metagalactic flux of photons will be able to ionize only a thin layer of matter on the surface. Near the surface, equilibrium between ionization and recombination will lead to a proton density of

$$n_p \cong \left(\frac{R}{\alpha}\frac{\Sigma}{l}\right)^{1/2} \cong 2\left(\frac{T}{10^4\,\mathrm{K}}\right)^{1/4}\left(\frac{\Sigma}{10^{22}\,\mathrm{cm}^{-2}}\right)^{1/2}\left(\frac{J_0 h}{l_0}\right)^{1/2}\,\mathrm{cm}^{-3} \tag{9.75}$$

(see equation (9.58)), which is about six times the neutral hydrogen density n_H of equation (9.72). Further, the penetration length λ of photons at such high column densities is low; we have

$$\lambda \cong \frac{1}{n_\mathrm{H}\sigma_\mathrm{PI}} \cong 0.14\,l_0 h^{-1}\,\mathrm{pc}\left(\frac{\Sigma}{10^{22}\,\mathrm{cm}^{-2}}\right)^{-1}. \tag{9.76}$$

(We have used (9.14) with $\omega \cong \omega_0$.) This length is much smaller than the size of the system. Note that $(\lambda/l) \cong (\Sigma\sigma_\mathrm{PI})^{-1}$; so $\lambda \cong l$ at column densities of $\Sigma \cong 10^{17}\,\mathrm{cm}^{-2}$. For $\Sigma \gg 10^{17}\,\mathrm{cm}^{-2}$, $\lambda \ll l$ and the system is mostly neutral. Thus, most of the mass in damped Lyman-α systems is contributed by the neutral hydrogen:

$$M_\mathrm{B} \cong M_\mathrm{H} \cong 10^{10}l_0^2 h^{-2}\,M_\odot. \tag{9.77}$$

(vi) The Ω contributed by damped Lyman-α systems is quite large and comparable to that of luminous matter in galaxies. We have, from (9.65),

$$\Omega_\mathrm{H} \cong 2 \times 10^{-3}\Omega^{1/2}h^{-1} \qquad (\text{for } \Sigma \cong 10^{22}\,\mathrm{cm}^{-2}; z \cong 3). \tag{9.78}$$

(b) The dark matter associated with such a cloud will be higher by a factor Ω_B^{-1}. If $\Omega_B \cong 0.06$, this factor will be about 18. Thus the total Ω associated with damped Lyman-α systems could be about (0.05–0.1) in the redshift range of $z \cong 2-3$. This region is also marked in figure 9.1(a). We see that this constraint is more severe than the one arising from quasars. The allowed models should pass to the right of the hatched line at $\sigma \cong 3$, $M \cong (10^{11} - 10^{12}) M_\odot$ in figure 9.1(b). In the extreme case, this requires $\Gamma \gtrsim 0.25$.

10
Very early universe

10.1 Flattening the Hubble radius

(a) As $t \to 0$, we can ignore the k/a^2 term in the Friedmann equations and write

$$d_{\mathrm{H}}^2 = \frac{a^2}{\dot{a}^2} = \frac{3}{8\pi G\rho}. \tag{10.1}$$

So

$$\frac{d\ln d_{\mathrm{H}}}{dt} = -\frac{1}{2}\frac{d\ln\rho}{dt} = \frac{3}{2}\left(1 + \frac{p}{\rho}\right)\frac{d\ln a}{dt}, \tag{10.2}$$

where we have used the relation $d(\rho a^3) = -p\,da^3$. The proper wavelengths scale as $\lambda \propto a(t)$, giving $d(\ln\lambda) = d(\ln a)$. Combining with (10.2), we obtain

$$\frac{d\ln d_{\mathrm{H}}}{d\ln\lambda} = \frac{3}{2}\left(1 + \frac{p}{\rho}\right). \tag{10.3}$$

For $(p/\rho) > 0$, the right-hand side is greater than unity. So d_{H} grows as λ^n with $n > 1$ during the early epochs of evolution. (Of course, n itself could vary with t if (p/ρ) is not a constant; but we will always have the condition $n > 1$.) It follows that $(d_{\mathrm{H}}/\lambda) < 1$ for sufficiently small a. In other words, the Hubble radius will be smaller than the wavelength of the perturbation at *sufficiently* early epochs (as $a \to 0$), provided $(p/\rho) > 0$.

(b) If $a(t) \propto \exp Ht$, the energy density ρ should be constant. From the equation $d(\rho a^3) = -p\,da^3$, it follows that $p = -\rho$. We saw in problem 6.3 that a scalar field dominated by potential energy can provide such an equation of state. One possible way of achieving this is as follows. We assume that the universe is populated by a scalar field with a potential $V(\phi)$. The equation of motion for such a field, if $\phi = \phi(t)$, is

$$\frac{1}{\sqrt{-g}}\partial_i\left(\sqrt{-g}\,\partial^i\phi\right) + V'(\phi) = \ddot{\phi} + 3\frac{\dot{a}}{a}\dot{\phi} + V'(\phi) = 0. \tag{10.4}$$

We now choose the potential such that $V(\phi)$ is very flat and constant (say, V_0) in some range $\phi_i < \phi < \phi_f$ with $\phi_i \cong 0$. In that case, the expansion of the universe

458

will be nearly exponential with

$$a(t) \cong a_i \exp \int_{t_i}^{t} H(t)\, dt, \quad H^2(t) \cong \frac{8\pi G}{3} V(\phi). \qquad (10.5)$$

When $V(\phi)$ is almost flat, ϕ varies slowly and we can ignore the $\ddot{\phi}$ term in (10.4). Then

$$\dot{\phi} \cong -\frac{1}{3}\frac{V'(\phi)}{H}. \qquad (10.6)$$

$$a(t) \cong a_i \exp \int_{\phi_i}^{\phi} H\frac{d\phi}{\dot{\phi}} \cong a_i \exp \int_{\phi_i}^{\phi} \frac{3H^2}{(-V')} d\phi$$

$$\cong a_i \exp 8\pi G \int_{\phi_i}^{\phi} \frac{V(\phi)}{[-V'(\phi)]} d\phi. \qquad (10.7)$$

By adjusting the parameters in $V(\phi)$, we can make sure that the inflation changes the scale factor by $[a(t)/a_i] \cong \exp 70 \cong 2.5 \times 10^{30}$ in a short time. For simplicity, we shall assume that H is a constant during inflation and $a(t) \propto \exp Ht$.

During inflation, physical wavelengths grow exponentially ($\lambda \propto a \propto \exp Ht$), while the Hubble radius remains constant. Therefore, a given lengthscale has the possibility of crossing the Hubble radius *twice* in the inflationary models. Consider, for example, a wavelength $\lambda_0 \cong 2\,\text{Mpc}$ today which contains a mass of a typical galaxy, $1.2 \times 10^{12}(\Omega h^2)\,M_\odot$. At the *end* of inflation, $t = t_f$, this scale would be

$$\lambda(t_f) = \lambda_0 \frac{a(t_f)}{a(t_0)} = 2\,\text{Mpc}\left(\frac{T_0}{T(t_f)}\right) \cong 1.9 \times 10^{-2}\,\text{cm}. \qquad (10.8)$$

This value, of course, is much *larger* than the typical Hubble radius at that epoch, $H^{-1} \cong 2 \times 10^{-24}\,\text{cm}$. But at the beginning of the inflation, this wavelength would correspond to a proper length of

$$\lambda(t_i) = \lambda(t_f) \cdot \frac{a(t_i)}{a(t_f)} = A^{-1}\lambda(t_i) = 1.8 \times 10^{-32}\,\text{cm}. \qquad (10.9)$$

This is much *smaller* than the Hubble radius. This is possible because the Hubble radius remains constant throughout the inflation while λ increases exponentially. In a time interval of about $\Delta t = t - t_i \cong 18H^{-1}$, λ will grow as big as the Hubble radius during the inflationary phase.

10.2 Horizon and flatness problems

(a) Consider the radiation dominated phase of the universe with the expansion law $a(t) \propto t^{1/2}$. The proper distance which a light signal could have travelled in the time interval $(0, t)$ is

$$R_H(t) \equiv a(t) \int_{0}^{t} \frac{dx}{a(x)} = 2t. \qquad (10.10)$$

Therefore, causal communication between two observers O and O' can exist only if they are within a distance $2R_H(t) = 4t$. This boundary is called the 'particle horizon'. Two observers O and O' separated by a proper distance larger than $2R_H(t)$ at an epoch t could never have influenced each other. Hence, there is no *a priori* reason to expect points O and O' to have similar physical environments.

If the present features of the universe were essentially determined at some early epoch – say, at $t = t_i$ when the temperature of the universe was $T \cong 10^{14}$ GeV – then we would expect a sphere, which had a radius $2R_H (t_i)$ at that epoch, to have expanded to encompass the presently observed universe. This will provide a natural explanation for the observed homogeneity of our universe. From the initial epoch $T \cong 10^{14}$ GeV to the present epoch with $T_0 = 2.75$ K $\cong 2.4 \times 10^{-4}$ eV the universe has expanded by a factor $(T/T_0) \cong 4 \times 10^{26}$. However, when $T \cong 10^{14}$ GeV the time was $t \cong 10^{-35}$ s, and $2R_H \cong 6 \times 10^{-25}$ cm. Thus the primordial sphere of homogeneity would have expanded only to a size of about 2.4×10^2 cm by today, a value far short of the size of the present universe.

The situation is worse as regards the isotropy of microwave background radiation. From the time t_i till the epoch of decoupling, t_{dec}, the universe would have expanded only by a factor $(T/T_{dec}) \cong 4 \times 10^{23}$. The primordial sphere of homogeneity would have expanded only to a size of 0.24 cm at $t = t_{dec}$. We saw in problem 8.8 that a one-degree scale in the sky probes a lengthscale of about 100 Mpc at $t = t_{dec}$. The MBR is extremely smooth at these lengthscales, even though only a tiny patch inside it could have evolved from a causally connected region at $t = t_i$.

To increase 0.24 cm to about 100 Mpc, one requires an additional expansion by a factor of about 10^{27}. This is easily achieved by inflation. For a more precise estimate we can proceed as follows. The coordinate size of the region in the LSS from where we receive signals today is

$$l(t_0, t_{dec}) = \int_{t_{dec}}^{t_0} \frac{dt}{a(t)} \cong \frac{3}{a_{dec}} \left(t_{dec}^{2/3} t_0^{1/3} \right), \qquad (10.11)$$

while the coordinate size of the horizon at $t = t_{dec}$ will be

$$l(t_{dec}, 0) = \int_0^{t_{dec}} \frac{dt}{a(t)} \cong \frac{4t_i}{a_{dec}} \left(\frac{t_{dec}}{t_f} \right)^{1/2} A. \qquad (10.12)$$

(We have used the facts $t_0 \gg t_{dec}$, $A \gg 1$, $t_i \cong H^{-1}$ and $a_{dec} = a_i A(t_{dec}/t_f)^{1/2}$.) The ratio

$$R = \frac{l(t_{dec}, 0)}{l(t_0, t_{dec})} \cong 2A \cdot \frac{t_i}{(t_f t_{dec})^{1/2}} \left[\frac{2}{3} \left(\frac{t_{dec}}{t_0} \right)^{1/3} \right] \cong 4 \times 10^4 \left(\frac{A}{10^{30}} \right), \qquad (10.13)$$

is much larger than unity for $A \cong 10^{30}$. Thus all the signals we receive today are from a causally connected domain in the LSS. Note that, in the absence of inflation, $l(t_{dec}, 0) = (2t_{dec}/a_{dec})$, so that $R = (2/3)(t_{dec}/t_0)^{1/3} \ll 1$. This value is amplified by a large factor in the course of inflation.

(b) We can write, at any time t, $\Omega(t)-1 = (k/\dot{a}^2)$. Assuming $k \neq 0$, a comparison with the present epoch, $t = t_0$, gives

$$\Omega(t) - 1 = \frac{\dot{a}_o^2}{\dot{a}^2} (\Omega_0 - 1).   \tag{10.14}$$

Expressed as a function of temperature, in the radiation dominated era, this relation becomes

$$\Omega(T) - 1 \cong 4 \times 10^{-15}(\Omega_0 - 1) \left(\frac{T}{1 \text{ MeV}} \right)^{-2}.   \tag{10.15}$$

The present astronomical observations imply that Ω_0 is between 0.1 to 5 (say) with liberal allowance given for systematic errors. Thus $| \Omega_0 - 1 |$ is of order unity. However, the above relation shows that at earlier epochs $|\Omega - 1|$ is far less than unity. If we assume that the initial conditions for the universe were 'set' at Planck time, when $T \cong 10^{19}$ GeV, then the a- and $\dot{a}$-terms should have been matched at that epoch to an accuracy of $|(\Omega - 1)| \cong 10^{-60}$ so as to result in the present values for a and $\dot{a}$. Had this fine-tuning not been resorted to, the universe would have contracted back to $a = 0$ (for $k = 1$) or diffused out to $a = \infty$ (for $k = -1$) *long before* the present epoch. In the absence of any physical mechanism, this fine-tuning has to be imposed in an *ad hoc* manner at some early epoch. (This is called the 'flatness problem'.)

Inflation of $a(t)$ by a factor A decreases the value of the k/a^2 term by a factor $A^{-2} \cong 10^{-60}$. Thus one can start with moderate values of (k/a^2) before inflation and bring it down to a very small value at $t \gtrsim 10^{-33}$ s. This solves the 'flatness problem', interpreted as the smallness of (k/a^2). Note that no classical process can change a $k \neq 1$ universe to a $k = 0$ universe, since it involves a change in topological properties. What inflation does is to decrease the value of (k/a^2) so much that, for all practical purposes, we can ignore the dynamical effect of the curvature term, thereby having the same effect as setting $k = 0$. This has the consequence that Ω at present must be unity to a high degree of accuracy.

10.3 Origin of density perturbations

(a) During inflation, the universe was assumed to be described by a Friedmann model with small inhomogeneities. This implies that the source – which should be some *classical* scalar field $\Phi(t, \mathbf{x})$ – can be split as $[\phi_0(t) + f(t, \mathbf{x})]$, where $\phi_0(t)$ denotes the average, homogeneous part and $f(t, \mathbf{x})$ represents the space-dependent, fluctuating part. Since the energy density due to a scalar field is $\rho \cong (1/2)\dot{\phi}^2 + V_0$, where V_0 is a constant, we obtain,

$$\delta\rho(t, \mathbf{x}) = \rho(\mathbf{x}, t) - \bar{\rho}(t) \cong \dot{\phi}_0(t)\dot{f}(t, \mathbf{x}),   \tag{10.16}$$

where we have assumed that $f \ll \phi_0$. Fourier transforming this equation, we find that

$$\delta\rho(\mathbf{k}, t) \cong \dot{\phi}_0(t)\dot{Q}_\mathbf{k}(t),   \tag{10.17}$$

with

$$f(t, \mathbf{x}) \equiv \int \frac{d^3\mathbf{k}}{(2\pi)^3} Q_{\mathbf{k}}(t) e^{i\mathbf{k}\cdot\mathbf{x}}. \tag{10.18}$$

Since the average energy density during inflation is dominated by the constant term V_0, the density contrast will be

$$\delta(\mathbf{k}, t) \cong \frac{\delta\rho}{V_0} = \frac{\dot{\phi}_0(t)\dot{Q}_{\mathbf{k}}(t)}{V_0}. \tag{10.19}$$

The 'mean' value ϕ_0 and the fluctuating field $f(t, \mathbf{x})$ appearing in this equation are supposed to be some *classical* objects *mimicking* the quantum fluctuations. It is not easy to devise and justify such quantities. What is usually done is to choose some convenient quantum mechanical measure for fluctuations and *define* ϕ_0 and $Q_{\mathbf{k}}$ in terms of this quantity.

In quantum theory, the field $\hat{\phi}(t, \mathbf{x})$ and its Fourier coefficients $\hat{q}_{\mathbf{k}}(t)$ will become operators related by

$$\hat{\phi}(t, \mathbf{x}) = \int \frac{d^3\mathbf{k}}{(2\pi)^3} \hat{q}_{\mathbf{k}}(t) e^{i\mathbf{k}\cdot\mathbf{x}}. \tag{10.20}$$

The quantum state of the field can be specified by giving the quantum state $\psi_{\mathbf{k}}(q_{\mathbf{k}}, t)$ of each of the modes $\hat{q}_{\mathbf{k}}$. One can think of $q_{\mathbf{k}}$ as coordinates of a particle and $\psi_{\mathbf{k}}(q_{\mathbf{k}}, t)$ as the wavefunction describing this particle. The fluctuations in $q_{\mathbf{k}}$ can be characterized by the dispersion

$$\sigma_{\mathbf{k}}^2(t) = \langle\psi|\hat{q}_{\mathbf{k}}^2(t)|\psi\rangle - \langle\psi|\hat{q}_{\mathbf{k}}(t)|\psi\rangle^2 = \langle\psi|\hat{q}_{\mathbf{k}}^2(t)|\psi\rangle \tag{10.21}$$

in this quantum state. (The mean value of the scalar field operator $\langle\hat{\phi}(t, \mathbf{x})\rangle$ is zero in the inflationary phase. Therefore, we have set the $\langle\hat{q}_{\mathbf{k}}\rangle$s to zero in the above expression. Note that we are interested only in the $\mathbf{k} \neq 0$ modes.) Expressing the $\hat{q}_{\mathbf{k}}$s in terms of $\hat{\phi}(t, \mathbf{x})$ it is easy to see that

$$\sigma_{\mathbf{k}}^2(t) = \int d^3\mathbf{x} \langle\psi|\hat{\phi}(t, \mathbf{x} + \mathbf{y})\hat{\phi}(t, \mathbf{y})|\psi\rangle e^{i\mathbf{k}\cdot\mathbf{x}}. \tag{10.22}$$

In other words, the 'power spectrum' of fluctuations $\sigma_{\mathbf{k}}^2$ is related to the Fourier transform of the two-point correlation function of the scalar field. Since $\sigma_{\mathbf{k}}^2(t)$ appears to be a good measure of quantum fluctuations, we may attempt to *define* $Q_{\mathbf{k}}(t)$ as

$$Q_{\mathbf{k}}(t) = \sigma_{\mathbf{k}}(t). \tag{10.23}$$

This is equivalent to *defining* the fluctuating classical field $f(t, \mathbf{x})$ to be

$$f(t, \mathbf{x}) \equiv \int \frac{d^3\mathbf{k}}{(2\pi)^3} \sigma_{\mathbf{k}}(t) e^{i\mathbf{k}\cdot\mathbf{x}}. \tag{10.24}$$

This leads to the result

$$\delta(\mathbf{k}, t) = \frac{\dot{\phi}_0(t)}{V_0} \dot{\sigma}_{\mathbf{k}}(t). \tag{10.25}$$

It should be stressed that the expressions chosen for $Q_{\mathbf{k}}$ and ϕ_0 are only two out of many possible choices available. Such an ambiguity cannot be avoided when semiclassical expressions have to be computed from quantum mechanical operators.

(b) Let us suppose that the particular mode (corresponding to the galactic scale, say) we are interested in leaves the Hubble radius at $t = t_1$ during the inflationary phase, with the amplitude $\epsilon(t_1) \equiv \delta(\mathbf{k}, t_1(\mathbf{k}))$. We have suppressed the $\mathbf{k}$-dependence in ϵ to simplify the notation. Let $t = t_f$ be the instant at which the universe makes a transition from the inflationary phase to the radiation dominated phase. During the time $t_1 < t < t_f$, the mode is outside the Hubble radius in an inflating universe. For modes bigger than the Hubble radius, we have derived a growth law in problem 7.6:

$$\left(\frac{\delta\rho}{\rho}\right) \propto \frac{1}{\rho_{\text{bg}} a^2} . \tag{10.26}$$

In the present context, $\rho_{\text{bg}} = $ constant and we obtain

$$\epsilon(t) \propto \exp(-2Ht). \tag{10.27}$$

This result can also be expressed as

$$\frac{\epsilon(t_f)}{\epsilon(t_1)} = \left[\frac{H(t_1)\,a(t_1)}{H(t_f)\,a(t_f)}\right]^2, \tag{10.28}$$

since $H(t)$ is a constant during this interval. If the transition from the inflationary phase to the radiation dominated phase is approximated as instantaneous, then the radiation density ρ_{rad} and the fluctuations in the radiation density $\delta\rho_{\text{rad}}$ originate directly from the corresponding terms in the inflationary phase: $\rho_{\text{rad}} \cong \dot{\phi}^2$ and $\delta\rho$ (radiation phase) $\cong \delta\rho$ (inflationary phase). Further, since the background energy density ρ in the inflationary phase is dominated by V_0, we can write

$$\left(\frac{\delta\rho}{\rho}\right)_{\text{rad}} \cong \left(\frac{\delta\rho}{\rho}\right)_{\text{deSitter}} \cdot \left(\frac{\rho_{\text{deSitter}}}{\rho_{\text{rad}}}\right) \cong \left(\frac{\delta\rho}{\rho}\right)_{\text{deSitter}} \cdot \left(\frac{V_0}{\dot{\phi}^2}\right). \tag{10.29}$$

During the time $t_f < t < t_{\text{enter}}$, the mode is bigger than the Hubble radius and is evolving in the radiation dominated universe. We know from problem 7.6 that such modes grow as $\delta \propto a^2 \propto t$, while the quantity $a(t)H(t) \propto t^{-1/2}$. Hence, at $t = t_{\text{enter}} \equiv t_2$,

$$\begin{aligned}
\epsilon(t_2) &= \left(\frac{V_0}{\dot{\phi}^2}\right)\left(\frac{\delta\rho}{\rho}\right)_{\text{deSitter}} \cdot \left[\frac{H(t_f)\,a(t_f)}{H(t_2)\,a(t_2)}\right]^2 \\
&= \epsilon(t_1)\left(\frac{V_0}{\dot{\phi}^2}\right)\left[\frac{H(t_1)\,a(t_1)}{H(t_f)\,a(t_f)}\right]^2\left[\frac{H(t_f)\,a(t_f)}{H(t_2)\,a(t_2)}\right]^2 \\
&= \epsilon(t_1)\left(\frac{V_0}{\dot{\phi}^2}\right)\left[\frac{H(t_1)\,a(t_1)}{H(t_2)\,a(t_2)}\right]^2 .
\end{aligned} \tag{10.30}$$

But for a mode which is leaving the Hubble radius at t_1 and entering it again at t_2, $H(t_1) a(t_1) = H(t_2) a(t_2)$. So

$$\epsilon(t_2) = \epsilon(t_1) \left(\frac{V_0}{\dot{\phi}^2} \right). \tag{10.31}$$

On using (10.25), we obtain

$$\delta(\mathbf{k}, t_{\text{enter}}) \cong \left(\frac{\dot{\sigma}_{\mathbf{k}}}{\dot{\phi}_0} \right)_{t=t_{\text{exit}}}. \tag{10.32}$$

More exact matching of energy densities at $t = t_{\text{f}}$ gives an additional factor of $(4/3)$.

10.4 Quantum fluctuations in inflationary phase

(a) Substituting the ansatz

$$\psi_{\mathbf{k}} = A_{\mathbf{k}}(t) \exp \left\{ -B_{\mathbf{k}}(t)[q_{\mathbf{k}} - f_{\mathbf{k}}(t)]^2 \right\} \tag{10.33}$$

into the Schrödinger equation

$$i \frac{\partial \psi_{\mathbf{k}}}{\partial t} = -\frac{1}{2a^3} \frac{\partial^2 \psi_{\mathbf{k}}}{\partial q_{\mathbf{k}}^2} + \frac{1}{2} a^3 \omega_{\mathbf{k}}^2 q_{\mathbf{k}}^2 \psi_{\mathbf{k}}, \quad \omega = |\mathbf{k}| \tag{10.34}$$

and equating the coefficients of various powers of $q_{\mathbf{k}}$, we obtain three equations:

$$i\dot{B} = \frac{2B^2}{a^3} - \frac{1}{2} ak^2, \tag{10.35}$$

$$i(\dot{B}f + B\dot{f}) = \frac{2B^2}{a^3} f, \tag{10.36}$$

$$i \frac{\dot{A}}{A} = i\dot{B}f^2 + 2iBf\dot{f} + \frac{B}{a^3} - \frac{2B^2 f^2}{a^3}, \tag{10.37}$$

where we have suppressed the index k, for simplicity. These equations can be transformed to a simpler form by introducing a variable Q defined by the relation: $B = -(i/2)a^3(\dot{Q}/Q)$. Simple algebra then shows that

$$f = (\text{const})(a^3 \dot{Q})^{-1}, \quad A = (\text{const})Q^{-1/2} \exp \left\{ -\frac{i}{2} \int k^2 af^2 dt \right\}, \tag{10.38}$$

while Q satisfies the linear equation:

$$\frac{1}{a^3} \frac{d}{dt} \left(a^3 \frac{dQ}{dt} \right) + \frac{k^2}{a^2} Q = 0. \tag{10.39}$$

From (10.33) we also see that

$$|\psi_{\mathbf{k}}|^2 = N_{\mathbf{k}} \exp \left\{ -\frac{(q_{\mathbf{k}} - R_{\mathbf{k}})^2}{2\sigma_{\mathbf{k}}^2} \right\}, \tag{10.40}$$

with

$$\sigma_k^2 = \frac{1}{2}(B_k + B_k^*)^{-1}, \quad R_k = \frac{B_k f_k + B_k^* f_k^*}{B_k + B_k^*}. \tag{10.41}$$

Equation (10.41) can be further simplified by noting that (10.39) implies the relation

$$\frac{d}{dt}\left\{a^3(Q_k^* \dot{Q}_k - \dot{Q}_k^* Q_k)\right\} = 0, \tag{10.42}$$

giving $Q_k^* \dot{Q}_k - \dot{Q}_k^* Q_k = i(\text{constant})a^{-3}$. Using this result in (10.41), we obtain

$$\sigma_k^2 = (\text{const})|Q_k|^2, \quad R_k = \text{Re}\,[(\text{const})\sigma_k^2 B_k f_k]. \tag{10.43}$$

Thus all relevant quantities can be expressed in terms of Q_k.

(b) To solve for $Q_k(t)$, it is convenient to transform the independent variable from t to another time coordinate τ with $d\tau = [dt/a(t)]$. Then

$$t = -H^{-1}\ln|(1 - H\tau)|, \quad a(\tau) = (1 - H\tau)^{-1} \tag{10.44}$$

and the equation for Q becomes

$$\frac{d^2 Q}{d\tau^2} - \frac{2H}{H\tau - 1}\frac{dQ}{dt} + k^2 Q = 0. \tag{10.45}$$

Changing the variable to $x = kH^{-1}(1 - H\tau) = kH^{-1}e^{-Ht}$, (10.45) can be written as

$$\frac{d^2 Q}{dx^2} - \frac{2}{x}\frac{dQ}{dx} + Q = 0. \tag{10.46}$$

The general solution to this equation is

$$Q(x) = a(1 + ix)e^{-ix} + b(1 - ix)e^{ix}, \tag{10.47}$$

where a and b are two constants which depend on the initial conditions imposed on $Q(x)$. Since the quantum state of the oscillator $\psi_k(q_k)$ is completely determined by the function $Q_k(t)$, we now have a *set* of quantum states, parametrized by the constants a_k, b_k. Expectation values of physical variables will, of course, depend on the quantum state in which they are evaluated. Therefore, to proceed further, we need to make a specific choice for the quantum state of the field.

Since the minimum amount of fluctuations will arise from the 'ground state' of the system, it seems natural to choose the quantum state to be the ground state of the system. It is, however, not easy to define a ground state in the expanding background. One way of doing this is the following. We know that when $a = 1$, and $H = 0$ we should recover the ordinary, flat space, field theory. As $H \to 0$, equation (10.47) becomes

$$Q_{\text{flat}}(\tau, \mathbf{k}) = \lim_{H \to 0}\left\{a\left(1 + \frac{ik}{H} - ik\tau\right)e^{i(k\tau - kH^{-1})} + b\left(1 - \frac{ik}{H} + ik\tau\right)e^{-i(k\tau - kH^{-1})}\right\}$$

$$= \alpha e^{ik\tau} + \beta e^{-ik\tau}, \tag{10.48}$$

provided we keep $\alpha = ikaH^{-1}\exp(-ik/H)$ and $\beta = -ikbH^{-1}\exp(ik/H)$ as *finite* constants when the limit is taken. But to obtain the 'standard vacuum' state in the flat spacetime each oscillator must be described by the wavefunction

$$\psi = N\exp\left(-\frac{i}{2}k\tau\right)\exp\left(-\frac{1}{2}kq^2\right) = (\text{const})Q^{-1/2}\exp\left(+\frac{i}{2}\frac{\dot{Q}}{Q}q^2\right). \quad (10.49)$$

Therefore, we must satisfy three conditions: (i) $(\dot{Q}/Q) = ik$; (ii) $\beta = 0$; and (iii) α is independent of k. This implies that our solution for Q_k must have the form

$$Q_k = a(1+ix)e^{-ix} = \alpha He^{ik/H}(ik)^{-1}(1+ix)e^{-ix}. \quad (10.50)$$

(Note that $\tau = t$ in the limit of $H \to 0$.) We can now compute all the physical quantities. Direct calculation gives

$$B_k = \frac{k^3 H^{-2}}{2(1+k^2/H^2a^2)}\left[1 - \frac{iHa}{k}\right], \quad \sigma_k^2 = \frac{1}{2}(B+B^*)^{-1} = \frac{H^2}{2k^3} + \frac{1}{2ka^2}. \quad (10.51)$$

To compute $\delta(k, t_{\text{enter}})$ we need the value of $\dot{\sigma}_k$ and $\dot{\phi}_0$ at $t = t_{\text{exit}}$. From the expression for $\sigma_k(t)$, we have

$$|\dot{\sigma}_k| = \frac{H}{\sqrt{2}k^{3/2}}\frac{1}{2}\left(1+\frac{k^2}{H^2G^2}\right)^{-1/2}\frac{2k^2\dot{a}}{H^2a^3} = \frac{H^2}{\sqrt{2}k^{3/2}}\left(1+\frac{k^2}{H^2a^2}\right)^{-1/2}\left(\frac{k^2}{H^2a^2}\right) \quad (10.52)$$

at any time t. At $t = t_{\text{exit}}$, $(k/Ha) = 2\pi$. So, at $t = t_{\text{exit}}$, we find

$$|\dot{\sigma}_k| = \frac{4\pi^2}{\sqrt{2}(1+4\pi^2)^{1/2}}\frac{H^2}{k^{3/2}} \cong 4\left(\frac{H^2}{k^{3/2}}\right) \quad (\text{at } t = t_{\text{exit}}). \quad (10.53)$$

We saw earlier that this spectrum of perturbation at the time of re-entry has the form $\delta_{\mathbf{k}}(t_{\text{enter}}) \cong (\dot{\sigma}_{\mathbf{k}}/\dot{\phi}_0)$. The $\mathbf{k}$-dependence of $\delta_{\mathbf{k}}$ arises solely from the $\mathbf{k}$-dependence of $\dot{\sigma}_{\mathbf{k}}$. The above analysis shows that $k^3|\delta_{\mathbf{k}}(t_{\text{enter}})|^2$ is independent of $\mathbf{k}$. This corresponds to the $n = 1$ spectrum discussed in chapter 7.

(c) For a potential with $V(\phi) \cong (V_0 - \lambda\phi^4/4)$ we have $V'(\phi) \cong -\lambda\phi^3$. During inflation, let us assume that ϕ varies from ϕ_i to ϕ_f. In the 'slow roll-over' phase $\dot{\phi} \cong -V'(\phi)/3H$, giving

$$\delta_k = \frac{\dot{\sigma}_k}{\dot{\phi}} = \frac{4H^2}{k^{3/2}}\left[\frac{3H}{-V'(\phi)}\right] = \frac{12H^3}{k^{3/2}[-V'(\phi)]}. \quad (10.54)$$

The expression on the right-hand side is to be evaluated at a time $t = t_{\text{exit}}$ when the mode leaves the Hubble radius. During the time $t_{\text{exit}} < t < t_f$ the universe inflates by the factor $\exp N$ with

$$N = \int_{t_{\text{exit}}}^{t_f}\frac{\dot{a}}{a}dt \cong 8\pi G\int_\phi^{\phi_f}\frac{V_0}{[-V']}d\phi = \frac{3H^2}{2\lambda}\left(\frac{1}{\phi^2} - \frac{1}{\phi_f^2}\right) \cong \frac{3H^2}{2\lambda\phi^2}, \quad (10.55)$$

since we may take $\phi \ll \phi_f$. Using this, we obtain

$$-V'(\phi) = \lambda\phi^3 = \left(\frac{3}{2}\right)^{3/2}\frac{H^3}{\lambda^{1/2}N^{3/2}}. \quad (10.56)$$

Using this in (10.54) we obtain

$$k^{3/2}\delta_k = 12H^3 \left(\frac{2}{3}\right)^{3/2} \frac{\lambda^{1/2}N^{3/2}}{H^3} \cong 12 \left(\frac{2}{3}\right)^{3/2} \lambda^{1/2}N^{3/2}. \tag{10.57}$$

In order to have sufficient inflation we need $N \cong 50$. Then $k^{3/2}\delta_k \cong 2.3 \times 10^3 \lambda^{1/2}$, which is far too large for $\lambda \cong 0.1\text{--}1$. To obtain $k^{3/2}\delta_k \cong 10^{-4}$ we need $\lambda \cong 10^{-15}$.

This has been the most serious difficulty faced by all the realistic inflationary models: they all produce too large an inhomogeneity. The qualitative reason for this result can be found from (10.32). To obtain *slow* roll-over and sufficient inflation we need to keep $\dot\phi_0$ small; this leads to an increase in the value of δ. The difficulty could have been avoided if it were possible to keep σ_k arbitrarily small; unfortunately, the inflationary phase induces a fluctuation of about $(H/2\pi)$ on any quantum field due to field theoretical reasons. This lower bound prevents us from obtaining acceptable values for δ unless we fine-tune the dimensionless parameters of $V(\phi)$. Several 'solutions' have been suggested in the literature to overcome this difficulty but none of them appear to be very compelling.

10.5 Cosmological defects

(a) The minima of the potential corresponds to $\phi = \pm\eta$. As the universe cools, the field in some regions will reach the value η and, in some other regions, $-\eta$. These regions will be separated by two-dimensional boundaries, where $\phi = 0$, with the value of the field changing from η to $-\eta$ across the surface. The field equation for ϕ in the static limit is given by

$$\frac{d^2\phi}{dl^2} = \frac{dV}{d\phi} = \lambda\phi\left(\phi^2 - \eta^2\right), \tag{10.58}$$

where the proper distance l is measured perpendicular to the local plane of the domain wall. Integrating this equation, with the boundary condition that $(d\phi/dl) = 0$ far away from the wall, we obtain

$$\left(\frac{d\phi}{dl}\right)^2 = 2V. \tag{10.59}$$

This has the solution

$$\phi(l) = \eta \tanh\left(\frac{l}{L}\right), \qquad L = \sqrt{\frac{2}{\lambda}}\,\eta^{-1}, \tag{10.60}$$

which changes from $(-\eta)$ to η over a lengthscale L. The stress-tensor for this configuration can be obtained using the general expression found in problem 5.9.

If the third axis is taken as normal to the domain wall, we obtain

$$\rho = T^{00} = -T^{11} = -T^{22} = \frac{1}{2}\left(\frac{d\phi}{dl}\right)^2 + V = 2V, \qquad (10.61)$$

$$T^{33} = \frac{1}{2}\left(\frac{d\phi}{dl}\right)^2 - V = 0. \qquad (10.62)$$

Most of the energy will be concentrated on the domain wall, which may be thought of as having an energy per unit area $\sigma \cong VL \cong \lambda^{1/2}\eta^3$. We saw in problem 5.10 that the source of gravitational acceleration in general relativity is $\rho_{\text{eff}} = (\rho + T^{11} + T^{22} + T^{33})$. In our case, $\rho_{\text{eff}} = -2V$ and the gravitational acceleration will be

$$\mathbf{g} = 2\pi G\sigma\mathbf{n}, \qquad (10.63)$$

where $\mathbf{n}$ is a unit vector along the third direction. Note that domain walls lead to gravitational repulsion.

(b) The minimum energy configuration will correspond to $\theta = $ constant. However, an arbitrary field configuration $[\phi_1(\mathbf{x}), \phi_2(\mathbf{x})]$ cannot always relax to such a configuration. To see this, consider a configuration $\theta(\mathbf{x})$ in which the value of θ increases by 2π while going around a closed path. Continuity demands that this result should hold even as the path is shrunk to zero area. But that would cause the field gradient to diverge. The minimum energy configuration, therefore, should contain a region of some width ξ, say, within which $\phi^2 = \phi_1^2 + \phi_2^2$ should pass through a zero, allowing the components ϕ_A to smoothly switch signs. The two equations $\phi_1(\mathbf{x}) = 0$ and $\phi_2(\mathbf{x}) = 0$ define two different two-surfaces which intersect in a line. This line is the location of a defect called the 'cosmic string'. It is clear from the above argument that such a defect is topologically stable.

(c) From the stress-tensor for the cosmic string we find that $(\rho + 3p) = \rho - \rho = 0$. Hence, the effective gravitational mass of the string vanishes.

To find the metric due to the cosmic strings, we shall start with an ansatz of the form

$$ds^2 = dt^2 - dr^2 - Q(r)\,d\theta^2 - dz^2. \qquad (10.64)$$

This form is chosen with hindsight. It has only two non-trivial components for G_k^i, which are G_0^0 and G_3^3. A simple computation using the formulas of chapter 5 gives

$$G_0^0 = G_3^3 = -\frac{Q''}{2Q} + \frac{Q'^2}{4Q^2}. \qquad (10.65)$$

Equating this expression to the corresponding source term and writing $Q = q^2$ we obtain

$$\frac{q''}{q} = -8\pi G\rho(r). \qquad (10.66)$$

For the $\rho(r)$ given in the question we can easily solve this equation for $r < r_0$. We obtain

$$q(r) = \begin{cases} k^{-1} \sin kr & \text{(for } r < r_0) \\ \alpha + \beta r & \text{(for } r > r_0) \end{cases}, \tag{10.67}$$

with $k^2 = 8\pi G\rho_0$. We have chosen the internal solution with the boundary condition that $Q = q^2 \to r^2$ near $r \to 0$, which is needed to make the geometry near $r = 0$ regular. The constants α and β are fixed by demanding the continuity of $q(r)$ and its first derivatives at $r = r_0$. This gives

$$q(r) = (r - r_0) \cos kr_0 + \frac{\sin kr_0}{k}, \tag{10.68}$$

thereby completely solving the problem. When $r \gg r_0$ and $kr_0 < 1$, the solution can be approximated as

$$q^2 = (1 - 8\pi G\rho_0 r_0^2)r^2 \tag{10.69}$$

and the line element is

$$ds^2 = dt^2 - dr^2 - (1 - 8\pi G\mu)r^2 d\theta^2 - dz^2, \tag{10.70}$$

with the mass per unit length of the string being $\mu = \pi r_0^2 \rho_0$.

In any local region, one can rescale θ, thereby reducing the above metric to that of flat spacetime. This agrees with the result that the cosmic string has no active gravitational mass. Such a rescaling, of course, is not a globally valid procedure. This is easily seen by calculating the circumference of a circle of radius $r = R$ in the $z =$ constant, $t =$ constant plane. We find that the circumference is

$$C = (2\pi - 8\pi G\mu)R, \tag{10.71}$$

showing that the geometry is that of a conical space with a 'deficit angle' given by $\Delta\theta = 8\pi G\mu$.

Useful Constants

1. Fundamental constants

Speed of light (c)	$= 3.0 \times 10^{10} \, \mathrm{cm \, s^{-1}}$
Planck's constant $(\hbar)$	$= 1.1 \times 10^{-27} \, \mathrm{erg \, s}$
Gravitational constant (G)	$= 6.7 \times 10^{-8} \, \mathrm{dyne \, cm^2 \, g^{-2}}$
Electronic charge (e)	$= 4.8 \times 10^{-10} \, \mathrm{esu}$
Electronic mass (m_e)	$= 9.1 \times 10^{-28} \, \mathrm{g}$
Proton mass (m_p)	$= 1.7 \times 10^{-24} \, \mathrm{g}$
Boltzmann's constant (k)	$= 1.4 \times 10^{-16} \, \mathrm{erg \, K^{-1}}$

2. Combinations of fundamental constants

Planck length $[(G\hbar/c^3)^{1/2}]$	$= 1.6 \times 10^{-33} \, \mathrm{cm}$
Planck time $[(G\hbar/c^5)^{1/2}]$	$= 5.4 \times 10^{-44} \, \mathrm{s}$
Planck mass $[(G/\hbar c)^{-1/2}]$	$= 2.2 \times 10^{-5} \, \mathrm{g}$
Gravitational structure constant $(Gm_p^2/\hbar c)$	$= 5.9 \times 10^{-39}$
Fine-structure constant $(e^2/\hbar c)$	$= 7.3 \times 10^{-3}$
Compton wavelength $(\hbar/m_e c)$	$= 3.8 \times 10^{-11} \, \mathrm{cm}$
Compton frequency $(m_e c^2/\hbar)$	$= 7.7 \times 10^{20} \, \mathrm{Hz}$
Classical electron radius $(e^2/m_e c^2)$	$= 2.8 \times 10^{-13} \, \mathrm{cm}$
Thomson cross-section	
$[\sigma_T = (8\pi/3)(e^2/m_e c^2)^2]$	$= 6.7 \times 10^{-25} \, \mathrm{cm^2}$
Square of electronic charge (e^2)	$= 15.1 \, \mathrm{eV \mathring{A}}$
Radius of hydrogen atom $[a_0 = (\hbar^2/m_e e^2)]$	$= 0.53 \times 10^{-8} \, \mathrm{cm}$

Ground state energy of hydrogen atom
$[E_0 = (m_e e^4/2\hbar^2)]$ $= 13.6 eV = 2.1 \times 10^{-11}$ erg
Ionization frequency $(E_0/\hbar)$ $= 2.1 \times 10^{16}$ Hz
Ionization wavelength $(2\pi c\hbar/E_0)$ $= 912 \,\mathring{A}$

Radiation constant $[a = (\pi^2 k^4/15\hbar^3 c^3)]$ $= 7.6 \times 10^{-15}$ erg s^{-1} cm^{-3}

3. Astrophysical quantities and combinations

Solar mass (M$_\odot$) $= 2.0 \times 10^{33}$ g $= 1.2 \times 10^{57} m_p$
Solar radius (R$_\odot$) $= 7.0 \times 10^{10}$ cm
Solar luminosity (L$_\odot$) $= 3.9 \times 10^{33}$ erg s^{-1}

Solar velocity $\equiv (GM_\odot/R_\odot)^{1/2}$ $= 436$ km s^{-1}
Solar density $\equiv (M_\odot/R_\odot^3)$ $= 5.8$ g cm^{-3}

4. Cosmological quantities and combinations

Megaparsec $= 3.1 \times 10^{24}$ cm

Hubble constant (H_0) $= 100\,h$ km s^{-1} Mpc^{-1}
Hubble time (H_0^{-1}) $= 3.1 \times 10^{17} h^{-1}$ s $= 9.8 \times 10^9 h^{-1}$ yr
Hubble radius (cH_0^{-1}) $= 3000\,h^{-1}$ Mpc $= 9.2 \times 10^{27}\,h^{-1}$ cm
Critical density $[\rho_{crit} = (3H_0^2/8\pi G)]$ $= 1.88 h^2 \times 10^{-29}$ g cm^{-3}
                                    $= 1.1 \times 10^4 h^2$ eV cm^{-3}
                                    $= 2.8 \times 10^{11} h^2$ M$_\odot$ Mpc^{-3}
                                    $= 1.1 \times 10^{-5} h^2$ protons cm^{-3}

Radiation energy density $= 4.8 \times 10^{-34} \left(T/2.75\,\text{K}\right)^4$ g cm^{-3}
Photon number density $= 421.8 \left(T/2.75\,\text{K}\right)^3$ cm^{-3}
Mass contained in a sphere of radius $\lambda/2$ $= 1.4 \times 10^{11}$ M$_\odot (\Omega h^2) \left(\lambda/1\,\text{Mpc}\right)^3$

5. Units and conversions (with $\hbar = c = k = 1$)

1 eV $= 1.3 \times 10^{-4}$ cm $= 12\,396.3 \,\mathring{A}$
                                    $= 1.6 \times 10^{-12}$ erg
                                    $= 1.2 \times 10^4$ K
1 g cm^{-3} $= 5.9 \times 10^{23}$ protons cm^{-3}
                                    $= 5.9 \times 10^{32}$ eV cm^{-3}
                                    $= 1.5 \times 10^{40}$ M$_\odot$ Mpc^{-3}
1 Jy $= 10^{-23}$ erg cm^{-2} s^{-1} Hz^{-1}
                                    $= 3.0 \times 10^{-7}$ photons cm$^{-3}(\ln \nu)^{-1}$

Notes and references

General references which are of relevence to an entire chapter are cited at the beginning. More specific comments are listed against each particular problem.

Chapter 1

The description of astrophysical processes for a student with a physics background is available in a few books. I found the following ones quite useful:

(a) F. H. Shu (1991), *Physics of Astrophysics*, vols. I and II (University Science Books, California).
(b) M. Harwitt (1988), *Astrophysical Concepts* (Springer-Verlag, New York).
(c) J. I. Katz (1987), *High Energy Astrophysics* (Addison-Wesley, California).

[1.1] This derivation for the radiation field was first given by J. J. Thomson (1907), in *Electricity and Matter* (Archibald Constable, London), chapter 3. It is repeated in F. S. Crawford (1968), *Waves*, Berkeley Physics Course, vol. III (McGraw-Hill, New York).

[1.3] Most textbooks on astrophysical processes discuss Bremsstrahlung and Synchrotron radiation with varying degrees of clarity. One of the best treatments is in L. D. Landau and E. M. Lifshitz (1975), *Classical Theory of Fields* (Pergamon Press, New York).

[1.11] This problem was originally discussed by Lord Kelvin. The derivation given here is based on T. Padmanabhan (1996), *Resonance* (to appear).

[1.14] An excellent discussion of gravitational dynamics can be found in J. Binny and S. Tremaine (1987), *Galactic Dynamics* (Princeton University Press).

Chapter 2

Several aspects of dynamics specifically geared for studying gravitating systems can be found in

(a) J. Binny and S. Tremaine (1987), cited in ref. 1.14 above.

Detailed discussion of statistical mechanics applied to gravitating systems can be found in

(b) T. Padmanabhan (1990), Statistical mechanics of gravitating systems, *Physics Reports*, **285**, 188.

[2.1(c)]   This problem is based on the unpublished work by D. Lynden-Bell and T. Padmanabhan (1993), Comment on Action Principle in General Relativity (preprint).

[2.4]   This problem and discussion is based on section 3.3 of J. Binny and S. Tremaine (1987), cited in ref. (a) above. The reader is referred to this text for more details.

[2.6(a)]   This example is from V. I. Arnol'd (1963), *Russ. Math. Surveys*, **18:6** (reprinted in R. S. Macay and J. D. Meiss (1987), *Hamiltonian Dynamical Systems: A reprint selection* (Princeton University Press)).

[2.8]   This example is from T. Padmanabhan (1990), cited in ref. (b) above.

[2.10]   The instability of the isothermal sphere for large values of R was first derived by V. A. Antonov (1962), *Vestn. Leningrad Gos. Univ.*, **7**, 135 (English translation is available in: Dynamics of globular clusters, *IAU Symp.*, **113** (eds., J. Goodman and P. Hut, Reidel, Dordrecht, 1985)). The simple derivation given here in part (c) is based on T. Padmanabhan (1990), cited in ref. (b) above.

[2.12]   The neat derivation of dynamical friction and diffusion in a combined fashion is due to L. D. Landau; see E. M. Liftshitz and L. P. Pitaevskii (1981), *Physical Kinetics* (Pergamon Press, Oxford). The original derivation was given for plasmas but adapting it to gravitational systems is straightforward.

[2.13(a)]   The process of 'violent relaxation' was discovered by D. Lynden-Bell. This problem is based on the original paper, D. Lyndel-Bell (1967), *MNRAS*, **136**, 101.

[2.13(b)]   This problem is based on the work by S. Tremaine, M. Henon and D. Lynden-Bell (1986), *MNRAS*, **219**, 285.

[2.16]   This problem is based on the work by A. Toomre (1982), *Ap. J.*, **259**, 535.

Chapter 3

An excellent discussion of most of these topics can be found in

(a) L. D. Landau and E. M. Liftshitz (1987), *Fluid Mechanics* (Pergamon Press, New York).
(b) E. M. Liftshitz and L. P. Pitaesvskii (1981), cited in ref. 2.12 above.

The following book is also very useful for topics related to accretion:

(c) J. Frank, A. King and D. Raine (1992), *Accretion Power in Astrophysics* (Cambridge University Press).

[3.7] The physical picture of turbulence based on the growth of successive instabilities is essentially based on chapter III of L. D. Landau and E. M. Liftshitz (1987), cited in ref. (a) above.

[3.12] This discussion is based on section 2.5 of J. Frank, A. King and D. Raine (1992), cited in ref. (c) above.

[3.14] This discussion is based on section 106 of L. D. Landau and E. M. Liftshitz (1987) cited in ref. (a) above.

Chapter 4

All the text books cited at the beginning of chapter 1 discuss radiative processes. The fundamental aspects of radiation theory are best discussed in L. D. Landau and E. M. Liftshitz (1975), cited in ref. 1.3 above.

[4.1(b)] This formula was originally derived by R. P. Feynmann and quoted and used in R. P. Feynmann (1964), *Feynmann Lectures in Physics*, vols. I and II (Addison-Wesley, USA). However, Feynmann has not given any derivation of the formula. The derivation given here is based on the author's work and the paper by A. R. Janah, T. Padmanabhan and T. P. Singh (1988), *Am. J. Phys.*, **56**, 1036.

[4.3(e)] The expression for the radiation reaction force in terms of the stress-tensor of the electromagnetic field was derived by the author. See T. Padmanabhan (1994), *Inverse Compton Scattering – revisited*, preprint IUCAA-25/94.

[4.8(c)] The derivation of the power radiated in the inverse Compton process is based on T. Padmanabhan (1994), cited in ref. 4.3(e) above.

[4.10] A direct derivation of Kompaneets' equations by judicious use of Lorentz transformations is given in P. J. E. Peebles (1993), *Principles of Physical Cosmology* (Princeton University Press).

[4.13(c)] This derivation is based on L. D. Landau and E. M. Liftshitz (1971), *Quantum Electrodynamics* (Pergamon Press, New York).

Chapter 5

The best place to learn general relativity is from the last four chapters of

(a) L. D. Landau and E. M. Liftshitz (1975), cited in ref. 1.3 above.

For a more detailed working practice, the following book is excellent:

(b) A. P. Lightman, W. H. Press, R. H. Price, S. A. Teukolsky (1975), *Problem Book in General Relativity and Gravitation* (Princeton University Press).

[5.9(c)] This derivation is based on the unpublished work by D. Lynden-Bell and T. Padmanabhan, cited in ref. 2.1(c) above.

Chapter 6

The topics in part 2 of this book (chapters 6–10) are discussed in far greater detail in

(a) T. Padmanabhan (1993), *Structure Formation in the Universe* (Cambridge University Press).
(b) P. J. E. Peeble (1993), *Principles of Physical Cosmology* (Princeton University Press).

The reader is referred to these two books for original references. Many of these topics are also covered in the classic:

(c) Ya. B. Zeldovich and I. Novikov (1983), *Relativistic Astrophysics – vol. II* (University of Chicago Press).

[6.12]   Many aspects of this problem are discussed in detail in B. J. T. Jones and R. F. G. Wise (1985), *Astron. Ap.*, **149**, 144.

Chapter 7

Many of these topics are discussed in detail in refs. (a) and (b) cited at the beginning of chapter 6. The reader is referred to these text books for original references.

[7.8(b)]   This is based on an unpublished work by the author, T. Padmanabhan (1992).

[7.13]   This is based on R. Nityananda and T. Padmanabhan (1994), *MNRAS*, **271**, 976. See also A. J. S. Hamilton *et al.* (1991), *Ap. J.*, **374**, L1. The fitting function is based on J. S. Bagla and T. Padmanabhan (1995), *Jour. Ap. Astron.* (in press). For an explanation of these scaling relations from first principles, see T. Padmanabhan (1996), *MNRAS* **278**, L29.

Chapter 8

The topics discussed here are all covered in detail in refs (a) and (b) cited at the beginning of chapter 6. The reader is referred to these text books for original references.

Chapter 9

The topics discussed here are covered in detail in refs (a) and (b) cited at the beginning of chapter 6. The reader is referred to these text books for original references.

[9.4(b)]　　　This problem is based on J. R. Bond *et al.* (1991), *Ap. J.*, **329**, 440. The solution to the diffusion equation is discussed in S. Chandrasekhar (1943), *Rev. Mod. Phys.*, **15**, 1.

[9.7]　　　The constraints on structure formation models from the abundance of the damped Lyman-alpha system is discussed, for example, in K. Subramanian and T. Padmanabhan, (1994), *Constraints on the Models for Structure Formation from the Abundance of Damped Lyman-alpha Systems*, preprint IUCAA-5/94.

Chapter 10

The topics discussed here are covered in detail in refs. (a) and (b) cited at the beginning of chapter 6. The reader is referred to these text books for original references. A more detailed discussion of several aspects related to the physics of the early universe can be found in

(a) E. W. Kolb and M. S. Turner (1990), *The Early Universe* (Addison-Wesley, USA).

Chapter 11

The solutions to some of the problems appearing in this chapter can be found in the references cited below.

[11.1]　　　J. Binney and S. Tremaine (1987), cited in ref. 1.14 above; see section 2.3.1

[11.2]　　　Most books on 'modern' classical mechanics discuss this issue. See, for example, V. I. Arnold (1989), *Mathematical Methods of Classical Mechanics* (Springer-Verlag).

[11.3], [11.4] For a pedagogical discussion, see T. Padmanabhan (1990), cited in ref. (b) of chapter 2 above. This review contains references to earlier literature.

[11.6]　　　See E. M. Liftshitz and L. P. Pitaevskii (1981), cited in ref. 2.12 above; see section 10.

[11.7]　　　A detailed discussion of this phenomena is available in F. H. Shu (1991), *Physics of Astrophysics, Vol. II* (University Science Books, California), chapter 20.

[11.8(a)]　　　This is derived in F. Rohrlich (1965), *Classical Charged Particles* (Addison-Wesley, Massachusetts).

[11.11]　　　This is discussed in P. J. E. Peebles (1993), cited in ref. 4.10 above.

[11.12]　　　Several books on general relativity discuss this problem. A clear and crisp discussion is available in L. D. Landau and E. M. Lifshitz, (1975) cited in ref. 1.3 above; see section 103.

[11.14]　　　All the standard books on cosmology discuss primordial nucleosynthesis. In the usual approach, equations governing nuclear reactions are solved numerically to obtain the results. Particularly noteworthy in this connection is the paper by R. Esmaizadeh *et al.* (1991), Primordial nucleosynthesis without a computer, *Ap. J.*, **378**, 504.

[11.17]   Some details regarding cosmological numerical simulations are available in, for example, G. Efstathiou *et al.* (1985), *Ap. J. Suppl.*, **57**, 241; F. R. Bouchet (1985), *Ap. J.*, **299**, 1. A general reference to numerical simulations is, R. W. Hockney and J. W. Eastwood (1988), *Computer Simulation using Particles* (Adam Hilger, New York).

Index

Printed in the United States
By Bookmasters